Benchmark Papers in Electrical Engineering and Computer Science

Series Editor: John B. Thomas
Princeton University

Volume

1 SYSTEM SENSITIVITY ANALYSIS / *José B. Cruz, Jr.*
2 RANDOM PROCESSES: Multiplicity Theory and Canonical Decompositions / *Anthony Ephremides and John B. Thomas*
3 ALGEBRAIC CODING THEORY: History and Development / *Ian F. Blake*
4 PATTERN RECOGNITION: Introduction and Foundations / *Jack Sklansky*
5 COMPUTER-AIDED CIRCUIT DESIGN: Simulation and Optimization / *S. W. Director*
6 ENVIRONMENTAL MODELING: Analysis and Management / *Douglas Daetz and Richard H. Pantell*
7 AUTOMATIC CONTROL: Classical Linear Theory / *George J. Thaler*
8 CIRCUIT THEORY: Foundations and Classical Contributions / *M. E. Van Valkenburg*
9 DATA COMMUNICATION: Fundamentals of Baseband Transmission / *L. E. Franks*
10 NONLINEAR SYSTEMS: Processing of Random Signals—Classical Analysis / *A. H. Haddad*
11 RANDOM PROCESSES, II: Poisson and Jump-Point Processes / *Anthony Ephremides*
12 DIGITAL FILTERS AND THE FAST FOURIER TRANSFORM / *Bede Liu*
13 DETECTION AND ESTIMATION: Applications to Radar / *Simon S. Haykin*
14 DATA COMPRESSION / *Lee D. Davisson and Robert M. Gray*
15 ACTIVE *RC* FILTERS: Theory and Application / *L. P. Huelsman*
16 NONLINEAR SYSTEMS: Stability Analysis / *J. K. Aggarwal and M. Vidyasagar*
17 LINEAR LEAST-SQUARES ESTIMATION / *Thomas Kailath*
18 OPTICAL COMMUNICATION THEORY / *R. O. Harger*
19 ERGODIC AND INFORMATION THEORY / *Robert M. Gray and Lee D. Davisson*
20 TWO-DIMENSIONAL DIGITAL SIGNAL PROCESSING / *Sanjit K. Mitra and Michael P. Ekstrom*
21 COMPUTER ARITHMETIC / *Earl E. Swartzlander, Jr.*
22 ARRAY PROCESSING: Applications to Radar / *Simon Haykin*
23 NONPARAMETRIC DETECTION: Theory and Applications / *Saleem A. Kassam and John B. Thomas*

Benchmark Papers in Electrical Engineering and Computer Science / 23

A BENCHMARK® Books Series

NONPARAMETRIC DETECTION
Theory and Applications

Edited by

SALEEM A. KASSAM

University of Pennsylvania

and

JOHN B. THOMAS

Princeton University

Dowden, Hutchinson & Ross, Inc.

STROUDSBURG, PENNSYLVANIA

Benchmark Papers in Electrical Engineering and Computer Science, Volume 23
Library of Congress Catalog Card Number: 79-22557
ISBN: 0-87933-359-6

82 81 80 1 2 3 4 5
Manufactured in the United States of America.

LIBRARY OF CONGRESS CATALOGING IN PUBLICATION DATA
Main entry under title:
Nonparametric detection.
(Benchmark papers in electrical engineering and computer science; 23)
Bibliography: p.
Includes indexes.
1. Nonparametric signal detection—Addresses, essays, lectures. I. Kassam, Saleem A., 1949- II. Thomas, John Bowman, 1925-
TK5102.5.N567 621.38'043 79-22557
ISBN 0-87933-359-6

Distributed world wide by Academic Press,
a subsidiary of Harcourt Brace Jovanovich,
Publishers.

SERIES EDITOR'S FOREWORD

The Benchmark Papers in Electrical Engineering and Computer Science series attempts to sift, organize, and make readily accessible to the reader the vast literature that has accumulated. Although the series is not intended as a complete substitute for a study of this literature, it serves at least three major critical purposes. First, it provides a practical point of entry into a given area of research. Each volume offers an expert's selection of the critical papers on a topic as well as his or her views on its structure, development, and present status. Second, the series provides a convenient and time-saving means for study in areas related to but not contiguous with one's principal interests. And last, the series allows the collection, in a particularly compact and convenient form, of the major works on which present research activities and interests are based.

Each volume in the series has been collected, organized, and edited by an authority in the area to which it pertains. In order to present a unified view of the area, the volume editor has prepared an introduction to the subject, has included comments on each article, and has provided a subject index to facilitate access to the papers.

I believe that this series will provide a manageable working library of the most important technical articles in electrical engineering and computer science. I hope that it will be equally valuable to students, teachers, and researchers.

This volume, *Nonparametric Detection: Theory and Applications,* has been edited by S. A. Kassam of the University of Pennsylvania and me. It contains thirty-five papers, a bibliography, and editorial comments on nonparametric techniques as applied to signal detection. Due to space limitation, it does not consider nonparametric estimation schemes. The volume is largely self-contained although some familiarity with classical signal detection will be useful. It should provide an introduction to and condensation of the literature in this area.

JOHN B. THOMAS

PREFACE

We have tried to collect in this volume most of the important papers that have appeared in the engineering literature on nonparametric signal detection. Lack of space has precluded an exhaustive survey and, in particular, we have omitted, reluctantly, almost all of the very extensive literature in the statistical journals. In an attempt to compensate for this necessary omission, we have listed a rather large bibliography.

We have included as many papers on applications as space permits, and, in keeping with the goals toward which this volume is aimed, a majority of these papers are concerned with the two-hypothesis testing problem: deciding between the alternatives of noise only and noise plus signal. However, particularly in Part IV, other applications, such as pattern classification and discrimination, are included.

SALEEM A. KASSAM
JOHN B. THOMAS

CONTENTS

CONTENTS BY AUTHOR

NONPARAMETRIC DETECTION

INTRODUCTION

Classical detection theory, as the term is usually understood in an electrical engineering context, is almost always concerned with the parametric detection of signals in additive Gaussian noise. The signals may be of known functional form with unknown or random parameters or may be assumed to be random processes. In the latter case, it is usually assumed that the signals are also Gaussian. This subject is ably treated by Helstrom[1] and, in greater detail, by Van Trees.[2]

The strong restriction that the random processes involved be Gaussian is necessary in order that the likelihood ratio be tractable—that is, be available in such a form that it can be interpreted as a realizable system operating on the input observation. At the root of the problem is the general difficulty in expressing N-dimensional joint densities.

In nonparametric procedures, an attempt is made to circumvent these problems by assuming that the statistics of input noises (and possibly signals) are not known exactly but obey some weak conditions, such as having known medians. Detectors that can guarantee a minimum value of some performance characteristic for such hypotheses are called nonparametric detectors. Such detectors are generally inferior to those where more specific assumptions are made but are often easily implemented and, in any case, may be optimal for the nonparametric hypotheses assumed. The ultimate justification for these detectors is that they are often not greatly inferior to detectors designed to more specific and more restrictive assumptions, and they will usually be less sensitive in performance deterioration when specific input assumptions are violated.

In most of the papers in this volume, a test between the null hypothesis H_0 and the alternative H_1 is treated where

$$H_0: x_i = \eta_i$$

and $\qquad i = 1, 2, \ldots, n$

$$H_1: x_i = \eta_i + s_i$$

Here the observation sequence $\{x_i\}_{i=1}^{n}$ consists, under H_0, of noise observations $\{\eta_i\}_{i=1}^{n}$ taken from a noise process $\{N_i\}_{i=1}^{n}$. Under H_1, the observation sequence also contains an additive signal $\{s_i\}_{i=1}^{n}$ where the s_i are usually assumed to be known but may in some cases be observations from a (random) signal process $\{S_i\}_{i=1}^{n}$. A detector is derived that operates on the $\{x_i\}$ and decide between H_0 and H_1.

Part I contains two survey papers (and a comment on one of these) that should be useful as a general introduction for those who are not too familiar with nonparametric techniques as applied to signal detection.

Part II consists of six papers treating detectors that use only the signs of the input observations. Such detectors are easy to implement and are often remarkably effective.

In Part III, more sophisticated detectors based on ranks are considered in seven papers. Improved performance over the sign detectors is obtained with a considerable increase in structural complexity and analytical difficulty.

In Part IV, ten papers consider other nonparametric methods, many of them designed to retain some of the advantages of rank tests while eliminating some of the complexities.

For reasons of tractability, it is usually necessary to assume that the sequence of random variables $\{X_i\}$ are independent. This is a strong restriction but one that is difficult to eliminate. The ten papers in Part V involve attempts to circumvent this restriction.

For readers unfamiliar with the field and, no less importantly, for others actively working in the theory and applications of nonparametric detection techniques, we have compiled a bibliography of publications in the field of nonparametric detection. The bibliography includes most of the developments reported in the various engineering journals and conferences since the late 1950s. The sources for this bibliography include all of the major IEEE professional group transactions, journals, and conferences and other major technical journals and conference proceedings. A list of the sources used for the bibliography precedes the alphabetically arranged entries. We have not included in this bibliography titles of dissertations or technical reports of limited availability. It is our belief that the bibliography is fairly complete, and most other references on topics related to nonparametric procedures, as well as much of the related statistical literature, can be obtained indirectly from the articles indexed.

The description *nonparametric* does not have a uniform meaning, and thus a clarifying note on its usage needs to be made. The sense in which most authors use the term *nonparametric procedure* is that which implies that the performance of the procedure, or test, is independent of the underlying probability distributions; that is, it is distribution free. In most cases, the performance measure implied is the probability of type I error (size), or the probabilities of both types of errors. Sometimes, however, particularly in pattern recognition and classification work, a nonparametric procedure is taken to be any procedure that has been formulated without a parametric model. This is not the meaning that we have accepted in compiling the bibliography. Sometimes a procedure is classified as being nonparametric when it can be more precisely described as being asymptotically nonparametric (having a distribution-free performance in the limit when sample sizes tend to infinity). Finally, the description *CFAR* (constant false-alarm rate, or constant probability of type I error) is often used synonymously with *nonparametric* in radar and sonar applications. According to our usage, a nonparametric test is often a CFAR test. However, for many parametric problems where some parameters are unknown, tests can be implemented whose performances are independent of the values of these parameters. Such tests would then also be called *CFAR* tests.

REFERENCES

1. C. W. Helstrom. *Statistical Theory of Signal Detection*. Pergamon Press, 1968.
2. H. L. Van Trees. *Detection, Estimation, and Modulation Theory*. John Wiley, 1968, 1971.

Part I

SURVEY ARTICLES ON NONPARAMETRIC DETECTION

Editors' Comments on Papers 1 and 2

1 **CARLYLE**
Nonparametric Methods in Detection Theory

2A **THOMAS**
Nonparametric Detection

2B **LYTLE, MOOSE, and SCHARF**
Comments on "Nonparametric Detection"

One important function of this book is to serve as a point of entry into the field for those who are not acquainted in depth with the methods of, and literature on, nonparametric detection. We thus felt it appropriate to include in this collection two review papers that survey the area without treating all of the details of specific schemes.

The survey article by Carlyle (Paper 1) was published as a chapter in a book devoted to various aspects of communication theory. Written by an early contributor in this field, it is one of the first extensive surveys of the main techniques of nonparametric detection and contains a description of the engineering developments up to 1968.

Paper 2A by Thomas is a short review of the main ideas in nonparametric detection and is useful as an introduction to the area. A brief correspondence article has been reprinted as Paper 2B and serves to clarify an important property of a very simple nonparametric detection scheme discussed in Paper 2A. Since 1970, when this latest review paper was published, new results, concepts, and application have been reported. These accomplishments, however, do not reduce the usefulness of these papers for potential contributors to the field and for others wishing to gain an understanding of the basic principles and methods of nonparametric detection.

1

Reprinted from pages 293–319 of *Communication Theory*, A. V. Balakrishnan, ed., New York: McGraw-Hill Book Company, 1968, 347pp. Used with permission of McGraw-Hill Book Company.

NONPARAMETRIC METHODS IN DETECTION THEORY†

J. W. Carlyle

1. Introduction

For a two-decision detection problem of the type "signal absent vs. signal present," a detector is said to be *nonparametric* or *distribution-free* if its false-alarm probability can be fixed in advance without knowledge of the functional form of the noise distributions. This will be made more precise in the following sections, but the objective is clear: maintenance of some system performance criterion at a minimum guaranteed level in an essentially unknown environment. The design of adaptive systems[1] has a similar goal, but in the nonparametric case we do not seek to optimize performance; the emphasis is on insensitivity to changes in environmental conditions rather than adaptation to them. The investigation of stability of detectors[2] is similarly motivated; however, in nonparametric detection theory we are primarily concerned with the gross insensitivities and extreme simplicities in implementation possible in systems based on some rather severe reduction of the data (e.g., to rank ordering and/or polarities).

Nonparametric statistical hypothesis-testing techniques are well known and are a principal subject of current research, and the associated mathematical literature is very extensive. However, engineering implications in signal-detection theory have only recently been examined; the results currently are therefore largely confined to analysis and extension of the role of existing nonparametric tests as signal detectors in relatively simple situations (rather than synthesis of essentially new forms). We consider an illustrative selection of such problems here.

We have not attempted to be exhaustive or historically complete in the bibliography, but most basic papers can be located at least indirectly from the references given. Certain results in mathematical statistics, particularly those available in recent textbooks (e.g., Refs. 3, 4, 5), are mentioned here without specific reference; some more detailed cita-

† The preparation of this chapter was supported in part by the Air Force Office of Scientific Research, Office of Aerospace Research, United States Air Force, under grant 700-67, and in part by the National Science Foundation under grant GP-3008.

tions of this nature are included in an earlier survey,[6] which covers portions of Secs. 3, 4, and 5 below.

2. Review of Basic Concepts; Definitions, Notations

In the problems discussed here w denotes the received data set to be processed by a detector, H is the no-signal hypothesis (the null hypothesis), and K is the signal-present hypothesis (the alternative). If P is the probability distribution of w, we recall that a hypothesis is, in effect, a statement that P belongs to a certain *class* of distributions (it is particularly important to keep this in mind in the nonparametric case where the classes are broad in scope); a hypothesis and the class to which it refers are here often regarded as identical for notational convenience. A hypothesis is simple if it consists of exactly one completely specified distribution, and is otherwise composite. A composite hypothesis is *parametric* if the indexing of its members P requires a finite number of continuous-range real parameters, and *nonparametric* if the membership is so large that no such indexing is possible. In this definition we avoid subtleties (connected with the indexing of classes of probability measures for random processes) by tacitly assuming that w is finite dimensional, e.g., a section of a discrete-time series, or a vector or matrix array of sampled data taken from a single or multiple process over a finite observation interval; this is the case for each of the examples discussed here.

If D or $D(w)$ is a detector, i.e., a statistical test or decision rule by which we agree to announce H or K (signal absent or signal present) based upon observation of w, then we let $\beta_D(P)$ denote the operating characteristic of D, defined to be the probability of announcing K when P is the distribution of w; thus $\beta_D(P)$ is a detection probability when P belongs to K and a false-detection or false-alarm probability when P belongs to H. We recall that the object of detector design from the Neyman-Pearson statistical hypothesis-testing viewpoint (where prior probabilities of hypotheses are not assumed) is to achieve suitably large detection probabilities for at least some members of K, subject to the constraint that no false-alarm probability is permitted to exceed the desired false-alarm level α preassigned. For given H, K, and α, the existence of an optimum detector (a detector which simultaneously maximizes each detection probability) is not typical, but if H and K are simple hypotheses, the well-known likelihood-ratio procedure is optimum. It may occasionally happen that the likelihood-ratio reduces to a test that remains optimum when H and K are composite, but in general it is necessary to seek a solution that is optimum in some restricted sense; e.g., optimum among those that are unbiased (all detection probabilities $\geq \alpha$) or among these which utilize only a reduced form of the data, or *weak-signal optimum* (locally optimum in the sense that the detection probability is maximized for signal-present distributions P "close" to H or as "$P \to H$" in a suitably defined manner.[3,24]

Let H be a nonparametric hypothesis; a detector D is said to be *nonparametric* or *distribution-free* (with respect to H) if the false-alarm probability $\beta_D(P)$ is constant for all P in H. (Conventional tests not enjoying this property, and designed to distinguish parametric hypotheses, are occasionally called "parametric detectors" below.) Thus a nonparametric detector is such that one of its performance criteria can be maintained at a guaranteed level in a largely unknown noise environment; this basic qualitative concept is evidently of interest in many detection and communication problems, but we have chosen to restrict attention here (as is the case in the recent engineering literature on nonparametric detection) to situations covered by the Neyman-Pearson two-hypothesis formulation so that some existing nonparametric statistical techniques can be examined directly in a signal-detection context. For the same reason, in Secs. 3, 4, and 5 we consider only white-noise sequences, i.e., situations where the components of w corresponding to different time instants are independent (when hypotheses reduce to statements about one-dimensional distributions); this case is of course not without interest, and in Sec. 6 we indicate how performance for correlated sequences can be evaluated for detectors that were designed to be nonparametric in white-noise situations. Various combinations of restricted optimality criteria (such as those mentioned in the preceding paragraph) have been used to synthesize, or to formally justify, existing tests nonparametric in white-noise situations; we refer the reader to recent statistical textbooks (e.g., Ref. 3) for such derivations. This material is significant as background, but we wish to concentrate on examples of analysis and engineering application or extension of some of these tests, to show that such devices (1) are very simple to implement operationally, (2) possess desirable properties of false-alarm level invariance (e.g., with respect to noise power level and predetection nonlinearities) even in nonwhite noise, and (3) exhibit weak-signal performance that may be quite acceptable, and can even be superior to optimum parametric detectors under departures from the parametric model assumptions.

A detector based on finite observation time, which fails to be distribution-free, may become so in the limit as the observation time grows. More precisely, suppose that P, H, and K pertain to the probability law of an infinite time series $\{w_1, w_2, \ldots\}$ and for each $n = 1, 2, \ldots$, let D_n be a detector that bases its decision on $\{w_k: 1 \leq k \leq n\}$, for example, on the first $T = n\Delta$ seconds of sampled data in the case where the w_i happen to be samples taken at Δ-second intervals from a continuous-time process $\{w(t):0 \leq t < \infty\}$. If the sequence $\{D_n\}$ of detectors has the property that for α preassigned,

$$\lim_{n\to\infty} \beta_{D_n}(P) = \alpha \tag{8-1}$$

when P is in H, and if H is a nonparametric hypothesis, then $\{D_n\}$ is said to be *asymptotically nonparametric*. This desirable property is often called *robustness* in the statistical literature. We also recall $\{D_n\}$ is said to be *consistent* (for K) if

$$\lim_{n\to\infty} \beta_{D_n}(P) = 1 \tag{8-2}$$

when P is in K, that is, if (as is typically the case) arbitrarily large detection probabilities can be achieved by $\{D_n\}$ through processing for sufficiently long time intervals [in (8-2) it is tacitly assumed that each D_n has the required false-alarm level α preassigned].

Let α and β be preassigned, $0 < \alpha < \beta < 1$, and let $\{D_n\}$, $\{D'_n\}$ be two sequences of detectors (each utilizing increasing observation times as defined above) with common false-alarm level α for a null hypotheses H. Let P be a fixed alternative ("signal present") distribution, let $N(\alpha,\beta,P)$ be the smallest n for which $B_{D_n}(P) \geq \beta$, and let $N'(\alpha,\beta,P)$ be defined in the same way for $\{D'_n\}$. Clearly the ratio

$$\frac{N'(\alpha,\beta,P)}{N(\alpha,\beta,P)} \tag{8-3}$$

is often a reasonable measure of the relative efficiency of $\{D_n\}$ with respect to $\{D'_n\}$; an *asymptotic relative efficiency* (ARE) of $\{D_n\}$ with respect to $\{D'_n\}$ is obtained by passing to the weak-signal limit (when it exists) in (8-3) as P approaches H in a properly restricted sense (and necessarily N, $N' \to \infty$ if the detectors are consistent). The ARE provides a useful figure of merit for $\{D_n\}$ which is ≤ 1 when $\{D'_n\}$ is optimum for the alternatives considered, but which may exceed 1 when other data distributions are examined; i.e., when an optimum detector is used in situations for which it is not designed to be optimum, it may be possible to obtain superior performance from a suboptimal detector that is less sensitive to departures from assumptions (it is in this sense that nonparametric detectors may be "superior" to conventional detectors when the nature of the noise environment is largely unknown). The ARE has been shown to be a well-defined constant, independent of α and β, in many cases of interest involving comparison of conventional statistical tests with some of their nonparametric competitors. We illustrate the calculation of ARE below in simple situations where one can proceed directly from the somewhat imprecise definition just given, and in other cases we merely quote ARE results; the references should be consulted for further details (see Refs. 3 to 5, 7, and 8).

Notations used frequently here include the following: μ and σ^2 are the mean and variance of a one-dimensional distribution; $\Phi(\cdot)$ is the standard gaussian cumulative distribution function with $\mu = 0$ and $\sigma^2 = 1$; sgn $(\cdot)$ and $u(\cdot)$ are the sign function and unit step function (for $x > 0$, sgn $(x) = u(x) = 1$, and for $x \leq 0$, sgn $(x) = -1$ and $u(x) = 0$). All detectors examined below are of the threshold-comparison type, where one announces H or K according to whether some statistic $V(w)$ (a real-valued function of the data) is less than or greater than a preselected constant C; this situation is summarized, in the usual fashion, by writing merely

$$V(w) \gtrless C$$

All threshold levels are denoted by the same letter C; the value of C for a particular detector can, of course, be deduced from the context and the desired false alarm level α.

3. Detection of a Sure Signal in White Noise

In this section we consider detectors utilizing received data

$$w = (x_1, x_2, \ldots, x_n)$$

where the x_i are independent scalar random variables with cumulative distribution functions $F_i(x) = P(x_i \leq x)$, assumed to be identical with common value $F(x)$ unless otherwise specified. The detection problems examined are those in which the alterations induced in the no-signal F by the presence of a signal include a shift in a location parameter (mean or median). In particular we have in mind the basic problem of detection of a constant signal (i.e., a fixed but unknown level) in an additive stationary white-noise sequence, and in generalizations of this problem when the requirement of simple additivity is relaxed or reinterpreted and some nonstationarities are permitted. We also briefly consider nonconstant signals and the phase-incoherent case.

Parametric Detectors for a Constant Signal. Let σ_0^2 be a known constant; under the preceding assumptions regarding the data w, the problem of testing the simple hypothesis

$$H_1\text{: } F(x) \text{ is gaussian} \qquad \mu = 0 \qquad \sigma^2 = \sigma_0^2$$

vs. the composite (parametric) alternative

$$K_1\text{: } F(x) \text{ is gaussian} \qquad \mu > 0 \qquad \sigma^2 = \sigma_0^2$$

corresponds to the problem of detecting a constant signal of known polarity and unknown amplitude in the presence of additive stationary white gaussian noise of known intensity. For testing H_1 vs. any fixed alternative distribution in K_1, the well-known optimum rule, obtained from the likelihood ratio, is of course

$$\sum_{i=1}^{n} x_i \gtrless \sqrt{n}\,\sigma_0 C \qquad \text{(8-4a)}$$

where the threshold C is determined by

$$\Phi(C) = 1 - \alpha \qquad \text{(8-4b)}$$

and since the result is identical for any distribution in K_1, (8-4) is in fact the optimum detector for H_1 vs. K_1 with false-alarm probability α. For brevity we refer to a rule of the form $\Sigma x_i \gtrless C$ as a linear detector.

If the noise power level σ_0^2 is not known, we have the parametric hypothesis and alternative

H_2: $F(x)$ is gaussian with zero mean and unknown variance
K_2: $F(x)$ is gaussian with positive mean and unknown variance

Note that signal and noise are not restricted to be simply additive here; an increase in the mean might be accompanied by unknown changes in the intensity σ^2 of the noise, for example. It can be shown that for practical false-alarm levels there does not exist an optimum detector in the unrestricted sense for H_2 vs. K_2, but when attention is restricted to the class of unbiased rules, there is an optimal solution within that class given by Student's t test

$$\frac{\sum_{i=1}^{n} x_i}{\sqrt{n}\left[\frac{1}{n-1}\sum_{i=1}^{n}\left(x_i - \frac{1}{n}\sum_{j=1}^{n} x_j\right)^2\right]^{1/2}} \gtrless C \qquad \text{(8-5)}$$

where C is determined approximately by (8-4b) for large n and exactly for any n through the use of Student's distribution. [Since the bracketed quantity in the denominator of (8-5) is an estimate of σ^2, this structure may of course be regarded as "adaptive" in a rudimentary sense.]

The removal of the assumption that the noise is gaussian leads to a nonparameteric problem such as

H_3: $\mu = 0, \sigma^2 < \infty$ $F(x)$ otherwise unspecified
K_3: $\mu > 0, \sigma^2 < \infty$ $F(x)$ otherwise unspecified

The bracketed quantity in (8-5) converges in probability to σ^2 as $n \to \infty$, and by the central limit theorem $\Sigma x_i/\sqrt{n}$ is asymptotically gaussian with variance σ^2 and zero mean under H_3, so that (applying the well-known convergence theorem of Cramer) the left-hand side of (8-5) converges in distribution to a gaussian random variable with unit variance and zero mean under H_3; hence if C is selected according to (8-4b), the detectors (8-5) for increasing observation time ($n \to \infty$) are asymptotically nonparametric for hypothesis H_3 with false-alarm probability α (and in the same way we can verify consistency for K_3). Thus, for large n, a detector that implements Student's t test is appropriate for a detection problem in which the signal appears as a mean value in stationary white noise whose characteristics are essentially unknown and may, in fact, depend upon the signal amplitude in some unspecified manner (e.g., as a result of nonlinearities in the transmission channel). It should be kept in mind, however, that (8-5) is only asymptotically nonparametric, so that for limited observation time (small n), a threshold C chosen according to (8-4b) (or according to Student's t distribution) can sometimes result in false-alarm probabilities greater than α when the input is significantly nongaussian; in addition, the squaring operations on numerical values required in (8-5) may be inconvenient to implement if analog processing is involved. These are among the reasons for consideration of some of the systems which (1) utilize only qualitative features of the input data, such as signs and ranks, rather than actual numerical values, and (2) are exactly nonparametric (for any n) for suitable modifications of H_3. We first examine what is perhaps the simplest useful scheme of this type, based solely upon polarity information; subsequently we consider the retention of ranks as well as signs.

The Sign Detector. If, as in the problem (H_3,K_3), the presence of a signal is accompanied by an increase in mean value, and if H_3 and K_3 are modified by restricting attention to symmetric data distributions so that mean and median are coincident, it is clear that the number of positive x_i among $(x_1,x_2, \ldots ,x_n)$ will tend to be near $n/2$ when a signal is absent and greater than $n/2$ when a signal is present. In this sense, so-called *sign test*

$$\Sigma u(x_i) \gtrless C \tag{8-6}$$

or its symmetrized equivalent based on Σ sgn (x_i) (a "hard limiter and accumulator" system) may be a reasonable solution (and obviously a very easily implemented solution) for a nonparametric problem such as

$$H_4: F(-x) = 1 - F(x) \qquad F \text{ otherwise unspecified}$$
$$K_4: F(2\mu - x) = 1 - F(x) \qquad \text{for some } \mu > 0$$

or, if we restrict attention to distribution functions F possessing probability densities f,

$$H_4': f(-x) = f(x) \qquad f \text{ otherwise unspecified}$$
$$K_4': f(2\mu - x) = f(x) \qquad \text{for some } \mu > 0$$

(Note that these hypothesis classes are not subsets of H_3 and K_3 unless it is also specified that means and variances exist; where convenient, we implicitly assume existence of moments as needed. Note also that no relationship is assumed between F under H_4 and F under K_4; that is, as in H_3 and K_3 we do not restrict attention to fixed-noise characteristics with additive signals.)

We first show that (8-6) is nonparametric for hypotheses such as H_4. Evidently the left-hand side of (8-6) has the binomial distribution with parameter

$$p = P(x_i > 0) = 1 - F(0) \tag{8-7}$$

so that the operating characteristic for (8-6) is

$$\beta(F) = \sum_{k=C+1}^{n} \binom{n}{k} p^k (1-p)^{n-k} \tag{8-8}$$

If C is determined through

$$\sum_{k=C+1}^{n} \binom{n}{k} \left(\frac{1}{2}\right)^n = \alpha \tag{8-9}$$

we see that (8-6) has constant false-alarm probability α for all distributions in H_4; that is, the sign detector (8-6) is nonparametric for H_4 (or H_4'). Also, for any fixed $p > \frac{1}{2}$ in (8-8), the detection probability $\beta \to 1$ as $n \to \infty$; that is, (8-6) is consistent for the class of F in K_4 and strictly increasing at 0.

The sign detector evidently remains nonparametric for the much more general hypothesis

$$H_5: p = \tfrac{1}{2} \qquad F \text{ otherwise arbitrary}$$

and is consistent for the corresponding alternative

$$K_5: p > \tfrac{1}{2} \qquad F \text{ otherwise arbitrary}$$

In fact, (8-6) is the optimum test of H_5 vs. K_5. This is one of the few nontrivial cases in which an unrestricted optimum solution exists for a problem in which H and K are nonparametric; therefore we indicate what seems to be the most direct of the several possible proofs, although this particular optimality property is not essential to our present discussion. For the proof, fix any signal-present distribution for x_i (i.e., any member of K_5) that possesses a probability density, say f, and write $f = pf^+ + (1-p)f^-$, where f^+ and f^- are the conditional densities given $x_i > 0$ and $x_i \leq 0$, respectively; then $f_0 = (f^+ + f^-)/2$ is a no-signal density (i.e., a distribution in H_5). The optimum test of the simple hypothesis f_0 vs. the simple alternative f is $L(w) \gtrless C$ where L is the likelihood ratio $\prod_i f(x_i)/f_0(x_i)$; here f/f_0 is $pf^+/\frac{1}{2}f_0^+$ when $x_i > 0$ and $(1-p)f^-/\frac{1}{2}f_0^-$ when $x_i \leq 0$, so that

$$L = 2^n p^{\Sigma u(x_i)}(1-p)^{n-\Sigma u(x_i)}$$

and the test $L \gtrless C$ immediately reduces to (8-6). Thus the structure based on (8-6) and (8-9) maximizes the detection probability for the simple alternative f subject to the condition that the false-alarm probability be α for the simple hypothesis f_0. The more stringent condition that the false-alarm probability be α for all H_5 is also satisfied by (8-6) and (8-9), so that the test is actually optimum for the nonparametric hypothesis H_5 vs. the simple alternative f. This conclusion holds for every distribution in K_5 possessing a density, and likewise for every other distribution in K_5 if the density is taken with respect to some suitable measure. Since the test structure obtained is the same in each case (independent of the alternative), we conclude that (8-6) and (8-9) yield the optimum test of H_5 vs. K_5.

Hypotheses H_5 and K_5 are equivalent to median $= 0$ and median > 0, respectively; the preceding proof thus establishes that the sign test is optimal for testing for the presence of a shift in the median of an otherwise unknown distribution. This optimality property raises the question of the significance of the median as an indicator of the presence of a signal; since an arbitrary distribution function can have zero median but positive mean (and vice versa), the concept of signal as a shift in average level, even though other signal-induced changes in the distribution are permitted, is not properly reflected by general hypotheses such as H_5 and K_5. Hence the optimality of (8-6) for H_5 vs. K_5 may be of less importance here than its performance for H_4 vs. K_4. However, it is useful to observe that since (8-6) is nonparametric for H_5, its false-alarm probability is invariant with respect to a wide class of instantaneous nonlinearities that might perturb the amplitude of the detector input; in fact, if detector (8-6) is applied to data

$$x_i' = \psi(x_i) \tag{8-10}$$

where the graph of ψ passes through the origin and is otherwise confined to the interior of the first and third quadrants, not only the false-alarm probability (8-9) but also the detection probability (8-8) are unaffected by the presence of ψ [since $P(x_i' > 0) = P(x_i > 0)$]. More generally, the same conclusions clearly remain valid if the nonlinearity varies with time, i.e., if $x_i' = \psi_i(x_i)$, or equivalently if the input-time series is nonstationary in this sense. We also see that (8-6) remains nonparametric for the hypothesis

$$H_6\colon F_i(0) = \tfrac{1}{2} \qquad i = 1, 2, \ldots, n$$

a nonstationary generalization of H_5 where F_i is the distribution function of x_i and we do not require that all F_i coincide with a common F. Because of similar invariance properties under largely unspecified time-variable input conditions, other systems based on hard limiting have been of interest in underwater acoustic signal detection. In Secs. 5 and 6 we comment further on some multichannel extensions of the sign detector suitable for such applications. False-alarm rate stabilization by hard limiting has also been considered in radar applications.[26]

We now turn to the calculation of some asymptotic relative efficiencies for the sign test with respect to the linear detector (8-4). Let f be a probability density that is symmetric about the origin (a member of H_4'). We calculate ARE for fixed f and the simple alternatives in K_4' obtained by translation of f; that is, we consider f to be the probability density of the noise, and the signal to be a positive constant μ to which the noise is added, and "$P \to H$" or "weak-signal limit" here means $\mu \to 0$. In (8-3), let the sample sizes N and N' pertain, respectively, to the sign detector (8-6) and the linear detector (8-4), which is known to be optimum when f is gaussian with zero mean and known variance $\sigma^2 = \sigma_0^2$. Let us first consider this gaussian case: from (8-4) we immediately have

$$\beta = \Phi\left[\sqrt{n}\,\frac{\mu}{\sigma} - \Phi^{-1}(1-\alpha)\right] \tag{8-11}$$

Clearly N becomes larger than any desired bound, for fixed α and β, by restricting attention to sufficiently small signals μ (and the same is true for N'), and then we may pass to the asymptotic forms for (8-8) and (8-9) given by the central limit theorem to obtain

$$\beta \simeq \Phi\left[\sqrt{\frac{n}{pq}}\left(p - \frac{1}{2}\right) - \frac{\Phi^{-1}(1-\alpha)}{2\sqrt{pq}}\right] \tag{8-12}$$

with $q = 1 - p$. Replacement of n with N' in (8-11) and N in (8-12) yields close approximations. Equating these two expressions for β, we may solve for the ratio (8-3):

$$\frac{N'}{N} = \frac{\sigma^2(p - \frac{1}{2})^2}{\mu^2 p(1-p)} \tag{8-13}$$

neglecting a term of order $1/N$. Now, recalling that $p = P_\mu(x_i > 0)$, f is symmetric, and μ is small,

$$p = \tfrac{1}{2} + \int_0^\mu f(x - \mu)\,dx \simeq \tfrac{1}{2} + \mu f(0) \tag{8-14}$$

Using (8-14) in (8-13), and passing to the limit $\mu \to 0$, we obtain

$$\text{ARE} = 4\sigma^2 f^2(0) \tag{8-15}$$

and with $f(0) = (\sqrt{2\pi}\,\sigma)^{-1}$ in the gaussian case, the numerical value of the ARE is $2/\pi$. In this sense, for long-time observation of a weak additive constant signal in a white gaussian noise sequence of known intensity, the sign detector is 64 percent as efficient as the detector known to be optimal for the problem. If the noise intensity σ^2 is not known, the sign test has the same ARE $2/\pi$ with respect to the t test (8-5) appropriate for this situation, since (8-11) is correct as an asymptotic approximation to the detection probability for (8-5). Our proof of the well-known result (8-15) is somewhat skeletonized since connections among the several concurrent limiting processes must be made more explicit; the details are easily supplied (alternatively appeal may be made to general ARE theorems in the references previously cited).

The preceding derivation of (8-15) is unchanged when f is not gaussian; (8-15) is the ARE of the sign detector with respect to the linear detector (or the t test) for signal level $\mu \to 0$ with noise density f, which is symmetric and continuous at the origin and has finite variance σ^2 (the symmetry condition can be relaxed by requiring only that f assign weight $\frac{1}{2}$ to the positive axis). Evaluation of (8-15) may be expedited by noticing that the result is invariant with respect to the scale of f. As an example, for the double-exponential density $f(x) = a \cdot \exp(-b|x|)$, substitution of $f(x) = \exp(-2|x|)$ in (8-15) shows that the ARE is 2; thus the sign detector is twice as efficient as the linear detector in this case (we note in passing that it can be shown directly, by inspection of the likelihood ratio, that the sign detector is weak-signal optimal for this problem). In fact, for suitable noise densities, arbitrarily large ARE values can be attained. Such results are not entirely academic; in particular it can be shown that for certain impulse-type noise processes, reasonable models for the probability density yield ARE values in (8-15) significantly greater than one.[25] On the other hand, (8-15) can be zero, e.g., for a double-humped density f with $f(0) = 0$ (these conditions might be encountered when amplification and clipping precede the detector). Under the additional assumption that f is unimodal, it is easily demonstrated that (8-15) cannot fall below $\frac{1}{3}$ (the latter is attained for the uniform or rectangular density).[7] Results of this type, together with invariance properties such as those accompanying (8-10) and H_6, suggest in summary form some of the motivations for considering the sign test as a signal detector.

The Wilcoxon Detector. By retaining and processing polarities *and* ranking information for the input data sequence, one can achieve some ARE results markedly superior to those quoted above for the sign detector based solely on polarities. We consider a detector implementing the well-known Wilcoxon test, in which "signal present" is announced when the sum of the ranks of the positive observations among the absolute values exceeds a threshold. This rule may be written as

$$\sum_{k=1}^{n} d_k \gtrless C \tag{8-16}$$

where
$$d_k = \begin{cases} 0 & \text{if } x_{i_k} \leq 0 \\ k & \text{if } x_{i_k} > 0 \end{cases} \tag{8-17}$$

and
$$|x_{i_1}| < |x_{i_2}| < \cdots < |x_{i_n}|$$

[We ignore the question of equalities among magnitudes, since such events have zero probability when $F(x)$ is continuous; of course in practice ties must be anticipated, and we may then agree, e.g., to assign the group average of the ranks to each member of a tied group.] The set of integers d_k defined by (8-17) is one representation of what is often called the set of "signed ranks," since we have

$$-n(n+1) + 2\sum_{k=1}^{n} d_k = \sum_{i=1}^{n} k_i \operatorname{sgn}(x_i)$$

where k_i is the rank of $|x_i|$. When hypothesis H_4' is true, each of the 2^n sets of possible values of $(d_1, d_2, \ldots, d_n)$ has the same probability of occurrence; therefore (8-16), or any other detector based only on the d's, is nonparametric for H_4' and the relationship between the threshold C and the false-alarm rate α is determined by noting that exactly m of the 2^n sums Σd_k must exceed C, where $m/2^n$ is the desired false-alarm probability α. For large n, it can be shown that Σd_k under H_4' is approximately gaussian with mean $\mu_W = n(n+1)/4$ and variance

$$\sigma_W{}^2 = \frac{n^2(n+1)}{12}$$

so that C can be determined from $\Phi[(C - \mu_W)/\sigma_W] = 1 - \alpha$. For the hypothesis K_4', it is clear that positive observations will tend to occur more frequently and will tend to have larger ranks among the absolute values; therefore it is reasonable to decide that a signal is present if Σd_k is large.

The performance of (8-16) is particularly noteworthy from the ARE viewpoint. For comparison of the Wilcoxon structure with either the linear detector or the t test, the analog of (8-15) can easily be shown to be

$$\text{ARE} = 12\sigma^2 \left[\int_{-\infty}^{\infty} f^2(x)\, dx\right]^2 \tag{8-18}$$

for any symmetric density f. Thus in the gaussian case the result is $3/\pi$ or 95 percent efficiency with respect to the best detector. A minimization of (8-18) shows that the ARE is bounded below by $^{108}\!/_{125} = 0.864$ in *all cases* [and the lower bound is attained when $f(x)$ is of the form $a(b^2 - x^2)$ for $x^2 \leq b^2$].[7] This striking result should be contrasted with (8-15) where no lower bound exists unless additional restrictions (e.g., unimodality) are imposed.

While the quantity Σd_k is a readily computed function of the signs and ranks of the input data, it can also be put into a form that does not involve ranks explicitly, but only pairwise sums:

$$\sum_{k=1}^{n} d_k = \sum_{j=1}^{n} \sum_{i=j}^{n} u(x_i + x_j) \tag{8-19}$$

The expression on the right-hand side of (8-19), or its symmetrized equivalent where u is replaced by sgn, is perhaps easier to interpret as the response of a physical system whose input is the time sequence of amplitudes $x_1, x_2, \ldots, x_n$. Observe that (in contrast to the sign detector) this system must include some data-storage technique, since all pairs of inputs must be examined before the desired response can be produced. In return for this additional system complexity, one obtains an efficient nonparametric detector for H_4'.

Clearly, α for the Wilcoxon detector is unaffected by the nonlinearity (8-10) if ψ is continuous, odd, and strictly increasing. Moreover, α is constant for the following nonparametric version of H_4':

$$H_7: f_i(x) = f_i(-x) \qquad i = 1, 2, \ldots, n$$

(Ref. 3, pp. 242–243). Thus α is unaffected by predetection nonlinearities $x_i' = \psi_i(x_i)$ if each ψ_i satisfies the above conditions.

Although the ARE properties of the Wilcoxon detector are impressive, it is possible to obtain further improvement with other schemes based on the signed-ranks data set (8-17). In particular, the one-channel version of the normal scores test (also known as the Fisher-Yates test and the c_1 test) has ARE ≥ 1 with respect to the linear detector or the t test for any symmetric noise density with additive constant signal, and the minimum ARE of unity is attained for the gaussian density. In this sense the test is guaranteed to perform as well as or better than the test optimum for gaussian noise when the noise distribution is unknown. The normal-scores test is given by

$$\sum_{k=1}^{n} h_n(d_k) \gtrless C \tag{8-20}$$

where $h_n(i)$ is the expectation of the absolute value of the ith smallest among n independent gaussian variables with zero means and unit variances [and for notational convenience we have set $h(0) = 0$].

Among all structures based only on the signed ranks of the input data, (8-20) is locally optimal for weak signals and gaussian noise density. This optimality property, together with the "ARE $\geq$ 1" result quoted above, suggests that detector (8-20) should be employed in preference to the Wilcoxon detector [the two detectors, and others based on data reduction (8-18), have the same false-alarm invariance properties; both are nonparametric for H_4', H_7, etc.]. However, from the viewpoint of system implementation, (8-20) may be of somewhat less interest because of the necessity of storing or generating the numerical function $h_n(\cdot)$; the same practical difficulties apply to the asymptotically equivalent test of the van der Waerden type, in which $h_n(i)$ is replaced by $\Phi^{-1}[(n+1+i)/(2n+2)]$.

Time-varying Signals. Let $(\lambda s_1, \lambda s_2, \ldots, \lambda s_n)$ be the signal vector in a detection problem of the form "$\lambda = 0$" vs. "$\lambda > 0$," where the s_i are known constants, generalizing previously treated situations where we assumed $s_i = 1$ for all i ($\lambda = \mu$); for example, a natural extension of H_4' vs. K_4' is

$$H_4'': \text{each } x_i \text{ has density } f(x) \text{ symmetric about } 0$$
$$K_4'': x_i \text{ has density } f(x - \lambda s_i) \qquad i = 1, 2, \ldots, n$$

for some $\lambda > 0$, where f is symmetric about 0. When the noise density f in this model is gaussian with known variance, we have the well-known optimum detector

$$\Sigma s_i x_i \gtrless C$$

but for the nonparametric problem H_4'' vs. K_4'' with unknown noise densities f, we might first consider reduction of the data to signs, in which case an appropriate detector is

$$\Sigma s_i \operatorname{sgn}(x_i) \gtrless C$$

If the s_i are themselves unknown except for polarity, we could use

$$\Sigma \operatorname{sgn}(s_i) \operatorname{sgn}(x_i) \gtrless C$$

or equivalently $\Sigma u(s_i x_i) \gtrless C$. Evidently these sign detectors are nonparametric and have useful invariance properties with respect to predetection nonlinearities, as discussed in the case of the sign detector for a constant signal. Further details and calculations of ARE values are left to the reader as exercises.

The above sign-detector designs are based on the obvious premise that when a signal is present, x_i will tend to have the sign of s_i and large $|s_i|$ will often entail large $|x_i|$, so that the proposed sums of products will tend to be large. For the same reason, the following variation of the Wilcoxon test is appropriate for H_4'' vs. K_4'' if the data storage required in determining signed ranks is available:

$$\Sigma s_i k_i \operatorname{sgn}(x_i) \gtrless C$$

where k_i is the rank of $|x_i|$ among the x magnitudes (and as usual the ranking convention is that the largest has rank n). In the same way we may formulate a variant of the normal-scores test.

A multiple-channel version of the normal-scores test has been proposed as a signal detector by Daly and Rushforth;[10] we describe the structure briefly. Let x_{ij} be the output of the jth channel at time t_i, and suppose that there are m mutually independent channels, to each of which H_4'' or K_4'' applies simultaneously (i.e., the overall hypotheses refer to the simultaneous presence or absence of the same signal vector in m channels perturbed by independent white-noise sequences); let the total set of mn data values x_{ij} be pooled and arranged in order, and let $a_{ik} = 1$ if the kth smallest of these is an ith coordinate from some channel ($a_{ik} = 0$ otherwise). Then the generalized normal-scores structure can be written in the form

$$\sum_{i=1}^{n} s_i v_i \gtrless C$$

with

$$v_i = \sum_{k=1}^{mn} e_k a_{ik}$$

where the numerical weight e_k is the mean of the kth smallest among mn independent standardized gaussian variables. It has been shown that this detector enjoys the ARE properties of the one-channel normal-scores test; i.e., for gaussian noise the ARE is unity with respect to the structure

$$\sum_{i=1}^{n} s_i \left(\sum_{j=1}^{m} x_{ij} \right) \gtrless C$$

known to be optimal in the case of gaussian noise of given intensity, and for unknown noise distributions the ARE is guaranteed to be ≥ 1.

Phase-incoherent Signals. The phase-incoherent narrow-band signal is an engineering concept with no direct counterpart in conventional statistical theory. Construction of nonparametric detectors suitable for such signals may therefore require a nonroutine extension of known results. In the multichannel situation just reviewed above, Daly and Rushforth[10] also adapted their fixed-signal normal-scores test to apply the case of narrow-band phase-incoherent signals; the resulting structure again has ARE ≥ 1 with respect to the usual parametric detector optimal for gaussian noise (the standard envelope detector in this case). We omit the description of this multichannel structure and indicate instead how a one-channel Wilcoxon-type scheme could be formulated in similar fashion. For notational simplicity we assume that the signal is a pure sinusoid, although amplitude and phase modulation can obviously be inserted.

The problem is a restatement of (H_4'',K_4'') incorporating a phase random variable θ:

$$H_4''' = H_4''$$

$$K_4''': x_i = \lambda \cos\left(\frac{\omega Ti}{n} - \theta\right) + \eta_i \qquad \text{for } \lambda > 0$$

where each η_i has symmetric density f with zero mean, and θ is an unobservable random variable uniformly distributed in $0 \leq \theta \leq 2\pi$. Here, ω is the known signal frequency and T is the observation time. The conventional detector (optimal in the gaussian case subject to the usual assumptions $T \gg 1/\omega$ and $T/n \ll 1/\omega$) can be written as

$$M = \sqrt{A^2 + B^2} \gtrless C$$

where

$$A = \sum x_i \cos \frac{\omega Ti}{n}$$

$$B = \sum x_i \sin \frac{\omega Ti}{n}$$

This suggests that a corresponding Wilcoxon-type structure should also use $M \gtrless C$ with

$$A = \sum k_i \operatorname{sgn} x_i \cos \frac{\omega Ti}{n}$$

$$B = \sum k_i \operatorname{sgn} x_i \sin \frac{\omega Ti}{n}$$

where k_i is the rank of $|x_i|$ as defined previously. If it is desired to avoid ranking determinations and use only a sign test, $k_i = 1$ may be substituted in A and B. We omit the details of system performance evaluations; the basic results are comparable with those obtained for constant signals.

4. Detectors with a Reference-noise Source

Suppose that, in addition to n_1 independent observations $x_1, x_2, \ldots, x_{n_1}$ of the type utilized by the one-channel detection schemes of the preceding section, we can also obtain n_2 independent observations $y_1, y_2, \ldots, y_{n_2}$ taken under no-signal conditions, and the detector input data w consists of these $n_1 + n_2$ random variables. If the x's and y's are samples from two independent stationary random processes with first-order cumulative distribution functions F and G, respectively, and if we wish to detect the presence of a signal component in the x's with aid of the y's, it is appropriate to seek some method of comparison of the x's and y's that is sensitive to differences between F and G.

If the signal appears as an increase in level, the problem is one of comparison of locations; for example,

$$H_8: F(x) = G(x) \qquad -\infty < x < \infty$$
$$K_8: F(x) = G(x - \lambda) \qquad \text{for some } \lambda > 0$$

Here, if attention is restricted to gaussian distributions F and G with the same (unknown) variance, Student's t test for comparison of means provides the optimum unbiased detector, and if the gaussian restriction is removed, Student's test is easily seen to be asymptotically nonparametric for H_8. Moreover, a location-comparison version of the Wilcoxon test is exactly nonparametric for H_8, and has 95 percent asymptotic relative efficiency with respect to the t test, for gaussian alternatives as $\mu \to 0$, just as in the previous section where tests of (absolute) location were surveyed. The detector based on the Wilcoxon test (also called the Mann-Whitney test) here can be written in the form

$$\sum_{j=1}^{n_2} \sum_{i=1}^{n_1} u(x_i - y_j) \gtrless C$$

[this should be compared with (8-19)].

Several discussions of the application of these and other location-comparison statistical techniques to detection problems with a reference-noise source have appeared in the engineering literature.[11,12,13] Further details will not be repeated here, but it may be useful to remark that the reference-noise source from which the y's are to be obtained could be somewhat difficult to realize in practice, since its characteristics must be identical with those of the signal-bearing channel under no-signal conditions. Of course the output of the latter channel might itself serve as the reference source during the time intervals when it is known that a signal is not being transmitted, provided that these intervals are long enough to permit the establishment of suitable noise records $y_1, y_2, \ldots, y_{n_2}$, and provided that the channel characteristics do not change when one passes from such a reference interval to an interval in which a signal may be present.

5. Correlation Techniques

In this section, we consider some detection schemes based on two input channels in which the signal to be detected is a random process whose presence perturbs both inputs simultaneously; in the absence of a signal, it is assumed that the two inputs are independent noise processes. Thus the relevant statistical techniques are those suitable for testing a hypothesis of independence of the two inputs vs. an alternative corresponding to the type of dependence introduced by the signal in its role as a random component common to both inputs. We also mention briefly a generalization to $m > 2$ input channels. (Note that the m-channel techniques in Sec. 3 are suitable for sure signals, while in the present section the signal is supposed to be random.) Let the received data be $w = (w_1, w_2, \ldots, w_n)$, with $w_i = (x_i, y_i)$ where x_i and y_i are interpreted as data obtained from the two channels at the same time t_i. If $\{x_i\}$ and $\{y_i\}$ are stationary white sequences, the no-signal hypothesis reduces to a statement of the form $Q(x,y) = F(x)G(y)$, where Q is the two-dimensional distribution function for the random point (x_i, y_i), and F and G are the one-dimensional distribution functions for x_i and y_i, respectively.

Parametric Systems. Suppose that in the absence of a signal, the noises in the two channels are independent and gaussian with known variances σ_1^2 and σ_2^2 and known means, and that the random signal is also gaussian (independent of the others), with known mean and unknown variance σ_3^2, and appears additively in both channels. Under these assumptions the problem of testing the simple hypothesis $\sigma_3 = 0$ vs. the one-parameter alternative $\sigma_3 > 0$ has an optimum solution (the likelihood ratio detector structure is independent of the alternative σ_3); in particular, when the noise variances are equal, so that the correlation coefficient $\rho = \sigma_3^2/(\sigma_1^2 + \sigma_3^2)$, and all mean values are zero, the problem is

$$H_9\colon Q(x,y) \text{ is gaussian} \qquad \rho = 0$$

$$\sigma_x^2 = \sigma_y^2 = \sigma_1^2 \qquad \mu_x = \mu_y = 0$$

$$K_9\colon Q(x,y) \text{ is gaussian} \qquad \rho > 0$$

$$\sigma_x^2 = \sigma_y^2 = \frac{\sigma_1^2}{1-\rho} \qquad \mu_x = \mu_y = 0$$

and it is easy to verify that the optimum detector reduces to

$$\sum_{i=1}^{n} (x_i + y_i)^2 \gtrless C \tag{8-21}$$

If the parameters μ_x, μ_y, σ_x, σ_y are unknown, but it is still assumed that the distributions are gaussian, the problem becomes

$$H_{10}\colon Q(x,y) \text{ is gaussian} \qquad \rho = 0 \qquad \mu_x, \mu_y, \sigma_x, \sigma_y \text{ unspecified}$$
$$H_{10}\colon Q(x,y) \text{ is gaussian} \qquad \rho > 0 \qquad \mu_x, \mu_y, \sigma_x, \sigma_y \text{ unspecified}$$

In this case it has been shown that an optimum unbiased test exists and is given by

$$r = \frac{\Sigma x_i^* y_i^*}{(\Sigma x_i^{*2} \Sigma y_i^{*2})^{1/2}} \gtrless C \tag{8-22}$$

where $$x_i^* = x_i - \sum \frac{x_i}{n} \qquad y_i^* = y_i - \sum \frac{y_i}{n}$$

The statistic r is usually called the sample correlation coefficient. Under H_{10}, $[(n-2)r^2/(1-r^2)]^{1/2}$ has Student's t distribution with $n-2$ degrees of freedom; for large n, r is asymptotically gaussian with mean ρ and variance $(1-\rho^2)^2/n$.

If we drop the specification that the distributions are gaussian, it is easy to verify (assuming the existence of second moments) that the detector (8-22) is asymptotically nonparametric for a general hypothesis of independence

$$H_{11}: Q(x,y) = F(x)G(y)$$

and consistent for the alternative

$$K_{11}: \rho > 0 \qquad Q \text{ otherwise unspecified}$$

The hypothesis H_{11} is a natural generalization of H_9 or H_{10}, since for gaussian distributions the condition $\rho = 0$ is equivalent to independence; however, there is a gap between H_{11} and K_{11}, for there exist nongaussian bivariate distributions for which $\rho = 0$ does not imply independence. Nevertheless, situations of practical interest are properly represented by H_{11} and K_{11}; in particular, if the signal and noise are still assumed to be additive and independent, Q is in H_{11} when the signal is absent and $\rho = \sigma_3{}^2/\sigma_x\sigma_y$ so that $\rho > 0$ is equivalent to signal present.

Rank Correlation. If each x_i is replaced by its rank among the x observations and each y_i is replaced with its rank among the y observations, the correlation coefficient r defined in (8-22) becomes the *rank correlation coefficient* r', and the corresponding detector is nonparametric for H_{11}. [From (8-22), the denominator of r' is constant, and the r' detector reduces to rank-product sum $\gtrless C$. There are equivalent forms for r', as well as other techniques for testing H_{11} based on ranks, which will not be reviewed here.] The ARE of the r' test with respect to the r test is 0.91 as $\rho \to 0$ for gaussian distributions.[4] Thus we again have a situation in which the desirable invariance properties afforded by data reduction to ranks do not entail a significant loss in detection efficiency in the gaussian case (and the efficiency may exceed 1 in other cases).

The Polarity-coincidence Correlator. We now examine the consequences of reduction of data to polarity information for the two-channel problem.[14] The total number of points (x_i,y_i) that fall in the first and third quadrants of the xy plane can be written in the form

$$\sum_{i=1}^{n} \operatorname{sgn}(x_i) \operatorname{sgn}(y_i) \tag{8-23}$$

The *polarity-coincidence correlator* (PCC) decides that a signal is present when (8-23) exceeds a threshold. An equivalent form for the PCC is

$$\sum_{i=1}^{n} u(x_iy_i) \gtrless C \tag{8-24}$$

Since the distribution of $\Sigma u(x_i y_i)$ is binomial (n,p), where

$$p = 1 - F(0) - G(0) + 2Q(0,0) \tag{8-25}$$

is the probability that x_i and y_i have the same sign, the PCC is nonparametric for any hypothesis expressed in terms of p only, such as "$p = \frac{1}{2}$." In fact, (8-24) is a bivariate generalization of the sign test considered in Sec. 3 (both count the number of points which fall into some region); (8-24) is therefore the optimum detector for testing "$p = 0$" vs. "$p > 0$." These classes are too general in our present signal-detection context; it is perhaps more pertinent to consider PCC performance in some specific nonparametric version of the situation of an additive random signal common to the two channels, such as

$$H_{12}: Q(x,y) = F_1(x)F_2(y)$$
$$K_{12}: Q(x,y) = \int_{-\infty}^{\infty} F_1(x - z)F_2(y - z)\,dF_3(z)$$

where F_1, F_2, and F_3 are the (unspecified) distribution functions for the x-channel noise, the y-channel noise, and the signal, respectively. From (8-25), using the K_{12} form for Q, we obtain

$$p = \tfrac{1}{2} + 2\int[F_1(z) - \tfrac{1}{2}][F_2(z) - \tfrac{1}{2}]\,dF_3(z) \tag{8-26}$$

which shows that $p \geq \frac{1}{2}$ when the noise distributions have zero medians. Then we also have $p = \frac{1}{2}$ under H_{12}, so that the PCC is nonparametric for H_{12}, unbiased for K_{12}, and consistent for K_{12} [if we exclude pathological cases for which the integral in (8-26) vanishes]. Thus, in particular, the PCC is an appropriate detector when we append to H_{12} and K_{12} the assumption that each of the three F_j has a probability density that is symmetric about the origin.

Under H_{12} and K_{12}, let f_j, σ_j^2, and γ_j^4 denote the density, variance, and fourth central moment of F_j, $j = 1, 2, 3$. For weak signals (σ_3 much less than σ_1 or σ_2) (8-26) becomes

$$p \simeq \tfrac{1}{2} + 2f_1(0)f_2(0)\sigma_3^2,$$

[compare with (8-14)], and proceeding as in the proof of (8-15) in Sec. 3, we obtain

$$\text{ARE}_{(8\text{-}24):(8\text{-}21)} = (\gamma_1^4 + \gamma_2^4 + 4\sigma_1^2\sigma_2^2 - \sigma_1^4 - \sigma_2^4)f_1^2(0)f_2^2(0) \tag{8-27}$$

and

$$\text{ARE}_{(8\text{-}24):(8\text{-}22)} = 16\sigma_1^2\sigma_2^2 f_1^2(0)f_2^2(0) \tag{8-28}$$

for the asymptotic relative efficiencies (as $\sigma_3 \to 0$) of the PCC (8-24) with respect to the parametric detector (8-21) and the r detector (8-22). Note that from the ARE viewpoint, the r detector may be replaced with the structually simpler correlator detector

$$\Sigma x_i y_i \gtrless C \tag{8-29}$$

but of course this device does not possess the asymptotically nonparametric behavior (or the optimality properties) of the r detector (8-22). In the particular gaussian situation for which detector (8-21) is optimal, the ARE's (8-27) and (8-28) are $\frac{1}{2}(2/\pi)^2$ and $(2/\pi)^2$ (0.202 and 0.405), respectively. These losses in detection efficiency, sustained upon reduction of the data to signs in the gaussian case, are of course offset by gains in structural simplicity and false-alarm probability invariance. Furthermore (as in the case of the one-channel sign test) nongaussian noises may result in efficiencies exceeding one; for example, if $f_1 = f_2 =$ a double exponential density, (8-27) and (8-28) are 3.5 and 4.0, respectively.[14,25]

Polarity-coincidence Arrays. Multichannel generalizations of the PCC have been used to process the outputs of large arrays of receivers, particularly in the case of multiple hydrophone arrays in sonar. In such situations one is interested in detecting the presence of a signal in an ill-defined background that may be nonstationary, where the transmission medium may introduce time-variable nonlinearities. It is then appropriate to reduce the data to signs and make use of the invariance properties of polarity-coincidence detection schemes. (We are concerned here with detectors suitable for stochastic signals rather than the sure-signal multichannel problems mentioned in Sec. 3.) If there are m channels, with output x_{ij} from channel j at time t_i, the following is an example of a polarity coincidence array (PCA) detector:

$$\sum_{i=1}^{n} \left[\sum_{j=1}^{m} u(x_{ij}) - \frac{m}{2} \right]^2 \gtrless C \tag{8-30}$$

For $m = 2$, (8-30) reduces to the PCC. The absolute value of the quantity inside the brackets in (8-30) is the excess over $m/2$ of the number of channels having the most prevalent sign at time t_i. If the m channels have independent noise sources with zero medians, these excesses tend to be small under no-signal conditions, while increases are expected in the presence of an additive signal common to all channels.

In the case of stationary white-noise and signal sequences, the detector that is weak-signal optimum in the gaussian case is given by

$$\sum_{i=1}^{n} \sum\sum_{j \neq k} x_{ij} x_{ik} \gtrless C \tag{8-31}$$

When each channel has the same noise characteristics, the ARE of the PCA detector (8-30) with respect to the parametric system (8-31) is $16\sigma_1^4 f_1^4(0)$ [compare with (8-28)] as implied by the work of Kanefsky,[15] who also observes that (8-30) is weak-signal optimum for the gaussian case among processors based only on signs, and studies the weak-signal performance of other PCA detectors, in particular the intuitively reasonable rule

$$\sum_{i=1}^{n} v_i \gtrless C$$

with
$$v_i = \begin{cases} 0 & \text{if } \frac{m}{2} - C_1 \leq \sum_{j=1}^{m} u(x_{ij}) \leq \frac{m}{2} + C_1 \\ 1 & \text{otherwise} \end{cases}$$

is shown to introduce only a slight loss in performance with respect to (8-30) for large arrays ($m \to \infty$) when the optimum value $\sqrt{m}/2$ is chosen for the intermediate threshold C_1 (the choice $C_1 = m/2 - 1$, or $v_i = 1$ only when all channels have the same sign at time t_i, has sometimes been made, but is very inefficient for large arrays).

6. Operation in Colored Noise

In each of the preceding sections, white-noise sequences have been assumed. This may be the case in practice for some communications problems, or in radio astronomy (where polarity-coincidence processing for large antenna arrays has been considered), or in any situation where the sampling rate in acquiring discrete-time data from a continuous-time process can be made sufficiently small in comparison with the reciprocal of the process bandwidth. In problems such as those encountered in sonar, the latter may be an unrealistic restriction. It may be argued that because of their structural simplicity and invariance properties, the nonparametric detectors suggested in the white-noise case are of interest in any event. This is often true, but it must be noted that these structures may fail to be nonparametric when nonwhite-noise assumptions are admitted in the no-signal hypothesis; e.g., it is clear that for fixed C, the false-alarm probability α for the sign test (8-6) will differ for, say, a gaussian first-order Markov sequence in comparison to α for white noise. Indeed, for a fixed-noise correlation function α can vary when the noise distributions are changed. For the sign test and Markov sequences this has been demonstrated for the gaussian case vs. the case where the one-dimensional density is double-exponential;[16] α varies here even asymptotically ($n \to \infty$). However, invariance with respect to transformations of the type (8-10) shows that α is constant for the class of all processes that can be derived from a given process through memoryless nonlinear transformations that preserve zero-crossings; this property of

α is of considerable engineering significance and is evidently not shared by (8-4) or any of the other conventional parametric detectors. Such an invariance property is useful when, e.g., the incoming processes are gaussian with known power spectrum but are subject to the action of an unknown nonlinearity prior to the point of data processing.

Since sign and rank structures fail to be generally nonparametric for correlated noise, the value of the threshold-comparison constant C required to achieve a preassigned false-alarm probability α is a function of the noise distributions. Thus C cannot in general be fixed in advance (except by simulation) if the distributions of the noise are actually unknown in the correlated case. The same is true of the conventional parametric detectors with respect to which we might calculate ARE values; indeed the threshold-adjustment problem for the corresponding sign or rank tests is much less severe (since their invariance properties yield protection against various nonstationarities and predetection nonlinearities). On the other hand, it has been shown that the threshold constants C for parametric detectors such as (8-29) or (8-31) can be replaced by functions of the data constructed in such a way that the resulting detectors are asymptotically nonparametric for the class of noise processes with given correlation function (and satisfying certain regularity conditions) and in some cases are asymptotically much more efficient than the corresponding polarity-coincidence devices.[9]

As a typical example of the calculations that can be carried out for detectors based on signs in colored-noise situations, we evaluate the ARE of the two-channel PCC (8-24) with respect to the correlator (8-29) when the two noises and the signal have normalized correlation functions $\rho_1(k)$, $\rho_2(k)$, and $\rho_3(k)$, respectively; other assumptions and notations are as in Sec. 5. We suppose[9] that the random sequences are sampled data from processes for which a form of central limit theorem for dependent variables can be invoked to yield asymptotic gaussian distributions for the test statistics $V = \Sigma u(x_i y_i)$ and $V' = \Sigma x_i y_i$; this is the case if the processes satisfy so-called strong mixing conditions,[17] which we shall not repeat here (roughly, events defined by widely separated sets of random variables are required to be independent asymptotically as the separation is increased indefinitely). Such conditions are met, e.g., by wide classes of Markov processes, gaussian processes, impulse-noise processes, and processes obtained from these through, say, finite-memory nonlinear operations. As in (8-11), (8-12), etc., we equate β for (8-24) and (8-29), using the asymptotic gaussian distributions for V and V', solve for the ratio N'/N and pass to the weak-signal limit; thus, equivalently,

$$\text{ARE} = \lim \frac{[\mu_+(V) - \mu_0(V)]^2/\sigma_0^2(V)}{[\mu_+(V') - \mu_0(V')]^2/\sigma_0^2(V')} \tag{8-32}$$

where the limit is $\sigma_3 \to 0$ (and $n \to \infty$) and the subscripts + and 0 refer respectively to the no-signal ($\sigma_3 = 0$) and signal-present (σ_3 fixed > 0) simple hypotheses chosen. The evaluation of (8-32) is straightforward, since only the σ_0^2 expressions are needed. The other terms are identical to those already calculated in the case of white sequences; hence (8-28) appears as a multiplicative factor, and the ratio of (8-32) to (8-28) is

$$\frac{\text{ARE}_{\substack{\text{colored}\\ \text{noise}}}}{\text{ARE}_{\substack{\text{white}\\ \text{noise}}}} = \frac{1 + 2\sum_{k=1}^{\infty} \rho_1(k)\rho_2(k)}{1 + 2\sum_{k=1}^{\infty} \Gamma_1(k)\Gamma_2(k)} \tag{8-33}$$

where Γ_1 and Γ_2 are the correlation functions of hard-limited transformations, $\text{sgn}(\cdot)$, of the noises in the x and y channels, respectively.

According to (8-33), the ARE for colored noise is not a function of the signal correlation $\rho_3(k)$; in particular, for detection of a colored-signal sequence in white noise, the ARE is the same as that obtained for a white-signal sequence. The absence of ρ_3 in (8-33) is obvious by inspection of (8-32) since neither $\mu_+(V)$ nor $\mu_+(V')$ contain product-moment terms for the V and V' being compared here. [Of course (8-24) and (8-29) are both suboptimal if ρ_3 is known and the processes are gaussian, for then the weak-signal optimal detector, a quadratic operation on $\{x_i\}$ and $\{y_i\}$ with numerical weights involving values of ρ_3, has superior performance as a function of ρ_3. This structure reduces to (8-29) for white gaussian signal and noise sequences, and in general may be inferior to the PCC in nongaussian cases.]

If the noises are gaussian, then $\Gamma_j(k) = (2/\pi) \arcsin \rho_j(k)$, $j = 1, 2$. In this case it is easy to see from (8-32) that when the channel-noise correlation functions are identical [or more generally when the signs of $\rho_1(k)$ and $\rho_2(k)$ agree for each fixed k] the ARE for colored noise is at least as large as the ARE for white noise. The interpretation of this result is that although the PCC is not as efficient as the correlator in white gaussian noise, since the ARE is then 0.405, this performance ratio is maintained or improved for colored gaussian noise; i.e., for fixed false-alarm probability the PCC cannot suffer a greater asymptotic degradation in detection probability than that sustained by the correlator for given $\rho_1(k)$ and $\rho_2(k)$ in the gaussian case.

For the polarity-coincidence array detector (8-30) with respect to the product device (8-31), in the case where all channels have the same correlation function $\rho(k)$, expression (8-33) also holds with

$$\rho_1(k) = \rho_2(k) = \rho(k)$$

and has been evaluated numerically for some representative noise spectra in the gaussian case.[15] Actually, (8-32) amounts to a quotient of system signal-to-noise ratios for weak signals, and array detection using polarity coincidences has also been studied from the traditional viewpoint of signal-to-noise ratios[18,19] (without reference to nonparametric interpretations). For certain input nonstationarities of the fluctuating-intensity type encountered in sonar, a polarity-coincidence technique has marked advantages[19] over its parametric competitor, as might be expected.

In the particular study of nonstationarity just cited, a continuous-time version of the polarity-coincidence scheme is examined, and the signal and noise correlation functions are assumed to be shaped by the action of a given narrow-band prefilter (as might be the case in a sonar application). In general, it is clear that such colored-noise model assumptions must be made when considering continuous-time analogs of detectors based on signs or ranks (for these have no operational meaning for continuous-time *white* noise) and identifying discrete-time results for fast sampling with corresponding continuous-time results. In a related matter, we note that continuous-time analogs of the stochastic-signal-in-noise problems in Sec. 5 are singular[2] in the gaussian case; i.e., as is well known, it is possible to make correct decisions with probability one for continuous-time observation over any interval when the signal and the two additive noises are white gaussian processes, or nonwhite gaussian with common covariance function. For fixed observation time and increasing sampling rate or continuous-time processing, structures which fail to attain $\beta = 1$ are suboptimal in this sense, but are still of practical interest since no implementable system has $\beta = 1$. These model anomalies are of course not present in the discrete-time problems of Sec. 5, including their asymptotic form for fixed sampling rate and infinite observation time.

7. Supplementary Remarks

Since the definition of ARE involves both of the limits "signal $\rightarrow 0$" and $n \rightarrow \infty$ (observation time $\rightarrow \infty$), one might question the utility of ARE as a performance indicator in practice when the record length n is finite. However, numerical studies have frequently shown that for small n there is no gross violation of the conclusions reached on an ARE basis, and often the latter are actually reinforced for finite n. Most of the work of this type has been in standard statistical situations with independent random variables (e.g., Refs. 7, 20), but unpublished correlated-noise simulation results known to the author again lead to the conclusion that ARE is a conservative performance criterion for finite n. Evidently much additional computation and/or simulation of this nature is needed to present a more complete picture of nonparametric detector performance.

As noted in Sec. 6, the calculation of ARE can often[4] be reduced to a quotient of signal-to-noise ratios as in (8-32). However, there is clearly much more to be said in the general theory, in view of the several interdependent limits involved. We shall not attempt to summarize the many pertinent results here, except to note that (1) obviously interchanges in the order in which the limits are taken are not always possible;[7] (2) other types of efficiency comparisons can be made by holding some of the variables fixed in (8-3), passing to limit in the others, and studying the rates of convergence;[7,21] (3) when the structure of a weak-signal optimum test is known, ARE's may sometimes be evaluated conveniently by noting that the ARE of any test with respect to the weak-signal optimum is (under certain regularity conditions) the square of the asymptotic correlation function between the two test statistics under no-signal conditions.[8]

Conclusions based on ARE for two-hypothesis Neyman-Pearson situations may have less obvious relevance to communication-systems problems (where multiple hypotheses, coding, data rates, etc., are significant) than to surveillance-data-processing problems involving detection of fixed sources (when long records are available). However, communication systems design can benefit from the general observations that systems utilizing such qualitative data features as signs and ranks may be easily implemented, efficient, and insensitive to departures from assumptions. A specific communication systems example is Slepian's "permutation modulation" schemes[22] based on rankings, with an optimal decoding algorithm that remains invariant for a nonparametric class of noise distributions. Another illustration is given by the transmission of quantiles to achieve data compression in a space telemetry system.[23]

REFERENCES

1. Chapter 9.
2. Chapter 4.
3. Lehmann, E. L.: "Testing Statistical Hypotheses," John Wiley & Sons, Inc., New York, 1959.
4. Kendall, M. G., and A. Stuart: "The Advanced Theory of Statistics," vol. 2, Hafner Publishing Company, New York, 1961.
5. Fraser, D. A. S.: "Nonparametric Methods in Statistics," John Wiley & Sons, Inc., New York, 1957.
6. Carlyle, J. W., and J. B. Thomas: On Nonparametric Signal Detectors, *IEEE Trans. Inform. Theory*, vol. IT-10, pp. 146–152, April, 1964.
7. Hodges, J. L., Jr., and E. L. Lehmann: The Efficiency of Some Nonparametric Competitors to the *t*-test, *Ann. Math. Stat.*, vol. 27, pp. 324–335, 1956.
8. Van Eeden, C.: The Relation between Pitman's Asymptotic Relative Efficiency of Two Tests and the Correlation Coefficient between Their Test Statistics, *Ann. Math. Stat.*, vol. 34, pp. 1442–1451, 1963.

9. Kanefsky, M., and J. B. Thomas: On Adaptive Nonparametric Detection Systems Using Dependent Samples, *IEEE Trans. Inform. Theory*, vol. IT-11, pp. 521–526, October, 1965.
10. Daly, R. F., and C. K. Rushforth: Nonparametric Detection of a Signal of Known Form in Additive Noise, *IEEE Trans. Inform. Theory*, vol. IT-11, pp. 70–76, January, 1965.
11. Capon, J.: A Nonparametric Technique for the Detection of a Constant Signal in Additive Noise, *IRE WESCON Conv. Record*, pt. 4, pp. 92–103, 1959.
12. Capon, J.: Locally Optimum Nonparametric Methods for the Detection of Signals in Noise, *Proc. of the 1963 IEEE Symposium on Adaptive Processes*, pp. 64–77, September, 1963.
13. Millard, J. B., and L. Kurz: Nonparametric Signal Detection—an Application of the Kolmogorov-Smirnov, Cramèr-von Mises Tests (abstract), *IEEE Trans. Inform. Theory*, vol. IT-12, p. 275, April, 1966.
14. Wolff, S. S., J. B. Thomas, and T. R. Williams: The Polarity-coincidence Correlator; a Nonparametric Detection Device, *IRE Trans. Inform. Theory*, vol. IT-8, pp. 5–9, January, 1962.
15. Kanefsky, M.: Detection of Weak Signals with Polarity Coincidence Arrays, *IEEE Trans. Inform. Theory*, vol. IT-12, pp. 260–268, April, 1966.
16. Wolff, S. S., J. L. Gastwirth, and H. Rubin: The Sign Test on Autoregressive Data (abstract), *IEEE Trans. Inform. Theory*, vol. IT-12, p. 279, April, 1966.
17. Rosenblatt, M.: Independence and Dependence, *1961 Proc. Fourth Berkeley Symposium on Mathematical Statistics*, vol. 2, pp. 431–433.
18. Ekre, H.: Polarity Coincidence Correlation Detection of a Weak Noise Source, *IRE Trans. Inform. Theory*, vol. IT-9, pp. 18–23, January, 1963.
19. Thomas, J. B., and T. R. Williams: On the Detection of Signals in Nonstationary Noise by Product Arrays, *J. Acoust. Soc. Am.*, vol. 31, no. 4, pp. 453–462, April, 1959.
20. Arnold, H. J.: Small Sample Power for the One Sample Wilcoxon Test for Non-normal Shift Alternatives, *Ann. Math. Stat.*, vol. 36, pp. 1767–1778, December, 1965.
21. Klotz, J.: Alternative Efficiencies for Signed Rank Tests, *Ann. Math. Stat.*, vol. 36, pp. 1759–1766, December, 1965.
22. Slepian, D.: Permutation Modulation, *Proc. IEEE*, vol. 54, pp. 228–236, March, 1965.
23. Eisenberger, I., and E. C. Posner: Systematic Statistics Used for Data Compression in Space Telemetry, *J. Am. Stat. Assn.*, vol. 60, pp. 97–133, March, 1965.
24. Capon, J.: On the Asymptotic Relative Efficiency of Locally Optimum Detectors, *IEEE Trans. Inform. Theory*, vol. IT-7, pp. 67–72, April, 1961.
25. Kanefsky, M., and J. B. Thomas: On Polarity Detection Schemes with Non-Gaussian Inputs, *J. Franklin Inst.*, vol. 280, pp. 120–138, August, 1965.
26. Bello, P., and W. Higgins: Effect of Hard Limiting on the Probabilities of Incorrect Dismissal and False Alarm at the Output of an Envelope Detector, *IEEE Trans. Inform. Theory*, vol. IT-7, pp. 60–66, April, 1961.

2A

Reprinted with permission from *IEEE Proc.* **58**:623–631 (1970)

Nonparametric Detection

JOHN B. THOMAS, FELLOW, IEEE

Invited Paper

Abstract—This paper considers some of the simpler nonparametric detection schemes and compares their asymptotic relative efficiencies to those of detectors which are optimal in the Neyman–Pearson sense. In the one-input case, the nonparametric sign and Wilcoxon detectors are compared to the linear detector which is optimal for the detection of a dc signal of unknown amplitude in Gaussian noise. For two-input systems the nonparametric polarity coincidence correlator is compared to the system which is optimal for the detection of a common random Gaussian component in two-input Gaussian noises. The nonparametric detectors are shown to offer advantages in simplicity of implementation and in insensitivity to changes in input statistics while performing moderately well compared to the parametric detectors. More impressive results can be obtained with more complicated detectors utilizing nonlinear rank statistics.

I. Introduction

IN the introductory paper of this issue [1] and in most treatments of detection theory, emphasis is placed on the determination of *optimal detectors* requiring an essentially complete statistical description of the input signals and noises. There may be compelling reasons which lead to consideration of other (nonoptimal)detectors:

1) a complete statistical description of the input may not be available when the detector must be designed,
2) the statistics of the input data set may vary with time or may change from one application to another, or
3) the optimal detector may be too complex to implement.

Adaptive and learning detectors have been developed to meet the first two conditions and can perform in a near-optimal sense in an unknown or changing environment by proper adaptation of the detector structure; however, such detectors tend toward considerable complexity. *Nonparametric* or *distribution-free* detectors are intended to exhibit *insensitivity* to environment and *simplicity* in implementation at a cost of some deterioration in performance compared to optimal detectors.

Although the terms nonparametric and distribution-free are often used interchangeably, they are not synonyms and, in a certain sense, neither is an accurate description of the detectors to be discussed here. Distribution-free is used in the statistical literature to describe tests which make no assumptions on the form of the distributions of the input data set. Nonparametric refers to the notion that the possible distributions of the input data set comprise such a large class that they cannot be indexed by a finite number of real parameters.

Manuscript received November 7, 1969. This work was supported in part by the USAF Office of Scientific Research under AFOSR Grant 1333-67, in part by the National Science Foundation under Grant GK-1439, and in part by the U. S. Army Research Office under Contract DAHC04-69-C-0012.

The author is with the Department of Electrical Engineering, Princeton University, Princeton, N. J. 08540.

In this treatment, attention will be restricted to the two-decision problem "signal absent versus signal present." A detector will be called nonparametric[1] (or distribution-free) if its false-alarm probability is fixed under weak assumptions on the statistical character of the input data set. These weak assumptions will usually be taken to be lack of knowledge of the functional form of the input noise distributions. This definition of a nonparametric detector is not precise, but the objectives of such detector design will be made clear: these are certain minimum guaranteed performance levels in unknown or changing environments, often coupled with the requirement that the detector structure be simple or easily implemented.

The literature of nonparametric statistics is very large. A bibliography [2] published in 1962 gives more than 3000 references, and a three-volume handbook [3] is required to list the principal nonparametric procedures. On the other hand, elementary textbooks devoted entirely to nonparametric statistics have appeared only recently [4]–[7]; other statistics texts have usually included a chapter treating some of the simpler nonparametric techniques with little organization or justification. As a consequence, the impression exists among engineers that nonparametric statistics consists of a collection of unrelated techniques which may be applied to data sets with vaguely defined statistical properties. Although the field of nonparametric statistics lacks the compact structure of classical hypothesis testing, a reasonably well-organized body of knowledge is available. This knowledge has been applied only rarely to detection problems and the engineering literature is sparse. To the author's knowledge, no exhaustive bibliography on nonparametric detection exists, although we will list a number of the more accessible papers.

II. Definitions and Terminology

The input data set will be denoted by w and will be either a random process $\{w(t);\ 0 \le t \le T\}$ or a finite sequence of random variables $\{w_i;\ i=1, 2, \cdots, n\}$. The probability distribution of w will be denoted by P and will comprise a complete statistical description of w. It will be assumed either 1) that w consists of "noise only" in which case it will be said that the *hypothesis* H is in effect, or 2) that w consists of "noise and signal" in which case it will be said that the alternative K is in effect. An hypothesis (or alternative) will be called *simple* if it consists of one (and only one) completely specified form for P; otherwise it will be said to be *composite*. A composite hypothesis (or alternative) will be called *nonparametric* if it cannot be described by a finite number of real parameters.

A detector D is a device or system which examines w and decides that either the hypothesis H or the alternative K is in effect corresponding to "noise only" or "signal and noise," respectively.[2] Thus, as shown in Fig. 1, the detector

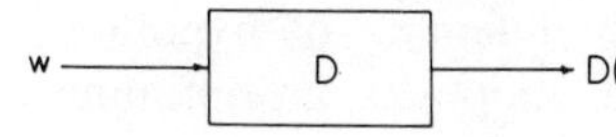

Fig. 1. The detector $D(w)$.

D can be represented as a system whose input is the data set w and whose output is the *decision function* (or *statistical test function*) $D(w)$. This function is a random variable with possible values 0 and 1; if $D(w)=0$, we accept H; if $D(w)=1$, we accept K.

The decision function $D(w)$ is a random variable with possible values 0 and 1. The mathematical expectation of $D(w)$ will be denoted by $\beta_D(P)$ where

$$\beta_D(P) = E\{D(w)\} \tag{1}$$

and will satisfy the inequality

$$0 \le \beta_D(P) \le 1. \tag{2}$$

The quantity $\beta_D(P)$ is the *power function* or *operating characteristic* of the detector D and is the probability of acceptance of K (or rejection of H). It is clear that it depends on the form of the detector and on the distribution P of the input data set w. In this treatment, we accept the viewpoint of the Neyman–Pearson criterion [1] that $\beta_D(P)$ describes the detector performance. Two detectors will be said to be *equivalent* if they have the same operating characteristic.

It is clear that the detector D can commit two kinds of errors: Type I errors where H is falsely rejected and Type II errors where K is falsely rejected. For obvious reasons, errors of Type I are called *false alarms* and errors of Type II are called *misses*. If $P \varepsilon H$, then $\beta_D(P)$ is the *false-alarm probability* while the quantity

$$\sup_{P \varepsilon H} \beta_D(P) = \alpha \tag{3}$$

is the *size*, *significance level*, or *maximum false-alarm rate* of the detector D.[3] If $P \varepsilon K$, then $\beta_D(P)$ is the *probability of (signal) detection* or *power* of the detector D for the alternative P. The object of most detector design is to make the false-alarm probability as small as possible while making the probability of detection as large as possible subject to constraints on realizability, cost, complexity, etc.

Let Δ be the class of all detectors satisfying

$$\beta_D(P) \le \alpha \quad \text{for all } P \varepsilon H. \tag{4}$$

A detector $D^* \varepsilon \Delta$ is said to be optimum, in the Neyman–Pearson sense (for hypothesis H, alternative K, and level α), if

$$\beta_{D^*}(P) \ge \beta_D(P) \quad \text{for all } P \varepsilon K \text{ and all } D \varepsilon \Delta; \tag{5}$$

that is, if D^* simultaneously maximizes all detection probabilities of interest subject to the condition that its false-alarm rate be no greater than the desired level α.

If H and K are simple, an optimum detector can always be constructed according to the procedure given by the Ney-

[1] It would be more accurate to call these *constant false-alarm rate* detectors.

[2] For brevity, these two situations will be denoted by $P \varepsilon H$ and $P \varepsilon K$, respectively. This notation is inconsistent (but conventional) since P is a mathematical function while H and K are hypotheses.

[3] The notation $\sup_{P \varepsilon H}$ refers to the *supremum* or *least upper bound* of $\beta_D(P)$ taken over all w with distributions $P \varepsilon H$.

man–Pearson fundamental lemma of hypothesis testing [1], [8]. If H and K are composite, an optimum detector usually does not exist. In such cases, one may search for a detector which is optimum within some class of detectors of special form. For example, a detector is said to be *unbiased* [9] (at level α) if (4) holds and if, in addition, $\beta_D(P) \geq \alpha$ for all $P \varepsilon K$; a detector D^* is said to be *optimum unbiased* if (5) holds when Δ is the class of unbiased detectors. One may also seek a detector which has an appropriate *local* optimum property [10], in the sense that (5) is satisfied not necessarily for all $P \varepsilon K$, but only for certain *weak-signal conditions*, that is, only in the limit as P becomes "close" to H in a suitably restricted sense. Finally, one may consider some reasonable preliminary reduction of the input data and then seek useful detectors among those which employ only the reduced data. Data reduction may be justified by invariance considerations [11], or it may be the result of such operations as frequency conversion, filtering, clipping, sampling of continuous waveforms, and ranking or grouping of discrete observations. In particular, all detectors discussed here, unless otherwise stated, base their decision upon a finite set $(w_1, w_2, \cdots, w_n)$ of *independent* samples of the input process. This assumption constitutes a strong restriction on the input but it will often be reasonable to suppose that sufficiently many such samples can be obtained from the processes of practical interest, if the total observation time is large in comparison to the reciprocal of the bandwidth.

Let $w_1, w_2, \cdots, w_n$ be independent observations of the input, and let $D(w_1, w_2, \cdots, w_n)$ be a detector based on the sample data; D is said to be *consistent* [12] for an alternative K if for any $P \varepsilon K$,

$$\lim_{n\to\infty} \beta_{D(w_1,w_2,\cdots,w_n)}(P) = 1 \tag{6}$$

that is, if the probability of detection for any fixed alternative can be made as large as desired when sufficiently many observations are available. It is tacitly assumed that for the hypothesis H of interest, D has the same false-alarm rate α for each n, or, if H is a nonparametric hypothesis, that D is at least *asymptotically nonparametric* (or *robust* [13] and [14]) in the sense that for all $P \varepsilon H$,

$$\lim_{n\to\infty} \beta_{D(w_1,w_2,\cdots,w_n)}(P) = \alpha. \tag{7}$$

As a general rule, each detection problem has many possible solutions; that is, given an hypothesis H and an alternative K, there are many (nonequivalent) detectors which can be implemented. It is therefore necessary to develop means for comparing the usefulness of various detectors. In this respect, two of the commonest situations encountered are 1) to compare two nonparametric detectors under the same hypothesis H, or 2) to compare a nonparametric detector and a parametric detector, both operating under a subclass of the hypothesis H (and/or alternative K).

The usual method of comparing two detectors (tests) for the same hypothesis H consists of computing the *relative efficiency* of one detector with respect to the other. Although it is possible to define many relative efficiencies, the one most commonly used is due to Pitman. Let $N_1 = n_{D_1}(\alpha, \beta, P)$ and $N_2 = n_{D_2}(\alpha, \beta, P)$ be the smallest number of observations required for detectors D_1 and D_2 to attain a probability of detection at least equal to β for the alternative $P \varepsilon K$ and for the same false-alarm probability α under the same hypothesis H. The relative efficiency $e_{2,1}$ of detector D_2 with respect to detector D_1 is defined as

$$e_{2,1} = \frac{n_{D_1}(\alpha, \beta, P)}{n_{D_2}(\alpha, \beta, P)}. \tag{8}$$

It should be clear that $e_{2,1}$ is a function of α, K, N_1, and N_2; and it may be difficult to evaluate for arbitrary α, K_1, N_1, and N_2. It is often simplest to let N_1 and N_2 approach infinity. Since in such a case the probability of detection (power) will approach unity (for consistent tests against fixed alternatives), it is necessary to allow the alternative K to approach the hypothesis H in such a way that the probability of detection (power) β remains constant as N_1 and $N_2 \to \infty$. We are led thus to the concept of *asymptotic relative efficiency* (ARE) $E_{1,2}$ where

$$E_{1,2} = \lim_{\substack{N_1\to\infty \\ N_2\to\infty \\ K\to H}} e_{1,2}(\alpha_1, K, N_1, N_2)|_{\beta=\beta_1} \tag{9}$$

and α_1 and β_1 are fixed values of α and β. In many cases, $E_{1,2}$ is just a number, independent of α.

Since the definition of ARE has been obtained by allowing K to approach H, it is basically a weak-signal measure of detector performance. It is clear that it may provide a useful figure of merit less than unity for a detector D_2 compared to a detector D_1 which is optimal for the (H, K) being considered. On the other hand, the ARE of D_2 compared to D_1 may exceed unity for some (H, K) for which D_1 is not optimal; that is, it may be possible to obtain superior performance from a suboptimal detector which is less sensitive to departures from assumptions and which is operating in an environment different from that for which the optimal detector was designed.

In subsequent sections, we will discuss and compare specific parametric and nonparametric detection schemes. The most disconcerting aspect of nonparametric detection is the lack of a simple constructive procedure corresponding to the Neyman–Pearson lemma for the parametric case. There would appear to be no way around this problem since it is an inherent aspect of nonparametric techniques. As mentioned earlier, however, elementary nonparametric statistics is relatively well organized. Also, it might be well to keep in mind that the structural simplicity and compactness of parametric detection theory is somewhat misleading. As a practical matter, the only parametric problem that has received extensive attention is the rather specialized one where the input under the hypothesis is stationary Gaussian noise.

In Section III, we will consider detectors with one input; that is, when $w = (w_1, w_2, \cdots, w_n)$. In Section IV, we will treat the two-input case where a signal appears as a component common to both inputs. In this case, the input data are the set of pairs $w = [(u_1, v_1), (u_2, v_2), \cdots, (u_n, v_n)]$ where the pair (u_i, v_i) are two inputs observed simultaneously at some time t_i. Some comments will be made and references given for extension to the n-input case. The so-called "one-and-a-half" input case [15]–[17] where a noise reference source is available will not be treated due to lack of space. Some readers may find two earlier surveys [18], [19] more comprehensive and somewhat more advanced.

III. One-Input Detectors

In this section we consider the case where the input data are a single sequence

$$w = (w_1, w_2, \cdots, w_n) \tag{10}$$

of independent random variables, each with density function $f(w_i)$ and cumulative distribution function $F_i(x) = P\{w_i \leq x\}$. The mean of each w_i will be denoted by μ_i and the variance by σ_i^2.

The one-input problem most often treated [20] is the detection of a sure (deterministic) signal $\{s_1, s_2, \cdots, s_n\}$ in white Gaussian noise $\{n_1, n_2, \cdots, n_n\}$ of zero mean and fixed variance σ_0^2. In this case, the simple hypothesis corresponding to $w_i = n_i$ will be called H_0 and the composite (parametric) alternative corresponding to $w_i = s_i + n_i$ will be called K_0 or, equivalently,

$$H_0: F_i(x) \text{ is Gaussian}, \quad \mu_i = 0, \sigma_i^2 = \sigma_0^2$$
$$K_0: F_i(x) \text{ is Gaussian}, \quad \mu_i = s_i, \sigma_i^2 = \sigma_0^2$$

The joint density functions of the w_i under the hypothesis H_0 and the alternative K_0 will be denoted by $f_0(w)$ and $f_1(w)$, respectively, where it is easy to see that [21]

$$f_0(w) = (2\pi\sigma_0^2)^{-n/2} \exp\left[-\sum_{k=1}^{n} w_k^2/2\sigma_0^2\right] \tag{11}$$

and

$$f_1(w) = (2\pi\sigma_0^2)^{-n/2} \exp\left[-\sum_{k=1}^{n} (w_k - s_k)^2/2\sigma_0^2\right]. \tag{12}$$

According to the Neyman–Pearson lemma [1], [8], the likelihood ratio

$$L(w) = f_1(w)/f_0(w) \tag{13}$$

is calculated for the input data w and is compared to a threshold L_0. If $L(w) < L_0$, then $D(w) = 0$ and we accept H_0, "noise only." If $L(w) > L_0$, then $D(w) = 1$ and we accept K_0, "signal and noise."

In this particular case, the likelihood ratio is easy to find by inspection:

$$L(w) = \exp\left[\sum_{k=1}^{n} (2s_k w_k - w_k^2)/2\sigma_0^2\right]. \tag{14}$$

Since σ_0^2 and each s_k are known, it is clear that comparing $L(w)$ to a threshold L_0 is equivalent to comparing

$$\sum_{k=1}^{n} s_k w_k$$

to some other number C_0 so that the detector D_0 becomes

$$\sum_{n=1}^{n} s_k w_k \begin{cases} < C_0 \Rightarrow D_0(w) = 0. \\ > C_0 \Rightarrow D_0(w) = 1. \end{cases} \tag{15}$$

Here C_0 (or L_0) is obtained by fixing the false-alarm probability α, as discussed in detail in [21].

The detector D_0 of (15) is often called a *correlation detector* or *correlator* since it calculates the time cross correlation function of the (known) signal s and the input data w. The correlation detector is implemented by a *matched filter;* thus the left side of (15) can be obtained by passing the input data sequence $w = (w_1, w_2, \cdots, w_n)$ through a filter with impulse response h_k given by

$$h_k = \begin{cases} s_{n-k}, & k = 1, 2, \cdots, n \\ 0, & \text{elsewhere.} \end{cases}$$

In the continuous-time case, the left side of (15) can be written as [21]

$$\int_0^T s(t)w(t)dt$$

corresponding to the output of a matched filter with input $w(t)$ and impulse response $h(t)$ given by

$$h(t) = \begin{cases} s(T-t), & 0 \leq t \leq T \\ 0, & \text{elsewhere.} \end{cases}$$

This last expression is the more familiar form.

A simpler case occurs when the signal is taken to be an unknown positive constant, corresponding to the detection of a positive dc signal of unknown amplitude in Gaussian noise. Now the simple hypothesis H_1 and the composite (parametric) alternative K_1 are

$$H_1: F_i(x) \text{ is Gaussian}, \quad \mu_i = 0, \sigma_i^2 = \sigma_0^2$$
$$K_1: F_i(x) \text{ is Gaussian}, \quad \mu_i = \mu > 0, \sigma_i^2 = \sigma_0^2.$$

Since s_k is constant for all k, the detector D_0 of (15) becomes the *linear detector* D_1 given by

$$\sum_{k=1}^{n} w_k \begin{cases} < C_1 \Rightarrow D_1(w) = 0 \\ > C_1 \Rightarrow D_1(w) = 1 \end{cases} \tag{16}$$

where, as before, C_1 is obtained by fixing α. In this case, it is easy to see that the density function of

$$\sum_{k=1}^{n} w_k$$

under H_1 is a Gaussian density with zero mean and variance $n\sigma_0^2$. Thus C_1 is obtained from

$$\alpha = \frac{1}{\sqrt{2\pi n}\sigma_0} \int_{C_1}^{\infty} e^{-x^2/2n\sigma_0^2} dx \tag{17}$$

or, equivalently,

$$C_1 = \sqrt{n}\sigma_0 K \tag{18}$$

where K is found from

$$\Phi(K) = 1 - \alpha \tag{19}$$

and $\Phi(\cdot)$ is the unit normal cumulative distribution function.

Let us now consider a much broader class of detection problems than those described by (H_0, K_0) and (H_1, K_1). In the process it will become clear that the median of a distribution is a natural generalization in the nonparametric case of the mean of a Gaussian distribution.[4]

Let the w_i of (10) be independent and identically distributed with the common cumulative distribution function $F(x)=P\{w_i \le x\}$. Define the probability p by

$$p = P\{w_i > 0\} = 1 - F(0). \tag{20}$$

Consider the *nonparametric hypothesis* H_2 and the *nonparametric alternative* K_2 where

$$H_2: p = 1/2, \quad F(x) \text{ otherwise arbitrary}$$
$$K_2: p > 1/2, \quad F(x) \text{ otherwise arbitrary.}$$

This is equivalent to assuming as the hypothesis that the w_i are independent noise observations with zero medians versus the alternative that the medians of the independent observations are greater than zero due to the presence of an additive positive signal.

A Neyman–Pearson optimal detector $D_2(w)$ exists for (H_2, K_2) and is easily found. Let f be the density function of w_i under any signal-present condition (i.e., under any member of K_2). Define f^+ and f^- as the conditional densities of w_i given $x_i > 0$ and $w_i \le 0$, respectively. From the definition of conditional density, it follows that

$$f = pf^+ + (1-p)f^-. \tag{21}$$

Define a new density function f_0 by

$$f_0 = (f^+ + f^-)/2. \tag{22}$$

It is clear that f_0 is a density function and that it belongs to a random variable with zero median since it can be obtained from (21) by setting $p = 1/2$. Thus f_0 is an acceptable density under the signal-absent condition, that is, under H_2.

The Neyman–Pearson lemma for testing the simple hypothesis f_0 versus the simple alternative f requires the formation of the likelihood ratio

$$L_2(w) = \prod_{i=1}^{n} f(w_i)/f_0(w_i). \tag{23}$$

The multiplicand in this last expression takes on one value if $w_i > 0$ and another if $w_i \le 0$:

$$\frac{f(w_i)}{f_0(w_i)} = \frac{pf^+}{\frac{1}{2}f^+} = 2p, \qquad w_i > 0 \tag{24}$$

and

$$\frac{f(w_i)}{f_0(w_i)} = \frac{(1-p)f^-}{\frac{1}{2}f^-} = 2(1-p), \qquad w_i \le 0. \tag{25}$$

Thus (23) becomes

$$L_2(w) = 2^n p^{\sum_{i=1}^{n} u(w_i)} (1-p)^{n - \sum_{i=1}^{n} u(w_i)} \tag{26}$$

where $u(w_i)$ is the unit step function given by

$$u(w_i) = \begin{cases} 1, & w_i > 0 \\ 0, & w_i \le 0. \end{cases} \tag{27}$$

It is clear that comparing L_2 to a threshold L_{20} is equivalent to comparing

$$\sum_{i=1}^{n} u(w_i)$$

to another constant C_2. Thus, the Neyman–Pearson detector $D_2(w)$ is the so-called *sign detector* and is given by

$$\sum_{i=1}^{n} u(w_i) \begin{cases} < C_2 \Rightarrow D_2(w) = 0 \\ > C_2 \Rightarrow D_2(w) = 1. \end{cases} \tag{28}$$

As usual, the constant C_2 is determined by fixing the false-alarm rate α. The quantity $\sum_i u(w_i)$ is the number k of posititive observations (successes) and is a random variable which is binomially distributed with parameter p. Thus k has a (discrete) density function under K_2 given by the binomial density

$$b(k; n, p) = \binom{n}{k} p^k (1-p)^{n-k}, \qquad k = 0, 1, \cdots, n \tag{29}$$

and under H_2 by

$$b(k; n, 1/2) = \binom{n}{k}\left(\frac{1}{2}\right)^n, \qquad k = 0, 1, \cdots, n \tag{30}$$

where $\binom{n}{k}$ is the binomial coefficient.[5] The operating characteristic of the detector D_2 is given by (1):

$$\beta_{D_2}(F) = \sum_{k=C_2+1}^{n} \binom{n}{k} p^k (1-p)^{n-k} \tag{31}$$

where C_2 is the largest integer satisfying

$$\alpha \le \sum_{k=C_2+1}^{n} \binom{n}{k}\left(\frac{1}{2}\right)^n \tag{32}$$

and α is the predetermined false-alarm rate.

Note from (32) that the value of C_2 (and the false-alarm rate) is the same for all distributions under H_2; that is, for all input noises with zero medians. Thus *the sign detector is a nonparametric detector for* H_2. Of more importance, it is clear from (32) that *the sign detector is nonparametric for any hypothesis which is a subclass of* H_2; for example, for

[4] The *median* m of a distribution function $F(x)=P\{X \le x\}$ is defined by $F(m)=\frac{1}{2}$. If the corresponding density function $f(x)$ is symmetric about the mean, then the mean and the median coincide.

[5] Note that (26) for the likelihood ratio can also be obtained by taking the ratio of (30) to (29).

the hypotheses H_0 and H_1. Of course, the sign detector is suboptimal for (H_0, K_0) and for (H_1, K_1) as well as for any other (H, K) where H is a subclass of H_2. However, it is easy to implement: it is nonparametric under very general conditions; and it will turn out that its asymptotic performance (as the number of input data sample points becomes large) is not too poor compared to the performance of optimal detectors. Also, as mentioned previously, the performance of the sign detector may be superior to that of an optimal detector operating under an inappropriate (H, K).

We proceed now to compare the sign detector of (28) to the linear detector of (16) using the concept of ARE given by (9). We shall evoke a simple form of the central limit theorem which states that sums of independent and identically distributed random variables tend in the limit to normal or Gaussian distributions. In particular, as n becomes large, the binomial density function $b(k; n, p)$ of (29) with mean np and variance $np(1-p)$ tends to the Gaussian density function

$$\eta(x) = \frac{1}{\sqrt{2\pi}\sigma} \exp\left[-\frac{1}{2}\left(\frac{x-\mu}{\sigma}\right)^2\right] \qquad (33)$$

with mean $\mu = np$ and variance $\sigma^2 = np(1-p)$. Actually, as pointed out by Feller [22], the convergence to the normal density is very rapid if p is not too far from one-half.

As before, let

$$k = \sum_{i=1}^{n} u(w_i)$$

be the number of positive observations from the input data set $w = (w_1, w_2, \cdots, w_n)$. In the limit as $n \to \infty$, (31) becomes the asymptotic probability of signal detection β_2 for the sign detector. On invoking the central limit theorem and using (33), we have

$$\beta_2 = \lim_{n\to\infty} \beta_{D_2}(F) = \int_{C_2}^{\infty} \eta(x)dx. \qquad (34)$$

After a change in variable, this last integral can be written as

$$\beta_2 = 1 - \Phi\left[\frac{C_2 - np}{\sqrt{np(1-p)}}\right] = \Phi\left[\frac{np - C_2}{\sqrt{np(1-p)}}\right] \qquad (35)$$

where, as before, $\Phi(\cdot)$ is the unit normal cumulative distribution function. It is clear that the limiting expression for (32) is obtained by setting $p = 1/2$ in (35); thus the asymptotic false-alarm rate for the sign detector is

$$\alpha_2 = \lim_{n\to\infty} \alpha = 1 - \Phi\left[\frac{2C_2 - n}{\sqrt{n}}\right]. \qquad (36)$$

Now the threshold C_2 may be eliminated from (35) and (36) to yield

$$\beta_2 = \Phi\left[\frac{\sqrt{n}(2p-1) - \Phi^{-1}(1-\alpha_2)}{2\sqrt{p(1-p)}}\right] \qquad (37)$$

where $\Phi^{-1}(\cdot)$ is the inverse of $\Phi(\cdot)$.

Let us now consider the linear detector of (16) operating on the input data set $w = (w_1, w_2, \cdots, w_m)$ consisting of m observations and obeying (H_2, K_2). If the w_i are Gaussian (as was assumed in deriving the linear detector from the Neyman–Pearson lemma), then $\sum_{i=1}^{m} w_i$ is Gaussian also. On the other hand, if the w_i are only identically distributed with means μ and variances σ^2, the sum $\sum_{i=1}^{m} w_i$ is asymptotically Gaussian with mean $m\mu$ and variance $m\sigma^2$. Thus the asymptotic probability of signal detection β_1 for the linear detector is, from comparison with (35),

$$\beta_1 = \Phi\left[\frac{m\mu - C_1}{\sqrt{m}\sigma}\right] \qquad (38)$$

The asymptotic false-alarm probability α_1 can be obtained by assuming the w_i have zero mean[6] under H_2. Now $\sum_{i=1}^{m} w_i$ is asymptotically Gaussian under H_2 with zero mean and variance $m\sigma^2$ and α_1 becomes

$$\alpha_1 = \frac{1}{\sqrt{2\pi m}\sigma}\int_{C_1}^{\infty} e^{-x^2/2m\sigma^2} dx = 1 - \Phi\left[\frac{C_1}{\sqrt{m}\sigma}\right]. \qquad (39)$$

Again the threshold C_1 may be eliminated from (38) and (39) to yield

$$\beta_1 = \Phi\left[\sqrt{m}\frac{\mu}{\sigma} - \Phi^{-1}(1-\alpha_1)\right]. \qquad (40)$$

An examination of (9) shows that the asymptotic relative efficiency $E_{2,1}$ of the sign detector with respect to the linear detector may be found by setting $\beta_1 = \beta_2$ and $\alpha_1 = \alpha_2$ and solving for the ratio of m to n. The result is

$$E_{2,1} = \lim_{\substack{n\to\infty \\ m\to\infty}} \frac{m}{n} = \frac{\sigma^2(p-1/2)^2}{\mu^2 p(1-p)}. \qquad (41)$$

We have already assumed that $p > 1/2$, that the noise distribution f is symmetric, and that μ is small (i.e., that a weak-signal condition holds). Thus p may be approximated by

$$p = 1/2 + \int_0^{\mu} f(x-\mu)dx \approx \frac{1}{2} + \mu f(0). \qquad (42)$$

Now (41) becomes

$$E_{2,1} \approx \frac{4\sigma^2 f^2(0)}{1 - 4\mu^2 f^2(0)}. \qquad (43)$$

In the limit of vanishingly small signals, $\mu \to 0$ and we have

$$E_{2,1} \approx 4\sigma^2 f^2(0). \qquad (44)$$

This last expression gives the ARE of the sign detector with respect to the linear detector for the detection of a dc signal in any noise having a symmetric density function f.

If the noise is Gaussian, then $f(0) = 1/\sqrt{2\pi}\,\sigma$ and the

[6] This is equivalent to assuming that the density function f of the w_i under H_2 is symmetric so that the mean and the median coincide and are zero.

right side of (44) becomes $2/\pi$. Recall that this corresponds to (H_1, K_1) for which the linear detector is optimal. Thus the sign detector is only about 64 percent as efficient as the optimal detector for a dc signal in Gaussian noise. On the other hand, suppose that the noise has a two-sided exponential distribution so that f is given by

$$f(x) = \frac{\alpha}{2} e^{-\alpha|x|}, \qquad -\infty < x < \infty. \tag{45}$$

This density function corresponds to a random variable with zero mean and variance $\sigma^2 = 2/\alpha^2$. In this case, (44) becomes $E_{2,1} = 2$ and the sign detector is twice as efficient as the linear detector. More generally, it is obvious that (44) will be greater than unity whenever

$$\frac{1}{f(0)} < 2\sigma = 2\int_{-\infty}^{\infty} x^2 f(x)dx. \tag{46}$$

(Recall that f not only is symmetric, but has zero mean.) Thus any noise with a symmetric density function which falls off slowly enough will yield an ARE greater than unity in comparing the sign detector to the linear detector.

Since the sign detector was derived as the (Neyman-Pearson) optimal detector for (H_2, K_2), it is clear that it is a natural test for a shift in median. However, it uses relatively little information from the input data set and, specifically, does not take account of the distance of an observation w_i from the origin. Instead of $\sum_{i=1}^{n} u(w_i)$, it might be more reasonable to consider some statistic $\sum_{i=1}^{n} \lambda_i u(w_i)$ where the λ_i are weights reflecting the distance of an observation w_i from the origin. Let us *rank* the observed data $w = (w_1, w_2, \cdots, w_n)$ by ordering the absolute values of the w_1 so that

$$|w_{k_1}| < |w_{k_2}| < \cdots < |w_{k_n}|. \tag{47}$$

Now consider the detector given by

$$\sum_{i=1}^{n} d_i \begin{cases} < C_2' \Rightarrow D_2'(w) = 0 \\ > C_2' \Rightarrow D_2'(w) = 1 \end{cases} \tag{48}$$

where d_i is given by

$$d_i = \begin{cases} 0, & w_{k_i} \le 0 \\ i, & w_{k_i} > 0. \end{cases} \tag{49}$$

The detector $D_2'(w)$ is called the *Wilcoxon detector* and is easily compared to the linear detector [6], [18], [19], [23]. Space precludes the calculations, but the ARE of the Wilcoxon detector with respect to the linear detector is given by

$$E_{2',1} = 12\sigma^2\left[\int_{-\infty}^{\infty} f^2(x)\,dx\right]^2 \tag{50}$$

where, as before, f is the noise density (that is, the density of one of the w_i under H) and σ^2 is the noise variance. For the Gaussian case, $E_{2',1}$ becomes

$$E_{2',1} = \frac{3}{\pi}\left[\frac{1}{\sqrt{2\pi}}\int_{-\infty}^{\infty} e^{-y^2/2}dy\right]^2 = \frac{3}{\pi} = 0.955. \tag{51}$$

For the two-sided exponential density (45) we have

$$E_{2',1} = \frac{3}{2}\left[2\alpha\int_{0}^{\infty} e^{-2\alpha x}dx\right]^2 = \frac{3}{2} = 1.5. \tag{52}$$

Thus the Wilcoxon detector is more than 95 percent as efficient as the linear detector for Gaussian noise and, as might be expected, considerably better for noise with a two-sided exponential distribution. A more surprising result [23] is that $E_{2',1}$ given by (50) cannot be less than a positive lower limit of 0.864 for any f.

The Wilcoxon detector of (48) is a *linear rank detector* as is the closely related Mann–Whitney detector [24]. Even more impressive results can be obtained with rank statistics which are not linear [19], [24]. The Kolmogorov-Smirnov detector is especially interesting [24] and can be expressed as a maximum of a finite number of linear rank detectors. Also included in this general type are the Rènyi and the Cramèr-von Mises detectors [24]. The complexity of implementation of such schemes grows rapidly, of course. However, with the increasing speed and memory size of digital computers, such detectors become more practical and merit further attention. Another fruitful approach [25], [26] is to rank small blocks $(w_{i+1}, w_{i+2}, \cdots, w_{i+k})$, $k << n$, of data and form "mixed statistics" or detectors based on naturally ordered blocks of ranked data. This approach reduces greatly the complexity of implementation and the memory and/or storage requirements of the detector. The analyses of most detectors employing ranking tend to be complicated and to rely critically on the applicability of the central limit theorem.

Lack of space precludes further discussion of the many one-input systems summarized and tabulated in the literature [3], [4], [7]. We now consider very briefly some two-input systems which may be considered of practical importance in the context of signal detection.

IV. Two-Input Systems

A class of two-input systems is often encountered where a common signal may exist in two channels perturbed by independent noises. Applications exist in the fields of radio astronomy [27]–[29], underwater sound detection [30]–[32], and geophysics [33], [34]. In this case, as mentioned in Section II, the input data are the set of pairs

$$w = [(u_1, v_1), (u_2, v_2), \cdots, (u_n, v_n)]; \tag{53}$$

when w consists of "noise alone," the $\{u_i\}$ and $\{v_i\}$ are independent noise processes. When w consists of "signal plus noise," a common random component $\{s_i\}$ is introduced into both $\{u_i\}$ and $\{v_i\}$.

We shall parallel to some degree the treatment of one-input detectors and begin by assuming that, under H, the processes $u = (u_1, u_2, \cdots, u_n)$ and $v = (v_1, v_2, \cdots, v_n)$ are independent sequences of independent noises with zero means and fixed variances $E_H\{u_i^2\} = \sigma_1^2$ and $E_H\{v_i^2\} = \sigma_2^2$. In contradistinction to the one-input case, we will assume that, under K, a common *random* signal $s = (s_1, s_2, \cdots, s_n)$ will be added to both u and v. The s_i will have zero means, a fixed variance $E\{s_i^2\}=\sigma_3^2$, and will be independent of each other and of the u and v processes.

Let us now consider the special case where the u, v, and s are Gaussian and require that $\sigma_1^2=\sigma_2^2=\sigma^2$; now the testing hypothesis and alternative become

$$H_3: Q(u, v) \text{ is Gaussian}, \quad \rho = 0, \sigma_u^2 = \sigma_v^2 = \sigma^2$$

$$K_3: Q(u, v) \text{ is Gaussian}, \quad \rho > 0, \sigma_u^2 = \sigma_v^2 = \frac{\sigma^2}{1-\rho}$$

where $Q(u, v)$ is the joint distribution of u and v, $\sigma_u^2=E\{u_i^2\}$, $\sigma_v^2=E\{v_i^2\}$, and ρ is the correlation coefficient given by

$$\rho = \frac{\sigma_3^2}{\sigma^2 + \sigma_3^2} \tag{54}$$

and obeying the inequalities $0 \le \rho \le 1$. For (H_3, K_3), a (Neyman-Pearson) optimum detector exists and is found by forming the likelihood ratio $L_3(w)=P_{K_3}(w)/P_{H_3}(w)$. The calculation is straightforward and the result is the detector $D_3(w)$ where

$$\sum_{i=1}^{n} (u_i + v_i)^2 \begin{cases} < C_3 \Rightarrow D_3(w) = 0 \\ > C_3 \Rightarrow D_3(w) = 1 \end{cases} \tag{55}$$

and C_3 is determined by fixing the false-alarm rate.

A much more general form of (H_3, K_3) arises on assuming only that the noises and signal are independent with density functions symmetric about the origin. Let the (arbitrary) distribution functions F_1, F_2, and F_3 be given by

$$F_1(x) = P_H\{u_i \le x\}, \quad \text{all } i \tag{56}$$

$$F_2(x) = P_H\{v_i \le x\}, \quad \text{all } i \tag{57}$$

$$F_3(x) = P\{s_i \le x\}, \quad \text{all } i$$

and let the corresponding symmetric density functions be denoted by f_1, f_2, and f_3. We have the nonparametric hypothesis and alternative

$$H_4: Q(u, v) = F_1(u)F_2(v)$$

$$K_4: Q(u, v) = \int_{\infty}^{\infty} F_1(u - x)F_2(v - x)dF_3(x)$$

corresponding to "noise only in each channel" for H_4 and "common signal present in both channels" for K_4.

The development of Section III suggests a generalization of the sign detector for (H_4, K_4). Such a generalization is given by the *polarity coincidence correlator* (PCC) which is the detector

$$\sum_{i=1}^{n} u(u_i v_i) \begin{cases} < C_4 \Rightarrow D_4(w) = 0 \\ > C_4 \Rightarrow D_4(w) = 1. \end{cases} \tag{58}$$

Here $u(\cdot)$ is the unit step function of (27). The detector calculates the number of *sign coincidences* among the paired observations and compares this number to a threshold. The random variable $k=\sum u(v_i v_i)$ is binomially distributed with density function $b(k; n, p)$ given by (29), where p is the probability of sign coincidence in a single observation and is given by

$$p = P\{u(u_i v_i) = 1\} = P\{u_i > 0 \text{ and } v_i > 0\} + P\{u_i < 0 \text{ and } v_i < 0\} \tag{59}$$

$$= \frac{1}{2} + 2\int_{-\infty}^{\infty} [F_1(x) - 1/2][F_2(x) - 1/2]\, dF_3(x).$$

Under H, $s \equiv 0$ and $p=1/2$ since u_i and v_i have zero medians; under K, $p>1/2$. Thus the PCC is *nonparametric* for H_4. It is also *unbiased* and *consistent* for K_4.

Under weak-signal conditions, (59) may be written [compare with (42)] as

$$p \approx 1/2 + 2f_1(0)f_2(0)\sigma_3^2 \tag{60}$$

and the ARE of the PCC with respect to the optimum detector D_3 may be shown [35] to be

$$E_{4,3} = (\gamma_1^4 + \gamma_2^4 + 4\sigma_1^2\sigma_2^2 - \sigma_1^4 - \sigma_2^4)f_1^2(0)f_2^2(0) \tag{61}$$

where γ_1^4 and γ_2^4 given by

$$\gamma_1^4 = E\{[u_i - Eu_i]^4\}, \quad \text{all } i \tag{62}$$

$$\gamma_2^4 = E\{[v_i - Ev_i]^4\}, \quad \text{all } i \tag{63}$$

are the fourth central moments of the noises.

In the Gaussian case given by (H_3, K_3) [where the detector D_3 is optimal], the ARE of the PCC with respect to detector D_3 becomes

$$E_{4,3} = 2/\pi^2 \approx 0.202 \tag{64}$$

since $f_1(0)=f_2(0)=1/\sqrt{2\pi}\,\sigma$, $\sigma_1^2=\sigma_2^2=\sigma^2$, and $\gamma_1^4=\gamma_2^4=3\sigma^4$. This comparatively poor performance arises from the large amount of information that the PCC ignores compared to the optimum detector D_3. On the other hand, let both f_1 and f_2 have the two-sided exponential distributions of (45) with variances $\sigma^2=2/\alpha^2$ and fourth moments $\gamma^4=24/\alpha^4$; then (64) becomes

$$E_{4,3} = 3.5 \tag{65}$$

indicating that the PCC may perform much better than detector D_3 if the Gaussian assumption on the noise is not valid.

V. Comments

A question of considerable engineering importance is the validity of the ARE as a measure of system performance for small sample sizes N_1 and N_2 where alternative and hypothesis are not necessarily close. Relatively little work appears to have been published along these lines. However, unpublished studies involving both computer simulation and actual implementation indicate that the relative efficiency converges rather rapidly to the ARE for detectors and inputs of the type discussed here.

The assumption made in this paper that the input data samples w_i are independent and identically distributed is frequently unrealistic under both hypothesis and alternative. If the noise violates this assumption, the detectors may not be nonparametric (that is, the false-alarm rate may vary) even asymptotically [36]. Furthermore, if the samples are dependent, it is much more difficult to invoke the central limit theorem and to insure asymptotic normality with the consequent simplification of calculations [37].

References

[1] W. L. Root, "An introduction to the theory of the detection of signals in noise," *Proc. IEEE*, this issue, pp. 610-623.
[2] I. R. Savage, *Bibliography of Nonparametric Statistics.* Cambridge, Mass.: Harvard University Press, 1962.
[3] J. E. Walsh, *Handbook of Nonparametric Statistics.* Princeton, N. J.: Van Nostrand, vol. 1, 1962; vol. 2, 1965, vol. 3, 1968.
[4] S. Siegel, *Nonparametric Statistics for the Behavioral Sciences.* New York: McGraw-Hill, 1956.
[5] D. A. S. Fraser, *Nonparametric Methods in Statistics.* New York: Wiley, 1957.
[6] G. E. Noether, *Elements of Nonparametric Statistics.* New York: Wiley, 1967.
[7] J. V. Bradley, *Distribution-Free Statistical Tests.* Englewood Cliffs, N. J.: Prentice-Hall, 1968.
[8] E. L. Lehmann, *Testing Statistical Hypotheses.* New York: Wiley, 1959, p. 65.
[9] *Ibid.*, ch. 4, 5.
[10] *Ibid.*, pp. 329, 342.
[11] *Ibid.*, ch. 6.
[12] M. G. Kendall and A. Stuart, *The Advanced Theory of Statistics.* New York: Hafner, vol. 2, 1962, p. 240.
[13] *Ibid.*, ch. 31.
[14] G. E. P. Box and S. L. Andersen, "Permutation theory in the derivation of robust criteria and the study of departures from assumptions," *J. Roy. Statist. Soc.*, ser. B, vol. 17, pp. 1–34, 1955.
[15] J. Capon, "A nonparametric technique for the detection of a constant signal in additive noise," *IRE WESCON Conv. Rec.*, pt. 4, pp. 92–103, 1959.
[16] ——, "Locally optimum nonparametric methods for the detection of Signals in Noise," *1963 Proc. IEEE Symp. on Adaptive Processes*, pp. 64–77, 1963.
[17] J. B. Millard and L. Kurz, "Nonparametric signal detection—An application of the Kolmogorov–Smirnov, Cramèr–von Mises tests (Abstract only), *IEEE Trans. Information Theory*, vol. IT-12, p. 275, April 1962.
[18] J. W. Carlyle and J. B. Thomas, "On nonparametric signal detectors," *IEEE Trans. Information Theory*, vol. IT-10, pp. 146–152, April 1964.
[19] J. W. Carlyle, "Nonparametric methods in detection theory," in *Communication Theory*, A. V. Balakrishnan, Ed. New York: McGraw–Hill, 1958, ch. 8.
[20] C. W. Helstrom, *Statistical Theory of Signal Detection.* London: Pergamon, 1968.
[21] *Ibid.*, ch. 4.
[22] W. Feller, *An Introduction to Probability Theory and Its Application*, vol. 1. New York: Wiley, 1968, ch.7.
[23] J. L. Hodges and E. L. Lehmann, "The efficiency of some nonparametric competitors to the *t*-test," *Ann. Math. Statist.*, vol. 27, pp. 324–335, 1956.
[24] J. Hajek and Z. Sidak, *Theory of Rank Tests.* New York: Academic Press, 1967.
[25] E. A. Feustel and L. D. Davisson, "The asymptotic relative efficiency of mixed statistical tests," *IEEE Trans. Information Theory*, vol. IT-13, pp. 247–255, April 1967.
[26] ——, "On the efficacy of mixed locally-most-powerful one- and two-sample rank tests," *IEEE Trans. Information Theory*, vol. IT-14, pp. 776–778, September 1968.
[27] R. C. Jones, "The ultimate sensitivity of radiation detectors," *J. Opt. Soc. Am.*, vol. 37, pp. 879–898, November 1947.
[28] R. H. Dicke, "The measurement of thermal radiation at microwave frequencies," *Rev. Sci. Instr.*, vol. 17, pp. 268–275, July 1946.
[29] S. J. Goldstein, Jr., "A comparison of two radiometer circuits," *Proc. IRE*, vol. 43, pp. 1663–1666, November 1955.
[30] J. J. Faran, Jr., and R. Hills, Jr., "Correlators for signal reception," Harvard University, Acoustics Res. Lab., Cambridge, Mass., Tech. Memo. 27, September 15, 1952.
[31] ——, "The application of correlation techniques to acoustic receiving systems," Harvard University, Acoustic Res. Lab., Cambridge, Mass., Tech. Memo. 28, November 1, 1962.
[32] J. B. Thomas and T. R. Williams, "On the detection of signals in nonstationary noise by product arrays," *J. Acoust. Soc. Am.*, vol. 31, pp. 453–462, April 1959.
[33] B. S. Melton and P. H. Karr, "Polarity coincidence scheme for revealing signal coherence," *Geophysics*, vol. 22, pp. 553–564, July 1957.
[34] H. Ekre, "Polarity coincidence correlation detection of a weak noise source," *IRE Trans. Information Theory*, vol. IT-9, pp. 18–23, January 1963.
[35] S. S. Wolff, J. B. Thomas, and T. R. Williams, "The polarity-coincidence correlator: A nonparametric detection device," *IRE Trans. Information Theory*, vol. IT-8, pp. 5–9, January 1962.
[36] S. S. Wolff, J. L. Gastwirth, and H. Rubin, "The sign test on autoregressive data" (Abstract only), *IEEE Trans. Information Theory*, vol. IT-12, p. 279, April 1966.
[37] M. Kanefsky and J. B. Thomas, "On adaptive nonparametric detection systems using dependent samples," *IEEE Trans. Information Theory*, vol. IT-11, pp. 521–526, October 1965.

2B

Reprinted with permission from *IEEE Proc.* **59**:1018–1019 (1971)

COMMENTS ON "NONPARAMETRIC DETECTION"

Dean W. Lytle, Paul H. Moose, and Louis L. Scharf

Abstract—The sign test is discussed in order to clarify the sense of its optimality and emphasize why it cannot be considered optimal for any commonly understood signal detection problem.

The above paper[1] is an excellent tutorial introduction to nonparametric detection. However, we feel the discussion of the optimality of the sign test may be misleading and deserves clarification. The development in

Manuscript received August 31, 1970.
[1] J. B. Thomas, "Nonparametric detection," *Proc. IEEE*, vol. 58, May 1970, pp. 623–

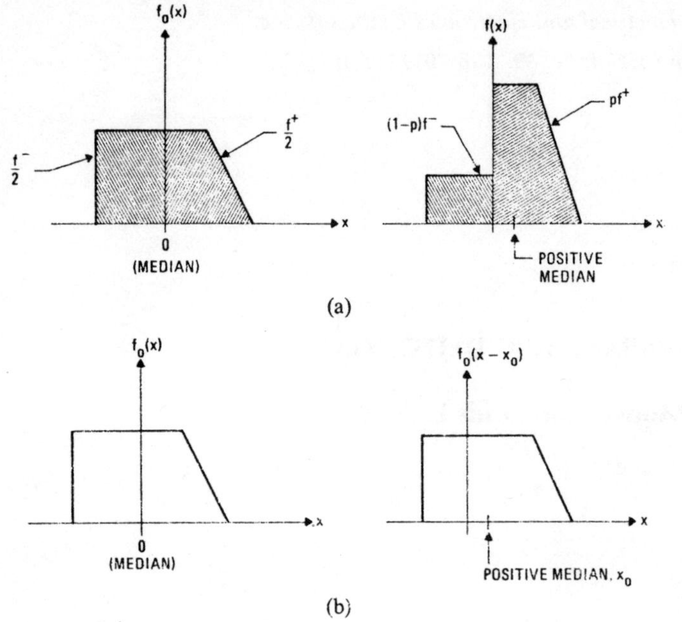

Fig. 1. Contrasting density functions. (a) Densities satisfying (21) and (22) of Thomas[1]. (b) Densities satisfying (H_2', K_2').

the previously mentioned paper[1] is likely, we feel, to leave the reader with the impression the sign test is the Neyman–Pearson optimal detector for detecting an additive positive signal in zero median noise with an arbitrary probability distribution function. This impression follows from Thomas' statement (see Thomas,[1] p. 627) that his nonparametric hypothesis H_2 and nonparametric alternative K_2 are equivalent to assuming that the data samples w_i are independent observations with zero medians under the hypothesis, but that they have positive medians under the alternative due to the presence of an additive positive signal. This statement of equivalence seems to imply that if the sign test is optimum for (H_2, K_2) then it must also be optimum for detecting an additive positive signal in zero median noise when the distribution is arbitrary.

In our opinion, the almost universal interpretation of the signal detection problem just stated is

$$H_2': \quad p = 1/2,\ F(x) \text{ arbitrary}$$
$$K_2': \quad F(x - x_0)$$

where x_0 is the positive signal level[2] and $p = 1 - F(0)$. Thomas' (H_2, K_2), where

$$H_2: \quad p = 1/2,\ F(x) \text{ otherwise arbitrary}$$
$$K_2: \quad p > 1/2,\ F(x) \text{ otherwise arbitrary}$$

clearly includes (H_2', K_2') as a subset. Therefore (H_2, K_2) and (H_2', K_2') should not be considered equivalent.

The Neyman–Pearson optimality of the sign test[3] for (H_2, K_2) depends critically upon the fact that for any distribution with density f which belongs to K_2 [see (21)[4]], the corresponding *least favorable* distribution within H_2 has density f_0 given by (22). As Thomas proved, the sign test is optimum for testing any alternative distribution versus its corresponding least favorable hypothesis distribution. Since the sign test is nonparametric and optimum for worst-case pairs in (H_2, K_2), it is the uniformly most powerful test (i.e., Neyman–Pearson optimum) for H_2 versus K_2. When the problem is constrained to (H_2', K_2'), the pairing of an f and its least favorable f_0 can no longer occur and thus the sign test cannot be Neyman–Pearson optimal for (H_2', K_2'). These remarks are clarified in Fig. 1, where densities satisfying (21) and (22) are contrasted with those satisfying (H_2', K_2'). It is noted that the densities in Fig. 1(a) bear little resemblance to any physically meaningful detection problem one might envision.

The conclusion is that the sign test is indeed Neyman–Pearson optimal (and nonparametric) for (H_2, K_2). However, it retains only its nonparametric claims for (H_2', K_2'); it is not Neyman–Pearson optimal for the commonly understood nonparametric detection problem (H_2', K_2'). Admittedly, this fact is implicit in the first sentence of Thomas' paper, p. 628.[1] However, this point is likely to be missed.

We hope this letter will clarify the reasons why the sign test is optimal for (H_2, K_2), but suboptimal for the signal detection problem (H_2', K_2').

[2] We are assuming for convenience that the signal level is fixed. This is not necessary.
[3] E. L. Lehmann, *Testing Statistical Hypotheses*. New York: Wiley, 1959, pp. 90–94.
[4] All equation numbers refer to Thomas' paper.[1]

Part II

NONPARAMETRIC DETECTION BASED ON SIGNS (HARD-LIMITING) OF OBSERVATIONS

Editors' Comments on Papers 3 Through 8

Perhaps the simplest nonparametric detector is the sign detector for a constant signal in additive zero-median noise, using a set of n independent observations $X_1, X_2, \ldots, X_n$. Viewed as a statistical hypothesis testing problem, we have under the null hypothesis H_0 independent observations from a common density function $f(x)$, and under the alternative hypothesis H_1 (signal present) independent observations from a common density function $f(x-\theta)$. Here $f(x)$ represents the probability density function of noise observations, and θ is the constant signal. As long as $f(x)$ is known to have zero median, the distribution of $T(X_1, X_2, \ldots, X_n) = \Sigma_{i=1}^{n} sgn\ (X_i)$ is completely specified under H_0, where

$$sgn\ (x) = \begin{cases} 1\ , x > 0 \\ 0\ , x = 0\ . \\ -1\ , x < 0 \end{cases}$$

Thus the test statistic T can form the basis for a nonparametric test of H_0 versus H_1. The result is an extremely simple test, which has been known to statisticians for a long time. It has been established that the test is a uniformly most powerful test for the location, or median, of the density function of observations (Fraser).[1]

The six papers in this part are concerned with signal detection schemes based on signs, or polarities, of the observations. They are variations of the above basic sign test suitable for different detection situations. Paper 3 by Kanefsky and Thomas analyzes the performance of the simple sign test for various types of noise density functions. The performance analysis is based on the concept of asymptotic relative efficiency (ARE) for a pair of tests. This is the ratio of sample sizes required by two tests for a hypothesis testing problem so that both have the same size and power asymptotically as sample sizes approach infinity and signal strength becomes vanishingly small. The asymptotic comparisons are formulated in a direct manner, which is appealing and which allows a study to be made of the applicability of the asymptotic performance results in practical finite-sample and finite-signal situations.

Both Paper 3 and Paper 4 by Wolff, Thomas, and Williams consider the detection of a random signal common to two channels and receivers, with a background of independent additive noise processes in each channel. The paper by Wolff, Thomas, and Williams was the first study of the polarity-coincidence-correlator (PCC) as a nonparametric detector, although the PCC had been previously considered for weak-signal detection (see, for example, Thomas and Williams).[2]

Kanefsky, in Paper 5, and Stoll and Kurz in Paper 6 extend the formulation of the two-input, hard-limiting PCC to detector structures for nonparametric detection of a random signal common to an array of k receivers. Kanefsky compares performances based on a definition of signal-to-noise ratio (SNR) for a test statistic under weak signal conditions, which is equivalent to a comparison of detector efficacies (Capon)[3] leading to the ARE. One scheme that emerges as a useful nonparametric detector is based on a test statistic formed as a sum of the $k(k-1)/2$ PCC-type test statistics for each of the pairs of channels in the array. Kanefsky also considers the effect of dependent sampling and obtains the SNRs; under conditions of dependent sampling, the polarity-coincidence-array (PCA) detectors do not remain nonparametric in their performance.

The paper by Hansen and Olson (Paper 7) illustrates the applicability of simple sign test detectors in practical radar systems. It describes a radar detector that operates by comparing received data with a set of reference noise observations. The sum of hard-limited differences of these two sets of observations forms the test statistic. The detector is analyzed for asymptotic performance, and numerical results are given for finite-sample, finite-signal power. Several other interesting extensions of this type of scheme have also been discussed in subsequent literature.

All of these papers deal with the fixed-sample-size detection problem. The sign test can also be applied for nonparametric sequential detection of signals, and Paper 8 by Tantaratana and Thomas describes and analyzes a sequential procedure based on signs of independent observations. This scheme is nonparametric in that it yields fixed probabilities of errors for a nonparametric class of symmetric noise distribution functions, which are required only to have known values $F(s)$ and $F(-s)$ at two points $\pm s$, which represent the levels of the signal in an antipodal signaling scheme. A sequential sign test with a constant size for detecting a shift in the median of a density function may also be shown similarly to achieve better performance than the fixed-sample-size sign detector, based on results for the average sample number for the sequential procedure.

REFERENCES

1. D. A. S. Fraser. *Nonparametric Methods in Statistics*. John Wiley, 1957, pp. 167–169.
2. J. B. Thomas and T. R. Williams. "On the Detection of Signals in Nonstationary Noise by Product Arrays." *Acoust. Soc. Am. J.* **31**:453–462 (1959).
3. J. Capon. "On the Asymptotic Efficiency of Locally Optimum Detectors." *IEEE Trans. Inf. Theory* **IT-7**:67–72 (1961).

3

Reprinted from *J. Franklin Inst.* **280**:120–138 (1965)

On Polarity Detection Schemes with Non-Gaussian Inputs

by M. KANEFSKY AND J. B. THOMAS

Department of Electrical Engineering
Princeton University, Princeton, New Jersey

ABSTRACT: *The design of signal detectors which are optimum (in the Neyman-Pearson sense) requires a complete statistical description of the input processes. Certain non-optimum detectors, the "nonparametric detectors," have the interesting property that their false alarm rates can be fixed for wide classes of inputs. These detectors are often easy to implement and are useful when the input statistics are not completely known.*

This paper treats two practical nonparametric detectors, the one-input sign detector and the two-input polarity coincidence correlator, both of which utilize only the polarity information about the input.

The performance of these nonparametric detectors is evaluated for a wide class of non-gaussian input noises. Their performance is compared to that of their Neyman-Pearson counterparts designed on the assumption of gaussian stationary noises. The comparisons are based on asymptotic relative efficiency (a.r.e.). While the a.r.e. gives a close approximation to the actual relative efficiency for stationary gaussian inputs, significant differences result when considering non-gaussian inputs, particularly of the impulse type.

For stationary gaussian inputs the sign detector is 64 per cent as efficient as its Neyman-Pearson counterpart, the mean detector. Under the same conditions the polarity coincidence correlator (PCC) is 20 per cent as efficient as its Neyman-Pearson counterpart.

It is demonstrated in this paper that these known results are relatively insensitive to small changes in the noise statistics. However, as the amplitude density of the noise processes become more sharply peaked and/or more asymmetrical, then the nonparametric detectors become more efficient than the Neyman-Pearson detectors designed for gaussian inputs. These types of noise processes resemble what is commonly called impulse noise.

Introduction

The usual detection problem reduces to determining whether or not a "signal" is present in a corrupting noise process. When the statistical properties of the noise (and of the signal if it is non-deterministic) are known, then detectors can be found that are optimum in a sense. For example, the Neyman-Pearson detector is the optimum (most powerful) detector in the sense of minimizing the "miss" probability for a fixed "false alarm" probability (1).

Such optimum detectors can be used as standards for the evaluation of less powerful detectors (non-optimum). Some of the non-optimum detectors have the interesting property that their false alarm rates can be fixed for a wide class of noise processes and are therefore useful when the statistical properties of the noise are not completely known. Such detectors are called nonparametric and are often easy to implement (**2**). A class of these detectors utilizes only the polarity information about the inputs. Two detectors belonging to this class and which are practical devices are the one-input sign detector (**3**) and the two-input polarity coincidence correlator (**4**).

In order to evaluate nonparametric detectors it is important to understand their behavior for wide classes of non-gaussian inputs. We shall consider a number of the types of non-gaussian noise that might be encountered in practice, and the performance of the nonparametric detectors for such noise is compared to that of their optimum (Neyman-Pearson) counterpart designed under the assumption of gaussian stationary noise. The sensitivity of the Neyman-Pearson detectors to the parametric assumption is also considered. All detectors are presumed to operate on sampled input data.

The techniques used to compare detectors will be that of asymptotic relative efficiency (a.r.e.) (**5**). The relative efficiency of two detectors is the ratio of the respective sample sizes which are required to achieve the same power of detection (probability of a hit) for a given false alarm rate. The a.r.e. is the limit of this ratio as the sample sizes approach infinity and the signal-to-noise power ratio decreases to zero. More specifically, if two detectors D_1 and D_2 require N_1 and N_2 samples, respectively, to attain the same signal detection rate for a given false alarm rate, then the asymptotic relative efficiency of D_1 relative to D_2 is defined as

$$e_{D_1/D_2} = \lim_{\substack{N_1 \to \infty \\ \sigma_s^2/\sigma_n^2 \to 0}} N_2/N_1,$$

where the signal-to-noise power ratio σ_s^2/σ_n^2 approaches zero in such a way that the signal detection rate is kept constant. While the a.r.e. gives a close approximation to the actual relative efficiency for stationary gaussian inputs, significant differences result when considering non-gaussian inputs, particularly of the impulse type. These differences are discussed and calculated here for practical conditions.

Description of the Problem

The two types of detection problems considered are illustrated in Fig. 1. In the one-input system, the input $x(t)$ consists either of a signal and an addi-

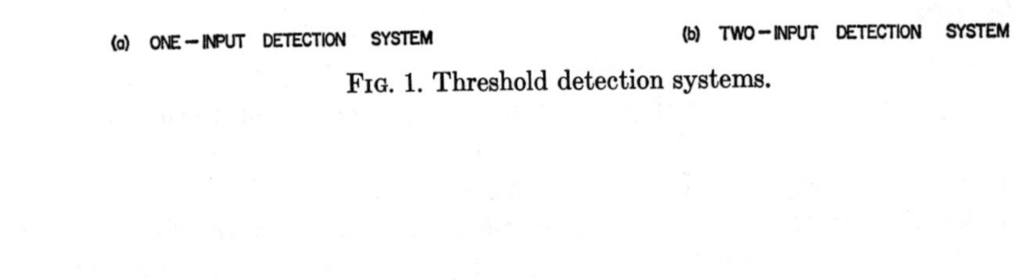

FIG. 1. Threshold detection systems.

tive noise, or of the noise alone. It is presumed that samples x_i spaced T seconds apart are used as input data and that the interval T is chosen such that the x_i are statistically independent. We are interested in testing an hypothesis H_1 against an alternative K_1 where

$$\begin{aligned} H_1: &\quad x(t) = n(t) \\ K_1: &\quad x(t) = n(t) + s(t), \end{aligned}$$

where the noise $n(t)$ and the signal $s(t)$ are statistically independent. The detector will calculate a test statistic S and compare it to a fixed threshold S_0. If $S > S_0$, K_1 will be chosen, *i.e.*, the signal will be considered present.

In the two-input system of Fig. 1(b), the signal will be common to both channels when it is present, therefore

$$\begin{aligned} H_2: &\quad x^{(j)}(t) = n^{(j)}(t) &\quad j = 1, 2, \\ K_2: &\quad x^{(j)}(t) = n^{(j)}(t) + s(t) &\quad j = 1, 2, \end{aligned}$$

where $n^{(1)}(t)$, $n^{(2)}(t)$, and $s(t)$ are statistically independent; otherwise the two-input scheme functions in the same manner as the one-input case.

The sign detector is a one-input device that is often considered for detecting a d.c. signal s in additive stationary noise with zero median (**3**). This detector calculates the statistic

$$S_s = \sum_{i=1}^{N_s} u(x_i),$$

where x_i is the value of the ith sample, $u(\cdot)$ is the unit step function, and N_s is the total number of samples taken. Thus S_s is the number of positive samples. The Neyman-Pearson counterpart (**6**) designed under the assumption that the x_i are independent samples from a gaussian population calculates the statistic

$$S_m = \sum_{i=1}^{N_m} x_i.$$

The polarity coincidence correlator (PCC) is a two-input device that is a reasonable system for detecting a random stationary signal with zero median in additive independent noises with the same statistical properties as the signal (**4**). The PCC calculates the statistic

$$S_{pcc} = \sum_{i=1}^{N_{pcc}} u(x_i^{(1)} x_i^{(2)}).$$

The Neyman-Pearson counterpart (**7**) designed under the assumption that the three random processes are all gaussian calculates the statistic

$$S_{np} = \sum_{i=1}^{N_{np}} [x_i^{(1)} + x_i^{(2)}]^2.$$

Both nonparametric detectors (*i.e.*, the sign detector and the PCC) convert the signal samples into binary samples by some hard limiting process. When dealing with physical systems, the procedure of limiting and then operating on binary functions is generally easier than operating on the value of the function, this being the case here. In addition, certain other properties of these nonparametric tests make them useful.

The following theorems (**8**) have been proved:

1) the sign test is uniformly most powerful with respect to the class of zero median distribution functions, and
2) for nonstationary statistics the sign test is uniformly most powerful invariant.

Because of its similarity to the sign test, equivalent statements can be made about the polarity coincidence correlator (**9**). It seems reasonable to assume, therefore, the existence of a class of distribution functions for which the nonparametric detectors are more powerful than their Neyman-Pearson counterparts designed for gaussian noise. In fact, we know (**10**) that the sign test is the asymptotically most powerful parametric test for the exponential distribution $f = \frac{1}{2}e^{-|x-\theta|}$. For this distribution we also know the PCC is asymptotically 3.5 times more powerful than the Neyman-Pearson detector (**11**). From such considerations we expect that the sign detector and the PCC may outperform, for a large class of non-gaussian processes, Neyman-Pearson detectors designed on the gaussian assumption.

Both nonparametric detectors have test statistics which are binomially distributed with "probability of success" depending on the noise statistics (**12**). The parametric detectors have test statistics which are gaussian distributed for the one-channel detector and chi-squared distributed with N_{np} degrees of freedom for the two-channel detector under the assumption of gaussian noise statistics (**13**). In the parametric case the test statistics are not generally simple when the assumption of gaussian noise is violated. As the detectors are compared asymptotically (*i.e.*, for very large sample sizes) the Central Limit Theorem can be applied. Thus, under reasonable restrictions, the distribution of all the test statistics can be considered gaussian. The Central Limit Theorem must be applied with considerable caution, but the following results can be proved (**14**):

1) For the statistics considered, the Central Limit Theorem can be applied as the sample size increases beyond bound if and only if the signal-to-noise power ratio decreases to zero "sufficiently" fast as the sample size increases.
2) The limiting processes mentioned in (1) can be performed in such a way that the power of the detectors approach unity while simultaneously the error decreases in the application of the Central Limit Theorem.
3) The errors involved for finite sample sizes and small but nonzero signal-to-noise powers can be computed and shown to be small.

Thus the detectors will be compared under the assumption of vanishingly small signal-to-noise power ratios and very large sample sizes. The necessary corrections involved for more practical conditions will be discussed later.

Calculation of Detector Performance

The level or false alarm rate α of a test is defined as the probability that the detector decides the signal is present when the signal is absent. As previously mentioned, all of the detectors discussed are of the threshold type; *i.e.*, they compare their test statistic S with some threshold level S_0 and decide that a signal is present when $S > S_0$. Thus the level is given asymptotically by

$$\alpha = \int_{S_0}^{\infty} f_H(S)dS = 1 - \Phi\left\{\frac{S_0 - E_H(S)}{\sqrt{\mathrm{Var}_H(S)}}\right\}, \tag{2}$$

where $f(\cdot)$ is the gaussian density function, Φ is the gaussian cumulative distribution with zero mean and unit variance, and the subscript H refers to the hypothesis (noise only). The power β of a test is defined as the probability that the detector decides the signal is present when indeed the signal is present. The power is given asymptotically by

$$\beta = \int_{S_0}^{\infty} f_K(S)dS = 1 - \Phi\left\{\frac{S_0 - E_K(S)}{\sqrt{\mathrm{Var}_K(S)}}\right\}, \tag{3}$$

where the subscript K refers to the alternative (signal and noise present.) Combining Eqs. 2 and 3, we obtain the following result for the power of any threshold detector whose test statistics are gaussian:

$$\beta = 1 - \Phi\left\{\frac{\sqrt{\mathrm{Var}_H(S)}\Phi^{-1}(1-\alpha) + E_H(S) - E_K(S)}{\sqrt{\mathrm{Var}_K(S)}}\right\}. \tag{4}$$

Consider two detectors with test statistics S_1 and S_2 and corresponding samples sizes N_1 and N_2. If the test statistics are sums of independent random variables (as are S_s, S_m, S_{pcc}, and S_{np}) then, under the assumption of stationarity, the means and variances of S_1 and S_2 are given by

$$E_\Psi(S_i) = N_i\mu_{i\Psi}$$

and

$$\mathrm{Var}_\Psi(S_i) = N_i\rho_i{}^2{}_\Psi,$$

where $i = 1, 2$; Ψ is either H or K depending on whether or not the signal is present; and $\mu_{i\Psi}$ and $\rho_i{}^2{}_\Psi$ are the mean and variance, respectively, of any one summand in the test statistic S_i. On equating the powers $\beta_1 = \beta_2$ for equal false alarm rates α, we obtain

$$\sqrt{\frac{N_2}{N_1}} = \left[\frac{\mu_{1K} - \mu_{1H}}{\mu_{2K} - \mu_{2H}}\right]\frac{\rho_{2K}}{\rho_{1K}} - \frac{\Phi^{-1}(1-\alpha)}{\sqrt{N_1}}\frac{\rho_{2K}}{(\mu_{2K} - \mu_{2H})}\left[\frac{\rho_{1H}}{\rho_{1K}} - \frac{\rho_{2H}}{\rho_{2K}}\right]. \tag{5}$$

It is clear from this equation that the a.r.e. can be given by

$$e_{D_1/D_2} = \lim_{\substack{N_1 \to \infty \\ \sigma_s^2/\sigma_n^2 \to 0}} N_2/N_1 = \lim_{\sigma_s^2/\sigma_n^2 \to 0} \left[\frac{\mu_{1K} - \mu_{1H}}{\mu_{2K} - \mu_{2H}}\right]^2 \frac{\rho_{2K}{}^2}{\rho_{1K}{}^2}. \tag{6}$$

Two errors are to be considered for practical cases ($N < \infty$). The first (to be called the Central Limit Theorem error) is implicit in Eq. 5 and is the result of the gaussian assumption. The second error (or computational error) is in the approximation of Eq. 5 by Eq. 6 for finite N Equation 5 can be written in a more convenient form if it is assumed that $\beta = 1 - \alpha$. Solving for $\Phi^{-1}(1 - \alpha)/\sqrt{N}$ from Eq. 4 and substituting in Eq. 5 yields

$$\frac{N_2}{N_1} = \left[\frac{\mu_{1K} - \mu_{1H}}{\mu_{2K} - \mu_{2H}}\right]^2 \left[\frac{\rho_{2K} + \rho_{2H}}{\rho_{1K} + \rho_{1H}}\right]^2. \tag{7}$$

Before considering the two errors, the detectors will be compared using the a.r.e. expression of Eq. 6.

A.R.E.—One-Channel Detectors

Consider the mean detector first where $E(S_m) = N_m E(x_i)$; therefore $E_H(x_i) = \mu_{mH} = E\{n(t)\}$ and $E_K(x_i) = \mu_{mK} = E\{n(t)\} + s$, where $n(t)$ is the noise process and s is the d.c. signal. It is also true that $\text{Var}(S_m) = N_m \text{Var}(x_i)$; therefore $\text{Var}_H(x_i) = \rho_m{}^2 = \sigma_n{}^2$ and $\text{Var}_K(x_i) = \rho_{mK}{}^2 = \sigma_n{}^2$, where $\sigma_n{}^2$ is the variance of the noise. The test statistic S_s of the sign detector is binominally distributed. Define $p_o = \text{Prob}\{n(t) > 0\} = 1 - F(0)$, and $p_+ = \text{Prob}\{n(t) + s > 0\} = 1 - F(-s)$, where $F(\cdot)$ is the cumulative distribution function of the noise. Then $E_H\{u(x_i)\} = \mu_{sH} = p_o$, $E_K\{u(x_i)\} = \mu_{sK} = p_+$, $\text{Var}_H\{u(x_i)\} = \rho_{sH}{}^2 = p_o(1 - p_o)$, and $\text{Var}_K\{u(x_i)\} = \rho_{sK}{}^2 = p_+(1 - p_+)$. Substituting these results into Eq. 6 yields

$$e_{s/m} = \lim_{s^2/\sigma_n^2 \to 0} \frac{(p_+ - p_o)^2}{s^2} \frac{\sigma_n{}^2}{p_+(1 - p_+)}. \tag{8}$$

Consider that

$$\lim_{s^2/\sigma_n^2 \to 0} p_+ = \lim_{s^2/\sigma_n^2 \to 0} [1 - F(-s)] = 1 - F(0) + s^2 f(0^-) = p_0 + s^2 f(0^-),$$

where $F(\cdot)$ is continuous at zero and $F'(\cdot)$ is $f(\cdot)$, the density function of the noise. It is apparent that Eq. 8 can be written as

$$e_{s/m} = \frac{\sigma_n{}^2 f^2(0^-)}{F(0)[1 - F(0)]}. \tag{9}$$

For gaussian processes $F(0) = \frac{1}{2}$, and $f(0) = 1/\sqrt{2\pi\sigma_n{}^2}$, which implies that the sign detector is asymptotically 64 per cent as efficient as the mean detector for gaussian inputs. The dependence of the a.r.e. on $f(\cdot)$ will be demonstrated in a later section.

A.R.E.—Two-Channel Detectors

Consider first the Neyman-Pearson detector. For this detector

$$\mu_{npK} - \mu_{npH} = E\{[n^{(1)}(t) + n^{(2)}(t) + 2s(t)]^2\} - E\{[n^{(1)}(t) + n^{(2)}(t)]^2\},$$

where $n^{(1)}(t)$ and $n^{(2)}(t)$ represent the two independent noise processes and $s(t)$ represents the signal process. It follows, therefore, for signals with zero means, that $\mu_{npK} - \mu_{npH} = 4\sigma_s^2$, where σ_s^2 is the variance of the signal $s(t)$. It is also apparent that

$$\rho^2_{npH} = E\{[n^{(1)}(t) + n^{(2)}(t)]^4\} - E^2\{[n^{(1)}(t) + n^{(2)}(t)]^2\}.$$

This expression easily yields

$$\rho^2_{npH} = 2\mu_4 + 2\sigma_n^4 + 16\mu\mu_3 + 32\mu^2\sigma_n^2.$$

where $\mu = E\{n(t)\}$ and $\mu_r = E\{[n(t) - \mu]^r\}$. An expression for ρ^2_{npK} is not required at the moment.

The test statistic S_{pcc} of the PCC detector is binomially distributed. Thus, if p_o and p_+ are the "probability of success" under the hypothesis and alternative, respectively, it follows that $\mu_{pccH} = p_o$, $\mu_{pccK} = p_+$, and $\rho^2_{pccH} = p_o(1-p_o)$. For this detector,

$$\begin{aligned} p_o &= \text{Prob. } \{n^{(1)}(t) > 0 \text{ and } n^{(2)}(t) > 0\} \\ &\qquad + \text{Prob. } \{n^{(1)}(t) < 0 \text{ and } n^{(2)}(t) < 0\} \\ &= [1 - F(0)]^2 + F^2(0) = 1 - 2F(0) + 2F^2(0), \end{aligned}$$

and

$$\begin{aligned} p_+ &= \text{Prob. } \{n^{(1)}(t) + s(t) > 0 \text{ and } n^{(2)}(t) + s(t) > 0\} \\ &\qquad + \text{Prob. } \{n^{(1)}(t) + s(t) < 0 \text{ and } n^{(2)}(t) + s(t) < 0\} \\ &= \int \text{Prob. } \{n^{(1)}(t) > -u \text{ and } n^{(2)}(t) > -u\} dF_s(u) \\ &\qquad + \int \text{Prob. } \{n^{(1)}(t) < -u \text{ and } n^{(2)}(t) < -u\} dF_s(u) \\ &= \int [1 - F(-u)]^2 dF_s(u) + \int F^2(-u) dF_s(u) \\ &= \int [1 - 2F(-u) + 2F^2(-u)] dF_s(u), \end{aligned}$$

where $F_s(\cdot)$ is the cumulative distribution function of the signal $s(t)$. Upon substituting all of the previous results into Eq. 6 we obtain

$$e_{pcc/np} = \frac{\left[1 + \frac{\mu_4}{\sigma_n^4} + 8\frac{\mu}{\sigma_n}\frac{\mu_3}{\sigma_n^3} + 16\frac{\mu^2}{\sigma_n^2}\right]}{16F(0)[1 - F(0)][1 - 2F(0) + 2F^2(0)]} \lim_{\sigma_s^2/\sigma_n^2 \to 0} \frac{p_+ - p_o}{\sigma_s^2/\sigma_n^2}. \qquad (10)$$

The following results can be obtained (**14**). For the types of noise statistics to be considered, p_+ can be expanded in a power series of the signal-to-noise power ratio of the form

$$p_+ = p_o + \sum_{n=0}^{\infty} [a_n(\sigma_s/\sigma_n)^{cn+1} + b_n(\sigma_s/\sigma_n)^{cn+2}],$$

for some $c > 1$. This series converges for "reasonable" signal-to-noise power ratios in such a way that, for signals with zero mean,

$$\lim_{\sigma_s^2/\sigma_n^2 \to 0} (p_+ - p_o)/\sigma_s^2/\sigma_n^2 = \sigma_n^2[f^2(0^+) + f^2(0^-)],$$

Thus Eq. 10 becomes

$$e_{pcc/np} = \frac{\sigma_n^4[f^2(0^+) + f^2(0^-)]^2 \left[1 + \frac{\mu_4}{\sigma_n^4} + 8\frac{\mu}{\sigma_n}\frac{\mu_3}{\sigma_n^3} + 16\frac{\mu^2}{\sigma_n^2}\right]}{F(0)[1 - F(0)][1 - 2F(0) + 2F^2(0)]}. \tag{11}$$

For gaussian processes $F(0) = \frac{1}{2}$, $f(0^+) = f(0^-) = 1/\sqrt{2\pi\sigma_n^2}$, $\frac{\mu}{\sigma_n} = 0$, and $\frac{\mu_4}{\sigma_n^4} = 3$, which results in $e_{pcc/np} = 2/\pi^2$. In other words, for gaussian inputs the PCC is asymptotically 20 per cent as efficient as its Neyman-Pearson counterpart.

Comparison of Detectors for Non-Gaussian Inputs

The following general class of density functions has the property that it can represent a wide variety of specific unimodal densities in a mathematically tractable form

$$f(x) = \begin{cases} f(0^+)e^{-\frac{1}{2}\left(\frac{x}{\sigma_+}\right)^c}; & 0 < x \leq \Lambda\sigma_+ \\ f(0^-)e^{-\frac{1}{2}\left(\frac{x}{\sigma_-}\right)^c}; & 0 > x \geq \Lambda\sigma_- \\ 0; & \text{elsewhere,} \end{cases} \tag{12}$$

where

$$\int_{-\infty}^{\infty} f(x)dx = 1 \quad \text{and} \quad \int_{-\infty}^{\infty} x^2 f(x)dx = \sigma_n^2. \tag{12a}$$

There are four subgroups of this class that depend on a single parameter and include the gaussian distribution as a special case:

I. Generalized Gaussian Statistics

$$f(x) = (a/\sigma_n e)^{-\frac{1}{2}\left(\frac{x}{\sigma'}\right)^c} \tag{13}$$

where a, σ', and c are positive real numbers, Λ approaches infinity, and $\sigma' = \sigma_+ = \sigma_-$. The constants a and σ' can be written as functions of c from the two normalizing equations. Then c becomes the only parameter. It is apparent that for $c = 2$, $f(x)$ is gaussian with variance σ_n^2. Also, for $c = 1$, the result is the double exponential distribution.

II. Truncated Gaussian Statistics

$$f(x) = \begin{cases} (K/\sqrt{2\pi\sigma'^2})e^{-\frac{1}{2}\left|\frac{x}{\sigma'}\right|^2}; & |x| \leq \Lambda\sigma' \\ 0; & \text{elsewhere.} \end{cases} \tag{14}$$

For this class of distribution functions, where Λ is finite, it is convenient to write the constants K and σ' as functions of Λ from the normalizing equations. Then the single parameter Λ is a measure of the amount of truncation. This type of distribution is similar to that encountered in practice when gaussian processes have been subjected to saturation and limiting effects. It will turn out that the presence or absence of delta functions at $|x| = \Lambda\sigma'$ will not affect the results.

III. Type I Asymmetrical Statistics

$$f(x) = \begin{cases} \dfrac{1}{\sqrt{2\pi\sigma_-^2}}\, e^{-\frac{1}{2}\frac{x^2}{\sigma_-^2}}; & x < 0 \\ \dfrac{1}{\sqrt{2\pi\sigma_+^2}}\, e^{-\frac{1}{2}\frac{x^2}{\sigma_+^2}}; & x > 0, \end{cases} \tag{15}$$

where $\sigma_+^2 = A^2\sigma_-^2$ and $\Lambda = \infty$. The constants σ_+^2 and σ_-^2 can be written as functions of A from the normalizing equations. This class of distribution functions depends on the single parameter A. It is apparent that $f(x)$ is gaussian for $A = 1$. The density given by Eq. 15 is obtained by the following simple instantaneous nonlinear operation on a gaussian process $x(t)$:

$$y(t) = \begin{cases} K_1 x(t); & x(t) \geq 0 \\ K_2 x(t); & x(t) < 0, \end{cases}$$

where $K_1/K_2 = A$.

IV. Type II Asymmetrical Statistics

$$f(x) = \begin{cases} Ke^{-\frac{1}{2}\frac{x^2}{\sigma_+^2}}; & x > 0 \\ Ke^{-\frac{1}{2}\frac{x^2}{\sigma_-^2}}; & x < 0, \end{cases} \tag{16}$$

where $\int_{-\infty}^{0} f(x)dx = F(0)$ and $\Lambda = \infty$. The constants K, σ_+, and σ_- can be written as functions of $F(0)$ from the constraint given and from the normalizing equations. It is apparent that $f(x)$ is gaussian for $F(0) = \frac{1}{2}$. This class of distribution functions represent a class of unimodal asymmetrical density functions that have no discontinuities.

The asymptotic relative efficiency expressions of Eqs. 9 and 11 will now be evaluated for the four classes of distributions just discussed.

Comparison of Detectors for Generalized Gaussian Statistics

From the density function of Eq. 13 and the two normalizing constraints of Eq. 12a we obtain the following results:

$$F(0) = \tfrac{1}{2}, \qquad f(0) = a/\sigma_n,$$

$$a = (c/2)\left[\frac{\Gamma(3/c)}{\Gamma^3(1/c)}\right]^{1/2},$$

$$\left(\frac{\sigma'}{\sigma_n}\right)^c = \frac{1}{2}\left[\frac{\Gamma(1/c)}{\Gamma(3/c)}\right]^{c/2},$$

$$\frac{\mu}{\sigma_n} = 0$$

$$\frac{\mu_4}{\sigma_n{}^4} = \frac{\Gamma(5/c)\Gamma(1/c)}{\Gamma^2(3/c)}$$

where $\Gamma(\cdot)$ is the Gamma function. Substituting these expressions into Eqs. 9 and 11 we obtain

$$e_{s/m} = c^2\,\frac{\Gamma(3/c)}{\Gamma^3(1/c)} \tag{17}$$

and

$$e_{pcc/np} = \frac{c^4}{8}\,\frac{\Gamma^2(3/c) + \Gamma(1/c)\Gamma(5/c)}{\Gamma^6(1/c)}. \tag{18}$$

These expressions are shown graphically in Fig. 2. From these curves it is apparent that both nonparametric detectors perform well relative to their

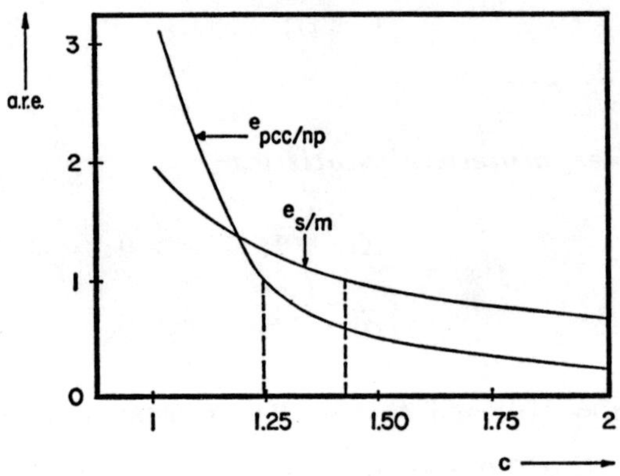

FIG. 2. Asymptotic relative efficiency—generalized gaussian statistics.

Neyman-Pearson counterparts for $c < 1.5$. We observe that $e_{s/m} \approx 1$ for $c = 1.40$, and $e_{pcc/np} \approx 1$ for $c = 1.24$. The density functions corresponding to these values of c are shown in Fig. 6 together with the gaussian density ($c = 2$). A discussion of these results will be given later.

Comparison of Detectors for Truncated Gaussian Statistics

From the density function of Eq. 14 and the normalizing constraints of Eq. 12a we obtain:

$$F(0) = \tfrac{1}{2}; \qquad f(0) = \frac{K}{\sqrt{2\pi\sigma'^2}}$$

$$K = \frac{1}{2\Phi(\Lambda) - 1}; \qquad \mu/\sigma_n = 0$$

$$(\sigma'/\sigma_n)^2 = \frac{2\Phi(\Lambda) - 1}{2\Phi(\Lambda) - 2\Lambda G(\Lambda) - 1}$$

$$\frac{\mu_4}{\sigma_n^{\,4}} = \frac{[2\Phi(\Lambda) - 1]\{3[2\Phi(\Lambda) - 1] - 2\Lambda G(\Lambda)[\Lambda^2 + 3]\}}{[2\Phi(\Lambda) - 2\Lambda G(\Lambda) - 1]^2}.$$

Substituting these expressions into Eq. 9 and 11 yields

$$e_{s/m} = 2/\pi \left[\frac{2\Phi(\Lambda) - 2\Lambda G(\Lambda) - 1}{[2\Phi(\Lambda) - 1]^3}\right], \tag{19}$$

and

$$e_{pcc/np} = 2/\pi^2 \left[\frac{[2\Phi(\Lambda) - \Lambda G(\Lambda) - 1]^2 - \frac{1}{2}[2\Phi(\Lambda) - 1]\Lambda G(\Lambda)[1 + \Lambda^2]}{[2\Phi(\Lambda) - 1]^6}\right]. \tag{20}$$

These expressions are shown graphically in Fig. 3. The following results are apparent. A small amount of truncation ($\Lambda > 3$) has little effect on the detectors considered. The performance of the nonparametric detectors deteriorate relative to the Neyman-Pearson detectors under severe truncation. Similar results would occur if delta functions were allowed at $|x| = \Lambda\sigma$. The Neyman-Pearson detectors designed for gaussian noise perform relatively well in the presence of noise with a flat amplitude distribution ($\Lambda \to 0$). For example, the sign detector is 33 per cent as efficient as the mean detector under these conditions and the PCC is 3.6 per cent as efficient as its parametric counterpart.

Comparison of Detectors for Type I Asymmetrical Statistics

From the density function of Eq. 15 and the normalizing constraints of Eq. 12a we obtain

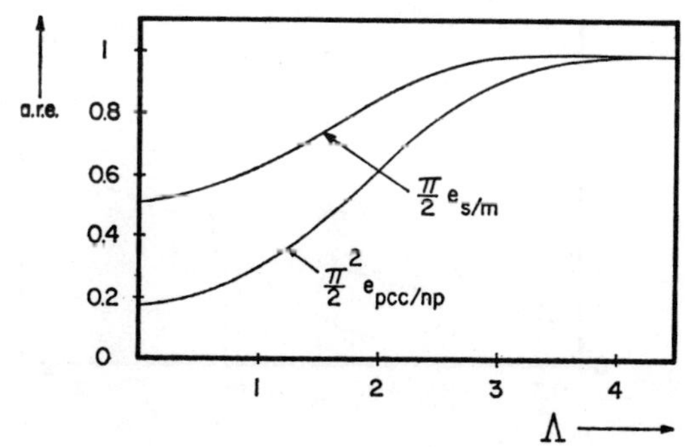

FIG. 3. Asymptotic relative efficiency—truncated gaussian statistics.

$$F(0) = \tfrac{1}{2}$$

$$f(0^+) = \frac{1}{\sqrt{2\pi\sigma_+^2}}; \qquad f(0^-) = \frac{1}{\sqrt{2\pi\sigma_-^2}}$$

$$\left(\frac{\sigma_+}{\sigma_n}\right)^2 = A^2/\xi; \qquad \left(\frac{\sigma_-}{\sigma_n}\right)^2 = 1/\xi$$

$$\frac{\mu}{\sigma_n} = \frac{A-1}{\sqrt{2\pi\xi}}$$

$$\frac{\mu_4}{\sigma_n^4} = \frac{3}{2}\frac{A^4+1}{\xi^2} - \frac{4}{\pi}\frac{(A-1)(A^3-1)}{\xi^2} + \frac{3}{2\pi}\frac{(A-1)^2(A^2+1)}{\xi^2} - \frac{3}{4\pi^2}\frac{(A-1)^4}{\xi^2}$$

$$\frac{\mu_3}{\sigma_n^3} = \sqrt{2/\pi}\,\frac{A^3-1}{\xi^{3/2}} - \frac{3}{\sqrt{8\pi}}\frac{(A-1)(A^2+1)}{\xi^{3/2}} + \frac{1}{\sqrt{2\pi^2}}\frac{(A-1)^3}{\xi^{3/2}},$$

where

$$\xi = \frac{\pi-1}{2\pi}A^2 + \frac{1}{\pi}A + \frac{\pi-1}{2\pi}.$$

Substituting these expressions into Eqs. 9 and 11 yields

$$e_{s/m} = \begin{cases} 2/\pi \quad \xi; & s > 0 \\ 2/\pi \quad \xi/A^2; & s < 0 \end{cases} \tag{21}$$

and

$$e_{pcc/np} = \frac{2}{\pi^2}\theta\left[\frac{\xi(A^2+1)}{2A^2}\right]^2, \tag{22}$$

where

$$\theta = \frac{1}{4} + \frac{3}{8}\frac{A^4+1}{\xi^2} + \frac{1}{\pi}\frac{(A-1)(A^3-1)}{\xi^2} - \frac{1}{4\pi}\frac{(A-1)^2}{\xi} - \frac{5}{16\pi^2}\frac{(A-1)^4}{\xi^2}.$$

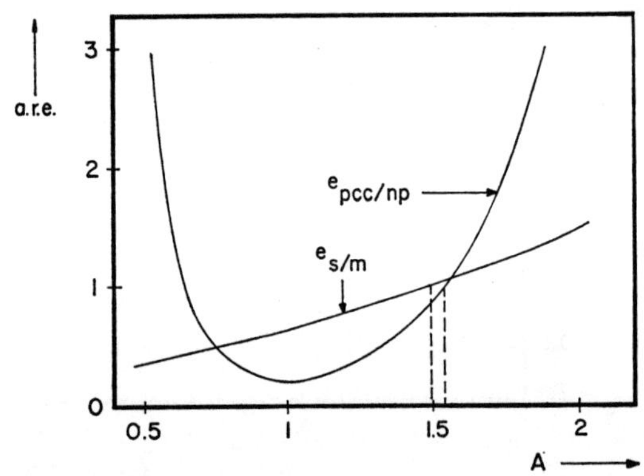

FIG. 4. Asymptotic relative efficiency—Type I asymmetrical statistics.

These expressions are shown graphically in Fig. 4. We observe that $e_{s/m} \approx 1$ for $A = 1.49$ and $e_{pcc/np} \approx 1$ for $A = 1.53$. The density functions corresponding to these values of A are shown together with the gaussian density ($A = 1$) in Fig. 7. It is worth considering how the a.r.e. would be affected by the inevitable smoothing of the discontinuity in any practical situation. It is clear from Eq. 9 that the power of the sign detector depends critically upon the value of the peak of the density function and hence its effectiveness would be reduced by smoothing. Thus the plot of $e_{s/m}$ in Fig. 4 overestimates the sign detector in the presence of smoothing. On the other hand, the effectiveness of the PCC depends upon the quantity $[f(0^+)^2 + f(0^-)^2]$ and hence a little smoothing would have a small effect on $e_{pcc/np}$. Thus these results have more practical meaning for the two-channel detectors.

Comparison of Detectors for Type II Asymmetrical Statistics

From the density function and constraint of Eq. 16 and the normalizing constraints of Eq. 12a we obtain

$$f(0) = K; \qquad K\sigma_n = \left[\frac{2\xi}{\pi}\right]^{1/2}$$

$$\left(\frac{\sigma_+}{\sigma_n}\right)^2 = \frac{[1 - F(0)]^2}{\xi}; \qquad \left(\frac{\sigma_-}{\sigma_n}\right)^2 = \frac{F^2(0)}{\xi}.$$

If

$$\theta \triangleq \left\{1 + \frac{\mu_4}{\sigma_n{}^4} + 8\frac{\mu}{\sigma_n}\left[\frac{\mu_3}{\sigma_n{}^3} + 2\frac{\mu}{\sigma_n}\right]\right\},$$

then

$$\theta = \chi/\xi^2,$$

where

$$\xi = (1 - 2/\pi) - (3 - 8/\pi)F(0) + (3 - 8/\pi)F^2(0),$$

and

$$\chi = \xi^2 + 3[[1 - F(0)]^5 + F^5(0)] + \frac{16}{\pi}[1 - 2F(0)][[1 - F(0)]^4 - F^4(0)] - \frac{4}{\pi}[1 - 2F(0)]^2\xi - \frac{20^2}{\pi^2}[1 - 2F(0)]^4.$$

Substituting these expressions into Eqs. 9 and 11 yields

$$e_{s/m} = 2/\pi \frac{\xi}{F(0)[1 - F(0)]}, \tag{23}$$

and

$$e_{pcc/np} = 2/\pi^2 \frac{\chi}{2F(0)[1 - F(0)][1 - 2F(0) + 2F^2(0)]}. \tag{24}$$

These expressions are plotted in Fig. 5 where it is observed that $e_{s/m} \approx 1$ for $F(0) = 0.2345$ and $e_{pcc/np} \approx 1$ for $F(0) = 0.27$. The density functions corresponding to these values of $F(0)$ are compared with the gaussian density function $[F(0) = 0.5]$ in Fig. 8. The results of the preceding calculations will now be discussed.

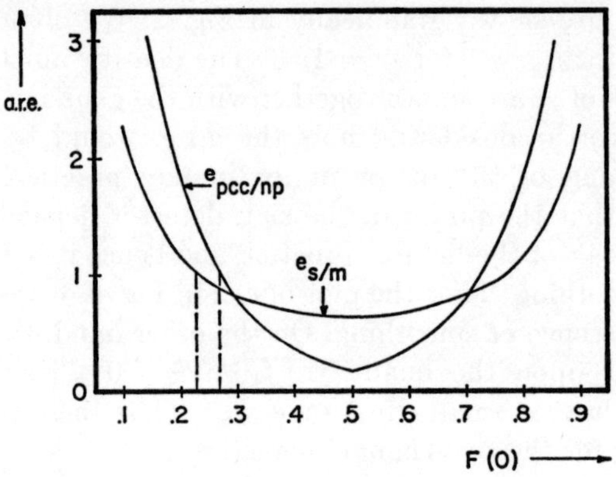

FIG. 5. Asymptotic relative efficiency—Type II asymmetrical statistics.

Discussion of Results

It has been demonstrated in the previous section that for some input processes the nonparametric detectors are more efficient than the parametric ones designed for gaussian statistics. While this class of inputs cannot be uniquely defined, definite characteristics can be observed. The nonparametric detectors are more efficient for noise processes whose amplitude densities are more peaked at the origin than the gaussian density and/or are asymmetrical. For processes whose amplitude densities are "flatter" than gaussian the nonparametric detectors are very inefficient relative to the parametric detectors The question now arises as to which of these classes of input processes occur more commonly in practice.

One of the most common and troublesome non-gaussian processes that occurs is known as impulse noise [**15**, **16**)). One model for this process is a train of independently occurring pulses with a relatively small amount of overlapping. The amplitude density of such a process depends heavily on the shape

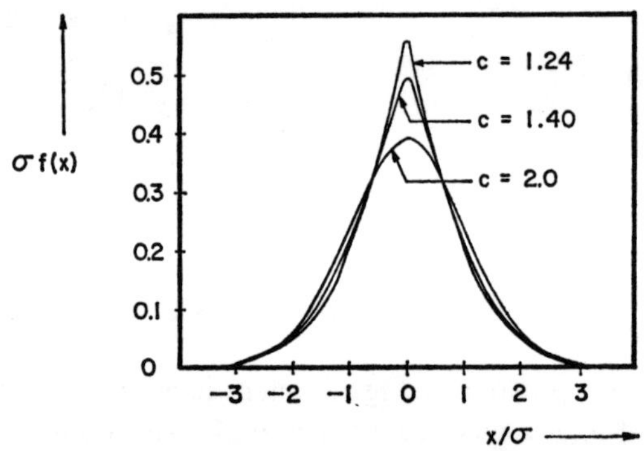

FIG. 6. Density functions of the form

$$\frac{a}{\sigma}\exp\left[-b\left|\frac{x}{\sigma}\right|^{c}\right].$$

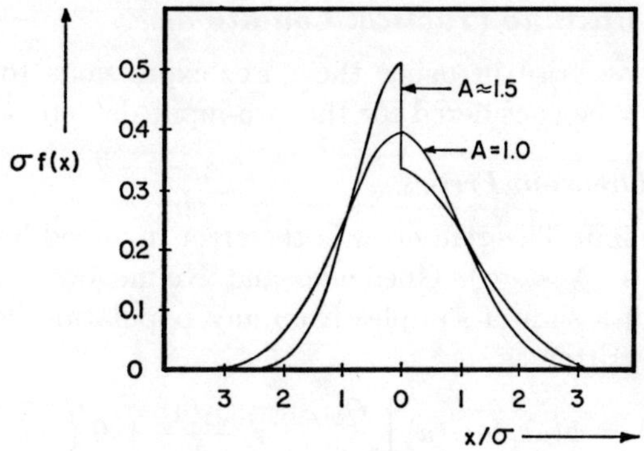

Fig. 7. Density functions of the form

$$[2\pi\sigma_+^2]^{-1/2}\exp\left[-\frac{1}{2}\left(\frac{x}{\sigma_+}\right)^2\right]; \qquad x > 0$$

$$[2\pi\sigma_-^2]^{-1/2}\exp\left[-\frac{1}{2}\left(\frac{x}{\sigma_-}\right)^2\right]; \qquad x < 0.$$

of the pulse (**17**). For this reason impulse noise is not well-defined. The amplitude densities of this impulse noise model have been calculated for a number of simple pulse shapes (**17**). They were shown as having the general properties of "peakiness" at the origin and/or asymmetry. The densities calculated (**17**) strongly resemble those shown in Figs. 6 and 8. It is therefore reasonable to assume that the nonparametric detectors are efficient (relative to their parametric counterparts designed under the gaussian assumption) in the presence of impulse noise.

The curves of Figs. 2 through 5 give an indication of the sensitivity of the parametric detectors to the gaussian assumption. We observe that for inputs that are nearly gaussian, the comparison of detectors under the gaussian assumption is reasonable.

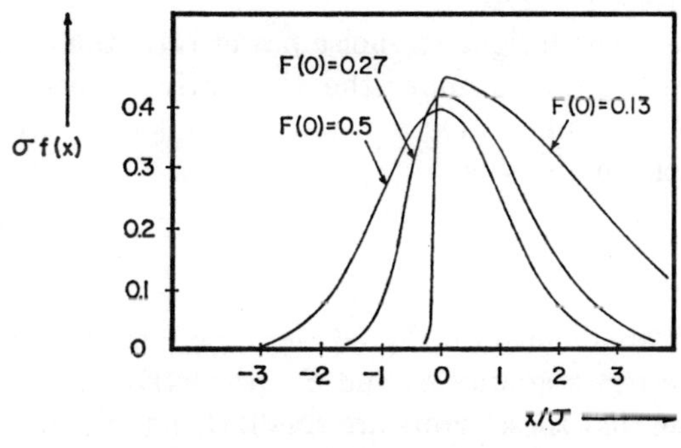

Fig. 8. Density functions of the form

$$\frac{a}{\sigma}\exp\left[-\frac{1}{2}\left(\frac{x}{\sigma_+}\right)^2\right]; \qquad x > 0$$

$$\frac{a}{\sigma}\exp\left[-\frac{1}{2}\left(\frac{x}{\sigma_-}\right)^2\right]; \qquad x < 0.$$

Application of A.R.E. to Practical Conditions

The errors that arise in using the a.r.e. expressions for practical cases ($N < \infty$) will now be considered for the two-input detectors only.

Central Limit Theorem Errors

The Central Limit Theorem error is the error involved in using Eqs. 5 or 7 for practical cases ($N < \infty$). Gnedenko and Kolmogorov (**18**) showed that the distribution of a sum of samples from any population density with finite moments can be written as

$$F(\mu) = \Phi(\mu) + G(\mu)\left[\frac{Q_1(\mu)}{N^{1/2}} + \frac{Q_2(\mu)}{N} + 0\left(\frac{1}{N^{3/2}}\right)\right], \tag{25}$$

where

$$Q_1(\mu) = \frac{\lambda_3}{6}(1 - \mu^2),$$

$$Q_2(\mu) = \frac{10\lambda_3^2}{6!}\mu^5 - \frac{1}{8}\left(\frac{\lambda_4}{3} - \frac{10\lambda_3^2}{9}\right)\mu^3 - \left(\frac{5\lambda_3^2}{24} - \frac{\lambda_4}{8}\right)\mu,$$

where

$$\lambda_3 = \mu_3/\sigma_n^3; \qquad \lambda_4 = \frac{\mu_4}{\sigma_n^4} - 3,$$

and where $G(\cdot)$ is the normalized gaussian density function. Since $F(u)$ and $\Phi(u)$ are both single-valued, monotonic functions that map the real line into the interval (0, 1), it is possible to define the error ϵ introduced in u, when replacing $F(u)$ by $\Phi(u)$, by the equation

$$\Phi[\mu(1 + \epsilon)] \triangleq F(\mu) = \Phi(\mu)\left[1 + \frac{G(\mu)}{\Phi(\mu)}\delta\right], \tag{26}$$

where δ is defined by Eq. 25. For $\epsilon \ll 1$,

$$\Phi[\mu(1 + \epsilon)] \approx \Phi(\mu) + \mu\epsilon\Phi'(\mu) = \Phi(\mu)\left[1 + \frac{G(\mu)}{\Phi(\mu)}\mu\epsilon\right].$$

Thus for small errors, $\epsilon \approx \delta/u$.

For a sufficiently small signal-to-noise power ratio the cumulative distribution of a given test statistic under the alternative is essentially equal to a shifted version of the cumulative distribution under the hypothesis. A more accurate expression for Eq. 7 is

$$\frac{N_2}{N_1} = \left[\frac{\mu_{1K} - \mu_{1H}}{\mu_{2K} - \mu_{2H}}\right]^2 \left[\frac{\rho_{2K} + \rho_{2H}}{\rho_{1K} + \rho_{1H}}\right]^2 \left[\frac{1 + \epsilon^{(1)}}{1 + \epsilon^{(2)}}\right]^2, \tag{27}$$

where $\epsilon^{(1)}$ and $\epsilon^{(2)}$ are defined by Eq. 26 and represent the errors introduced in assuming that the test statistics S_1 and S_2, respectively, are normally distributed. If we assume that these errors are small, then the Central Limit Theorem error ϵ_1 can be defined as $[2\epsilon^{(1)} - 2\epsilon^{(2)}]$ or

$$\epsilon_1 = \frac{2}{\mu}[\delta^{(1)} - \delta^{(2)}]. \tag{28}$$

This error represents the relative difference between Eq. 7 and Eq. 27.

For near-gaussian inputs, the PCC test statistic S_{pcc} is approximately binomially distributed and its Neyman-Pearson counterpart S_{np} is approximately chi-squared. The chi-squared distribution approaches the normal distribution much more slowly than does the binomial distribution. This can be easily verified by substituting the suitable numbers into Eq. 25. For this reason, only the Central Limit Theorem error due to theapproximation of the Neyman-Pearson test statistic will be considered. For a given level α and power β (which determines u in Eq. 28), the error can be tabulated as a function of signal-to-noise power ratio (**14**), where only a bound is given for the non-gaussian input processes. Thus, for $\alpha \geq 1.27$ per cent and $\beta \leq 98.73$ per cent or $u \leq 2.236$, bounds on the Central Limit Theorem error ϵ_1 are given in Table I. These errors lead to an over-estimation of the efficiency of the PCC detector for practical conditions. They are relatively small and will be shown to be smaller than the computational errors by a considerable amount for the non-gaussian inputs.

TABLE I

Central Limit Theorem Error Bounds

σ_s^2/σ_n^2 \ Inputs	Gaussian	Generalized Gaussian	Type II Asymmetrical Process
10^{-1}	-0.05	-0.10	-0.05
10^{-2}	-0.005	-0.010	-0.005
10^{-3}	-0.0005	-0.0010	-0.0005

Computational Errors

The computational error was previously defined to be the relative difference between the exact expression of Eq. 7 and the a.r.e. approximation of Eq. 11. If this error ϵ_2 is defined as

$$\epsilon_2 = \frac{N_{np}/N_{pcc} - e_{pcc/np}}{e_{pcc/np}},$$

then considering Eqs. 7 and 11 we obtain

$$\epsilon_2 = \left[\frac{1 + \rho_{2K}/\rho_{2H}}{1 + \rho_{1K}/\rho_{1H}}\right]^2 \frac{[(p_+ - p_o)/\sigma_s^2]^2}{\lim\limits_{\sigma_s^2/\sigma_n^2 \to 0} [(p_+ - p_o)/\sigma_s^2]^2} - 1. \tag{29}$$

Thus it is necessary to compute ρ_{npK} and p_+ as functions of the signal-to-noise ratio in order to calculate the computational error ϵ_2. These terms will be evaluated for gaussian signal processes. By definition

$$\rho^2_{npK} = E\{[n^{(1)}(t) + n^{(2)}(t) + 2s(t)]^4\} - E^2\{[n^{(1)}(t) + n^{(2)}(t) + 2s(t)]^2\}.$$

For gaussian signals this expression becomes

$$\rho^2_{npK} = \rho^2_{npH} + 32\sigma_n^4 \left[\frac{\sigma_s^2}{\sigma_n^2} + \frac{\sigma_s^4}{\sigma_n^4}\right].$$

For gaussian signals and all of the noise processes considered here, p_+ can be evaluated (**14**) as

$$p_+ = p_o + [1 - 2F(0)] \sum_{n=0}^{\infty} \left[\frac{\sigma_n{}^{cn+1}}{\sigma_-{}^{cn}} f(0^-) - \frac{\sigma_n{}^{cn+1}}{\sigma_+{}^{cn}} f(0^+) \right]$$

$$\times \frac{(-1)^n}{n!} \frac{2^{(cn+1)/2}}{2^n\sqrt{\pi}} \frac{\Gamma\left(\frac{cn+2}{2}\right)}{(cn+1)} \left(\frac{\sigma_s}{\sigma_n}\right)^{cn+1}$$

$$+ \sum_{n=0}^{\infty} \left[\frac{\sigma_n{}^{cn+2}}{\sigma_-{}^{cn}} f^2(0^-) + \frac{\sigma_n{}^{cn+2}}{\sigma_+{}^{cn}} f^2(0^+) \right]$$

$$\times \frac{(-1)^n}{2^n} \frac{2^{(cn+2)/2}}{\sqrt{\pi}} \Gamma\left(\frac{cn+3}{2}\right) P(n, c) \left(\frac{\sigma_s}{\sigma_n}\right)^{cn+2},$$

where

$$P(n, c) = \sum_{r=0}^{n} \frac{1}{r!(n-r)!(cr+1)(cn-cr+1)}.$$

For the inputs considered, the range of convergence is:

Gaussian:	$\sigma_s^2/\sigma_n^2 < 0.5$
Generalized Gaussian:	$\sigma_s^2/\sigma_n^2 < 1.0$
Type II Asymmetrical Process $[F(0) = 0.27]$:	$\sigma_s^2/\sigma_n^2 < 0.133$

The computational error given by Eq. 29 can now be evaluated, and is given in Table II. This error also leads to an over-estimation of the efficiency of the PCC detector for practical conditions.

TABLE II

Computational Error

σ_s^2/σ_n^2 \ Inputs	Gaussian	Generalized Gaussian (c = 1.24)	Type II Asymmetrical Process $F(0) = 0.27$
10^{-1}	0.002	−0.398	−0.43
10^{-2}	−0.0004	−0.1605	−0.122
10^{-3}	−0.000004	−0.0446	−0.0333

The following conclusions are reached: The a.r.e. comparison can be accurately applied to the small signal-to-noise case ($\sigma_s^2/\sigma_n^2 \leq 0.1$) when considering gaussian inputs. For the non-gaussian processes considered, the computational errors are much larger than the Central Limit Theorem errors and lead to significant overestimation of the efficiency of the PCC detector for practical conditions ($N < \infty$ and $\sigma_s^2/\sigma_n^2 > 10^{-3}$). It should be realized that in order to attain a power $\beta \geq 90$ per cent at a level $\alpha \leq 10$ per cent for $\sigma_s^2/\sigma_n^2 \leq 10^{-3}$, it is necessary to have a sample size $N \geq 4.6 \times 10^6$ for the generalized gaussian process considered and $N \geq 13 \times 10^6$ for the Type II asymmetrica process considered. Thus, in order to apply the a.r.e. comparison to many practical conditions, corrections for these two types of errors must be made. These corrections are cumbersome even for mathematically tractable density functions.

References

(1) E. L. Lehmann, "Testing Statistical Hypothesis," New York, John Wiley and Sons, Inc., p. 64, 1959.
(2) M. G. Kendall and A. Stuart, "The Advanced Theory of Statistics-Inference and Relationship," Vol. 2, New York, Hafner Publishing Co., p. 162, 1961.
(3) Kendall and Stuart (2), pp. 513–517.
(4) S. S. Wolff, J. B. Thomas, and T. R. Williams, "The Polarity Coincidence Correlator: A Nonparametric Detection Device," *IRE Trans. Infor. Theory*, Vol. IT-8, No. 1, pp. 5–9, Jan. 1962.
(5) Kendall and Stuart (2), p. 265.
(6) Kendall and Stuart (2), p. 162.
(7) J. W. Carlyle and J. B. Thomas, "On Nonparametric Signal Detectors," *IEEE Trans. Infor. Theory*, Vol. IT-10, No. 2, pp. 146–152, Apr. 1964.
(8) Lehmann (1), pp. 93, 242; 1959.
(9) Lehmann (1), pp. 93–94.
(10) Lehmann (1), p. 344.
(11) Wolff, Thomas, and Williams (4), p. 9.
(12) Carlyle and Thomas (7), pp. 148, 151.
(13) Carlyle and Thomas (7), p. 150.
(14) M. Kanefsky, "On Sign Tests and Adaptive Techniques for Non-parametric Detection," Ph.D. Dissertation, Princeton University, 1964.
(15) A. A. Alexander, R. M. Gryb and D. W. Nast, "Capabilities of the Telephone Network for Data Transmission," *Bell System Tech. Jour.*, Vol. 39, No. 3, pp. 431–476, May, 1960.
(16) A. B. Bodonyi, "Effects of Impulse Noise on Digital Data Transmission," *IRE Trans. Comm. Systems*, Vol. 9, pp. 355–361, Dec. 1961.
(17) E. N. Gilbert and H. O. Pollak, "Amplitude Distribution of Shot Noise," *Bell System Tech. Jour.*, Vol. 39, No. 2, pp. 333–350, Mar. 1960.
(18) B. V. Gnedenko and A. N. Kolmogorov, "Limit Distributions for Sums of Independent Random Variables," Cambridge, Mass., Addison-Wesley Pub. Co., p. 195, 1954.

4

Reprinted with permission from *IRE Trans. Inf. Theory* **IT-8**:5–9 (1962)

The Polarity-Coincidence Correlator: A Nonparametric Detection Device*

S. S. WOLFF†, MEMBER, IRE, J. B. THOMAS†, SENIOR MEMBER, IRE, AND T. R. WILLIAMS†, MEMBER, IRE

I. Introduction

TWO-INPUT detection systems in which the inputs are correlated to determine the presence or absence of a signal common to both inputs are widely employed in radio-astronomical[1-3] and underwater sound[4] applications. The specifications of a frequently used mathematical model are as follows:

1) In the absence of signal, the two inputs are given by $f^{(1)}(t) = x^{(1)}(t)$, $f^{(2)}(t) = x^{(2)}(t)$, and $0 \leq t \leq T$;

2) $x^{(1)}(t)$ and $x^{(2)}(t)$ are independent stationary random processes having identical Gaussian probability distributions with zero means;

3) The signal is additive in both channels so that $f^{(1)}(t) = x^{(1)}(t) + s(t)$, $f^{(2)}(t) = x^{(2)}(t) + s(t)$, when signal is present;

4) $s(t)$ is a stationary random process independent of both $x^{(1)}(t)$ and $x^{(2)}(t)$ but with the same probability distributions of all orders (the last condition implies that all three processes have the same spectral density function).

For the purposes of this paper, the model will be modified in the following two respects.

1) The inputs are assumed to have been sampled every Δ seconds, and only the sampled values (two sequences of random variables) are available to the detector. Furthermore, it is assumed that Δ has been chosen so that the samples of each process are independent among themselves. The following notation is introduced:

$$\varphi_t^{(n)} \triangleq f^{(n)}(t\,\Delta)$$

$$\xi_t^{(n)} \triangleq x^{(n)}(t\,\Delta) \qquad n = 1, 2$$

$$\gamma_t \triangleq s(t\,\Delta);$$

where t is a positive integral-valued index.

2) Specifications 2) and 4) are relaxed in that $x^{(1)}(t)$, $x^{(2)}(t)$ and $s(t)$ are no longer supposed to be Gaussian, although the requirement of stationarity is retained. The precise model for the *PCC* appears in Section II.

With the mathematical model outlined, several systems may be used for detection of the signal. One of these, termed *correlation detection*, consists in calculating for each pair of received sequences the statistic

$$C_1 \triangleq \sum_{t=1}^{N_1} \varphi_t^{(1)} \varphi_t^{(2)}, \tag{1}$$

where N_1 is the number of samples per sequence. A signal is considered to be present if C_1 exceeds a threshold value; determination of the threshold, together with other characteristics of the correlation detector, will be outlined in Section III.

A second method of detection arises from applying the fundamental lemma of statistical hypothesis testing to the Gaussian model and then employing the resulting system for the generalized model.[5] This (Neyman-Pearson) detector calculates the statistic

$$C_2 \triangleq \sum_{t=1}^{N_2} (\varphi_t^{(1)} + \varphi_t^{(2)})^2, \tag{2}$$

where N_2 is the number of samples per sequence. A signal is said to be present if C_2 exceeds a threshold. Again, further discussion will be postponed to Section III.

The polarity-coincidence correlator (*PCC*), a third-detection device, calculates

$$C_3 \triangleq \sum_{t=1}^{N_3} \operatorname{sgn}(\varphi_t^{(1)}) \operatorname{sgn}(\varphi_t^{(2)}), \tag{3}$$

* Received by the PGIT, February 2, 1961.

† Dept. of Elec. Engrg., Princeton University, Princeton, N. J.

1 R. C. Jones, "The ultimate sensitivity of radiation detectors," *J. Opt. Soc. Am.*, vol. 37, pp. 879–898; November, 1947.

2 R. H. Dicke, "The measurement of thermal radiation at microwave frequencies," *Rev. Sci. Instr.*, vol. 17, pp. 268–275; July, 1946.

3 S. J. Goldstein, Jr., "A comparison of two radiometer circuits," Proc. IRE, vol. 43, pp. 1663–1666; November, 1955.

4 J. B. Thomas and T. R. Williams, "On the detection of signals in nonstationary noise by product arrays," *J. Acoust. Soc. Am.*, vol. 31, pp. 453–462; April, 1959. See references 1–3.

5 E. L. Lehmann, "Testing Statistical Hypotheses," John Wiley and Sons, Inc., New York, N. Y., p. 64; 1959.

where N_3 is the number of samples per sequence and sgn $(\cdot)$ is the sign function

$$\text{sgn}(u) = \begin{cases} 1 & u > 0 \\ 0 & u = 0 \\ -1 & u < 0. \end{cases}$$

Faran and Hills[6,7] have analyzed and tested a *PCC* within the framework of the first mathematical model given above (the operations were in continuous time and the summation was replaced by averaging in a low-pass filter). Thomas and Williams[4] analyzed the *PCC* and concluded that while it was somewhat inferior to a correlation detector within the mathematical model, it became markedly superior if the restriction of noise stationarity were removed. Melton and Karr[8] have considered multiple-coincidence schemes, but without specific consideration of nonparametric aspects.

It is the purpose of this paper to show that the *PCC*, as a statistical technique, possesses nonparametric properties, and that it can in certain cases outperform the correlation detector. Also, being nonparametric, the *PCC* may be used when the noise distributions are unknown, and it is therefore impossible to apply the fundamental lemma to the design of a Neyman-Pearson detector.

II. Analysis of the Polarity-Coincidence Correlator (PCC)

The mathematical model appropriate to the *PCC* is less confining than that initially given in Section I. Specifically, the restrictions placed on the probability distribution functions P_ξ of the random variables $\xi_t^{(1)}$ and $\xi_t^{(2)}$ and P_γ of the γ_t are:

1) P_ξ and P_γ have zero median, $P_\xi(0) = P_\gamma(0) = \frac{1}{2}$;
2) $P_\xi(u) = 1 - P_\xi(-u)$, $P_\gamma(u) = 1 - P_\gamma(-u)$;
3) $\int u^2\, dP_\xi(u) < \infty$, $\int u^2\, dP_\gamma(u) < \infty$;
4) P_ξ and P_γ are absolutely continuous with respect to Lebesgue measure.

According to (3), the *PCC* calculates the statistic

$$C_3 = \sum_{t=1}^{N_3} \text{sgn}\,(\varphi_t^{(1)})\,\text{sgn}\,(\varphi_t^{(2)}).$$

For the case "no signal present" (denoted hereafter by H), a summand is -1 or $+1$, each with probability $\frac{1}{2}$, neglecting values of $\varphi_t^{(1)}$ and $\varphi_t^{(2)}$ having probability zero. When "signal is present" (denoted by K), the probability of a summand taking the value $+1$ is

$$\begin{aligned} Pr_K\,(\text{sgn}\,\varphi_t^{(1)}\,\text{sgn}\,\varphi_t^{(2)} = +1) &= Pr_K\,(\varphi_t^{(1)} > 0\ \&\ \varphi_t^{(2)} > 0) \\ &\quad + Pr_K\,(\varphi_t^{(1)} < 0\ \&\ \varphi_t^{(2)} < 0) \\ &= \int Pr\,(\xi_t^{(1)} > -u\ \&\ \xi_t^{(2)} > -u)\, dP_\gamma(u) \\ &\quad + \int Pr\,(\xi_t^{(1)} < -u\ \&\ \xi_t^{(2)} < -u)\, dP_\gamma(u) \\ &= \int [1 - P_\xi(-u)]^2\, dP_\gamma(u) + \int P_\xi^2(-u)\, dP_\gamma(u) \\ &= \int [1 - 2P_\xi(u) + 2P_\xi^2(u)]\, dP_\gamma(u) \\ &\triangleq p_+. \end{aligned} \tag{4}$$

It is to be expected that $p_+ > \frac{1}{2}$ for any nonzero signal γ_t common to both inputs; this is indeed true. Since P_ξ has zero median, $P_\xi(u)$ can be written as $\frac{1}{2} + R$, with $R \gtrless 0$ for $u \gtrless 0$; then the integral defining p_+ is

$$\begin{aligned} p_+ &= \int [1 - 1 - 2R + \tfrac{1}{2} + 2R + 2R^2]\, dP_\gamma(u) \\ &= \tfrac{1}{2} + 2 \int R^2\, dP_\gamma(u) > 1/2. \end{aligned}$$

Since the terms in the sum defining C_3 are independent, it is evident that C_3 is binomially distributed[9] with "probability of success" $\frac{1}{2}$ in H and p_+ in K. Letting the symbol ρ denote $\frac{1}{2}$ or p_+ indifferently, the expectation and variance of C_3 are

$$\begin{aligned} E\,C_3 &= N_3(2\rho - 1) \\ \text{var}\,C_3 &= 4N_3\rho(1 - \rho). \end{aligned}$$

Although the binomial distribution is well known and extensively tabulated, it is convenient for analytic purposes here (as it will be elsewhere in this paper) to pass immediately to the limit of large sample sizes ($N_3 \gg 1$) and replace the binomial distribution by its normal approximation.[10] Thus, for very large N_3, the quantity

$$C_3' \triangleq \frac{C_3 - N_3(2\rho - 1)}{[4N_3\rho(1 - \rho)]^{1/2}} \tag{5}$$

is normally distributed with zero mean and variance 1.

To use the *PCC* as a statistical test, C_3 is compared to a threshold value c_3 at the termination of each sequence of N_3 inputs. If $C_3 > c_3$, statement K is made; if $C_3 < c_3$,

[6] J. J. Faran, Jr., and R. Hills, Jr., "Correlators for Signal Reception," Acoustics Res. Lab., Harvard University, Cambridge, Mass., Tech. Memo. No. 27; September 15, 1952.

[7] J. J. Faran, Jr., and R. Hills, Jr., "The Application of Correlation Techniques to Acoustic Receiving Systems," Acoustics Res. Lab., Harvard University, Cambridge, Mass., Tech. Memo. No. 28; November 1, 1952.

[8] B. S. Melton and P. H. Karr, "Polarity coincidence scheme for revealing signal coherence," *Geophysics*, vol. 22, pp. 553–564; July, 1957.

[9] W. Feller, "An Introduction to Probability Theory and Its Applications," John Wiley and Sons, Inc., New York, N. Y., 2nd ed., p. 136; 1957.

[10] *Ibid.*, p. 168. See also p. 170ff. for the error involved in the approximation. Note that even for small N_3 the approximation may be surprisingly accurate.

statement H.[11] In any statistical test two parameters of interest are α, defined as the probability of stating K when H is in fact true (the "false-alarm" probability), and β, defined as the probability of stating K when K is true ("probability of detection"). The designer of a test sets the value of α he is willing to accept the judges the effectiveness of the test by how large a β results. For the *PCC*,

$$\alpha = Pr_H(C_3 > c_3) = 1 - \Phi\left\{\frac{c_3}{\sqrt{N_3}}\right\} \tag{6}$$

$$\beta = Pr_K(C_3 > c_3) = 1 - \Phi\left\{\frac{c_3 - N_3(2p_+ - 1)}{[4N_3p_+(1-p_+)]^{1/2}}\right\}, \tag{7}$$

using (5). $\Phi(\cdot)$ is the unit normal distribution function. Since $\Phi(\cdot)$ is continuous and monotonic, (6) can be inverted to find c_3 uniquely as a function of α, and the result can be substituted in (7) to yield

$$\beta = 1 - \Phi\left\{\frac{\frac{1}{2}\Phi^{-1}(1-\alpha) - N_3^{1/2}(p_+ - 1/2)}{[p_+(1-p_+)]^{1/2}}\right\}. \tag{8}$$

Eq. (6) shows that α does not depend on the signal and noise probability distributions; the *PCC* is said to be nonparametric[12,13] for the class of distributions given at the beginning of this section. That is, the *PCC* has the desirable feature that its false-alarm rate is independent of the noise distributions at its input. Eq. (8) specifies the detection probability of the *PCC* explicitly as a function of α and the number $p_+ > \frac{1}{2}$ dependent upon both signal and noise probability distributions. In no event can β be less than α; the *PCC* is unbiased.[14] Furthermore, $\lim_{N_3\to\infty} \beta = 1$ for any $\alpha > 0$; the *PCC* is consistent.[15]

If a *PCC* is designed for a particular value of α (by choosing c_3), β can be expected to vary as the input distributions change, even though the input SNR may remain constant. However, the following theorem shows that as long as the sample size is large enough, an increasing SNR can never cause β to decrease, provided that the input distributions change only in variance but not in form.

Theorem: If $N^{1/2} \geq \Phi^{-1}(1-\alpha)$ and the input probability distributions remain unchanged except for scale changes, β is a nondecreasing function of input SNR.

Proof: The proof follows directly from two lemmas which are stated and proved in the Appendix.

III. Large Sample Analysis of the Neyman-Pearson and Correlation Detectors for Gaussian Imputs with Zero Mean

According to (1), the correlation detector calculates

$$C_1 \triangleq \sum_{t=1}^{N_1} \varphi_t^{(1)}\varphi_t^{(2)}.$$

Each summand is a continuously-distributed random variable with a distribution function which is the Mellin convolution of the φ_t distribution with itself.[16] The term C_1, being the sum of N_1 such independent random variables, has a distribution function which is the N_1-fold Fourier convolution of the summand distribution with itself. However, for large sample sizes, recourse to the central limit theorem[17] shows that, asymptotically, $(C_1 - EC_1)/[\text{var } C_1]^{1/2}$ has the unit normal distribution. The expectation and variance of C_1 are given by

$$E_H C_1 = 0$$
$$\text{var}_H\, C_1 = N_1\nu^4$$
$$E_K C_1 = N_1\sigma^2$$
$$\text{var}_K\, C_1 = N_1(\nu^4 + 2\nu^2\sigma^2),$$

where σ^2 and ν^2 are the variances of the signal and noise distributions, respectively. Following the development of (6)–(8), it is readily found that, for the correlation detector,

$$\alpha = Pr_H(C_1 > c_1) = 1 - \Phi\left\{\frac{c_1}{\nu^2 N_1^{1/2}}\right\} \tag{9}$$

$$\beta = 1 - \Phi\left\{\frac{c_1 - N_1\sigma^2}{[N_1(\nu^4 + 2\nu^2\sigma^2)]^{1/2}}\right\} \tag{10}$$

$$= 1 - \Phi\left\{\frac{\Phi^{-1}(1-\alpha) \quad \frac{\sigma^2}{\nu^2}[N_1]^{1/2}}{\left[1 + 2\frac{\sigma^2}{\nu^2}\right]^{1/2}}\right\}. \tag{11}$$

According to (2), the Neyman-Pearson detector calculates

$$C_2 \triangleq \sum_{t=1}^{N_2} (\varphi_t^{(1)} + \varphi_t^{(2)})^2.$$

Again the distribution of C_2 is calculable, but the central limit theorem is applied at once to show that $(C_2 - EC_2)/[\text{var } C_2]^{1/2}$ is asymptotically normal with zero

[11] Having passed to a limiting continuous distribution for C_3, the event $C_3 = c_3$ has zero probability. Nevertheless, in a practical situation, C_3 has a discrete distribution and may in fact equal c_3; statement H or K is then made dependent upon the outcome of a Bernoulli trial with specified probability of success. This technique is well known and will not be discussed further here. See D. A. S. Fraser, "Nonparametric Methods of Statistics," John Wiley and Sons, Inc., New York, N. Y., p. 73; 1957.

[12] *Ibid.*, p. 125.

[13] Lehmann, *op. cit.*, p. 261.

[14] Fraser, *op. cit.*, p. 93.

[15] *Ibid.*, p. 108.

[16] B. Epstein, "Some applications of the mellin transform in statistics," *Ann. Math. Stat.*, vol. 19, p. 370–000; Month, 1948.

[17] H. Cramér, "Mathematical Methods of Statistics," Princeton University Press, Princeton, N. J., p. 213; 1958.

mean and unit variance. The expectation and variance of C_2 are easily calculated to be

$$E_H C_2 = 2\nu^2 N_2$$
$$\mathrm{var}_H\, C_2 = 8\nu^4 N_2$$
$$E_K C_2 = 2(\nu^2 + 2\sigma^2) N_2$$
$$\mathrm{var}_K\, C_2 = 8(\nu^2 + 2\sigma^2)^2 N_2.$$

Once more paralleling the development in Section II, the Neyman-Pearson detector is found to produce

$$\alpha = Pr_H\,(C_2 > c_2)$$
$$= 1 - \Phi\left\{\frac{c_2 - 2\nu^2 N_2}{[8\nu^4 N_2]^{1/2}}\right\} \quad (12)$$
$$\beta = 1 - \Phi\left\{\frac{c_2 - 2N_2(\nu^2 + 2\sigma^2)}{(\nu^2 + 2\sigma^2)[8N_2]^{1/2}}\right\} \quad (13)$$
$$= 1 - \Phi\left\{\frac{\Phi^{-1}(1-\alpha) - [2N_2]^{1/2}\sigma^2/\nu^2}{1 + 2\sigma^2/\nu^2}\right\}. \quad (14)$$

For both the correlation detector and the Neyman-Pearson detector, the event $C_j = c_j$, $j = 1, 2$, continues to have zero probability in the limit of large sample sizes and will continue to be ignored.

IV. Comparison of Detectors

In order to compare the *PCC* with the other two detectors under the same conditions of operation, it is necessary to calculate p_+ for Gaussian inputs to the *PCC*. Eq. (4) is used:

$$p_+ = \int [1 - 2P_\xi(u) + 2P_\xi^2(u)]\, dP_\gamma(u)$$

with

$$P_\xi(u) = \int_{-\infty}^{u} \epsilon^{-x^2/2\nu^2}(2\pi\nu^2)^{-1/2}\, dx$$

and

$$P_\gamma(u) = \int_{-\infty}^{u} \epsilon^{-x^2/2\sigma^2}(2\pi\sigma^2)^{-1/2}\, dx.$$

The resulting expression is considerably simplified if attention is restricted to the important case of small signals, $\sigma^2/\nu^2 \ll 1$. The expressions in P_ξ and P_ξ^2 may be expanded in a Maclaurin series, and only those terms leading to the first order in σ^2/ν^2 are retained. The result is

$$p_+ = \frac{1}{2} + \frac{1}{\pi}\frac{\sigma^2}{\nu^2}.$$

Now β for the *PCC* (8) may be evaluated for this p_+, yielding

$$\beta = 1 - \Phi\left\{\Phi^{-1}(1-\alpha) - \frac{2}{\pi} N_3^{1/2}\frac{\sigma^2}{\nu^2}\right\}, \quad (15)$$

once again retaining only first-order terms in σ^2/ν^2.

Comparison of the three detectors discussed is made using the concept of asymptotic relative efficiency e. If two statistical tests T_1 and T_2 require sample sizes n_1 and n_2 to achieve the same β, given the same α, then the asymptotic relative efficiency of T_1 with respect to T_2 is defined[18] by

$$e_{T_1,T_2} \triangleq \lim_{\substack{n_1\to\infty\\ n_2\to\infty}} \frac{n_2}{n_1}.$$

In what follows, the subscripts C, NP, PCC will refer in an obvious fashion to the three detectors. To find $e_{C,NP}$, for example, β_C [(11)] is set equal to β_{NP} [(14)], and the limiting ratio of N_2/N_1 solved for

$$1 - \Phi\left\{\frac{\Phi^{-1}(1-\alpha) - \frac{\sigma^2}{\nu^2}N_1^{1/2}}{\left[1 + 2\frac{\sigma^2}{\nu^2}\right]^{1/2}}\right\}$$
$$= 1 - \Phi\left\{\frac{\Phi^{-1}(1-\alpha) - [2N_2]^{1/2}\sigma^2/\nu^2}{1 + 2\frac{\sigma^2}{\nu^2}}\right\}$$

$$\frac{N_1}{1 + 2\frac{\sigma^2}{\nu^2}} = \frac{2N_2}{\left(1 + 2\frac{\sigma^2}{\nu^2}\right)^2}$$

$$\lim_{\substack{N_1\to\infty\\ N_2\to\infty}} \frac{N_2}{N_1} = e_{C,NP} = \frac{1}{2}.$$

Vanishingly small signals are assumed. Thus for small SNR's and large sample sizes, the correlation detector requires twice as many samples to achieve the same performance as the Neyman-Pearson detector, when the input distributions are Gaussian. In the same way there results

$$e_{PCC,C} = \left(\frac{2}{\pi}\right)^2 = 0.405$$

$$e_{PCC,NP} = \frac{1}{2}\left(\frac{2}{\pi}\right)^2 = 0.202.$$

Thus for Gaussian noises and signal, the *PCC* is inferior both to the correlator and the Neyman-Pearson device.

Suppose that the assumption of normality is in fact not valid. By confining attention to vanishingly small SNR's, a general expression for $e_{PCC,C}$ may be derived. By generalizing the development for p_+ at the beginning of this section, it is found that

$$p_+ = \tfrac{1}{2} + 2P_\xi'^2(0)\sigma^2$$

as $\sigma^2/\nu^2 \to 0$ for any distributions P_ξ and P_γ satisfying the restrictions of Section II. (P_ξ' is the derivative of P_ξ with respect to its argument.) Substituting this p_+ in (8) results in

$$\beta_{PCC} = 1 - \Phi\left\{\frac{\frac{1}{2}\Phi^{-1}(1-\alpha) - N_3^{1/2}2P_\xi'^2(0)\sigma^2}{[\frac{1}{4} - 4P_\xi'^4(0)\sigma^4]^{1/2}}\right\}.$$

[18] Lehmann, *op. cit.*, p. 239.

Similarly paralleling the development of (11) of the correlator for unspecified P_ξ and P_γ,

$$\beta_C = 1 - \Phi\left\{\frac{\Phi^{-1}(1-\alpha) - N_1^{1/2}\sigma^2/\nu^2}{[1 + 2\sigma^2/\nu^2]^{1/2}}\right\}.$$

By setting $\beta_C = \beta_{PCC}$, allowing N_1 and N_3 to increase without limit and solving for their ratio, the following general result is obtained:

$$e_{PCC,C} = [4\nu^2 P_\xi'^2(0)]^2. \tag{16}$$

Clearly, $e_{PCC,C}$ given by (16) can take on all values from zero (when the noise is bimodally distributed with zero density at the origin) upwards.

For example, consider a double exponential distribution

$$P_\xi'(u) = \frac{\exp(-2^{1/2}\mu/\nu)}{2^{1/2}\nu}$$

for the noise. Substitution in (16) yields

$$e_{PCC,C} = 4.$$

Furthermore, if the Neyman-Pearson detector specified by (2) for Gaussian noise were used in this double exponential noise, a simple calculation shows that

$$e_{PCC,NP} = 7/2.$$

V. Conclusions

It has been shown that the *PCC* is nonparametric in its false-alarm rate with respect to a wide class of signal and noise distributions. Operating in fixed distributions, the probability of detection is a nondecreasing function of input SNR. Although the *PCC* performs less well than an ordinary correlator for Gaussian inputs, it can be superior both to the correlator and to a Neyman-Pearson device operating in noise for which it was not designed.

Appendix

Proof of the theorem in Section II consists in applying the following two lemmas in turn.

Lemma 1: If $N^{1/2} \geq \Phi^{-1}(1-\alpha)$, β is an increasing function of p_+.

Proof: Differentiation of (8) results in

$$d\beta/dp_+ = M[(2p_+ - 1)\Phi^{-1}(1-\alpha) - N_3^{1/2}],$$

where M is a negative number. Since

$$0 < (2p_+ - 1) \leq 1, \text{ if } N_3^{1/2} \geq \Phi^{-1}(1-\alpha),$$

$d\beta/dp_+$ is always positive.

Lemma 2: If the input probability distributions remain unchanged except for scale changes, p_+ is a nondecreasing function of input SNR defined as var (γ_t)/var $(\xi_t^{(n)})$.

Proof: Referring to (4), the function

$$F(u) \triangleq 1 - 2P_\xi(u) + 2P_\xi^2(u)$$

is positive, even, and nondecreasing for positive arguments; hence, $F(au) \leq (\geq)\ F(u)$ for $|a| < (>)\ 1$. If $P_{\xi'}(u) \triangleq P_\xi(au)$, $a > 1$; var $\xi_t'^{(n)} <$ var $\xi_t^{(n)}$; and

$$p_+' = \int F_{\xi'}(u)\, dP_\gamma(u)$$

$$= \int F_\xi(au)\, dP_\gamma(u)$$

$$\geq p_+.$$

If $P_{\gamma'}(u) = P_\gamma(au)$, $a < 1$; var $\gamma_t' >$ var γ_t; and

$$p_+' = \int F_\xi(u)\, dP_{\gamma'}(u)$$

$$= \int F_\xi(u)\, dP_\gamma(au)$$

$$= \int F_\xi(u/a)\, dP_\gamma(u)$$

$$\geq p_+.$$

5

Detection of Weak Signals with Polarity Coincidence Arrays

MORTON KANEFSKY, MEMBER, IEEE

***Abstract*—Polarity Coincidence Array detectors (PCA) are considered for testing the hypothesis that a random signal is common to an array of receivers which contain noise processes that are independent representations of a given class of stochastic processes. A standard procedure is to reduce the received data by sampling and then hard limiting. Hard limiting is shown to introduce an inherent loss in input signal power of 1.96 dB when the input data is a sequence of independent samples from a stationary Gaussian process. However, when the stationary and/or Gaussian assumptions are violated, the relative efficiencies of the PCA detectors can greatly improve. When the input samples are dependent, it is necessary to assume Gaussian inputs in order to analyze the PCA detectors. However, these devices are still unaffected by a nonstationary noise level that is slowly varying relative to the inverse bandwidth of the pre-filter. Furthermore, the loss due to clipping is considerably reduced as the sample dependence (i.e., sampling rate) increases. For rapid sampling rates, the spectral shapes of the inputs must be known accurately in order to fix the false-alarm rate at some preassigned value.**

I. Introduction

IN UNDERWATER sound detection (SONAR) one typically decides whether or not a random signal is common to an array of receivers (hydrophones) which contain noise processes that are independent but statistically identical [1]–[5]. As the noise power is often nonstationary, the input data is sometimes reduced by sampling and then hard limiting [1]–[4]. A detector which operates on the polarity coincidence of two channels is called a polarity coincidence correlator (PCC) [1], [6]. This device has interesting nonparametric properties when the inputs are a sequence of independent samples. The false-alarm probability can be set for any sequence of samples whose amplitude densities have zero median [6]. For certain non-Gaussian noise processes, particularly of the impulse type, the PCC can be more efficient than the optimum two-channel detector based on stationary Gaussian inputs [6], [7]. Furthermore, the PCC is exceedingly simple to implement. In sonar practice, however, it is desirable to consider an array of receivers that may be quite large. Polarity coincidence techniques can be extended to the array case, but suggested procedures [1] are very inefficient for large arrays. Efficient hard limiting procedures have been utilized (i.e., the DIMUS system [4]), but they are not regarded as logical extensions of the PCC.

In this paper a general class of polarity coincidence array detectors (PCA), which are logical extensions of the two-channel PCC, are discussed. The implementations considered include the locally optimum[1] detector based on stationary Gaussian inputs [10], as well as some suboptimum devices that may be more easily implemented and are still quite efficient.

II. Terminology

Under the hypothesis H (noise with zero median and signal absent), approximately half the channels will have the same sign. Under the alternative K (signal present), the number of channels with the same sign as the signal will increase. Let $\delta(t)$ be the difference between the

This research was supported by the Office of Naval Research, under Contract NOnr 2512 (00) with General Dynamics/Electric Boat Company as part of the SUBIC (Submarine Integrated Control) program.
The author is with Yale University, New Haven, Conn.

[1] A locally optimum detector is one that is optimum in the Neyman-Pearson sense [8] in the limit as the input signal-to-noise ratio in each channel approaches zero [9].

number of channels having the most prevalent sign at time t and half the number of channels ($M/2$). Hence,

$$\delta(t) = \left| \frac{1}{2} \sum_{j=1}^{M} \operatorname{sgn}(x_j(t)) \right|, \quad (1)$$

where $x_j(t)$ are the M inputs, and

$$\operatorname{sgn}(\xi) = \begin{cases} 1; & \xi > 0 \\ -1; & \xi < 0 \end{cases}.$$

The PCA detectors sample the inputs and then perform some memoryless monotonic operation on δ_i ($\delta_i = \delta(t + i\tau)$ where τ is the sampling interval). Thus, the PCA test statistics are of the form

$$S_{pca} = \sum_{i=1}^{N} g(\delta_i), \quad (2)$$

where N is the total sample size and $g(\ )$ is some arbitrary monotonic function. This test statistic is then compared with a threshold c and the detector decides that the signal is present if $S_{pca} > c$. (See Fig. 1) for a schematic diagram of these detectors.) With little loss in generality the function $g(\ )$ can be normalized so that $g(\delta) = \delta$ when $\delta = 0$ or 1. With this normalization the PCA detectors reduce to the PCC detector when there are two inputs.

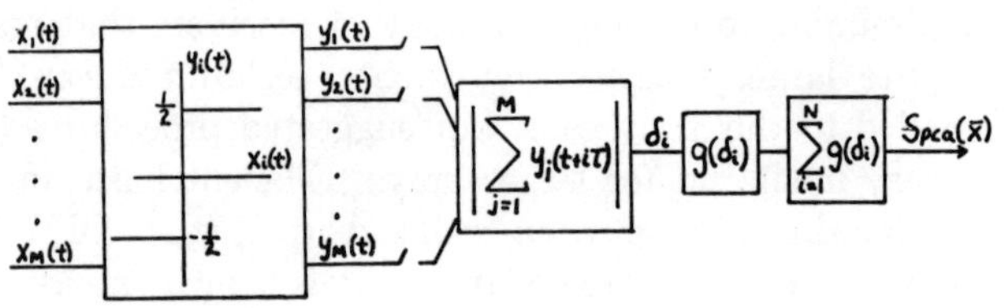

Fig. 1. Schematic for polarity coincidence array detectors.

For large N, S_{pca} is approximately normally distributed, and hence, c can be set equal to

$$c = E_H\{S_{pca}\} + \Phi^{-1}(1 - \alpha)\sqrt{\operatorname{Var}_H\{S_{pca}\}}, \quad (3)$$

where Φ is the normalized Gaussian cumulative distribution function, α the false-alarm probability, $E_H\{\ \}$ the expected value with respect to the hypothesis, and $\operatorname{Var}_H\{\ \}$ the variance under the hypothesis. Thus, $\alpha = \operatorname{Prob}(S_{pca} > c/H)$.

The output signal-to-noise ratio of a detector D, of the form $S_D \gtrless c$, is defined as

$$\mathrm{SNR}_D \triangleq \frac{E_K\{S_D\} - E_H\{S_D\}}{[\operatorname{Var}_K\{S_D\}]^{1/2}}. \quad (4)$$

For many practical applications the input signal-to-noise power ratio σ_s^2/σ_n^2 in each channel is small ($M\sigma_s^2/\sigma_n^2 \ll 1$). This is of course the justification for considering locally optimum detectors. Hence, we will compare detectors by calculating the limit of SNR_D as $\sigma_s^2/\sigma_n^2 \to 0$. Thus,

$$\overline{\mathrm{SNR}}_D \triangleq \lim_{\sigma_s^2/\sigma_n^2 \to 0} \frac{E_K\{S_D\} - E_H\{S_D\}}{[\operatorname{Var}_H\{S_D\}]^{1/2}}. \quad (5)$$

Under the assumptions that the cumulative distribution function of the noise can be expanded in a Taylor series near the origin and that the signal is amplitude limited with zero mean, it is shown in Appendix I that

$$\lim_{\sigma_s^2/\sigma_n^2 \to 0} [E_K\{S_D\} - E_H\{S_D\}]$$

$$= 8N\sigma_n^2 f^2(0) \frac{\sigma_s^2}{\sigma_n^2} \frac{2}{2^M} \cdot \sum_{r=M/2+\epsilon}^{M} \binom{M}{r} \left[\left(r - \frac{M}{2}\right)^2 - \frac{M}{4}\right] g\left(r - \frac{M}{2}\right) \quad (6)$$

$$= 8N\sigma_n^2 f^2(0) \frac{\sigma_s^2}{\sigma_n^2} \left[E_H\{\delta^2 g(\delta)\} - \frac{M}{4} E_H\{g(\delta)\} \right], \quad (7)$$

where $f(0)$ is the value of the noise amplitude density at the origin and

$$\epsilon = \begin{cases} 0, & M \text{ even} \\ \frac{1}{2}, & M \text{ odd} \end{cases}$$

The relationship between (6) and (7) follows from the fact that $\delta_i = (r - M/2)$ with probability $2/2^M \binom{M}{r}$ under the hypothesis, where r is the number of channels having the most prevalent sign. Hence, $E_H\{f(\delta_i)\}$ is given by

$$\sum_{r=M/2+\epsilon}^{M} \frac{2}{2^M} \binom{M}{r} f\left(r - \frac{M}{2}\right).$$

Thus, for independent samples

$$\overline{\mathrm{SNR}}_D = 8\sqrt{N}\, \sigma_n^2 f^2(0) \frac{\sigma_s^2}{\sigma_n^2} \frac{\left[E_H\{\delta^2 g(\delta)\} - \frac{M}{4} E_H\{g(\delta)\} \right]}{[E_H\{g^2(\delta)\} - E_H^2\{g(\delta)\}]^{1/2}}. \quad (8)$$

We will now consider particular implementations of the PCA, all of which reduce to the PCC for $M = 2$.

III. Polarity Coincidence Array Detectors

Consider the detector $D_{pca.1}$ which calculates the test statistic

$$S_{pca.1} = \sum_{i=1}^{N} \delta_i^2. \quad (9)$$

Since $g(\delta_i) = \delta_i^2$, $E_H\{\delta_i^2\} = M/4$, and $E_H\{\delta_i^4\} = 3M^2/16 - 2M/16$ (see Appendix III), it follows that

$$E_H\{S_{pca.1}\} = N\frac{M}{4}, \qquad \operatorname{Var}_H\{S_{pca.1}\} = \frac{N}{8} M(M - 1), \quad (10)$$

and from (8),

$$\overline{\mathrm{SNR}}_{pca.1} = \sqrt{N}\sqrt{8}\, \sigma_n^2 f^2(0) \frac{\sigma_s^2}{\sigma_n^2} \sqrt{M(M - 1)}. \quad (11)$$

But this is exactly the output signal-to-noise ratio determined for the locally optimum detector for independent clipped Gaussian samples [10]. Therefore, $D_{pca.1}$ is the optimum PCA detector and the others will be compared to it. Note that, when $M = 2$, (11) holds for all the PCA detectors to be considered.

Let us next analyze the suboptimum PCA, $D_{pca.2}$, which calculates the following test statistic:

$$S_{pca.2} = \sum_{i=1}^{N} \delta_i. \tag{12}$$

It is observed in Appendix II that sums of the type

$$\sum_{r=M/2+\epsilon}^{M} \binom{M}{r} \frac{2}{2^M} f\left(r - \frac{M}{2}\right)$$

approach

$$2 \int_{(2\epsilon-1/\sqrt{M})}^{\infty} \frac{1}{\sqrt{2\pi}} e^{-1/2y^2} f\left(\frac{\sqrt{M}}{2} y\right) dy$$

for large M ($M > 10$). The later integrals are easily evaluated. Therefore, (8) can be evaluated for large M using the fact that $g(\delta) = \delta$, resulting in

$$\overline{\text{SNR}}_{pca.2} \overset{M\to\infty}{\to} \sqrt{N} \sqrt{8}\, \sigma_n^2 f^2(0) \frac{\sigma_s^2}{\sigma_n^2} \frac{M}{\sqrt{\pi - 2}}, \tag{13}$$

where the limit is rapidly approached, particularly for M odd.

Thus, $D_{pca.2} = D_{pca.1}$ for $M = 2$, and for large arrays $D_{pca.2}$ introduces a loss of 0.29 dB in the input signal power over and above the loss inherent in hard limiting. It follows that the squaring of δ_i does not improve the detectability significantly. Note, however, that δ_i is defined as the absolute value of some quantity and squaring may not be more difficult to implement than the absolute value operation.

Finally we wish to consider the suboptimum PCA that introduces an intermediate threshold. Thus,

$$S_{pca.3} = \sum_{i=1}^{N} u(\delta_i - \bar{\delta}), \tag{14}$$

where

$$u(\xi) = \begin{cases} 1, & \xi > 0 \\ 0, & \xi \le 0, \end{cases}$$

and $\bar{\delta}$ is some constant. This intermediate threshold reduces the test statistic to a sum of 0's and 1's and may possibly be more easily implemented. In fact, $D_{pca.3}$ is probably the most natural extension of the two-channel PCC detector.[2] $S_{pca.3}$ is binomially distributed under the hypothesis with probability of success p_0 given by

$$p_0 = \text{Prob}_H(\delta_i > \bar{\delta}) = \frac{2}{2^M} \sum_{r=[M/2+\bar{\delta}]}^{M} \binom{M}{r}, \tag{15}$$

where $[x]$ is the smallest integer greater than x. The false-alarm probability α can be set exactly for any N if the threshold c is the smallest integer such that

$$\sum_{r=c}^{N} \binom{N}{r} p_0^r (1 - p_0)^{N-r} \le \alpha. \tag{16}$$

[2] Faran and Hills [1] consider this implementation where they set $\bar{\delta} = M/2 - 1$. Thus, the output is 1 only if all the channels have the same sign.

We wish to choose that $\bar{\delta}$ which maximizes the output signal-to-noise ratio. Since $\text{Var}_H[S_{pca.3}] = Np_0(1 - p_0)$, the output signal-to-noise ratio is seen from (5), (6), and (14) to be

$$\overline{\text{SNR}}_{pca.3} = \sqrt{N} \frac{8\sigma_n^2 f^2(0)}{\sqrt{p_0(1 - p_0)}} \frac{\sigma_s^2}{\sigma_n^2} \frac{2}{2^M} \cdot \sum_{r=[M/2+\bar{\delta}]}^{M} \binom{M}{r} \left[\left(r - \frac{M}{2}\right)^2 - \frac{M}{4}\right]. \tag{17}$$

As $\sqrt{p_0(1 - p_0)}$ is a relatively slowly varying function, we pick $r' = [M/2 + \bar{\delta}]$ to be the smallest integer that satisfies

$$\left(r' - \frac{M}{2}\right)^2 - \frac{M}{4} > 0. \tag{18}$$

It follows that

$$\bar{\delta}_{opt} = \frac{\sqrt{M}}{2}. \tag{19}$$

From Appendix II it is seen that

$$p_0 = \frac{2}{2^M} \sum_{r=[M/2+\sqrt{M/2}]}^{M} \binom{M}{r} \overset{M\to\infty}{\to} 2[1 - \Phi(1)] \cong 0.3174 \tag{20}$$

and

$$\frac{2}{2^M} \sum_{r=M/2+\sqrt{M/2}}^{M} \binom{M}{r} \left[\left(r - \frac{M}{2}\right)^2 - \frac{M}{4}\right] \overset{M\to\infty}{\to} \frac{M}{\sqrt{8\pi}} e^{-1/2} \approx 0.1210\, M. \tag{21}$$

Hence it follows that

$$\overline{\text{SNR}}_{pca.3} \overset{M\to\infty}{\to} \sqrt{N} \sqrt{8}\, \sigma_n^2 f^2(0) \frac{\sigma_s^2}{\sigma_n^2} \frac{M}{1.360}. \tag{22}$$

Thus, $D_{pca.3} = D_{pca.1}$ for $M = 2$, and for large arrays $D_{pca.3}$ introduces a loss of 1.34 dB in the input signal power over and above the loss inherent in hard limiting. Table 1 gives p_0, $\text{SNR}_{pca.3}/\text{SNR}_{pca.1}$, and the intermediate threshold $[M/2 + \sqrt{M}/2]$ for different values of M. Observe the fluctuations caused by the discreteness of the intermediate threshold.

IV. Optimum Detection of Unclipped Gaussian Samples

The array detector that is locally optimum, when the samples have Gaussian amplitude densities with unknown but stationary variance, calculates the following test statistic [11]:[3]

$$S_{opt} = \sum_{i=1}^{N} \left[\sum_{j \neq l}^{M} \sum^{M} x_j(t + i\tau) x_l(t + i\tau) \right]. \tag{23}$$

[3] To fix the false-alarm probability, the test statistic S_{opt} must be compared with a suitable estimate of the noise power [11]. In practice, however, S_{opt} as given in (23) is displayed as a function of possible target direction and the output is compared with the off-target output by inspection.

TABLE I
PERFORMANCE DATA FOR $D_{pca.3}$

M	2	3	4	5	6	7	8	14	20	26	32	38	44	50	$\rightarrow\infty$
$[M/2 + \sqrt{M}/2]$	2	3	4	4	5	5	6	9	13	16	19	23	26	29	$\rightarrow\infty$
p_0	0.500	0.250	0.125	0.375	0.219	0.453	0.289	0.424	0.263	0.327	0.377	0.256	0.291	0.322	$\rightarrow$ 0.317
$\overline{\text{SNR}}_{pca.3}/\overline{\text{SNR}}_{pca.1}$	1	1	0.926	0.817	0.878	0.719	0.820	0.700	0.794	0.749	0.712	0.784	0.762	0.744	$\rightarrow$ 0.735

The output signal-to-noise ratio of this locally optimum detector D_{opt} for arbitrary stationary inputs with zero mean is given by

$$\overline{\text{SNR}}_{opt} = \sqrt{\frac{N}{2}}\frac{\sigma_s^2}{\sigma_n^2}\sqrt{M(M-1)}. \tag{24}$$

If we define J_{D_1/D_2} as $\overline{\text{SNR}}_{D_1}/\overline{\text{SNR}}_{D_2}$, then

$$J_{pca.1/opt} = 4\sigma_n^2 f^2(0). \tag{25}$$

It also follows that

$$J_{pca.3/opt} = 4\sigma_n^2 f^2(0) J_{pca.3/pca.1}. \tag{26}$$

It is therefore observed that for Gaussian inputs $[\sigma_n f(0) = 1/\sqrt{2\pi}]$, $J_{pca.1/opt} = 2/\pi$, and thus, the process of hard limiting introduces an inherent loss in input signal power of 1.96 dB. Note that these losses are all based on the assumption of independent samples.

Let us now compare the PCA detectors with D_{opt} when the stationary Gaussian assumption is violated. For input probability densities that are peaked at the origin (impulse noise), the PCA detectors may become more efficient relative to D_{opt}. For example, consider input noise processes that have the double exponential density

$$\left[f(x) = \frac{1}{\sqrt{2\sigma_n^2}} e^{-\sqrt{2}|x/\sigma_n|}\right].$$

For these inputs $J_{pca.1/opt} = 2$, and therefore, D_{opt} introduces a loss of 3.01 dB in the input signal power relative to $D_{pca.1}$. Thus, the PCA detectors can improve greatly relative to D_{opt} when the Gaussian assumption is violated.

The analysis of the PCA detectors depends only on the sample median, and hence, the PCA is invariant with respect to any nonstationarity in the noise level [3]. This is not true of the optimum detector based on stationary Gaussian inputs. If the noise level varies sinusoidally about some mean, then $\text{Var}_H\{S_{opt}\}$ increases by an amount which approaches $(1 + k^2/2)$ asymptotically, where k is the modulation index. Thus, $\overline{\text{SNR}}_{opt}$ decreases by the square root of this amount, and the false-alarm rate increases from its assumed value if the threshold was set on the assumption of stationarity. Hence, $J_{pca/opt}$ is increased by as much as 1.225 for 100 percent modulation (or a 0.88 dB gain in signal power for the PCA detectors). The increase in false-alarm rate is also significant for these large modulations.

V. Operation of PCA Detectors with Dependent Samples

The nonparametric properties of the PCA array detectors follow from the assumption that independent samples are available. For realizable narrow-band signals there is no sampling rate for which the independent sample assumption is valid, although a reasonable approximation may be obtained with sufficiently slow sampling. This procedure, however, can be very inefficient given a finite decision time, and it may therefore be desirable to sample as fast as possible. The false-alarm rate of the PCA detectors can be fixed with dependent samples (and therefore, continuously) only if the shape of the noise spectrum and the distribution of zero crossings are known [12], [13]. However, there are still definite advantages, other than simplicity of implementation, in using these detectors instead of some optimum device.

Let us assume that the shape of the noise spectrum is essentially determined by prefilters that are based on knowledge of the signal spectrum. If any instantaneous nonstationarity in the noise level is slow enough relative to the inverse of the bandwidth of the prefilter, then the other spectral properties of the noise are essentially stationary. Hence, the PCA detectors, while no longer nonparametric in the strict sense, are still unaffected by any sufficiently slow nonstationary noise level. For a sinusoidally varying noise level that is sufficiently slow but nevertheless fast relative to the inverse of the decision time (not at all unreasonable), $J_{pca/opt}$ is still increased by the factor $[1 + k^2/2]^{1/2}$.

It will now be shown that, under the assumption that the zero crossings are the same as those from a Gaussian process, $J_{pca/opt}$ increases as the samples become more dependent. This improvement is a function of the sampling rate and the spectral shape of the noise. It will be seen that for sufficiently fast sampling, or continuous operation, much of the cost of clipping is recovered.

When the input samples are dependent, the only change in the output signal-to-noise ratios of the detectors considered is the result of an increase in the variance of the test statistics. For any test statistic of the form $S = \sum_{i=1}^{N} h(\bar{x}_i)$, where $\bar{x}_i = x_j(t + i\tau)$, $j = 1, \cdots, M$, and h is any memoryless operation, the variance for independent samples is given by $\text{Var}\{S\} = N\,\text{Var}\{h[\bar{x}(t)]\}$. When the samples are dependent, it is easily shown (see Appendix III) that

$$\mathrm{Var}_H \{S\} = N \,\mathrm{Var}_H \{h[\bar{x}(t)]\} \cdot \left[1 + 2 \sum_{k=1}^{N} \left(1 - \frac{k}{N}\right) R(k\tau)\right], \tag{27}$$

where

$$R(k\tau) = \frac{E_H\{h[\bar{x}(t)]h[\bar{x}(t + k\tau)]\} - E_H^2\{h[\bar{x}(t)]\}}{\mathrm{Var}_H \{h[\bar{x}(t)]\}}. \tag{28}$$

It follows that the output signal-to-noise expressions of (6) or (8) hold for dependent samples if N is replaced by

$$N_{eq} \triangleq \frac{T}{\tau + 2\tau \sum_{k=1}^{N} \left(1 - \frac{k}{N}\right) R(k\tau)} \underset{\tau \to 0}{\longrightarrow} \frac{T}{2 \int_0^T \left(1 - \frac{\xi}{T}\right) R(\xi)\, d\xi}, \tag{29}$$

where T is the decision time. The function $R(k\tau)$ depends on the detector considered and on the spectral shape, or normalized correlation function $\rho_n(\tau)$, of the noise inputs.

The following results are given in Appendix III. For arbitrary stationary inputs, R_{opt} is given by

$$R_{\mathrm{opt}}(k\tau) = \rho_n^2(k\tau). \tag{30}$$

When the zero crossings are those of a Gaussian process, $R_{\mathrm{pca.1}}$ is given by

$$R_{\mathrm{pca.1}}(k\tau) = \left(\frac{2}{\pi} \sin^{-1} [\rho_n(k\tau)]\right)^2. \tag{31}$$

In the limit as M approaches infinity, $R_{\mathrm{pca.3}}$ is given by

$$R_{\mathrm{pca.3}}(k\tau) \overset{M \to \infty}{\longrightarrow} \frac{1}{p_0(1 - p_0)} \int_{-1}^{1} G(\bar{\rho}, y) \frac{1}{\sqrt{2\pi}} e^{-1/2y^2}\, dy - \frac{(1 - p_0)}{p_0}, \tag{32}$$

where

$$G(\bar{\rho}, y) = \int_{(-1-\bar{\rho}y)/\sqrt{1-\bar{\rho}^2}}^{(1-\bar{\rho}y)/\sqrt{1-\bar{\rho}^2}} \frac{1}{\sqrt{2\pi}} e^{-1/2\xi^2}\, d\xi,$$

and

$$\bar{\rho} = \frac{2}{\pi} \sin^{-1} [\rho_n(k\tau)].$$

These functions are plotted vs. $\rho_n^2(\tau)$ in Fig. 2.

Since $R_{\mathrm{pca}} < R_{\mathrm{opt}}$ for $\rho(k\tau) \neq 0$, it follows that, for a given decision time, the equivalent number of uncorrelated samples $[N_{\mathrm{eq}}]$ for the PCA detectors is larger than that of D_{opt}. Therefore, when the input samples are dependent, $J_{\mathrm{pca/opt}}$ is larger than when they are independent. Observe from Fig. 2 that $R_{\mathrm{pca.1}} > (2/\pi)^2 \rho^2$ and

$$R_{\mathrm{pca.3}}(M \to \infty) > \left(\frac{2}{\pi} \frac{1}{1.360}\right)^2 \rho^2.$$

Thus it can be determined from (29) that even in the limit as $\tau \to 0$, $J_{\mathrm{pca/opt}} < 1$. This is as it should be, since

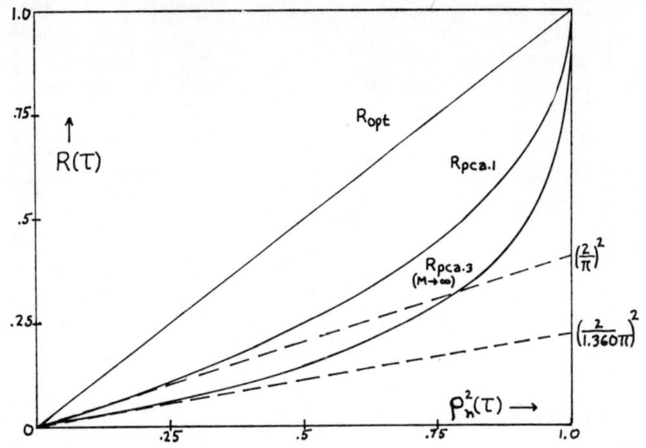

Fig. 2. Normalized correlation function of test statistic argument.

we are assuming essentially Gaussian inputs. Nevertheless, much of the loss due to clipping should be recovered in the limit as $\tau \to 0$.

If we define I as the ratio of $J_{\mathrm{pca/opt}}$ with dependent samples to that with independent samples, then from (8) and (29) one obtains

$$I^2 = \frac{1 + 2 \sum_{k=1}^{N} \left(1 - \frac{k}{N}\right) \rho_n^2(k\tau)}{1 + 2 \sum_{k=1}^{N} \left(1 - \frac{k}{N}\right) R_{\mathrm{pca}}(k\tau)}. \tag{33}$$

Under the assumption that the decision time is much larger than the width of $\rho_n(t)$, this expression can be accurately approximated by

$$I^2 \cong \frac{1 + 2 \sum_{k=1}^{\infty} \rho_n^2(k\tau)}{1 + 2 \sum_{k=1}^{\infty} R_{\mathrm{pca}}(k\tau)}. \tag{34}$$

This improvement in the PCA detectors relative to D_{opt} depends of course on the spectral shape.

In underwater sound detection one is normally looking for acoustic signals, and hence, the prefilters are often low-pass filters. Let us consider the single pole low-pass filter that falls off at a rate of 6 dB/octave ($\rho_n(k\tau) = e^{-\omega_c k\tau}$), and the ideal low-pass filter

$$\left(\rho_n(k\tau) = \frac{\sin \omega_c k\tau}{\omega_c k\tau}\right),$$

where $\omega_c/2\pi$ is the cutoff frequency or the bandwidth.[4] Given a particular sampling rate f_s, I can be evaluated for either case. The improvement of $D_{\mathrm{pca.1}}$ relative to D_{opt} vs. the sampling rate is given in Fig. 3, and the improvement of $D_{\mathrm{pca.3}}(M \to \infty)$ relative to $D_{\mathrm{pca.1}}$ is given in Fig. 4.

Observe first that for a sufficiently fast sampling rate, or continuous operation, much of the loss due to clipping (as well as the loss due to an intermediate threshold) is

[4] The cutoff frequency is partially determined by the white noise that is generated locally in the hydrophones and associated circuitry. The presence of this noise makes the detection problem nonsingular even in the limit of continuous operation [10].

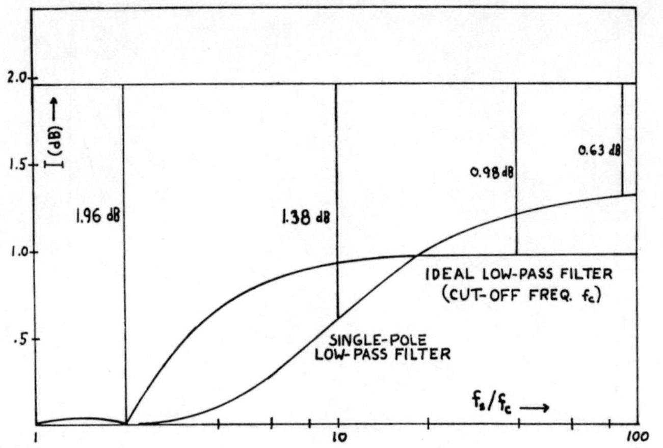

Fig. 3. Comparison of $D_{pca.1}$ to D_{opt} as a function of sampling rate.

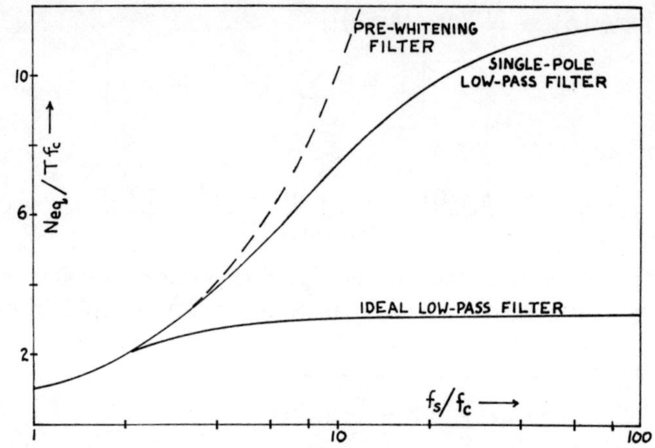

Fig. 5. Equivalent number of uncorrelated samples for $D_{pca.1}$.

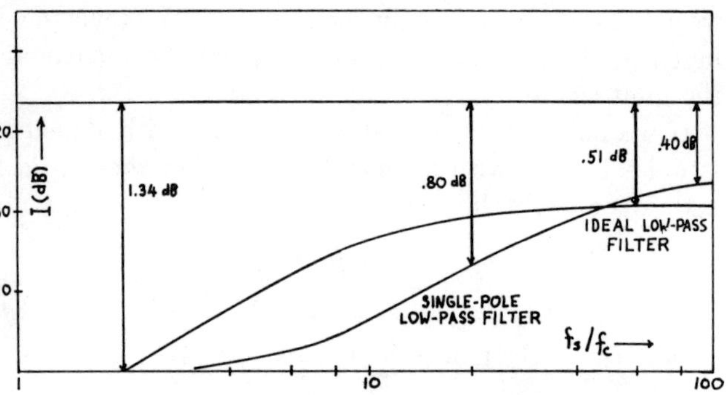

Fig. 4. Comparison of $D_{pca.3}$ to $D_{pca.1}$ for large arrays as a function of sampling rate.

recovered. Thus, for essentially Gaussian inputs and continuous operation, $D_{pca.1}$ introduces a loss in the signal power as low as 0.63 dB for the single pole prefilter. The loss depends somewhat on the filter shape. In addition, some of the remaining loss can be recovered if the noise level is nonstationary.

It should be pointed out, however, that when using D_{opt} there is little point in sampling faster than $2f_c$ for the ideal filter (Nyquist rate) or $6f_c$ for the single-pole filter. For these sampling rates, the loss due to clipping is almost that amount calculated for independent samples. Thus, one would have to sample at a rate faster than typical to recover some of the loss due to clipping. It is sometimes common in practice to filter the inputs sharply at some frequency f_2 that is significantly larger than f_c without prior prewhitening and then to sample at the Nyquist rate of the processing filter. This type of operation would give the rapid sampling needed to recover the loss due to clipping.

One way of looking at this phenomenon is that the process of clipping spreads thé spectrum of the input processes. In order to receive all the available information, one has to sample faster than necessary for the unclipped inputs. Thus, the 2 dB loss due to clipping is due in part to this spreading of the spectrum and only in part to an actual loss in information. Figure 5 is a plot of the equivalent number of uncorrelated samples for a fixed decision time vs. the sampling rate for $D_{pca.1}$. The results indicate that, when sampled rapidly, it is *necessary* to know the spectral shape accurately in order to set the threshold properly. If the spectral shape is not known and it is desirable to set the threshold with some degree of accuracy, then one cannot sample significantly faster than $2f_c$.

VI. Conclusions

Polarity coincidence array detectors have been analyzed and compared with the optimum parametric device based on stationary Gaussian inputs. While devices based on clipped data are inherently less efficient than the optimum detector, their nonparametric properties can make them more efficient when the stationary Gaussian assumption is violated. In addition, when the PCA detectors are operated as a parametric device with dependent samples, the loss due to clipping can be greatly reduced with sufficiently fast sampling. Even under these conditions, the PCA detectors are unaffected by certain types of nonstationarities in the noise level. The advantages for either mode of operation coupled with the obvious ease of implementation could make the PCA array detectors useful devices for underwater sound detection.

Appendix I

Output Signal-to-Noise Ratio

If $p(r)$ is the probability that r channels have the samd sign, then

$$E\{S_{pca}\} = N \sum_{r=[M/2]}^{M} g\left(r - \frac{M}{2}\right)p(r),$$

where $[x]$ is the smallest integer greater than or equal to x. Under the hypothesis, the m inputs are independent and have zero median, and hence,

$$p_H(r) = 2\binom{M}{r}\left(\frac{1}{2}\right)^M.$$

Thus,

$$E_H\{S_{pca}\} = N \sum_{r=[M/2]}^{M} \frac{2}{2^M}\binom{M}{r} g\left(r - \frac{M}{2}\right).$$

Under the alternative, there is an additive signal common to the M channels, and hence,

$$p_K(r) = \binom{M}{r}\left[\int_{-\infty}^{\infty} \{[1 - F(-u)]^r[F(-u)]^{M-r} + [F(-u)]^r[1 - F(-u)]^{M-r}\}f_s(u)\, du\right],$$

where $F(\ )$ is the cumulative distribution function of the noise and $f_s(u)$ is the amplitude density of the signal. It will now be assumed that $F(\ )$ can be expanded in a Taylor series about the origin over a finite range, and thus,

$$F(-u) = F(0) - uf(0) + \frac{u^2}{2!}f'(0) - \frac{u^3}{3!}f''(0) + 0[u^3].$$

Substituting this expression into the previous equation and expanding leads to the result, where $F(0) = \frac{1}{2}$,

$$p_K(r) = \binom{M}{r}\frac{2}{2^M}\left\{1 + 2[(2r - M)^2 - M]f^2(0) \cdot \int_{-\infty}^{\infty} u^2 f_s(u)\, du + 0\left[\int u^3 f_s(u)\, du\right]\right\},$$

where convergence is assumed. If the signal is amplitude limited, then for some sufficiently small imput signal-to-noise ratio (σ_s^2/σ_n^2) convergence is guaranteed and

$$p_K(r) = p_H(r) + 2\sigma_n^2 f^2(0)\frac{\sigma_s^2}{\sigma_n^2}\frac{2}{2^M}\binom{M}{r}[(2r - M)^2 - M] + 0\left[\frac{\sigma_s^3}{\sigma_n^3}\right].$$

This result follows from the fact that

$$f^n(0)\int_{-A\sigma_s}^{A\sigma_s} u^n f_s(u)\, du < f^n(0)(A\sigma_s)^n = (f(0)\sigma_n)^n\left(\frac{\sigma_s}{\sigma_n}\right)^n A^n,$$

and hence, $\sigma_s/\sigma_n < 1/A$ is a sufficient condition. Therefore,

$$E_K\{S_{pca}\} - E_H\{S_{pca}\} \underset{\sigma_s^2/\sigma_n^2 \to 0}{\longrightarrow} \delta N\sigma_n^2 f^2(0)\frac{\sigma_s^2}{\sigma_n^2}\frac{2}{2^M} \cdot \sum_{r=[M/2]}^{M}\binom{M}{r}\left[\left(r - \frac{M}{2}\right)^2 - \frac{M}{4}\right]g\left(r - \frac{M}{2}\right).$$

Appendix II

Limiting Results for Certain Sums

We are concerned with evaluating, for large values of M, sums of the type

$$\sum_{r=L}^{M} p_H(r) f\left(r - \frac{M}{2}\right),$$

where

$$p_H(r) = \frac{1}{2^M}\binom{M}{r}.$$

The term $p_H(r)$ corresponds to the binomial distribution with probability of "success" equal to 0.5 and is therefore closely approximated by the normal distribution with mean $M/2$ and variance $M/4$. Thus, the sums can be approximated by [14]:

$$\int_{L-1/2}^{\infty} \frac{1}{\sqrt{2\pi\frac{M}{4}}}\exp\left\{-\frac{1}{2}\frac{\left(r - \frac{M}{2}\right)^2}{\frac{M}{4}}\right\}f\left(r - \frac{M}{2}\right) dr.$$

Changing variables by setting $(r - M/2)/\sqrt{M}/4 = y$, one obtains

$$\int_{(2L-M-1/M)}^{\infty} \frac{1}{\sqrt{2\pi}}\exp\{-\tfrac{1}{2}y^2\}f\left(\frac{M}{2}y\right) dy.$$

In (15) the term p_0 was defined as

$$p_0 = 2\sum_{r=[M/2+\sqrt{M}/2]}^{M} \frac{1}{2^M}\binom{M}{r}.$$

It follows that for large M,

$$p_0 \cong 2\int_{1'}^{\infty} \frac{1}{\sqrt{2\pi}} e^{-1/2y^2}\, dy = 2[1 - \Phi(1')],$$

where $1'$ is equal to $(2[M/2 + \sqrt{M}/2] - M - 1)/\sqrt{M}$ which equals $(1 + \epsilon/\sqrt{M})$, where $-1 < \epsilon < 1$. In the limit as $M \to \infty$,

$$p_0 \overset{M\to\infty}{\longrightarrow} 2[1 - \Phi(1)] = 0.3174.$$

In (17) we were concerned with the term

$$p' \triangleq 2\sum_{r=[M/2+\sqrt{M}/2]}^{M} \frac{1}{2^M}\binom{M}{r}\left[\left(r - \frac{M}{2}\right)^2 - \frac{M}{4}\right].$$

It follows that for large M

$$p' \approx 2\frac{M}{4}\int_{1'}^{\infty}\frac{1}{\sqrt{2\pi}}e^{-1/2y}(y^2 - 1)\, dy.$$

Integrating by parts, it is easily shown that

$$p' \approx \frac{M}{\sqrt{8\pi}}(1')e^{-1/2(1')^2}.$$

Hence,

$$p' \overset{M\to\infty}{\longrightarrow} \frac{M}{\sqrt{8\pi}}e^{-1/2} = 0.1210M.$$

Appendix III

Test Statistic Variance for Dependent Samples

Consider an arbitrary test statistic of the form

$$S = \sum_{i=1}^{N} h[\bar{x}(t + i\tau)],$$

where $h(\ )$ may be any memoryless operation on the vector $\bar{x}(t + i\tau)$ whose elements are samples (with uniform sampling interval τ) of specific independent representations of the same stochastic process $\{x(t)\}$. The variance of S is given by

$$\text{Var}(S) = E\{S^2\} - E^2\{S\} = E\left\{\sum_i^N \sum_j^N [h[\bar{x}(t+i\tau)]h[\bar{x}(t+j\tau)] - Eh[\bar{x}(t)]]\right\}$$

$$= \sum_{=1}^N [E\{h^2[\bar{x}(t+i\tau)]\} - E^2\{h[\bar{x}(t)]\}] + \sum_{i\neq j}^N \sum^N [E\{h[\bar{x}(t+i\tau)]h[\bar{x}(t+j\tau)]\} - E^2\{h[\bar{x}(t)]\}]$$

$$\text{Var}\{h[\bar{x}(t)]\} + \sum_{i\neq j}^N \sum^N [E\{h[\bar{x}(t+i\tau)]h[\bar{x}(t+j\tau)]\} - E^2\{h[\bar{x}(t)]\}],$$

where the second term is zero if the samples are independent, assuming $\{\bar{x}(t)\}$ is stationary, and since there are $2(N-k)$ terms where $|j-i| = k$, one obtains

$$\text{Var}(S) = N\,\text{Var}\{h[\bar{x}(t)]\}\left[1 + 2\sum_{k=1}^N \left(1 - \frac{k}{N}\right)R(k\tau)\right],$$

where

$$R(k\tau) \triangleq \frac{E\{h[\bar{x}(t)]h[\bar{x}(t+k\tau)]\} - E^2\{h[\bar{x}(t)]\}}{\text{Var}\{h[\bar{x}(t)]\}}.$$

The function $R(k\tau)$ depends only not on the function $h(\ )$, but also on the normalized correlation function of $\{\bar{x}(t)\}$ denoted by $\rho(\tau)$.

Let us now consider the test statistic S_{opt} defined by (23). For this detector

$$h[\bar{x}(t)] = \sum_{i\neq j}^M \sum^M x_i(t)x_j(t),$$

where $x_j(t+i\tau)$, $j = 1, \cdots, M$, are the M components of the vector $\bar{x}(t+i\tau)$ and where it is assumed that $E\{x_j(t)\} = 0$ for all j. It follows that

$$E\{h[\bar{x}(t)]\} = \sum_{i\neq j}^M \sum^M E\{x_i(t)x_j(t)\} = 0,$$

and

$$\text{Var}\{h[\bar{x}(t)]\} = E\{h^2[\bar{x}(t)]\}$$
$$= \sum_{i\neq j}\sum \sum_{r\neq s}\sum E\{x_i(t)x_j(t)x_r(t)x_s(t)\}$$
$$= 2\sum_{i\neq j}\sum E\{x_i^2(t)x_j^2(t)\} = 2(M^2 - M)\,\text{Var}^2\{x(t)\},$$

and

$$E\{h[\bar{x}(t)]h[\bar{x}(t+k\tau)]\}$$
$$= \sum_{i\neq j}\sum \sum_{r\neq s}\sum E\{x_i(t)x_j(t)x_r(t+k\tau)x_s(t+k\tau)\}$$
$$= 2\sum_{i\neq j}\sum E\{x_i(t)x_i(t+k\tau)x_j(t)x_j(t+k\tau)\}$$
$$= 2(M^2 - M)\,\text{Var}^2[x(t)]\rho^2(k\tau).$$

Hence, by the definition of $R(k\tau)$, it is seen that

$$R_{\text{opt}}(k\tau) = \rho^2(k\tau).$$

Next let us consider the detector $D_{\text{pca.1}}$ in which $h[\bar{x}(t)] = \delta^2[\bar{x}(t)]$, where $\delta[\bar{x}(t)] = |\frac{1}{2}\sum_{i=1}^M \text{sgn}[x_i(t)]|$. For this test statistic

$$E\{h[\bar{x}(t)]\} = \frac{1}{4}\sum_i^M \sum_j^M E\{\text{sgn}[x_i(t)]\,\text{sgn}[x_j(t)]\}$$
$$= \frac{1}{4}\sum_{i=1}^M E\{\text{sgn}^2[x_i(t)]\} = \frac{M}{4},$$

and

$$\text{Var}\{h[\bar{x}(t)]\} = \frac{1}{16}\sum_i\sum_j\sum_r\sum_s E\{\text{sgn}[x_i(t)]\,\text{sgn}[x_j(t)]\,\text{sgn}[x_r(t)]\,\text{sgn}[x_s(t)]\} - \frac{M^2}{16}$$

$$= \frac{1}{16}\sum_{i=1}^M E\{\text{sgn}^4[x_i(t)]\} - \frac{3}{16}\sum_{i\neq j}\sum E\{\text{sgn}^2[x_i(t)]\,\text{sgn}^2[x_j(t)]\} - \frac{M^2}{16} = \frac{M}{16} + \frac{3}{16}(M^2 - M) - \frac{M^2}{16} = \frac{1}{8}M(M-1),$$

and

$$E\{h[\bar{x}(t)]h[\bar{x}(t+k\tau)]\} = \frac{1}{16}\sum_i\sum_j\sum_r\sum_s E\{\text{sgn}[x_i(t)]\,\text{sgn}[x_j(t)]\,\text{sgn}[x_r(t+k\tau)]\,\text{sgn}[x_s(t+k\tau)]\}$$

$$= \frac{1}{16}\sum_i\sum_j E\{\text{sgn}^2[x_i(t)]\,\text{sgn}^2[x_j(t+k\tau)]\} + \frac{2}{16}\sum_{i\neq j}\sum E\{\text{sgn}[x_i(t)]\,\text{sgn}[x_i(t+k\tau)]\,\text{sgn}[x_j(t)]\,\text{sgn}[x_j(t+k\tau)]\}$$

$$= \frac{M^2}{16} + \frac{1}{8}(M^2 - M)E^2\{\text{sgn}[x(t)]\,\text{sgn}[x(t+k\tau)]\} = \frac{M^2}{16} + \frac{1}{8}M(M-1)\left(\frac{2}{\pi}\sin^{-1}[\rho(k\tau)]\right)^2.$$

The result just given for the correlation function of a hard-limited process was shown by Van Vleck [15] and is based on the assumption that the distribution of zero crossings of $[x(t)]$ is the same as that of a Guassian process with zero mean. From the definition of $R(k\tau)$, it is seen that

$$R_{\text{pca.1}}(k\tau) = \left(\frac{2}{\pi}\sin^{-1}[\rho(k\tau)]\right)^2.$$

Finally we wish to consider the detector $D_{\text{pca.3}}$ in which $h[\bar{x}(t)] = u\{\delta[\bar{x}(t)] - \sqrt{M}/2\}$. For this statistic

$$E\{h[\bar{x}(t)]\} = \text{Prob}\left[\delta[\bar{x}(t)] > \frac{\sqrt{M}}{2}\right]$$

$$= \frac{2}{2^M} \sum_{r=[M/2+\sqrt{M}/2]}^{M} \binom{M}{r} = p_0,$$

$$\text{Var}\{h[\bar{x}(t)]\} = p_0(1-p_0),$$

and

$$E\{h[\bar{x}(t)]h[\bar{x}(t+k\tau)]\}$$

$$= \text{Prob}\left[\delta[\bar{x}(t+k\tau)] > \frac{\sqrt{M}}{2}, \delta[\bar{x}(t)] > \frac{\sqrt{M}}{2}\right].$$

This last probability can be evaluated conveniently for large M if we define $\delta[\bar{x}(t+k\tau)] \triangleq |\xi_k\sqrt{M}/2|$. Then ξ_k is a random variable with zero mean and unit variance that is asymptotically normally distributed as M increases beyond bound. Furthermore, $E\{\xi_k\xi_0\} = 2/\pi \sin^{-1}[\rho(k\tau)] \triangleq \bar{\rho}$. It follows that

$$R_{\text{pca.3}}(k\tau) = \frac{1}{p_0(1-p_0)}\left[\int_1^\infty \int_1^\infty W_2(\xi_k, \xi_0)\, d\xi_k\, d\xi_0 + \int_{-\infty}^{-1}\int_{-\infty}^{-1} W_2(\xi_k, \xi_0)\, d\xi_k\, d\xi_0 + \int_1^\infty \int_{-\infty}^{-1} W_2(\xi_k, \xi_0)\, d\xi_k\, d\xi_0 + \int_{-\infty}^{-1}\int_1^\infty W_2(\xi_k, \xi_0)\, d\xi_k\, d\xi_0\right] - \frac{p_0}{1-p_0},$$

or by symmetry

$$R_{\text{pca.3}}(k\tau) = \frac{1}{p_0(1-p_0)} \iint_{-1}^{1} W_2(\xi_k, \xi_0)\, d\xi_k\, d\xi_0 - \frac{(1-p_0)}{p_0},$$

where

$$W_2(\xi_k, \xi_0) \underset{M\to\infty}{\longrightarrow} \frac{1}{2\pi\sqrt{1-\bar{\rho}^2}} \cdot \exp\left[-\frac{1}{2(1-\bar{\rho}^2)}(\xi_k^2 - 2\bar{\rho}\xi_k\xi_0 + \xi_0^2)\right].$$

By completing the square and changing variables, one obtains

$$R_{\text{pca.3}}(k\tau) \underset{M\to\infty}{\longrightarrow} \frac{1}{p_0(1-p_0)} \int_{-1}^{1} \frac{1}{\sqrt{2\pi}} e^{-1/2x^2} \cdot \int_{(-1-\bar{\rho}x/\sqrt{1-\bar{\rho}^2})}^{(1-\bar{\rho}x/\sqrt{1-\bar{\rho}^2})} \frac{1}{\sqrt{2\pi}} e^{-1/2y^2}\, dy\, dx - \frac{(1-p_0)}{p_0},$$

where

$$(1-p_0) \underset{M\to\infty}{\longrightarrow} \int_{-1}^{1} \frac{1}{\sqrt{2\pi}} e^{-1/2x^2}\, dx.$$

These normalized covariance functions are plotted in Fig. 2.

References

[1] J. J. Faran, and R. Hills, "The application of correlation techniques to acoustic receiving systems," Acoustic Research Lab., Tech. Memo. 28, Harvard University, Cambridge, Mass., 1952.

[2] V. C. Anderson, "Arrays for the investigation of ambient noise in the ocean," *J. Acoust. Soc. Am.*, vol. 30, p. 470, 1958.

[3] J. B. Thomas and T. R. Williams, "On the detection of signals in non-stationary noise by product arrays," *J. Acoust. Soc. Am.*, vol. 31, p. 453, 1959.

[4] P. Rudnick, "Small signal detection in the dimus array," *J. Acoust. Soc. Am.*, vol. 32, p. 867, 1960.

[5] F. Bryn, "Optimal signal processing by three-dimensional arrays operating on Gaussian signals and noise," *J. Acoust. Soc. Am.*, vol. 34, p. 289, March 1962.

[6] S. S. Wolff, J. B. Thomas, and T. Williams, "The polarity coincidence correlator: a nonparametric detection device," *IRE Trans. on Information Theory*, vol. IT-8, pp. 1–19, January 1962.

[7] M. Kanefsky, "On sign tests and adaptive techniques for nonparametric detection," Ph.D. dissertation, Princeton University, Princeton, N. J., 1964.

[8] E. L. Lehmann, *Testing Statistical Hypotheses.* New York: Wiley, 1959, p. 60.

[9] ——, *ibid.* [8], p. 329.

[10] P. M. Schultheiss and F. B. Tuteur, "Optimum and suboptimum detection of directional Gaussian signals in an isotropic Gaussian noise field, part II: degradation of detectability due to clipping," *IEEE Trans. on Military Electronics*, vol. MIL-9, pp. 208–211, July/October 1965.

[11] F. B. Tuteur, "Optimum detection of a stochastic signal in a Gaussian noise of unknown strength," *Proc. IEEE (Correspondence)*, vol. 53, pp. 487–488, May 1965.

[12] M. Kanefsky and J. B. Thomas, "Polarity coincidence correlation using dependent samples," *Proc. 1964 Nat'l Electronics Conf.*, vol. 20, p. 714.

[13] H. Ekre, "Polarity coincidence correlation detection of a weak noise source," *IEEE Trans. on Information Theory*, vol. IT-9, pp. 18–23, January 1963.

[14] W. Feller, *An Introduction to Probability Theory and Its Applications.* New York: Wiley, 1957, p. 172.

[15] J. H. Van Vleck, "The spectrum of clipped Noise," Radio Research Lab., Harvard University, Cambridge, Mass., Rept. 51; 1943.

6

Reprinted with permission from *IEEE Trans. Commun. Tech.* **COM-16**:676–682 (1968)

A Contribution to the Multiple-Polarity Coincidence Detection Schemes

ERIC D. STOLL, MEMBER, IEEE, AND LUDWIK KURZ, MEMBER, IEEE

I. Introduction

Two-input detection schemes in which the inputs are correlated in order to determine the presence of a signal common to both have been advanced by Jones [1], Dicke [2], and Thomas and Williams [3]. Others including Faran and Hills [4], [5], Melton and Karr [6], Knudsen *et al.* [7], Anderson [8], and Sponsler [9] have considered polarity correlation where the two inputs are hard-limited and then correlated. Further, restriction of the noise process to white Gaussian has been common.

Recently, sampled systems with hard limiters have been analyzed and Ekre [10], Kanefsky and Thomas [11], and others have considered the problem allowing dependent but Gaussian noise. Wolff *et al.* [12] investigated the nonparametric properties of a two-input sampled detector called the polarity coincidence correlator (PCC). In what follows, multiple-input polarity coincidence schemes will be presented and analyzed from the standpoint of their nonparametric qualities.

II. Background Material

Following Wolff *et al.* [12], a brief description of the two-input PCC will be given to show the concepts and notation that will be used later.

The two-input PCC is shown in Fig. 1(a). As can be seen from the figure, the detector calculates the statistic

$$T_{2,N} = \sum_{i=1}^{N} Z_i \tag{1}$$

where

$$Z_i = \text{sgn}(s_i + n_{1i})\text{sgn}(s_i + n_{2i})$$

with

$s_i = s(t_i)$ signal sample at time t_i for $i = 1,2,\cdots,N$

$n_{ji} = n_j(t_i)$ noise sample in jth channel at time t_i for $j = 1,2$, and $i = 1,2,\cdots,N$

[1] The work in this paper was supported in part under Grant NSF-GP-2942 and in part by the Doctoral Support Plan of Bell Telephone Laboratories, Inc.

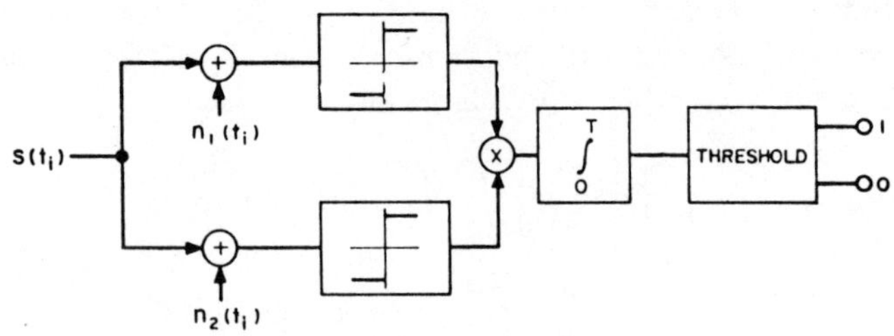

(a)

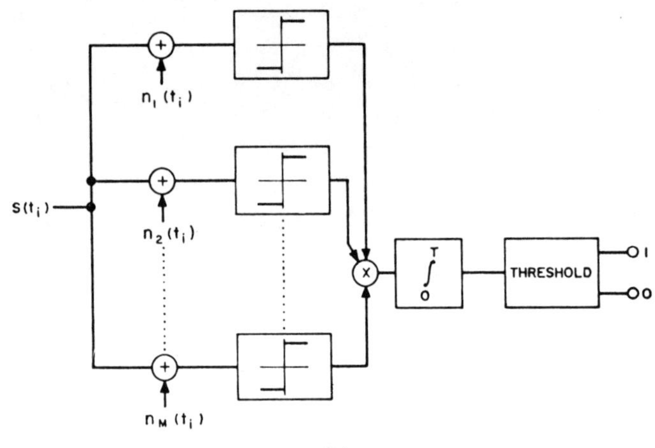

(b)

Fig. 1. (a) Two-input polarity coincidence correlator. (b) "Natural" generalization of (a).

$$\text{sgn}(x) = \begin{cases} 1 & \text{if } x > 0 \\ 0 & \text{if } x = 0 \\ -1 & \text{if } x < 0. \end{cases}$$

It will be assumed that all noise samples are statistically independent and come from populations with the same cumulative distribution function (cdf) $F_n(x)$. Further, it will be required that $F_n(x)$ be continuous for $-\infty < x < \infty$ and in most cases symmetry about $x = 0$ will be demanded.

The signals considered will be random in nature such that all signal samples are statistically independent from each other and the noise samples. The signal cdf $F_s(x)$ will not be required to be continuous.

Due to the continuity restriction on the noise cdf, the event $\{s_i + n_{ji} = 0\}$ has measure zero so that Z_i takes on one of two values, namely +1 or −1. By the central limit theorem, the modified statistic

$$\frac{T_{2,N} - N(2p_+ - 1)}{[4Np_+(1 - p_+)]^{1/2}} = T_{2,N}' \quad (2)$$

is approximately normally distributed with mean zero and variance one [13]–[15]. Here, p_+ denotes the probability of the event $\{Z_i = +1\}$.

In the language of hypothesis testing, the problem is expressed in the following way. It is desired to test the null hypothesis H_0 (no signal present) against the alternative hypothesis H_1 (signal present). Then, under

$$H_0: \quad Z_i = \text{sgn}(n_{1i})\text{sgn}(n_{2i}), \quad i = 1,2,\cdots,N$$

and under

$$H_1: \quad Z_i = \text{sgn}(s_i + n_{1i})\text{sgn}(s_i + n_{2i}), \quad i = 1,2,\cdots,N.$$

There will be a threshold K to which the random variable $T_{2,N}$ will be compared. The decision function D is

$$D = \begin{cases} 1 & \text{if } T_{2,N} > K \\ \xi & \text{if } T_{2,N} = K \\ 0 & \text{if } T_{2,N} < K \end{cases}$$

where ξ is a random variable that takes on the value 0 or 1 depending on the outcome of a Bernoulli trial with a specified frequency of success.[2] The threshold K depends upon the desired false alarm probability α. A brief discussion of detection philosophy when the distributions of the detector statistic under H_0 and H_1 are known is given in Appendix I.

Under H_0, $T_{2,N}$ is Gaussian (approximately) with mean zero and variance N. Thus

$$\Phi\left\{\frac{K}{N^{1/2}}\right\} = 1 - \alpha \quad (3)$$

or

$$K = N^{1/2}\Phi^{-1}(1 - \alpha) \quad (4)$$

where $\Phi\{\cdot\}$ is the cumulative distribution function for the standard normal distribution.

Under

$$H_1: \quad p_+ = \int_{-\infty}^{\infty} [1 - 2F_n(-x) + 2F_n^2(-x)]dF_s(x) \quad (5)$$

where $F_n(\cdot)$ and $F_s(\cdot)$ are the cdf's of the noise and signal, respectively. For the case where the noise distribution is symmetric about the origin

$$F_n(x) = \frac{1}{2} + R(x) \quad (6)$$

[2] For a complete discussion of this subject see Fraser [16].

where

$$R(x) = -R(-x)$$

$$R(0) = 0, \quad R(\infty) = \frac{1}{2}$$

which leads to

$$p_+ = \frac{1}{2} + 2\int_{-\infty}^{\infty} R^2(x)dF_s(x) > \frac{1}{2}. \quad (7)$$

Under H_1, $T_{2,N}$ is Gaussian (approximately) with mean $N(2p_+ - 1)$ and variance $4Np_+(1 - p_+)$. Thus the power of the test β is given by

$$\beta = 1 - \Phi\left\{\frac{K - N(2p_+ - 1)}{[4Np_+(1 - p_+)]^{1/2}}\right\} \quad (8)$$

which yields

$$\beta = 1 - \Phi\left\{\frac{\frac{1}{2}\Phi^{-1}(1 - \alpha) - N^{1/2}\left(p_+ - \frac{1}{2}\right)}{[p_+(1 - p_+)]^{1/2}}\right\}. \quad (9)$$

Inspection of (9) shows that β will always be greater than α, i.e., the test is unbiased.[3] Further, for fixed p_+ and α, $\beta \to 1$ as $N \to \infty$ so the test is consistent.[4] Wolff *et al.* [12] have shown that for $N^{1/2} \geq \Phi^{-1}(1 - \alpha)$, β increases as p_+ increases. For those signal and noise distributions whose variances exist, p_+ increases as the signal-to-noise ratio (SNR) increases where the SNR is defined as the signal variance divided by the noise variance. Thus the average probability of error decreases as the SNR increases.

III. Multiple-Input Polarity Schemes

A. "Natural" Generalization

An immediate generalization of the two-input PCC to M inputs is shown in Fig. 1(b). Here, the M hard-limited samples are all multiplied together and the results accumulated as in the two-input case. It is clear that the performance of this detector can be found simply by finding the quantity $p_+ = \Pr\{Z_i = +1\}$ and applying the results stated in the previous section.

In Appendix II the following expression for p_+ is derived:

$$p_+ = \frac{1}{2} + (-1)^M 2^{M-1} \int_{-\infty}^{\infty} R^M(x)dF_s(x) \quad (10)$$

where $F_n(x) = 1/2 + R(x)$ as before. Clearly, for an odd number of inputs and a symmetric signal distribution, $p_+ = 1/2$. For an unsymmetric distribution p_+ could even be less than 1/2 due to the negative sign before the integral. This detector will not be considered further for M odd due to its obvious faults.

[3] An unbiased statistical test is one where the power of the test is never less than the false alarm probability [16].

[4] A consistent statistical test is one where the power tends asymptotically to unity with increasing sample size for a fixed false alarm probability [16].

For M even and any signal distribution (except the degenerate one at the origin), $p_+ > 1/2$. However, a closer look reveals that $p_+ \to 1/2$ as M increases for even the most favorable signal-to-noise conditions. For example, assume that the signal is a bipolar one that is large compared to the noise in the following manner:

$$dF_s(x) = \begin{cases} \frac{1}{2}, & x = \pm S \\ 0, & \text{otherwise} \end{cases}$$

$$F_n(S) = 1 - \epsilon$$

$$F_n(-S) = \epsilon.$$

Then

$$\int R^M(x)dF_s(x) = 2\left[\frac{1}{2} - \epsilon\right]^M \tag{11}$$

and

$$p_+ = \frac{1}{2} + [1 - 2\epsilon]^M. \tag{12}$$

Substitution of this quantity into (9) produces

$$\beta \cong 1 - \Phi\{\Phi^{-1}(1 - \alpha) - 2N^{1/2}[1 - 2\epsilon]^M\} \tag{13}$$

for ϵ small. Although for fixed M and increasing N the performance improves, the fact that increasing M degrades performance certainly makes this detector of doubtful usefulness. Thus a "natural" generalization of the two-input PCC has been shown to be of questionable value as a detector.

B. Type I Modified PCC

A more useful detector that uses the same components as the PCC is shown in Fig. 2(a). For reference purposes, this detector will be called the Type I modified PCC. In this detector, the hard-limited samples are multiplied together pairwise and the results jointly accumulated. The quantities Z_{ji} are under

$$H_0: \quad Z_{ji} = \text{sgn}(n_{ji})\text{sgn}(n_{j+1,i})$$

and under

$$H_1: \quad Z_{ji} = \text{sgn}(s_i + n_{ji})\text{sgn}(s_i + n_{j+1,i})$$

for $j = 1,2,\cdots,M$ and $i = 1,2,\cdots,N$.

This detector calculates the statistic

$$T_{M,N}^{\text{I}} = \sum_{j=i}^{M} \sum_{i=1}^{N} Z_{ji}. \tag{14}$$

From the application of the central limit theorem [13] again, the modified statistic

$$T_{M,N}^{\text{I}'} = \frac{T_{M,N}^{\text{I}} - (M - 1)N(2p_+ - 1)}{[4(M - 1)Np_+(1 - p_+)]^{1/2}} \tag{15}$$

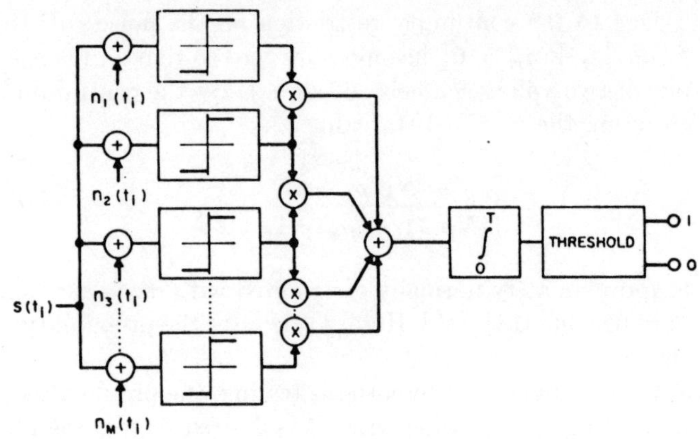

(a)

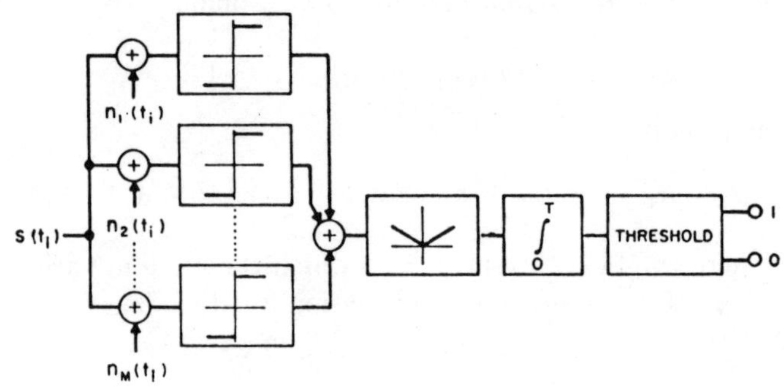

(b)

Fig. 2. (a) Type I modified PCC. (b) Type II modified PCC.

is approximately Gaussian with mean zero and variance one. Again, $p_+ = \Pr\{Z_{ji} = +1\}$ for $j = 1,\cdots,M$ and $i = 1,\cdots,N$. Using arguments similar to those presented earlier, a suitable threshold K is found from

$$K = [(M - 1)N]^{1/2}\Phi^{-1}(1 - \alpha) \tag{16}$$

and the power β from

$$\beta = 1 - \Phi\left\{\frac{\frac{1}{2}\,\Phi^{-1}(1 - \alpha) - [(M - 1)N]^{1/2}\left(p_+ - \frac{1}{2}\right)}{[p_+(1 - p_+)]^{1/2}}\right\} \tag{17}$$

where p_+ is precisely that quantity given for the two-input PCC in (5).

The formula for the power above shows that the multiple inputs contribute in a manner similar to taking a larger number of samples with the two-input detector. The Type I detector thus has all the desirable properties of the two-input detector with the added feature of increased power for a fixed sample size.

C. Type II Modified PCC

The truth tables below show that, for two samples of $+1$'s or -1's, multiplication is equivalent to algebraically adding, full-wave rectifying, and subtracting 1 from the result.

Truth Table A
(multiplication)

	+1	−1
+1	+1	−1
−1	−1	+1

Truth Table B
(adding and rectifying)

	+1	−1
+1	+2	0
−1	0	+2

Herein lies the basic idea for the Type II modified PCC that is shown in Fig. 2(b). The hard-limited samples are added algebraically, rectified, and accumulated.

The analysis of this detector follows. The quantity Z_{ji} is defined as follows: under

$$H_0:\ Z_{ji} = \operatorname{sgn}(n_{ji})$$

and under

$$H_1:\ Z_{ji} = \operatorname{sgn}(s_i + n_{ji})$$

and

$$Z_i = \sum_{j=1}^{M} Z_{ji}$$

for $j = 1,2,\cdots,M$ and $i = 1,2,\cdots,N$.

The quantity Z_i represents the input to the full-wave rectifier. The distribution of Z_i is discrete and ranges from $-M$ to M with finite probability at the points

$$-M, -M+2, -M+4,\cdots,M-4, M-2,M.$$

The output of the rectifier is denoted by Z_{Ri} and this has a discrete distribution with finite probabilities at the points

$$0,2,4,\cdots,M-2,M, \quad M \text{ even}$$

$$1,3,5,\cdots,M-2,M, \quad M \text{ odd}.$$

Since the Z_{Ri} $(i = 1,2,\cdots,N)$ are independent, the central limit theorem [13] can be used to obtain the approximate distribution of the statistic

$$T_{M,N}^{\text{II}} = \sum_{i=1}^{N} Z_{Ri}. \tag{18}$$

However, the exact distribution of Z_{Ri} is required. Unless the Z_{ji} $(j = 1,2,\cdots,M)$ are independent, computational problems arise in the form of unusable expressions for the probabilities.

To circumvent these problems, the signal will be restricted to a small amplitude bipolar one and a "reasonable" assumption will be made about the noise. Specifically the distribution of the signal is given by

$$\Pr\{S_i\} = A\delta(\epsilon) + (1-A)\delta(-\epsilon) \tag{19}$$

where $\delta(x)$ is the discrete distribution of unit amplitude at point x, ϵ is an arbitrarily small positive number, and A satisfies $0 \leq A \leq 1$. As for the noise distribution, it will be assumed that $F_n(x)$ and its derivative $F_n'(x)$ are everywhere continuous and that $F_n'(x) = F_n'(-x)$. Further, the only distributions from the above class that will be considered are those whose second derivatives $F_n''(x)$ exist in the neighborhood of the origin that contains the interval $[-\epsilon,\epsilon]$ and such that $F_n''(0) = 0$.

First, A will be set equal to 0 and then it will be shown by symmetry that A can have a value between 0 and 1 and not affect the results.

$$p = \Pr\{Z_{ji} = +1\}, \quad j = 1,2,\cdots,M, \quad i = 1,2,\cdots,N.$$

The probability distribution of Z_i is given by

$$\Pr\{Z_i\} = \sum_{\alpha=0}^{M} \binom{M}{\alpha} p^\alpha (1-p)^{M-\alpha}\, \delta(M-2\alpha). \tag{20}$$

The probability distribution of Z_{Ri} is given by

$$\Pr\{Z_{Ri}\} = \begin{cases} \displaystyle\sum_{\alpha=0}^{(M/2)-1} \binom{M}{\alpha} [p^\alpha(1-p)^{M-\alpha} + p^{M-\alpha}(1-p)^\alpha]\, \delta(M-2\alpha) + \frac{\binom{M}{M/2}}{2} p^{M/2}(1-p)^{M/2}\delta(0), & M \text{ even} \\ \displaystyle\sum_{\alpha=0}^{(M-1)/2} \binom{M}{\alpha} [p^\alpha(1-p)^{M-\alpha} + p^{M-\alpha}(1-p)^\alpha]\, \delta(M-2\alpha), & M \text{ odd} \end{cases} \tag{21}$$

where

$$p = \Pr\{s_i > -n_{ji}\} = \int F_n(-x)dF_s(x)$$

$$= \frac{1}{2} + \epsilon F_n'(0) + \frac{\epsilon^2 F_n''(0)}{2} + 0(\epsilon^3)$$

$$\cong \frac{1}{2} + \epsilon F_n'(0). \tag{22}$$

Only the calculations for M odd will be continued since similar expressions would be obtained for M even.

The function $G(\alpha,\epsilon,F_n)$ is defined as

$$G(\alpha,\epsilon,F_n) = p^\alpha(1-p)^{M-\alpha} + p^{M-\alpha}(1-p)^\alpha. \tag{23}$$

Substitution of (22) into (23) yields

$$G(\alpha,\epsilon,F_n) = \sum_{l=0}^{\alpha} \sum_{k=0}^{M-\alpha} \left(\frac{1}{2}\right)^{l+k} \binom{\alpha}{l} \binom{M-\alpha}{k} \cdot [\epsilon F_n'(0)]^{M-l-k}[(-1)^{M-\alpha-k} + (-1)^{\alpha-l}]$$

$$= \left(\frac{1}{2}\right)^{M-1} + \left(\frac{1}{2}\right)^{M-2} [\epsilon F_n'(0)]^2 \cdot [M^2 - (4\alpha+1)M + 4\alpha^2] + 0(\epsilon^3),$$

$$\alpha = 0,1,\cdots,\frac{M-1}{2}. \tag{24}$$

To use the central limit theorem the mean and variance of Z_{Ri} are needed. These are, under

$$H_0:\ E\{Z_{Ri}\} = \mu_{R0} = \sum_{\alpha=0}^{(M-1)/2} (M - 2\alpha)\binom{M}{\alpha} G(\alpha,0,F_n) \cong \frac{M}{2^{M-1}} \frac{\binom{M-1}{(M-1)/2}}{2} \tag{25}$$

$$\operatorname{var}\{Z_{Ri}\} = \sigma_{R0}^2 \cong M - \mu_{R0}^2 \tag{26}$$

and under

$$H_1:\ E\{Z_{Ri}\} = \mu_{R1} = \sum_{\alpha=0}^{(M-1)/2} (M - 2\alpha)\binom{M}{\alpha} G(\alpha,\epsilon,F_n) \cong \mu_{R0} + \frac{[\epsilon F_n'(0)]^2}{2^{M-2}}[\text{sum}] \tag{27}$$

where

$$[\text{sum}] = \sum_{\alpha=0}^{(M-1)/2} (M - 2\alpha)\binom{M}{\alpha} \cdot [M^2 - (4\alpha + 1)M + 4\alpha^2] \tag{28}$$

$$\operatorname{var}\{Z_{Ri}\} = \sigma_{R1}^2 \cong \sigma_{R0}^2 + \frac{[\epsilon F_n'(0)]^2}{2^{M-2}} \cdot [[2\ \text{sum}] - 2\mu_{R0}[\text{sum}]] \tag{29}$$

where

$$[2\ \text{sum}] = \sum_{\alpha=0}^{(M-1)/2} (M - 2\alpha)^2\binom{M}{\alpha} - [M^2 - (4\alpha + 1)M + 4\alpha^2]. \tag{30}$$

The above are given in terms of order ϵ^2 or less.

Using the central limit theorem, the threshold K is given by

$$\Phi\left(\frac{K - N\mu_{R0}}{N^{1/2}\sigma_{R0}}\right) = 1 - \alpha \tag{31}$$

or

$$K = N^{1/2}\sigma_{R0}\Phi^{-1}(1 - \alpha) + N\mu_{R0}. \tag{32}$$

The power is given by

$$\beta = 1 - \Phi\left\{\frac{K - N\mu_{R1}}{N^{1/2}\sigma_{R1}}\right\} \tag{33}$$

$$= 1 - \Phi\left\{\frac{\sigma_{R0}}{\sigma_{R1}}\Phi^{-1}(1 - \alpha) - \frac{N^{1/2}}{\sigma_{R1}}(\mu_{R1} - \mu_{R0})\right\}. \tag{34}$$

Examination of the expressions for μ_{R1} and σ_{R1}^2 shows that only constant terms and those in ϵ^2 are present. This means that a constant positive signal would have produced the same expressions. Further, since the mean and variance of Z_{Ri} and the rectified sample at time t_i were calculated, the polarity of the signal at any other time say t_{i+j} will have no effect on the mean and variance of $Z_{R(i+j)}$. Thus the signal can often change polarity arbitrarily and the operation of the detector will be unaffected.

D. Comparison of Type I and Type II Modified PCC's

The comparison of two detectors is frequently carried out on a relative efficiency basis [16]. The relative efficiency of detector A with respect to detector B is defined as

$$e_{A,B} = \frac{N_B(\alpha,\beta,F_n^*)}{N_A(\alpha,\beta,F_n^*)} \tag{35}$$

where N_A and N_B are the number of samples required by each detector to produce the same performance under the same conditions, i.e., the same false alarm probability α, the same power β, and the same noise distribution F_n^*. This will be the basis for comparison between the Type I and Type II modified detectors [16].

For the Type I detector then,

$$N_{\text{I}} = \frac{1}{M - 1} \cdot \left[\frac{\Phi^{-1}(1 - \alpha) - 2\sqrt{p_+(1 - p_+)}\,\Phi^{-1}(1 - \beta)}{2p_+ - 1}\right]^2. \tag{36}$$

With low-level bipolar signals p_+ is found to be approximately

$$p_+ \cong \frac{1}{2} + \epsilon^2\left[2\,[F_n'(0)]^2 - \frac{1}{2}F_n''(0)\right]. \tag{37}$$

From the previous assumptions about the noise

$$p_+ \cong \frac{1}{2} + 2[\epsilon F_n'(0)]^2 \tag{38}$$

and for very small ϵ the following expression is obtained:

$$N_{\text{I}} \cong \frac{1}{M - 1}\left[\frac{\Phi^{-1}(1 - \alpha) - \Phi^{-1}(1 - \beta)}{4[\epsilon F_n'(0)]^2}\right]. \tag{39}$$

For the Type II detector

$$N_{\text{II}} = \sigma_{R0}^2\left[\frac{\Phi^{-1}(1 - \alpha) - \frac{\sigma_{R1}}{\sigma_{R0}}\Phi^{-1}(1 - \beta)}{\mu_{R1} - \mu_{R0}}\right]^2 \tag{40}$$

TABLE I
TABULATION OF RELATIVE EFFICIENCY OF TYPE II MODIFIED PCC WITH RESPECT TO TYPE I MODIFIED PCC

M	[sum]	$[\sigma_{R0}^2]$	$e_{2,1}$
2			1.0
3	12	0.75	1.5
5	120	1.48	2.37
7	840	2.21	3.25
9	5040	2.94	4.11
11	2.77×10^4	3.67	5.0
13	1.44×10^5	4.40	5.86
15	7.21×10^5	5.13	6.68
17	3.50×10^6	5.85	7.62
19	1.66×10^7	6.58	8.43
21	7.76×10^7	7.31	9.31
23	3.57×10^8	8.04	10.2
25	1.62×10^9	8.76	11.05
27	7.30×10^9	9.49	12.02
29	3.26×10^{10}	10.2	13.0
31	1.44×10^{11}	10.9	13.7
33	6.35×10^{11}	11.7	14.6

Assumed noise distribution cdf $= F_n(x)$
1) $F_n(x)$, $F_n'(x)$—continuous
2) $F_n'(x) = F_n'(-x)$
3) F_n'' exists in neighborhood of origin and $F_n''(0) = 0$.

which, for small ϵ becomes

$$N_{II} \cong \sigma_{R0}^2 \left[\frac{\Phi^{-1}(1-\alpha) - \Phi^{-1}(1-\beta)}{\frac{[\epsilon F_n'(0)]^2}{2^{M-2}}\,[\text{sum}]} \right]^2. \tag{41}$$

The relative efficiency of detector II with respect to detector I is

$$e_{II,I} = \frac{N_I}{N_{II}} = \frac{1}{(M-1)\sigma_{R0}^2} \left[\frac{[\text{sum}]}{2^M} \right] \tag{42}$$

which is valid for very low signal levels.

The quantities [sum] and σ_{R0}^2 were programmed on an IBM 7094 computer and values were obtained corresponding to odd M's in the range of 3 to 33.

A tabulation of these quantities including the relative efficiency is given in Table I. Briefly, the results show that for input channels numbering 3 or greater, the Type II modified detector is more efficient that the Type I by a margin that seems to grow with increasing number of inputs.

IV. CONCLUSIONS

Three multiple-input polarity detectors have been shown and analyzed. The last, called the Type II modified PCC was shown to be more efficient that the others when dealing with very low signal levels. Only some general assumptions were made about the noise and thus any member of a large class of noise will produce the same performance under low signal conditions.

APPENDIX I

In general, a detector used in a statistical test will combine the input samples in a certain manner and produce an output quantity that will itself be a random variable. In order to use a threshold for comparison purposes the distribution of this random variable under both H_0 and H_1 is required. Suppose under H_0, the distribution is given by the cdf $F_0(\cdot)$. For purposes of this discussion the cdf of the statistic under H_1, $F_1(\cdot)$, will be assumed to be to the "right" of $F_0(\cdot)$. (See Fig. 3.) Knowing this general behavior, a threshold K will be set and used in the following manner: 1) if the output statistic is less than K, the decision that H_0 exists will be made; 2) if the output statistic is greater than K, the decision that H_1 exists will be made.

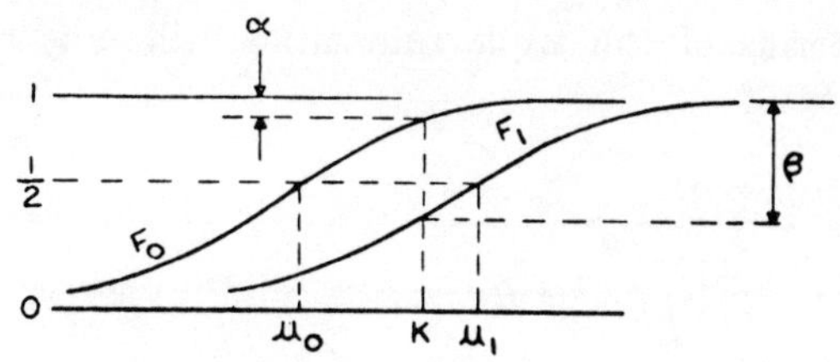

Fig. 3. Detector output statistic distributions.

The threshold K is set so that under H_0 the probability that K will be exceeded is at most the false alarm probability α. Hence

$$F_0(K) = 1 - \alpha. \tag{43}$$

The power of the test β is the probability that K will be exceeded given that H_1 is true. Hence

$$F_1(K) = 1 - \beta. \tag{44}$$

Equations (43) and (44) are demonstrated in Fig. 3 for the case of continuous cdf's which is the case for the work in this paper.

APPENDIX II

Here, the formula for p_+ for the natural generalization of the PCC is derived. There are two cases. Let $r_{ji} = s_i + n_{ji} \triangleq r_j$.

Case 1: $M = 2k - 1$, for some $k > 0$

$$p_+ = \sum_{l=0}^{k-1} \binom{2k-1}{2} \Pr\{r_1 < 0, r_2 < 0, \cdots, r_{2l} < 0, r_{2l+1} > 0, r_{2l+2} > 0, \cdots, r_{2k-1} > 0\}$$

$$= \sum_{l=0}^{k-1} \binom{2k-1}{2l} \int [F_n(x)]^{2l} [1 - F_n(x)]^{2k-2l-1} dF_s(x).$$

Now $F_n(x) = (1/2) + R(x)$ so the integrand becomes

$$[F_n(x)]^{2l}[1 - F_n(x)]^{2k-2l-1} = \sum_{j=0}^{2l} \sum_{i=0}^{2k-2l-1} \binom{2l}{j} \binom{2k-2l-1}{i} \left(\frac{1}{2}\right)^{i+j} \cdot (-1)^{2k-2l-i-1} R(x)^{2k-(i+j)-1}.$$

The orders of summation and integration are interchanged to yield an expression of the form

$$p_+ = \int \sum_{l=0}^{k-1} \sum_{j=0}^{2l} \sum_{i=0}^{2k-2l-1} \{\cdot\} dF_s(x).$$

The terms of the triple summation will now be investigated.

1) $i + j = 0$:

$$\sum_{l=0}^{k-1} \binom{2k-1}{2l} (-1) R^{2k-1}(x) = -2^{(M-1)} R^M(x).$$

2) $i + j = 2k - 1$:

$$\sum_{l=0}^{k-1} \binom{2k-1}{2l} \binom{2l}{2l} \binom{2k-1-2l}{2k-1-2l} \left(\frac{1}{2}\right)^{2k-1} \cdot (-1)^{2k-1-2l-(2k-1-2l)} R^0(x) = \left(\frac{1}{2}\right).$$

3) $i + j = \mu, \quad 0 < \mu < 2k - 1$:

$$\sum_{l=0}^{k-1} \binom{2k-1}{2l} \sum_{j=0}^{2l} \binom{2l}{j} \binom{2k-2l-1}{\mu - j} \left(\frac{1}{2}\right)^{\mu} \cdot (-1)^{2k-2l-1-\mu+j} R^{2k-1-\mu}(x)$$

$$= \left(\frac{1}{2}\right)^{\mu} R^{2k-1-\mu}(x)(-1)^{\mu} \cdot \left[-\sum_{j=0}^{k-1} \binom{2k-1}{\mu} \binom{\mu}{2j} 2^{2k-1-\mu-1} + \sum_{j=0}^{k-2} \binom{2k-1}{\mu} \binom{\mu}{2j+1} 2^{2k-1-\mu-1} \right]$$

$$= \left(\frac{1}{2}\right)^{\mu} R^{2k-1-\mu}(x)(-1)^{\mu} [-2^{\mu-1} + 2^{\mu-1}] = 0.$$

Thus

$$p_+ = \frac{1}{2} + (-1)^M 2^{M-1} \int_{-\infty}^{\infty} R^M(x) dF_s(x), \quad M = 2k - 1.$$

Case 2: $M = 2k$

$$p_+ = \sum_{l=0}^{k} \binom{2k}{2l} \Pr\{r_1 < 0, r_2 < 0, \cdots, r_{2l} < 0, r_{2l+1} > 0, r_{2l+2} > 0, \cdots, r_{2k} > 0\}$$

$$p_+ = \sum_{l=0}^{k} \binom{2k}{2l} \int_{-\infty}^{\infty} [F_n(x)]^{2l} [1 - F_n(x)]^{2k-2l} dF_s(x).$$

Using $F_n(x) = (1/2) + R(x)$ the integrand becomes

$$[F_n(x)]^{2l}[1 - F_n(x)]^{2k-2l} = \sum_{j=0}^{2l} \sum_{i=0}^{2k-2l} \binom{2l}{j} \binom{2k-2l}{i} \left(\frac{1}{2}\right)^{i+j} \cdot (-1)^{2k-2l-i} R^{2k-(i+j)}(x).$$

Summing terms as before:

1) $i + j = 0$:

$$\sum_{l=0}^{k} \binom{2k}{2l} R^{2k}(x) = 2^{2k-1} R^{2k}(x).$$

2) $i + j = 2k$:

$$\sum_{l=0}^{k} \binom{2k}{2l} \left(\frac{1}{2}\right)^{2k} R^0(x) = \left(\frac{1}{2}\right).$$

3) $i + j = \mu, \quad 0 < \mu < 2k$:

$$(-1)^{\mu} \left(\frac{1}{2}\right)^{\mu} R^{2k-\mu}(x) \left\{ \sum_{l=0}^{k} \binom{2k}{2l} \sum_{j'=0}^{l} \binom{2l}{2j'} \binom{2k-2l}{\mu - 2j'} - \sum_{l=0}^{k} \binom{2k}{2l} \sum_{j''=0}^{l-1} \binom{2l}{2j''+1} \binom{2k-2l}{\mu - 2j'' - 1} \right\}$$

$$= (-1)^{\mu} \left(\frac{1}{2}\right)^{\mu} R^{2k-\mu}(x) \binom{2k}{\mu} 2^{2k-\mu-1} \{2^{2\mu-1} - 2^{2\mu-1}\}$$

$$= 0.$$

Thus

$$p_+ = \frac{1}{2} + (1)^M 2^{M-1} \int_{-\infty}^{\infty} R^M(x) dF_s(x), \quad M = 2k.$$

References

[1] R. C. Jones, "The ultimate sensitivity of radiation detectors," *J. Opt. Soc. Amer.*, vol. 27, pp. 879–898, November 1947.

[2] R. H. Dicke, "The measurement of thermal radiation at microwave frequencies," *Rev. Sci. Instrum.*, vol. 17, pp. 268–275, July 1946.

[3] J. B. Thomas and T. R. Williams, "On the detection of signals in nonstationary noise by product arrays," *J. Acoust. Soc. Amer.*, vol. 31, pp. 453–462, April 1959.

[4] J. J. Faran, Jr., and R. Hills, Jr., "Correlators for signal reception," Acoustics Research Lab., Harvard University, Cambridge, Mass., Tech. Memo. 27, September 15, 1952.

[5] ——, "The application of correlation techniques to acoustic receiving systems," Acoustics Research Lab., Harvard University, Cambridge, Mass., Tech. Memo. 28, November 1, 1952.

[6] B. S. Melton and P. H. Karr, "Polarity coincidence scheme for revealing signal coherence," *Geophysics*, vol. 22, pp. 553–564, July 1957.

[7] V. O. Knudsen, R. S. Alford, and S. W. Emling, "Underwater ambient noise," *J. Marine Res.*, vol. 7, pp. 410–429, 1948.

[8] V. C. Anderson, "Arrays for the investigation of ambient noise in the ocean," *J. Acoust. Soc. Amer.*, vol. 30, pp. 470–477, 1958.

[9] G. C. Sponsler, "Note on coincidence detection optimal thresholds," *J. Appl. Phys.*, vol. 33, pp. 1889–1891, May 1962.

[10] H. Ekre, "Polarity coincidence correlation detection of a weak noise source," *IEEE Trans. Information Theory*, vol. IT-9, pp. 18–23, January 1963.

[11] M. Kanefsky and J. B. Thomas, "A comparison of some detectors and non-Gaussian inputs," Communications Lab., Princeton University, Princeton, N.J., Tech. Rept. 3, February 1964.

[12] S. S. Wolff, J. B. Thomas, and J. R. Williams, "The polarity-coincidence correlator: a nonparametric detection device," *IRE Trans. Information Theory*, vol. IT-8, pp. 5–9, January 1962.

[13] W. Feller, *An Introduction to Probability Theory and Its Applications*, vol. 1. New York: Wiley, 1957.

[14] H. Cramér, *Mathematical Methods of Statistics*. Princeton, N.J.: Princeton University Press, 1963.

[15] M. Fisz, *Probability Theory and Mathematical Statistics*. New York: Wiley, 1963.

[16] D. A. S. Fraser, *Nonparametric Methods in Statistics*. New York: Wiley, 1957.

Reprinted with permission from *IEEE Trans. Aerosp. Electr. Syst.* **AES-7**:942–950 (1971)

Nonparametric Radar Extraction Using a Generalized Sign Test

VILHELM GREGERS HANSEN, Member, IEEE
Raytheon Company
Wayland, Mass. 01778

BJARNE ALLAN OLSEN
Technical University of Denmark
Lyngby, Denmark

Abstract

A nonparametric procedure used in a constant false alarm rate (CFAR) radar extractor for detecting targets in a background of noise with unknown statistical properties is described. The detector is based on a generalization of the well-known two-sample sign test and thus requires a set of reference noise observations in addition to the set of observations being tested for signal presence.

The detection performance against Gaussian noise is determined for a finite number of observations and asymptotically, for both nonfluctuating and pulse-to-pulse Rayleigh fluctuating target statistics.

It is noted that the performance loss, as compared to the optimum parametric detector, depends critically on the number of reference noise observations available when the number of hits per target is not large. In the same case a much larger loss is also found for a pulse-to-pulse fluctuating target even though the asymptotic loss is the same as for a nonfluctuating target.

A comparison is finally made with a detector based on the Mann-Whitney test, which usually is considered to be one of the better nonparametric procedures for the two-sample case.

I. Introduction

The use of automatic signal extraction and digital data processing in search radar systems requires a very good regulation of the false alarm rate out of the extractor under varying conditions of noise, clutter, and interference at the input to the receiver, if overloading of subsequent data processors is to be prevented.

In the past various receiver and extractor techniques have been used in order to obtain such constant false alarm rate (CFAR) performance. In many cases, however, these designs have been based on the restricting assumption of Gaussian statistics of the disturbances.

Recently much effort has been directed towards the study of nonparametric methods of signal detection, due to their property of ensuring a CFAR for wide classes of input noise distributions. However, with a few exceptions quantitative studies of the performance of nonparametric tests have been restricted to the determination of their asymptotic relative efficiency, which is totally insufficient when an application to radar is contemplated.

The three most important characteristics of a nonparametric procedure which is considered for use in an automatic radar extractor are:

1) the CFAR properties of the nonparametric extractor against the types of disturbances that are likely to occur;
2) the detection performance against Gaussian noise for a finite number of observations (hits per target) as compared to the optimum parametric procedure;
3) the complexity of the resulting implementation from a practical point of view.

In this paper a nonparametric procedure which can be regarded as a generalization of the well-known two-sample sign test is considered for use in a radar extractor.[1] Mainly the two latter of the preceding characteristics are considered, as the first one is closely related to the specific radar parameters, the signal processor preceding the extractor, and the operational use of the radar system.

The analysis of the performance against Gaussian noise is carried out for both nonfluctuating and pulse-to-pulse Rayleigh fluctuating signals. Asymptotic results are obtained on the basis of the concept of asymptotic relative efficiency (ARE), whereas the performance for a finite number of observations has been obtained through analytical and numerical computations.

A comparison is finally made with a detector based on the Mann-Whitney (or two-sample Wilcoxon) test, which is usually considered to be one of the more effective

[1] A similar procedure was recently proposed and analyzed by Dillard and Antoniak [1]. A somewhat related but less practical detector was described in [2]. In this paper the approach is more general, and a more comprehensive set of results has been obtained. This detector was first analyzed in [3], and some results were presented in a recent survey of nonparametric radar detection principles [4].

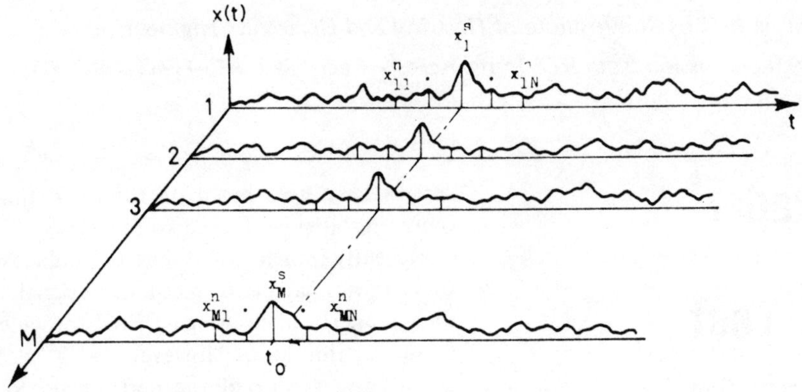

Fig. 1. Radar video output during M consecutive sweeps.

nonparametric tests in the two-sample case. An improved performance is indeed noted for this test but it is only obtained at the expense of a considerable increase in the complexity of the resulting extractor design.

II. Nonparametric Radar Extraction Based on a Generalized Sign Test

The procedure most frequently used in the past for obtaining CFAR performance in a radar detection system operating in an environment of nonstationary noise has been to sample the background noise in the time and/or frequency domain, in a region around the possible position of the target, in order to estimate the noise intensity and subsequently control receiver gain or decision threshold in an appropriate manner. This approach is usually based on the assumption that the noise is Gaussian, and the design problems have been to find strategies for collecting reference noise observations which are representative of the noise competing with the signal (if present) and in turn to use these observations in such a way that CFAR performance is achieved while an acceptable detection performance is maintained.

In many cases of practical importance the assumption of Gaussian noise can not be justified and no alternative distribution can usually be substituted with sufficient confidence for it to be used as a basis for the receiver design. Thus detection procedures are required which are invariant to the actual distribution of the noise; such procedures can be obtained and adapted from the theory of nonparametric statistics [5], [6].

In this paper the conventional approach to the design of CFAR radar detectors, as outlined previously, is extended by introduction of a nonparametric procedure for processing the probable signal-plus-noise observations and the reference noise observations. For the purpose of illustration a specific configuration of the nonparametric radar detector is considered in which the reference noise observations are obtained in the time domain. However, in practice any combination of time and frequency-domain sampling could be used.

In Fig. 1 the radar video output received during M repetition periods is shown. The presence of a target at range R_0 will cause a signal to appear with a corresponding time delay $t_0 = 2R_0/c$, and the signal-to-noise ratio is assumed to be the same in all M sweeps. This output corresponds to the usual assumption of a rectangular antenna beam shape for a scanning radar with M hits per beamwidth and a known target azimuth. Alternatively and more precisely, Fig. 1 will represent the output of an electronically scanned array radar transmitting M pulses in a given direction.

For the radar receiver a matched filter followed by an envelope detector is assumed. The justification for this choice is that the matched filter maximizes the output signal-to-noise ratio whenever the input interference has a spectral density, which is constant across the signal bandwidth. Lacking a more precise description of the statistical properties of the noise, this appears to be the most reasonable strategy to follow. An envelope detector is used because incoherent processing is assumed to be required.

The nonparametric test for target presence at a certain range R_0, corresponding to the video amplitudes $x_1{}^s, \cdots, x_M{}^s$ (Fig. 1), is carried out by comparison with a set of reference noise observations. These reference noise observations are taken in a region around the possible signal position but far enough away to ensure that they are not affected by the presence of a signal, and with sufficient individual spacing to make them approximately uncorrelated.[2] During each sweep N such reference noise observations are taken and in the jth sweep they are denoted $x_{j1}{}^n, x_{j2}{}^n, \cdots, x_{jN}{}^n$. On each successive sweep a statistic z_j is then computed as

$$z_j = \sum_{k=1}^{N} u(x_j{}^s - x_{jk}{}^n) \tag{1}$$

[2] As discussed later, independence between these observations is not required in order for the detection procedure to be nonparametric. Independence is, however, assumed in obtaining the quantitative results of Sections III and IV.

where $u(x)$ is the unit step function. The nonparametric test for target presence is then

$$T_{GS} = \sum_{j=1}^{M} z_j \begin{array}{l} > T_0 \text{ signal present} \\ < T_0 \text{ only noise present} \end{array} \quad (2)$$

where the threshold T_0 is set to yield the required false alarm probability.[3]

In order for this test to be nonparametric it is required that when only noise is present the first-order probability density functions (pdf) of all observations obtained on a single sweep are the same, and that, further, all the statistics $z_1, z_2, \cdots, z_M$ are independent. It is not required, however, that the observations on a single sweep are independent, nor that the first-order pdf is the same from sweep to sweep.

An example of the block diagram of the resulting extractor is shown in Fig. 2 for $N = 4$. A logarithmic compression could in practice be inserted after the matched filter in order to maintain a large dynamic range, but it is without effect on the theoretical performance.[4] In Fig. 2 it is assumed that the sum in (2) would have to be computed for a number of adjacent range bins and that a sweep-to-sweep updating is required as for a scanning radar. The implementation is completely analogous to that required for a moving window dig tal extractor, except that the usual (multilevel) amplitude quantizer is replaced by a circuit, which computes the statistic z_j as a function of range for each radar sweep. The synchronous switches S_1 and S_2 are assumed to perform the necessary range gating of this output into the corresponding registers and simultaneously to sample the summed outputs for examination by an appropriate detection logic. These more practical aspects are, however, not taken into account in this paper, and performance will only be determined in terms of the statistic T_{GS} for a specific range bin. The results can then be directly compared with standard curves describing radar detection performance.

For $N = 1$ the nonparametric test (2) is identical to the well-known two-sample sign test [5, p. 43]. For $N \geqslant 2$ the test can be considered as a generalized sign test or as a special case of a two-sample problem for randomized blocks [5, p. 4], where for each block the Mann-Whitney statistic z_j is computed and the test statistic T_{GS} is formed as in (2). Finally, the test could, in accordance with the terminology introduced in [7], be classified as a "mixed" Mann-Whitney test. This interpretation is, however, not entirely justified in the present case because the test specified by (2) will be nonparametric against a larger class of noise inputs than a

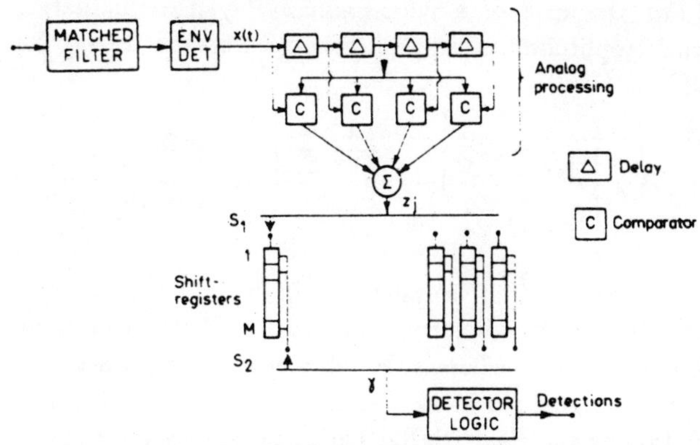

Fig. 2. Example of radar extractor based on generalized sign test for $N = 4$.

Mann-Whitney test designed to process the complete set of data collected during M sweeps.

In view of the fact that the Mann-Whitney test usually is considered to be one of the more efficient nonparametric tests in the two-sample case it was decided to obtain some results for the Mann-Whitney test for comparison. In terms of the observations specified in Fig. 1, the Mann-Whitney test statistic would be [8], [9]

$$T_{MW} = \sum_{j=1}^{M} \sum_{i=1}^{M} \sum_{k=1}^{N} u(x_j{}^s - x_{ik}{}^n). \quad (3)$$

A comparison of this expression with that for the nonparametric extractor [(2) and (1)] immediately reveals the considerable increase in complexity of the Mann-Whitney test.

As mentioned previously the detection performance of the two nonparametric detectors will be carried out for the specific case where the background noise is stationary, white Gaussian. The reason for selecting this case as the reference is that even though non-Gaussian noise may occur sufficiently often to be a serious problem, Gaussian noise will normally constitute the predominant background.

Assuming for convenience the input power to be normalized to unity the pdf at the output of the envelope detector has a Rayleigh pdf,

$$p_0(x) = \begin{cases} x \exp[-\tfrac{1}{2}x^2], & x \geqslant 0 \\ 0, & \text{otherwise.} \end{cases} \quad (4)$$

When a constant signal is present the pdf of the signal-plus-noise observation is Rician, i.e.,

$$p_1(x) = \begin{cases} x \exp[-\tfrac{1}{2}x^2] \exp[-S]\, I_0(x\sqrt{2S}), & x \geqslant 0 \\ 0, & \text{otherwise} \end{cases} \quad (5)$$

where $S = \bar{E}/N_0$ is the signal-to-noise ratio, $\bar{E}$ the (average) signal energy for each received signal pulse, and N_0 the (one-sided) spectral density of the white noise.

[3] The detection procedure studied in [1] is obtained by quantizing z_j, so that $z_j = 1$ if the sum in (1) is equal to N, and $z_j = 0$ otherwise. As suggested in [1] a better approach would be to perform this quantization relative to a threshold selected between 1 and N.

[4] In a digital system the possibility of ties (equality between inputs) in the comparators might also have to be taken into account.

For the case of a pulse-to-pulse Rayleigh fluctuating signal amplitude the signal-plus-noise observations have the pdf

$$p_1(x) = \begin{cases} \frac{x}{1+S} \exp\left[-\frac{x^2}{2(1+S)}\right], & x \geqslant 0 \\ 0, & \text{otherwise.} \end{cases} \quad (6)$$

For the quantitative evaluations in the following sections the pdf of (4) is therefore assumed for all reference noise observations x^n, whereas the pdf's of (5) and (6) are used for the signal-plus-noise observations x^s in the nonfluctuating and Rayleigh fluctuating cases, respectively.

III. Asymptotic Performance of Detectors

The asymptotic performance of a detector relates to a situation where the number of observations M is very large and the required signal-to-noise ratio per observation therefore approaches zero. The relative performance of two detection procedures against a specified noise background can then be characterized by their asymptotic relative efficiency [10].

If the number of observations required by a detector to yield specified values of detection probability P_d and false alarm probability P_f for a given signal-to-noise ratio S is denoted $n(P_d,P_f,S)$, the asymptotic relative efficiency of detector B, as compared to detector A, is defined as the limit

$$E_{B,A} = \lim_{S\to 0} \frac{n_A(P_d,P_f,S)}{n_B(P_d,P_f,S)}. \quad (7)$$

In many cases the detection statistics of the two detectors will be asymptotically normal and $E_{B,A}$ will then be independent of the specified values of P_d and P_f.

In engineering applications the asymptotic loss in signal-to-noise ratio of detector B relative to detector A is a more directly meaningful quantity. Defining $S(P_d,P_f,n)$ to be the signal-to-noise ratio required by a detector to yield specified values of P_d and P_f for a given number of observations n, the asymptotic loss of detector B as compared to detector A can be defined as the limit

$$L_\infty = \lim_{n\to\infty} \frac{S_B(P_d,P_f,n)}{S_A(P_d,P_f,n)}. \quad (8)$$

Depending on the definition of signal-to-noise ratio a simple relation usually holds between E and L_∞.

In Appendix I, E and L_∞ are determined for the case of a Gaussian noise background. Both the generalized sign detector and the Mann-Whitney detector are evaluated relative to the optimum parametric detector for stationary white Gaussian noise. The results obtained are identical for both nonfluctuating and Rayleigh fluctuating signals; a summary is given in Table I. It is seen that both detectors lead to similar asymptotic losses and that they become identical when the number of reference noise observations N is large.[5]

Table I
Asymptotic Performance of Nonparametric Detectors

N	Generalized Sign Detector		Mann-Whitney Detector	
	E	L_∞(dB)	E	L_∞(dB)
1	0.25	3.0	0.375	2.1
2	0.375	2.1	0.5	1.5
4	0.5	1.5	0.6	1.1
8	0.6	1.1	0.667	0.9
16	0.667	0.9	0.706	0.8
∞	0.75	0.6	0.75	0.6

A drawback of the asymptotic performance results as presented in Table I is that unless the number of observations actually available is large a rather unreliable indication of expected performance is obtained. In search radar applications the number of hits per target will often be relatively small and an analysis of the performance for a finite number of observations is therefore mandatory.

IV. Detection Performance of the Generalized Sign Detector

The detection performance of the generalized sign detector for a finite number of observations was determined through analytical and numerical computations. Again, the case of stationary white Gaussian noise was considered with signals of constant or pulse-to-pulse Rayleigh fluctuating amplitude. The pdf of the output z_j in (1) was first determined for the cases of signal-plus-noise and noise only present at the input, as shown in Appendix II. In deriving these results all reference noise observations were assumed to be independent.

These (discrete) densities were subsequently subjected to numerical convolutions in order to determine the distribution of the detector output T_{GS} as given by (2). From the results with only noise present the thresholds required to yield specified values of false alarm probability could be determined.[6] Due to the discreteness of the distributions, randomized decision rules were used to obtain the required value of P_f with equality.[7]

[5] For the nonparametric detector studied in [1] it is not difficult to show that the asymptotic relative efficiency is

$$E = \frac{1}{N}\left(\sum_{i=1}^{N} \frac{1}{1+i}\right)^2$$

which is maximum for $N = 6$ corresponding to $L_\infty = 1.9$ dB. For $N = 10$ the asymptotic loss is 2.0 dB which is close to the loss found in [1] for $M = 50$.

[6] For each M and N there is a lower limit on the achievable false alarm probability corresponding to a threshold $T_0 = M \cdot N$. This limit is easily seen to be $(P_f)_{min} = (1/N)^M$.

[7] Randomization would hardly ever be required in a practical system.

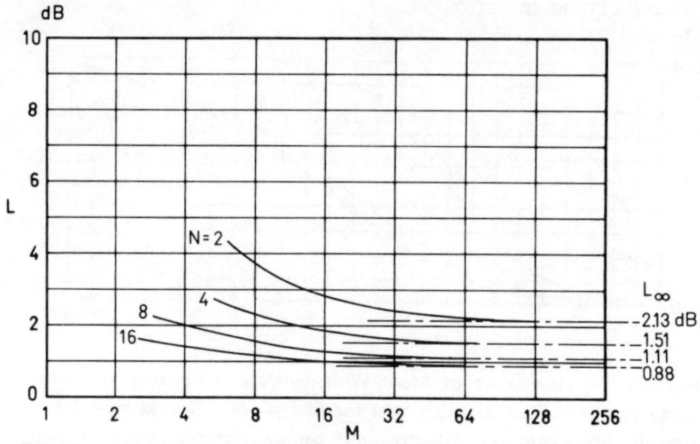

Fig. 3. Loss of generalized sign detector as a function of number of signal-plus-noise observations with N as parameter. Nonfluctuating signal for $P_d = 0.5$ and $P_f = 10^{-3}$.

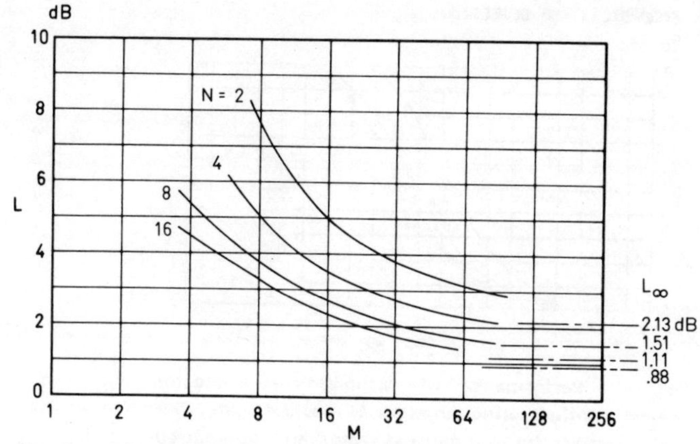

Fig. 5. Loss of generalized sign detector as a function of number of signal-plus-noise observations with N as parameter. Pulse-to-pulse Rayleigh fluctuating signal for $P_d = 0.5$ and $P_f = 10^{-3}$

Fig. 4. As in Fig. 3, except $P_f = 10^{-6}$.

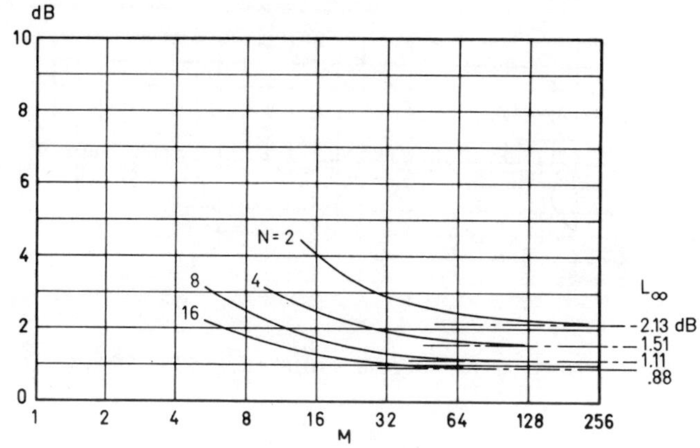

Fig. 6. As in Fig. 5, except $P_d = 10^{-6}$.

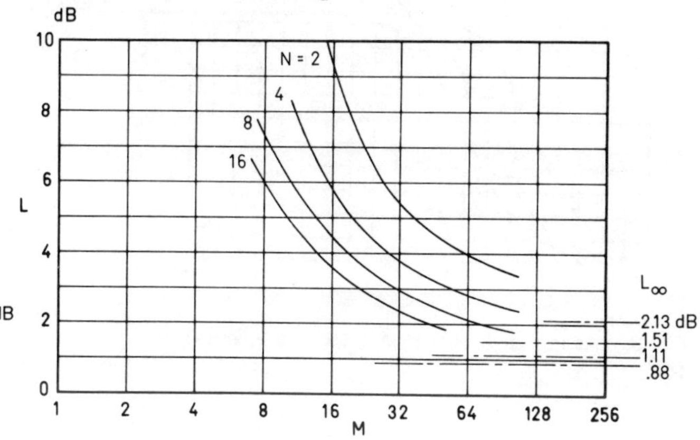

Subsequently curves of detection probability versus the signal-to-noise ratio $S = \bar{E}/N_0$ were determined on the basis of these thresholds and the convolutions of the signal-plus-noise distributions.

For the nonfluctuating case the results obtained are summarized in Figs. 3 and 4. These show the loss in signal-to-noise ratio of the nonparametric radar extractor as compared to the optimum parametric procedure for P_d = 0.5 and $P_f = 10^{-3}$ and 10^{-6}, respectively. The loss is shown as a function of the number of hits M and with the number of reference noise observations N per signal-plus-noise observation as parameter. The asymptotes are the asymptotic losses from Table I.

For a pulse-to-pulse Rayleigh fluctuating signal amplitude similar results are shown in Figs. 5 and 6. Note in this case the significantly larger loss for small values of M even though the asymptotic performance is the same in the two cases.

In the no-fading case the losses were 10-20 percent (of the decibel value) larger than shown in these figures when a detection probability of P_d = 0.9 was considered. For a pulse-to-pulse Rayleigh fluctuating signal amplitude, a somewhat larger increase in the loss for P_d = 0.9 was found. A quantitative indication of the loss for various detection probabilities can be obtained from the results described in the following section.

V. Comparison with the Mann-Whitney Test

The Mann-Whitney (or two-sample Wilcoxon) test is usually considered to be one of the more effective nonparametric tests for use in two-sample detection situations. It was therefore decided to make a quantitative comparison between the performance obtained with a detector based on the Mann-Whitney principle and that obtained with the generalized sign detector analyzed in the previous sections.

The operations performed by a Mann-Whitney detector are specified by (3). The asymptotic performance has already been determined and these results were shown in Table I.

For a finite number of observations the threshold settings required to give the specified false alarm probability first had to be determined. This was accomplished by using a recursive procedure for determining the distribution of T_{MW} when only noise is present. The recurrence relation is [9],

$$p_{L,K}(T) = p_{L-1,K}(T-N) + p_{L,K-1}(T). \qquad (9)$$

Here $p_{L,K}(T)$ is the probability of obtaining a statistic of value T when the number of reference noise observations is K and there are L signal-plus-noise observations (but with no signal present). With reference to (3) we wish to

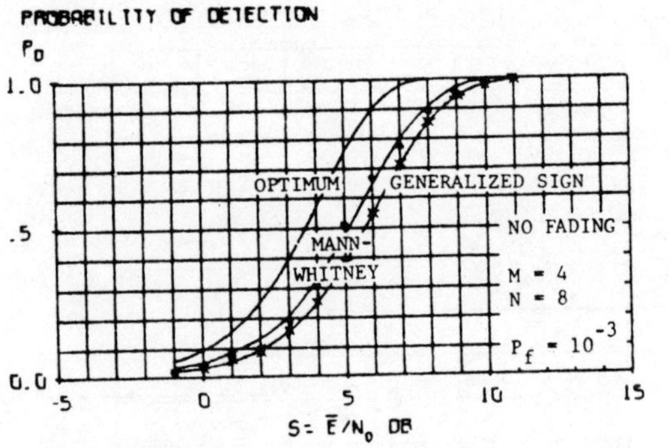

Fig. 7. Performance of Mann-Whitney detector against nonfluctuating target for $M = 4$, $N = 8$, and $P_f = 10^{-3}$. Results for optimum detector and generalized sign detector are shown for reference.

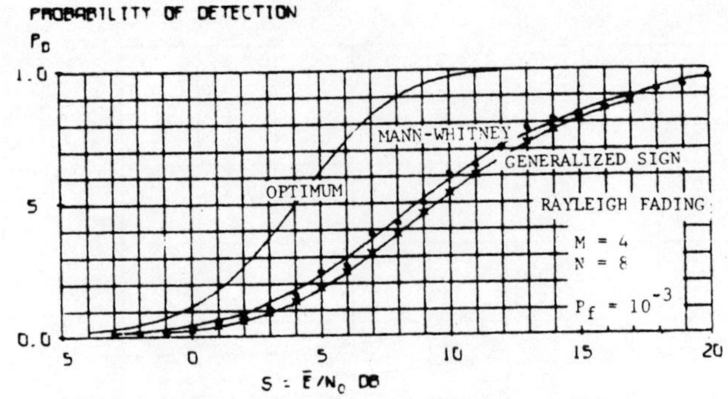

Fig. 10. Performance of Mann-Whitney detector against pulse-to-pulse Rayleigh fluctuating signal for $M = 4$, $N = 8$, and $P_f = 10^{-3}$. Results for optimum detector and generalized sign detector are shown for reference.

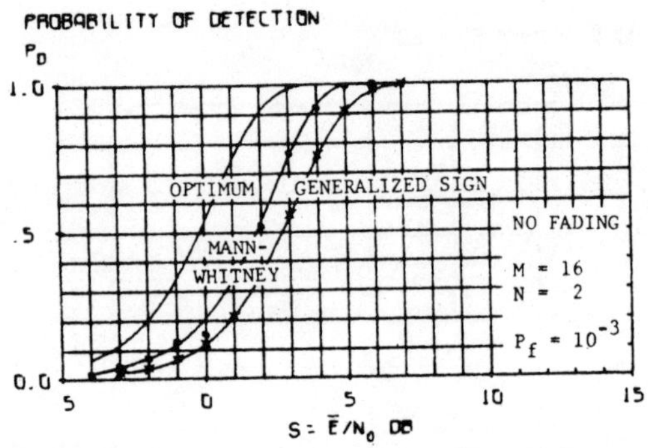

Fig. 8. As in Fig. 7, except $M = 16$, $N = 2$, and $P_f = 10^{-3}$

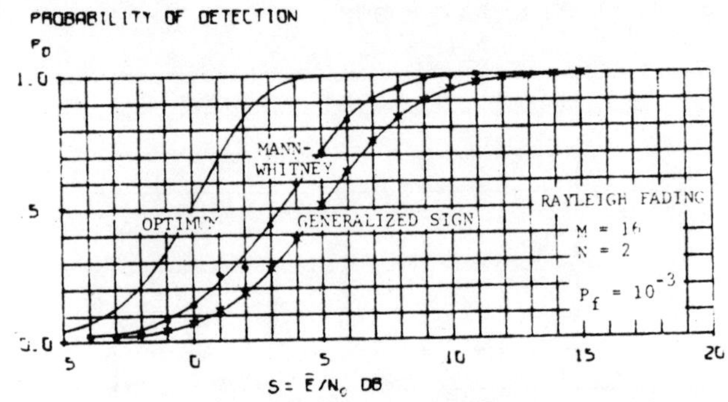

Fig. 11. As in Fig. 10, except $M = 16$, $N = 2$, and $P_f = 10^{-3}$.

Fig. 9. As in Fig. 7, except $M = 16$, $N = 2$, and $P_f = 10^{-6}$

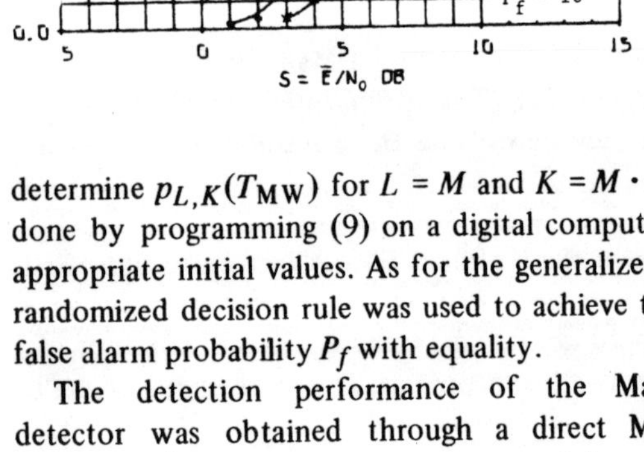

Fig. 12. As in Fig. 10, except $M = 16$, $N = 2$, and $P_f = 10^{-6}$.

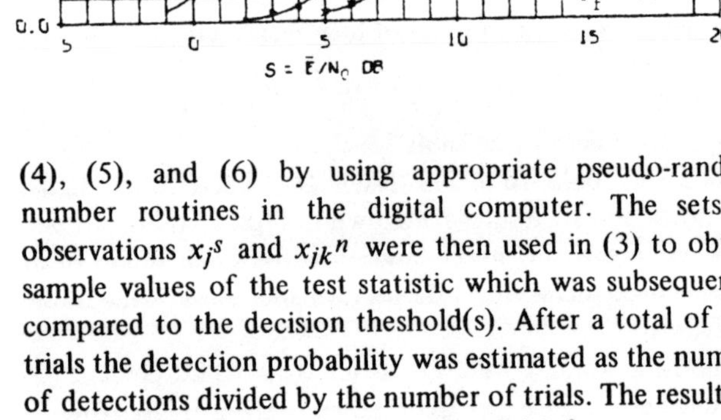

determine $p_{L,K}(T_{MW})$ for $L = M$ and $K = M \cdot N$. This was done by programming (9) on a digital computer using the appropriate initial values. As for the generalized sign test a randomized decision rule was used to achieve the specified false alarm probability P_f with equality.

The detection performance of the Mann-Whitney detector was obtained through a direct Monte Carlo simulation.[8] Observations were generated from the pdf's of (4), (5), and (6) by using appropriate pseudo-random number routines in the digital computer. The sets of observations x_j^s and x_{jk}^n were then used in (3) to obtain sample values of the test statistic which was subsequently compared to the decision theshold(s). After a total of 500 trials the detection probability was estimated as the number of detections divided by the number of trials. The results of these simulations are shown in Figs. 7-12 for some typical cases. The points shown in these figures are the actual results of the simulation. In addition, the results for the nonparametric radar extractor, obtained as described in Section IV, are shown in these figures together with the optimum curves for incoherent integration. A comparison of these curves shows that the Mann-Whitney detector,

[8] For the case of a Rayleigh pulse-to-pulse fluctuating signal, it appears to be possible to use the recursive procedure described in [11] to determine the detection performance, but this was not attempted since Monte Carlo techniques would in any case have been required for the no-fading case.

as expected, gives a better performance than the nonparametric radar extractor, in particular, for the smaller value of N. It does, however, seem doubtful that the improvement obtained would warrant the considerable increase in complexity which is required for a practical implementation of the Mann-Whitney detector.

VI. Conclusions

A nonparametric CFAR detector for use in radar systems which have to operate against a background of statistically incompletely specified noise has been described. A detailed study of the detection performance of this generalized sign detector for constant and fluctuating signals in a background of stationary white Gaussian noise has been carried out. In addition, some results were obtained for a Mann-Whitney detector, which is an interesting alternative.

The fact that a performance analysis for the case of white Gaussian noise constitutes the major part of the paper does not mean that this is the most important consideration. On the contrary, a CFAR, against all likely forms of interference and noise, is the primary design aim in an automatized detection system. Among the detectors satisfying this requirement one would then look for the one giving the best overall detection performance. As the background noise would be expected to be Gaussian in a large number of situations, our interest for this case follows.

In examining the losses of the generalized sign detector as presented in Figs. 3-6, it should be kept in mind the optimum parametric detector was used as a reference and that realistic conventional receivers such as LOG/FTC, Dicke-Fix, etc., also exhibit losses of up to several decibels as compared to the optimum. Furthermore, an extractor loss would normally have to be taken into account for these receivers, while for the generalized sign test this type of loss is already included.

Several conclusions can be made on the basis of Figs. 3 through 6.

1) For given values of M and N there exists a lower limit of false alarm probability which can be obtained.

2) For small values of M the loss for a pulse-to-pulse Rayleigh fluctuating target is appreciably larger than in the no-fading case.

3) The asymptotic losses are approached rather slowly. Up to several hundred hits are required in some cases before the asymptotic results apply.

An obvious simplification of the detector shown in Fig. 2 would be to subject the output z_j to a binary quantization. This would be particularly useful when N is large. Some results presented in [1] for a particular case of binary quantization suggest that this simplification can be introduced with little or no loss in detection performance.[9] Certain results in [2] show that such a quantization can in some cases even lead to an improved performance.

[9] Numerical results should be easily obtainable for this case from the pdf's given by (30) and (32) and tables of the binomial distribution.

Appendix I

Asymptotic Relative Efficiency of Detectors

The efficiency of a detector is defined [10] as

$$e = \lim_{M\to 0} \frac{1}{M} \frac{\left[\frac{\partial}{\partial S} E\{T|S\}\right]^2_{S=0}}{\sigma_0^2(T)} \tag{10}$$

where $\sigma_0^2(T)$ is the variance of the test statistic T in the noise only case and $E\{T|S\}$ is the expected value of T when the signal-to-noise ratio is S. The asymptotic relative efficiency of two detectors A and B is then given by the ratio of their efficiencies, i.e., [10]

$$E_{A,B} = \frac{e_A}{e_B}. \tag{11}$$

Both the generalized sign and the Mann-Whitney detectors are compared to the optimum parametric detector for white Gaussian noise. For small signal-to-noise ratio the optimum detector bases its decision on the test statistic[10]

$$T_{\text{opt}} = \sum_{i=1}^{M} (x_i^s)^2 \tag{12}$$

for a no-fading target as well as a pulse-to-pulse Rayleigh fluctuating target. From the distribution of x_i^s as given by (5) and (6) it can then be shown easily that $e_{\text{opt}} = 1.0$ for both of the target models considered.

For the Mann-Whitney detector as defined by (3) we have [5]

$$\sigma_0^2(T_{MW}) = \frac{M \cdot N \cdot M}{12}(M + N \cdot M + 1). \tag{13}$$

For the numerator in (10) we find

$$\begin{aligned} E\{T_{MW}|S\} &= \sum_{k=1}^{M}\sum_{j=1}^{M}\sum_{i=1}^{M} E\{u(x_k^s - x_{ji}^n)\} \\ &= M^2 N \cdot E\{u(x^s - x^n)\} \\ &= M^2 N \,\text{Prob}\{x^s > x^n\}. \end{aligned} \tag{14}$$

In general,

$$\text{Prob}\{x^s > x^n\} = \int_{-\infty}^{\infty} p(x^s)\left\{\int_{-\infty}^{x^s} p(x^n)dx^n\right\} dx^s \tag{15}$$

[10] A matched filter for the individual signal pulses is assumed for all detectors.

which for the pdf's given by (4) and (5) gives the result

$$\frac{\partial}{\partial S} E\left\{T_{MW}|S\right\} = M^2 N \, \tfrac{1}{4} \exp\left[-\tfrac{1}{2}S\right] \tag{16}$$

and thus, from (10),

$$e_{MW} = \lim_{M\to\infty} \frac{1}{M} \frac{(\tfrac{1}{4}M^2 N)^2}{\sigma_0^2(T_{MW})}$$

$$= \frac{3}{4} \frac{1}{1+N^{-1}} . \tag{17}$$

The asymptotic relative efficiency is then

$$E_{MW,\text{opt}} = \frac{e_{MW}}{e_{\text{opt}}} = \frac{3}{4} \frac{1}{1+N^{-1}} . \tag{18}$$

Exactly the same result is obtained if the pdf (6) for the pulse-to-pulse Rayleigh fluctuating signal is used for $p(x^s)$ in (15).

For the generalized sign detector it is easily shown that[11]

$$\sigma_0^2(T_{GS}) = M \cdot N \frac{N+2}{12} . \tag{19}$$

Further, by direct analogy to (14) we find

$$E\left\{T_{GS}|S\right\} = N \cdot M \,\text{Prob}\left\{x^s > x^n\right\} \tag{20}$$

and therefore the efficacy is

$$e_{GS} = \lim_{M\to\infty} \frac{1}{M} \frac{(\tfrac{1}{4}M \cdot N)^2}{\sigma_0^2(T_{GS})} = \frac{3}{4} \frac{1}{1+2N^{-1}} . \tag{21}$$

The asymptotic relative efficiency is therefore

$$E_{GS,\text{opt}} = \frac{e_{GS}}{e_{\text{opt}}} = \frac{3}{4} \frac{1}{1+2N^{-1}} . \tag{22}$$

For the case of incoherent processing and the signal-plus-noise pdf's (5) and (6) it can easily be shown that the asymptotic loss is given by

$$L_\infty = E^{-1/2} \tag{23}$$

or, in decibels,

$$L_\infty(\text{dB}) = -5 \log E . \tag{24}$$

Equations (18), (22), and (24) were used to compute the results shown in Table I.

Appendix II

pdf for Single-Sweep Statistic

From (1) the single-sweep statistic is given by

$$z = \sum_{k=1}^{N} u(x^s - x_k^n) \tag{25}$$

[11] Use the result (13) for the single-sweep statistic z_j.

where for simplicity the common subscript j has been left out. The statistic z can assume all integer values from 0 to N. Denoting the pdf for x^s, $p_1(x^s)$, and the pdf for x^n, $p_0(x^n)$, the probability that $z = \ell$ can be written

$$p(z=\ell) = \binom{N}{\ell} \int_{-\infty}^{\infty} p_1(x^s) \left[\int_{-\infty}^{x^s} p_0(x^n)\, dx^n\right]^{\ell}$$

$$\cdot \left[\int_{x^s}^{\infty} p_0(x^n)\, dx^n\right]^{N-\ell} dx^s. \tag{26}$$

Introducing the complementary distribution function of x^n,

$$F_0{}^c(x^n) = \int_{x^n}^{\infty} p_0(u)\, du \tag{27}$$

and expanding $[1 - F_0{}^c(x^s)]^{\ell}$ we find that the discrete pdf of z can be written

$$p(z=\ell) = \binom{N}{\ell} \sum_{i=0}^{\ell} (-1)^i \binom{\ell}{i}$$

$$\cdot \int_{-\infty}^{\infty} p_1(x^s)\, [F_0{}^c(x^s)]^{N-\ell+i}\, dx^s. \tag{28}$$

This pdf has to be evaluated for $p_0(x^n)$ as in (4) and with $p_1(x^s)$ given by either (5) or (6).

For a constant signal amplitude [$p_1(x^s)$ as in (5)] we find for the integral in (28)

$$\int_{-\infty}^{\infty} p_1(x^s)\, [F_0{}^c(x^s)]^{N-\ell+i}\, dx^s = \exp(-S) \int_0^{\infty} x^s$$

$$\cdot \exp\left[-\tfrac{1}{2} {x^s}^2 (N-\ell+i+1)\right] I_0(x^s \sqrt{2S})\, dx^s$$

$$= \frac{\exp(-S)}{N-\ell+i+1} \exp\left[\frac{S}{N-\ell+i+1}\right] \tag{29}$$

and (28) becomes

$$p(z=\ell) = \exp(-S) \binom{N}{\ell} \sum_{i=0}^{\ell} (-1)^i \binom{\ell}{i} \frac{1}{N-\ell+i+1}$$

$$\cdot \exp\left[\frac{S}{N-\ell+i+1}\right], \quad \ell = 0,1,\cdots,N. \tag{30}$$

When only noise is present $S = 0$ and (30) is simplified to

$$p(z=\ell) = \binom{N}{\ell} \sum_{i=0}^{\ell} (-1)^i \binom{\ell}{i} \frac{1}{N-\ell+i+1}$$

$$= \frac{1}{N+1}, \quad \ell = 0,1,\cdots,N \tag{31}$$

as may also be found more directly from simpler considerations.

For a fading signal the pdf given by (6) is used in (28) and the pdf of z becomes[12]

$$p(z = \ell) = \binom{N}{\ell} \sum_{i=0}^{\ell} (-1)^i \binom{\ell}{i} \frac{1}{(1 + S)(N - \ell + i) + 1} \, . \tag{32}$$

Again, for $S = 0$ this pdf reduces to (31).

The discrete pdf's given by (30)-(32) were evaluated for the required values of N and S, and the distribution of the test statistic T_{GS} [see (2)] was subsequently obtained through an M-fold numerical convolution followed by a numerical integration to obtain false alarm and detection probabilities as a function of the decision threshold T_0.

[12] As shown in [12, eq. (7)], this result can also be written as a recursive relation.

References

[1] G.M. Dillard and C.E. Antoniak, "A practical distribution-free detection procedure for multiple-range-bin radars," *IEEE Trans. Aerospace and Electronic Systems*, vol. AES-6, pp. 629-635, September 1970.

[2] C.E. Antoniak, G.M. Dillard, and R.A. Shorack, "Distribution-free detection in radar with multiple resolution elements," *1967 South Western IEEE Conf. and Exhibition (SWIEEECO) Rec.*, pp. 17-2-1 - 17-2-8.

[3] B.A. Olsen, "Nonparametric methods in radar and sonar systems" (in Danish), M.S. thesis, Technical University of Denmark, Copenhagen, Denmark, October 1968.

[4] V.G. Hansen, "Nonparametric detection procedures applied to radar," *Proc. UMR-Mervin I. Kelly Communications Conf.*, pp. 24-3-1 - 24-3-6, October 1970.

[5] G.E. Noether, *Elements of Nonparametric Statistics*. New York: Wiley, 1967.

[6] J. Hajek, *Nonparametric Statistics*. San Francisco: Holden-Day, 1969.

[7] E.A. Feustal and L.D. Davisson, "The asymptotic relative efficiency of mixed statistical tests," *IEEE Trans. Information Theory*, vol. IT-13, pp. 247-255, April 1967.

[8] J. Capon, "A nonparametric technique for the detection of a constant signal in additive noise," *WESCON '59 Rec.*, pt. 4, pp. 92-103.

[9] H.B. Mann and D.R. Whitney, "On a test of whether one of two random variables is stochastically larger than the other," *Ann. Math. Stat.*, vol. 18, pp. 50-60, 1947.

[10] J. Capon, "Optimum coincidence procedures for detecting weak signal in noise," *1960 IRE Internatl. Conv. Rec.*, pt. 4, pp. 154-166.

[11] R.A. Shorack, "Recursive generation of the distribution of the Mann-Whitney U-statistic under generalized Lehman alternatives," *Ann. Math. Stat.*, vol. 37, pp. 284-286, February 1966.

[12] C.E. Antoniak and G.M. Dillard, "A distribution-free sequential probability-ratio test for multiple-resolution element radars," *IEEE Trans. Information Theory*, vol. IT-14, pp. 822-825, November 1968.

8

Reprinted with permission from *IEEE Trans. Inf. Theory* **IT-23**:304–315 (1977)

On Sequential Sign Detection of a Constant Signal

SAWASD TANTARATANA, MEMBER, IEEE, AND JOHN B. THOMAS, FELLOW, IEEE

***Abstract*—In fixed sample size signal detection, the sign detector is easier to implement and requires much less knowledge of the noise density than does the optimum fixed sample size detector. These advantages accrue at a cost of more observations needed to achieve the same performance. By utilizing the sign detector sequentially, it is shown that considerable saving in the (average) number of observations is obtained over the corresponding optimal fixed sample size detector. The asymptotic relative efficiencies, as the signal-to-noise ratio and both probabilities of error approach zero, of the sequential sign detector compared to fixed sample size detectors are derived. The possibility of truncating the test to avoid excessively large numbers of observations is considered.**

I. INTRODUCTION

THE SEQUENTIAL probability ratio test (SPRT) introduced by Wald [1] is known to minimize the expected sample size in testing a simple hypothesis against a simple alternative [2]. In signal detection, considerable saving, on the average, in the number of observations over the optimum fixed sample size (FSS) detector can be achieved by using a sequential detector which has the same level α and the same power β. However, the randomness of the sample size of a sequential detector introduces some difficulty in implementation. In communication systems, a feedback link is needed to inform the transmitter when a decision is made [3]. In radar systems, an electronically steerable antenna is required to accommodate the random time spent at each beam position [4].

The sign detector is known for its robustness [5], for its ease of implmentation, and for requiring less knowledge of the noise density than does the corresponding optimum detector [6], although more samples are needed to achieve the same performance. Abel and Betten [7] used the sign test in the sequential detection of an FM signal. Worley [8] studied sequential tests on multilevel quantized data with Rice, Rayleigh, and Gaussian noises. Even with the reduced data, a saving in the number of observations over the optimum FSS detector is possible if the reduced data are tested sequentially.

We study in this paper the utilization of the sequential sign detector in detecting antipodal signals. The only knowledge needed to set up the test, under the assumption of a symmetric noise density, is the probability of a positive

Manuscript received May 10, 1976; revised August 23, 1976. This research was supported by the National Science Foundation under Grant GK-24187 and by the U.S. Army Research Office under Grant DAAG29-75-G-0192.

The authors are with the Department of Electrical Engineering, Princeton University, Princeton, NJ 08540.

observation when either the positive signal or the negative signal is present. The resulting SPRT is a sequential testing on the sum of the signs of the observations. The expected sample size of this detector is compared to the sample size of corresponding FSS detectors, and a significant saving is observed. However, the advantages of the FSS sign detector still remain. The asymptotic relative efficiency (ARE) is defined as the ratio of the expected sample size of a sequential detector to the sample size of an FSS detector as the signal-to-noise ratio and the probabilities of error approach zero in such a way that the probabilities of error are linearly related. The ARE is then calculated for some noise densities. In Section VI, the probability of a sample size exceeding some number N_T is obtained for this sequential sign test. The result shows that this probability is small for proper values of N_T.

II. Problem Formulation

Let $x_1, x_2, \cdots$ be observations on a sequence of independent identically distributed (i.i.d.) random variables taken either from the hypothesis H or the alternative K, where

$$H: \quad X_i \text{ has density } f(x+S)$$

$$K: \quad X_i \text{ has density } f(x-S).$$

Here, $S > 0$ is a constant signal and $f(x)$ is a symmetric noise density. Hence, we decide whether the signal imbedded in the noise is $-S$ or $+S$.

In FSS signal detection, where the number of observations is fixed before the experiment, the detector which is optimal in the Neyman–Pearson sense is [9]

$$\prod_{i=1}^{N} \frac{f(x_i - S)}{f(x_i + S)} \underset{H}{\overset{K}{\gtrless}} T,$$

where N is the (fixed) number of observations and T is the threshold chosen to satisfy the required probability of error under H.

The Wald SPRT requires the smallest expected sample size [2] among all tests (sequential or nonsequential) with probability of error under H, $P_{eH} \le \alpha$ and with probability of error under K, $P_{eK} \le 1 - \beta$. (Note that Wald [1] uses β to denote the probability of error under K.) At the nth observation, the SPRT is defined by

$$\prod_{i=1}^{n} \frac{f(x_i - S)}{f(x_i + S)} \begin{cases} \ge A \Rightarrow K \\ \le B \Rightarrow H \\ \text{otherwise} \Rightarrow \text{take an additional sample,} \end{cases}$$

where the thresholds A and B are set to give the desired values of P_{eH} and P_{eK}.

Let us assume that we know only the probability of a positive observation under either H or K. This suggests that we quantize the data into two levels by the operation [6]

$$y_i = \text{sgn}\,(x_i), \qquad i = 1,2,\cdots,$$

where

$$\text{sgn}\,(x) = \begin{cases} 1, & x \ge 0 \\ -1, & x < 0. \end{cases}$$

Now the problem of testing $Y_1, Y_2, \cdots$ becomes, by the symmetry of $f(x)$,

$$H: \quad \text{Prob}\,(Y_i = 1) = q \text{ and Prob}\,(Y_i = -1) = p$$

$$K: \quad \text{Prob}\,(Y_i = 1) = p \text{ and Prob}\,(Y_i = -1) = q,$$

where $p = \text{Prob}\,(X_i \geqslant 0 | K \text{ is true})$ and $q = 1 - p$. The SPRT for these binomially distributed $Y_1, Y_2, \cdots$ is [1, chap. 5], at the nth observation, to test

$$(p/q)^k (q/p)^{n-k} \cdot \begin{cases} \ge A \Rightarrow K \\ \le B \Rightarrow H \\ \text{otherwise} \Rightarrow \text{take an additional sample,} \end{cases}$$

where k = number of positive y_i and the constants A and B are chosen to guarantee $P_{eH} \le \alpha$ and $1 - P_{eK} \ge \beta$. An equivalent test is

$$T_n = \sum_{i=1}^{n} y_i = (2k - n) \cdot \begin{cases} \ge \ln A / \ln (p/q) \Rightarrow K \\ \le \ln B / \ln (p/q) \Rightarrow H \\ \text{otherwise} \Rightarrow \text{take an additional sample.} \end{cases}$$

This test statistic is an integer; therefore the thresholds might as well be integers, say a and $-b$, with a and b being positive integers; now the test is

$$T_n = \sum_{i=1}^{n} y_i \begin{cases} \ge a \Rightarrow K \\ \le -b \Rightarrow H \\ \text{otherwise} \Rightarrow \text{take an additional sample.} \end{cases}$$

We will discuss in the next section the choice of the thresholds a and $-b$ such that $P_{eH} \le \alpha$ and $1 - P_{eK} \ge \beta$.

The same results follow in testing, for symmetric noise density $f(x)$,

$$H_1: \quad X_i \text{ has density } f(x)$$

$$K_1: \quad X_i \text{ has density } f(x - S)$$

on defining $y_i = \text{sgn}\,(x_i - S/2)$, where S is a constant and $X_1, X_2, \cdots$ are i.i.d. random variables.

III. Distribution and Moments of the Sample Size

Recognizing the test to be the same as the simple random walk problem [10], we see that the resulting P_{eH} and P_{eK} are given by

$$1 - P_{eK} = \text{Prob (termination at } a | K) = p^a \frac{p^b - q^b}{p^{a+b} - q^{a+b}}$$

TABLE I
EXAMPLES OF THRESHOLDS

p	r	α	1-β	ℓn(β/α)/ℓn(r)	a	ℓn((1-α/k)/(1-β))/ℓn(r)	b	M
0.52	1.083	10^{-4}	10^{-4}	115.07	116	115.07	116	116
0.52	1.083	10^{-6}	10^{-6}	172.60	173	172.60	173	173
0.52	1.083	10^{-10}	10^{-6}	287.67	288	172.60	173	230
0.52	1.083	10^{-12}	10^{-8}	345.20	346	230.01	231	288
0.6	1.5	10^{-4}	10^{-4}	22.72	23	22.72	23	23
0.6	1.5	10^{-6}	10^{-6}	34.07	35	34.07	35	35
0.6	1.5	10^{-10}	10^{-6}	56.79	57	34.07	35	46
0.6	1.5	10^{-12}	10^{-8}	68.15	69	35.43	36	52
0.85	5.67	10^{-4}	10^{-4}	5.31	6	5.31	6	6
0.85	5.67	10^{-6}	10^{-6}	7.96	8	7.96	8	8
0.85	5.67	10^{-10}	10^{-6}	13.27	14	7.96	8	11
0.85	5.67	10^{-12}	10^{-8}	15.93	16	10.62	11	13

and

$$P_{eH} = \text{Prob (termination at } a|H) = q^a \frac{p^b - q^b}{p^{a+b} - q^{a+b}}.$$

We also have

$$\text{Prob (termination at } -b|K) = q^b \frac{p^a - q^a}{p^{a+b} - q^{a+b}}$$

and

$$\text{Prob (termination at } -b|H) = p^b \frac{p^a - q^a}{p^{a+b} - q^{a+b}}.$$

The first moments of the sample size $N = \min\{n: \sum_{i=1}^{n} y_i \notin (-b,a)\}$ are

$$E(N|H) = \frac{b}{p-q} - \frac{a+b}{p-q} \cdot \frac{(1-(p/q)^b)}{(1-(p/q)^{a+b})}$$

and

$$E(N|K) = \frac{b}{q-p} - \frac{a+b}{q-p} \cdot \frac{(1-(q/p)^b)}{(1-(q/p)^{a+b})}.$$

Of course, Wald's equations [1] also give the exact answers since a and b are integers so that there is no overshoot of the thresholds in terminating the test.

To find the second moments of N, we have the following probabilities of terminating at the kth observation [10],

$$\text{Prob } (N = k|H) = G(k,a,p) + G(k,b,q)$$

and

$$\text{Prob } (N = k|K) = G(k,a,q) + G(k,b,p),$$

where

$$G(k,c,p) = \begin{cases} \dfrac{2^{k+1}}{a+b} p^{(k-c)/2}(1-p)^{(k+c)/2} \sum_{i=1}^{M} \cos^{k-1}\left(\dfrac{\pi i}{a+b}\right) \sin\left(\dfrac{\pi i}{a+b}\right) \sin\left(\dfrac{\pi c i}{a+b}\right), & \text{if } (k-c) \text{ is even and } \geq 0 \\ 0, & \text{otherwise} \end{cases}$$

and $M = \lfloor (a+b)/2 \rfloor$, the largest integer smaller than or equal to $(a+b)/2$. Using these expressions, one can show the second moments of N to be

$$E(N^2|H) = V(q,p)$$

and

$$E(N^2|K) = V(p,q),$$

where

$$V(p,q) = \frac{2^{b+1}}{a+b} q^b \sum_{i=1}^{M} \left(\frac{b^2}{1-c} + \frac{4bc}{(1-c)^2} + \frac{4c(1+c)}{(1-c)^3}\right) \cos^{b-1}\left(\frac{\pi i}{a+b}\right) \sin\left(\frac{\pi i}{a+b}\right) \sin\left(\frac{\pi b i}{a+b}\right)$$
$$+ \frac{2^{a+1}}{a+b} p^a \sum_{i=1}^{M} \left(\frac{a^2}{1-c} + \frac{4ac}{(1-c)^2} + \frac{4c(1+c)}{(1-c)^3}\right) \cos^{a-1}\left(\frac{\pi i}{a+b}\right) \sin\left(\frac{\pi i}{a+b}\right) \sin\left(\frac{\pi a i}{a+b}\right)$$

and $c = 4pq \cos^2(\pi i/(a+b))$. The variances can now be computed from the first and second moments.

The value of M is normally not too large (See Table I); therefore, the second moments and the variances can be evaluated. These values are shown for a practical (α,β) pair in Table II.

We now consider choosing the thresholds a and $-b$, given α and β, such that the resulting test guarantees P_{eH}

TABLE II
$E(N)$, σ_N, AND $P(N \geqslant N_T)$ FOR $\alpha = 10^{-3}$ AND $1 - \beta = 10^{-5}$

Designed p	$E(N\|H)$	$\sigma_{N\|H}$	$E(N\|K)$	$\sigma_{N\|K}$	$E_{\theta=0}(N)$	$\sigma_{N\|\theta=0}$
0.93	5.8	1.4	8.1	1.6	40.0	34.9
0.84	10.3	3.4	17.6	4.5	84.0	75.1
0.69	39.2	15.1	60.1	18.7	360.0	317.3
0.54	916.4	378.8	1456.2	477.5	8280.0	7303.0

$P(N \geq N_T \| H)$

p \ $N_T = E(N\|H)+$	$\sigma_{N\|H}$	$2\sigma_{N\|H}$	$3\sigma_{N\|H}$	$4\sigma_{N\|H}$	$5\sigma_{N\|H}$	$6\sigma_{N\|H}$
0.93	.292	.072	.072	.0167	.0038	.00086
0.84	.127	.035	.018	.0067	.0012	.00062
0.69	.148	.050	.014	.0042	.0011	.00034
0.54	.145	.044	.013	.0036	.0010	.00027

$P(N \geq N_T \| K)$

p \ $N_T = E(N\|K)+$	$\sigma_{N\|K}$	$2\sigma_{N\|K}$	$3\sigma_{N\|K}$	$4\sigma_{N\|K}$	$5\sigma_{N\|K}$	$6\sigma_{N\|K}$
0.93	.384	.115	.031	.0079	.0020	.0020
0.84	.192	.068	.012	.0037	.0011	.0003
0.69	.156	.047	.011	.0028	.0007	.0001
0.54	.149	.042	.010	.0024	.0006	.0001

$\leq \alpha$ and $1 - P_{eK} \geq \beta$; that is, we require

$$1 - P_{eK} = \frac{r^{a+b} - r^a}{r^{a+b} - 1} \geq \beta \tag{1}$$

and

$$P_{eH} = \frac{r^b - 1}{r^{a+b} - 1} \leq \alpha, \tag{2}$$

where $r = p/q > 1$. The smallest integer pair (a,b) satisfying inequalities (1) and (2) will be the desired thresholds. It is shown in Appendix A that the necessary conditions to guarantee (1) and (2) are

$$r^a \geq \beta/\alpha \tag{3}$$

and

$$r^b \geq (1 - \alpha)/(1 - \beta). \tag{4}$$

We can now choose one of the thresholds, either a satisfying (3) or $-b$ satisfying (4). Then the other threshold has to satisfy a sufficiency condition for $P_{eH} \leq \alpha$ and $1 - P_{eK} \geq \beta$. Let us assume a is chosen first such that $a = \lceil \ln (\beta/\alpha)/\ln (r) \rceil$. This implies $\ln (\beta/\alpha)/\ln (r) \leq a < \ln (\beta/\alpha)/\ln (r) + 1$; thus $\beta/\alpha \leq r^a < r\beta/\alpha$ and (3) is satisfied. Then the sufficiency condition on b, given in Appendix B, is

$$\frac{1 - (\alpha/k)}{1 - \beta} \leq r^b \leq \frac{1 - \alpha}{1 - k\beta}, \quad \text{if } k\beta < 1 \tag{5}$$

or

$$\frac{1 - (\alpha/k)}{1 - \beta} \leq r^b, \quad \text{if } k\beta \geq 1, \tag{6}$$

where $k = r^a(\alpha/\beta) \geq 1$. Since smaller b gives smaller expected sample size, we choose b as the smallest integer satisfying (5) or (6). On the other hand, if we set b first with $b = \lceil \ln ((1 - \alpha)/(1 - \beta))/\ln (r) \rceil$, then a is chosen as the smallest integer satisfying the following sufficiency condition for $P_{eH} \leq \alpha$ and $1 - P_{eK} \geq \beta$,

$$\frac{1 - (1 - \beta)/k}{\alpha} \leq r^a \leq \frac{\beta}{1 - k(1 - \alpha)}, \quad \text{if } k(1 - \alpha) < 1$$

or

$$\frac{1 - (1 - \beta)/k}{\alpha} \leq r^a, \quad \text{if } k(1 - \alpha) \geq 1,$$

where $k = r^b(1 - \beta)/(1 - \alpha) \geq 1$.

In most cases, both choices give the same threshold pair $(a, -b)$. We list in Table I some examples of the calculations of the thresholds for various values of p, α, and β.

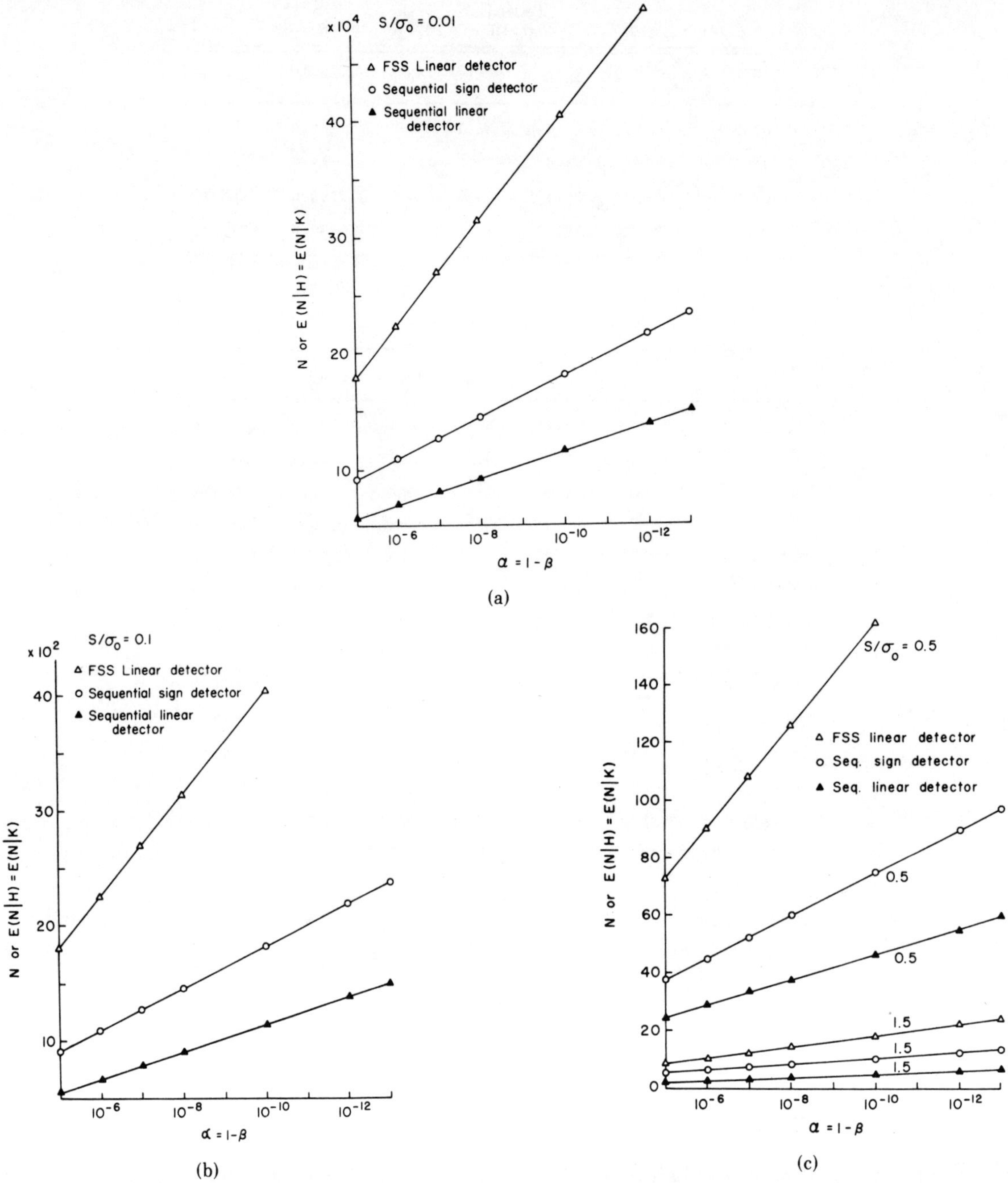

Fig. 1. Sample size and expected sample size with Gaussian noise.

IV. Performance of Sequential Sign Test

To evaluate the performance of this sequential sign detector, we compare its expected sample size to the expected sample size of the sequential linear detector and to the sample size of the FSS linear detector satisfying $P_{eH} \le \alpha$ and $1 - P_{eK} \ge \beta$ under Gaussian noise.

It is well-known that the linear detector is the Neyman–Pearson optimum detector for a constant signal in Gaussian noise, that is, to test

$$\sum_{i=1}^{N} x_i \underset{H}{\overset{K}{\gtrless}} T.$$

Here the sample size N and threshold T satisfy

$$\beta = \int_T^\infty f_G(x - NS, N\sigma_0^2)\,dx$$

and

$$\alpha = \int_T^\infty f_G(x + NS, N\sigma_0^2)\,dx$$

where σ_0^2 is the noise variance and

$$f_G(x,\sigma^2) = \frac{1}{\sqrt{2\pi}\sigma} e^{-x^2/2\sigma^2}, \qquad -\infty < x < \infty.$$

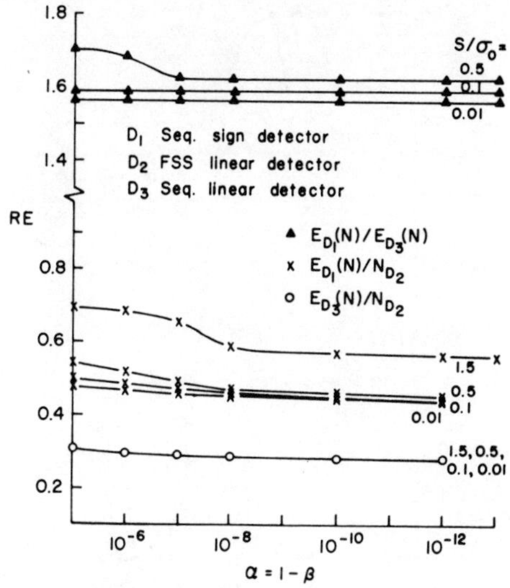

Fig. 2. Relative efficiencies with Gaussian noise.

Thus N is constrained to satisfy

$$N = \left[\frac{\Phi^{-1}(1-\alpha) + \Phi^{-1}(\beta)}{2(S/\sigma_0)}\right]^2$$

where $\Phi(x)$ is the standard Gaussian distribution function and $\Phi^{-1}(\cdot)$ is its inverse.

The expected sample size of the linear sequential detector is approximated by Wald's equations [1] so that

$$E(N|H) \cong \frac{\alpha \ln(\beta/\alpha) + (1-\alpha)\ln((1-\beta)/(1-\alpha))}{-2(S/\sigma_0)^2}$$

and

$$E(N|K) \cong \frac{\beta \ln(\beta/\alpha) + (1-\beta)\ln((1-\beta)/(1-\alpha))}{2(S/\sigma_0)^2},$$

for the Gaussian noise $f_G(x, \sigma_0^2)$. The errors in the approximations approach zero as the signal-to-noise ratio approaches zero.

We plot in Fig. 1 the sample size of the FSS linear detector and the expected sample sizes of the sequential sign detector and the sequential linear detector as functions of the probabilities of error α and $1-\beta$. Fig. 2 shows their ratios for comparison. The expected sample size of the sequential sign detector lies between those of the sequential linear detector and the FSS linear detector. At small signal-to-noise ratio and for α and $1-\beta$ in the range of practical interest, the number of samples needed by the sequential sign detector on the average is less than that needed by the FSS linear detector. These ratios, plotted in Fig. 2, converge as α and $1-\beta$ approach zero. The asymptotic results will be discussed in the next section. When the signal-to-noise ratio becomes larger, there is less saving in the number of samples. In Fig. 3, the expected sample sizes are plotted as functions of the signal-to-noise ratio with α and $1-\beta$ as parameters. Their ratios are shown in Fig. 4. As the signal-to-noise ratio approaches zero, these ratios converge. The graphs show that these ratios are very close to their limits at a signal-to-noise ratio smaller than 0.5. Observe that the ratio of the (expected)

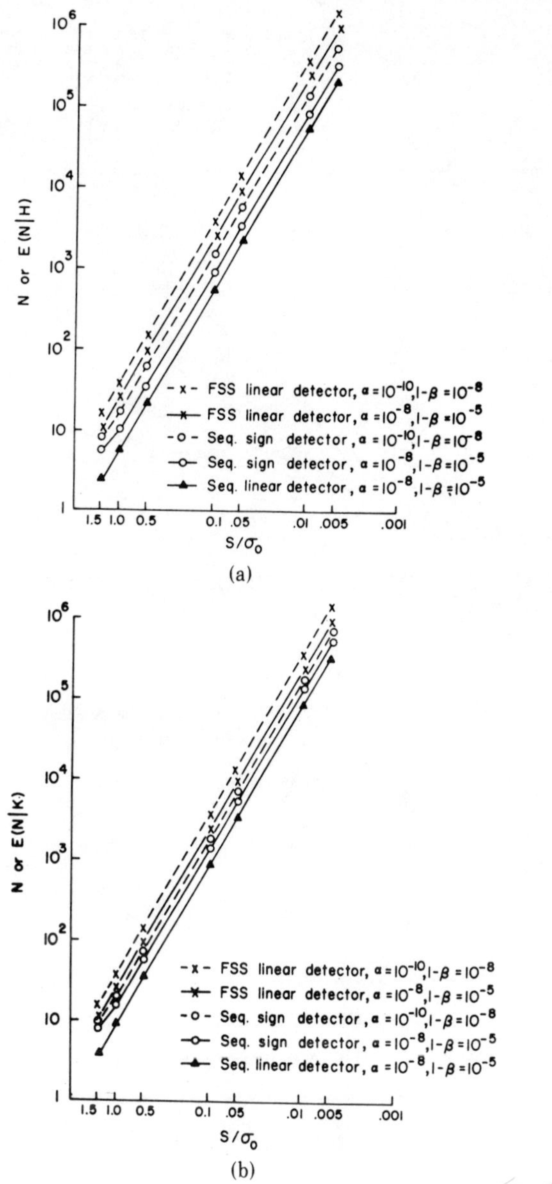

Fig. 3. (a) Sample size and expected sample size under H, Gaussian noise. (b) Sample size and expected sample size under K, Gaussian noise.

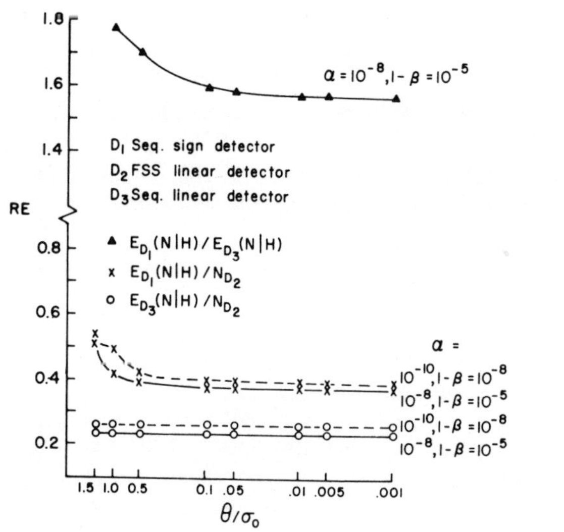

Fig. 4. Relative efficiencies under H, Gaussian noise.

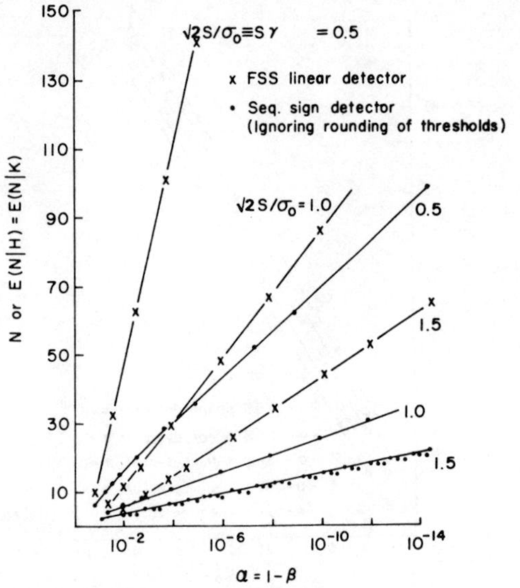

Fig. 5. Sample size and expected size with Laplace noise.

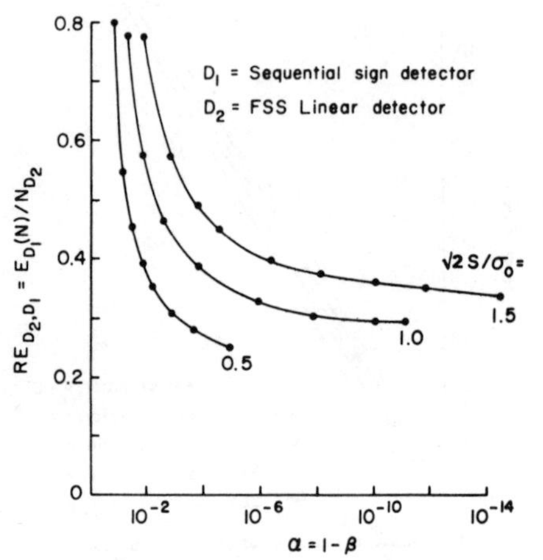

Fig. 6. Relative efficiencies with Laplace noise.

sample sizes for the FSS linear detector and the sequential linear detector is independent of the signal-to-noise ratio. This is generally not true for non-Gaussian noise distributions.

In Fig. 5 and Fig. 6, the performance of the sequential sign detector is compared to the FSS linear detector with Laplace noise. The values of P_{eH} and P_{eK} for the FSS linear detector in Laplace noise with density $f_L(x) = (\gamma/2) \exp(-\gamma|x|)$ are given by [11]

$$P_{eH} = e^{-NS\gamma} \sum_{k=0}^{N-1} a_k^{(N-1)} (NS\gamma)^k = P_{eK},$$

where we have set the threshold to be zero and where the coefficients $a_k^{(n)}$ are given recursively by

$$a_0^{(n)} = \tfrac{1}{2}$$

$$a_k^{(n)} = (\tfrac{1}{2}n) a_{k-1}^{(n-1)} + (1 - k/2n) a_k^{(n-1)}, \quad k = 1,2,\cdots,n-1$$

$$a_n^{(n)} = (\tfrac{1}{2}n) a_{n-1}^{(n-1)}.$$

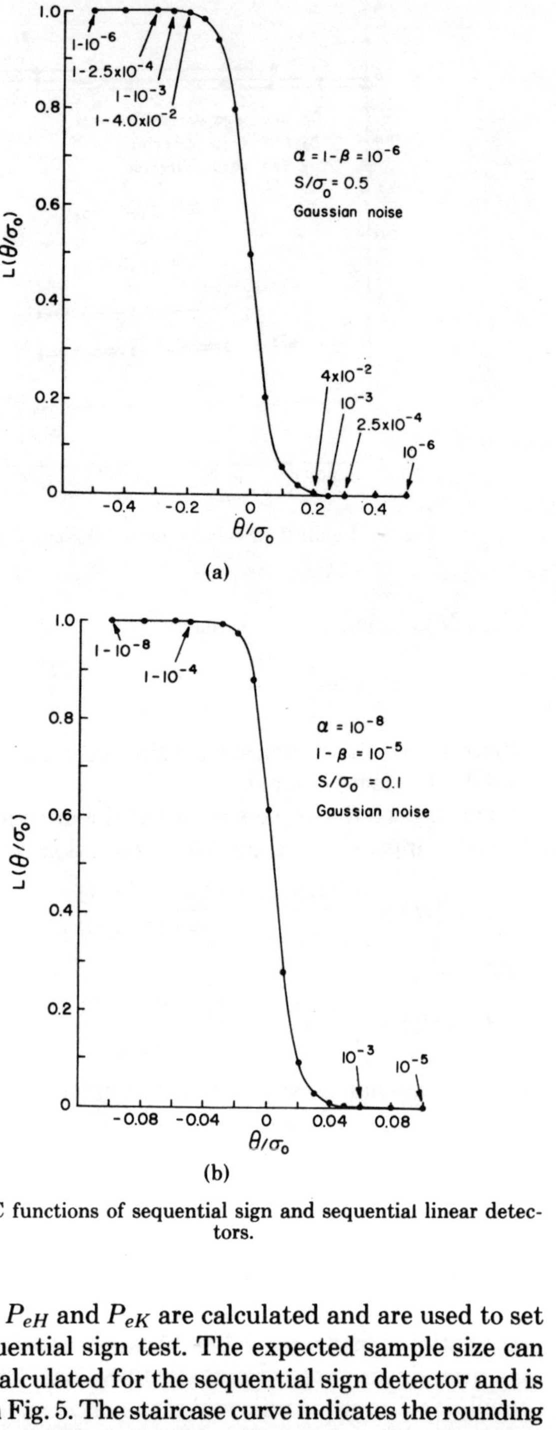

Fig. 7. OC functions of sequential sign and sequential linear detectors.

Given N, P_{eH} and P_{eK} are calculated and are used to set up a sequential sign test. The expected sample size can then be calculated for the sequential sign detector and is plotted in Fig. 5. The staircase curve indicates the rounding of the thresholds to integers. The ratio of the expected sample size to the FSS sample size is shown in Fig. 6. It converges as the errors α and $1 - \beta$ approach zero. When the signal-to-noise ratio $S\gamma/\sqrt{2}$ is smaller, this ratio is also smaller; hence there is more saving in the sample size. This saving is more than for the case of Gaussian noise, which is to be expected since the relative efficiency of an FSS sign detector with respect to an FSS linear detector is higher in Laplace noise than in Gaussian noise [6]. The sequential sign detector is expected to outperform the FSS linear detector also in other symmetric noise densities.

When the observations are samples of i.i.d. random variables with density $f(x - \theta)$, $\theta \neq S$ or $-S$, the perfor-

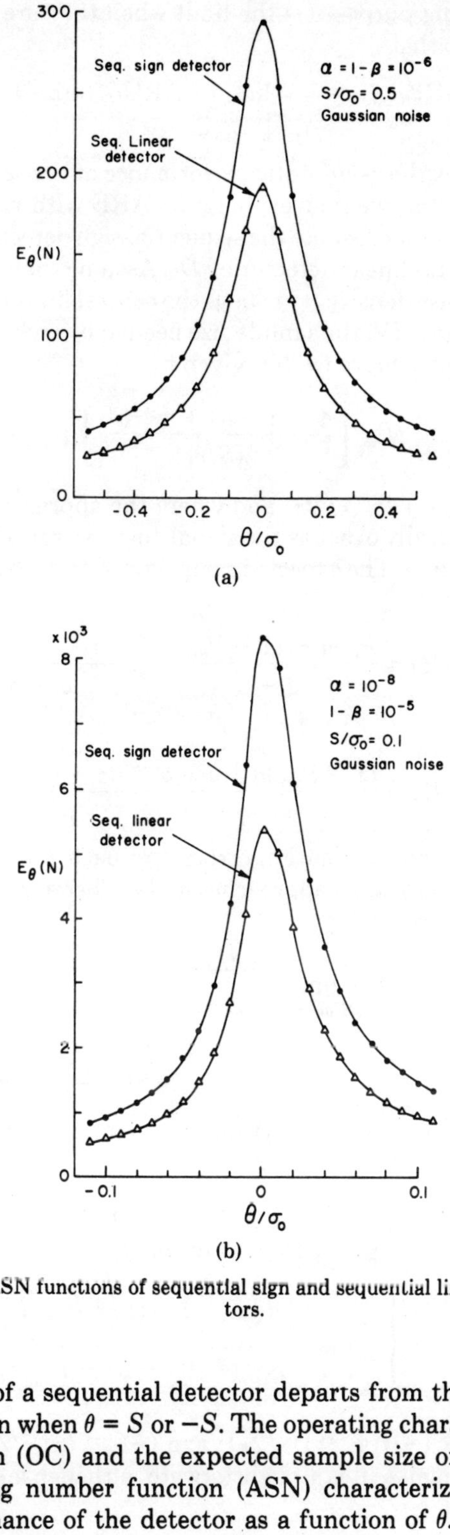

Fig. 8. ASN functions of sequential sign and sequential linear detectors.

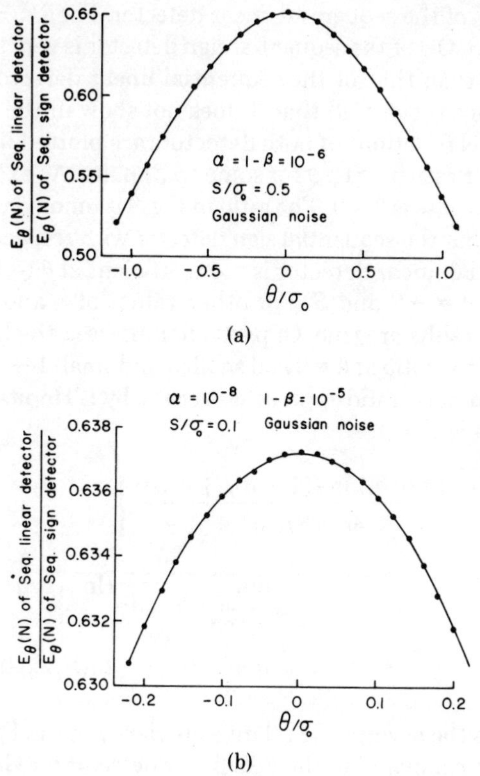

Fig. 9. Ratio of ASN functions of sequential sign and sequential linear detectors.

mance of a sequential detector departs from that of the situation when $\theta = S$ or $-S$. The operating characteristic function (OC) and the expected sample size or average sampling number function (ASN) characterize [1] the performance of the detector as a function of θ. The OC function $L(u)$ of the sequential sign detector is approximated by [1]

$$L(u) \cong \begin{cases} \dfrac{\ln(\beta/\alpha)}{\ln(\beta/\alpha) - \ln((1-\beta)/(1-\alpha))}, & u = 0.5 \\ \dfrac{(\beta/\alpha)^h - 1}{(\beta/\alpha)^h - ((1-\beta)/(1-\alpha))^h}, & u \neq 0.5, \end{cases}$$

where u denotes the probability of a positive observation when the signal is θ, namely $u = \int_0^\infty f(x-\theta)\,dx$, and h is determined from

$$u = \frac{1 - ((1-p)/p)^h}{(p/(1-p))^h - ((1-p)/p)^h}$$

and the ASN function is given by

$$E_u(N) = \begin{cases} \dfrac{(1 - L(u))a - L(u)b}{2u - 1}, & u \neq 0.5 \\ ab, & u = 0.5. \end{cases}$$

When $f(x)$ is a Gaussian density with variance σ_0, the OC and ASN functions for the sequential linear detector are [1]

$$L(\theta) \cong \begin{cases} \dfrac{\ln(\beta/\alpha)}{\ln(\beta/\alpha) - \ln((1-\beta)/(1-\alpha))}, & \theta = 0 \\ \dfrac{(\beta/\alpha)^h - 1}{(\beta/\alpha)^h - ((1-\beta)/(1-\alpha))^h}, & \theta \neq 0, \end{cases}$$

where $h = -\theta/S$, and

$$E_\theta(N) \cong \begin{cases} \dfrac{-\ln(\beta/\alpha)\ln((1-\beta)/(1-\alpha))}{4(S/\sigma_0)^2}, & \theta = 0 \\ \dfrac{(1 - L(\theta))\ln(\beta/\alpha) + L(\theta)\ln((1-\beta)/(1-\alpha))}{2(S/\sigma_0)(\theta/\sigma_0)}, & \theta \neq 0. \end{cases}$$

We plot in Fig. 7 the OC functions of the sequential sign detector and sequential linear detector. They have the same values at $\theta = S$, 0, and $-S$. For $-S > \theta > 0$ and $\theta > S$, OC of the sequential sign detector is very slightly larger than OC of the sequential linear detector. For $\theta < -S$ and $0 < \theta < S$, OC of the sequential sign detector is very slightly smaller than that of the sequential linear detector. The difference is so small that it does not show in the figure. The ASN functions of both detectors are plotted in Fig. 8 and their ratio in Fig. 9 for some (α,β) pairs. As expected, the worst case is $\theta = 0$. The ratio in Fig. 9 is unimodal which means that the sequential sign detector with respect to the sequential linear detector is more efficient at $\theta \in (-S,S)$ than at $\theta = -S$ and S. For other values of α and $1-\beta$, similar results are true. Of particular interest, the limiting value of the ratio at $\theta = 0$ and small α and small $1-\beta$ as the signal-to-noise ratio approaches zero is, by L'Hopital's rule and $p = \Phi(S/\sigma_0)$,

$$\lim_{S/\sigma_0 \to 0} \left[\frac{-\ln(\beta/\alpha)\ln((1-\beta)/(1-\alpha))}{4ab(S/\sigma_0)^2} \right] \cong \frac{\lim_{S/\sigma_0\to 0} \frac{\partial^2}{\partial(S/\sigma_0)^2}(\ln^2(p/q))}{\lim_{S/\sigma_0\to 0} \frac{\partial^2}{\partial(S/\sigma_0)^2}(4(S/\sigma_0)^2)} = 2/\pi$$

which is the asymptotic relative efficiency of the FSS sign detector compared to the FSS linear detector for Gaussian noise.

V. Relative Efficiency

The relative efficiency of a sequential detector with respect to another sequential detector or an FSS detector may be defined as [12], [13]

$$RE_{D_3,D_1} = \frac{E_{D_1}(N)}{E_{D_3}(N)}$$

or

$$RE_{D_2,D_1} = \frac{E_{D_1}(N)}{N_{D_2}},$$

where $E_{D_1}(N)$ and $E_{D_3}(N)$ are expected sample sizes of sequential detectors D_1 and D_3, and where N_{D_2} is the sample size of an FSS detector, each achieving the given (α,β). Thus the relative efficiency depends on α and β, and, in most cases, it also depends on the signal-to-noise ratio. Letting the signal-to-noise ratio approach zero, we obtain the asymptotic value of the relative efficiency ARE(α,β) as

$$\text{ARE}_{D_3,D_1}(\alpha,\beta) = \lim_{\text{SNR}\to 0} \frac{E_{D_1}(N)}{E_{D_3}(N)}$$

or

$$\text{ARE}_{D_2,D_1}(\alpha,\beta) = \lim_{\text{SNR}\to 0} \frac{E_{D_1}(N)}{N_{D_2}},$$

which still depends on α and β. We can further let the errors α and $1-\beta$ approach zero to obtain the ARE of singular detectors. Unfortunately, this ARE is not unique since it depends on the functional relationship of α and $1-\beta$ as they approach zero. A reasonable definition for engineering purposes is the limit when they are linearly related so that

$$\text{ARE}_{D_2,D_1} = \lim_{\substack{\alpha = c^*(1-\beta)\to 0\\ c^*=\text{constant}>0}} \text{ARE}_{D_2,D_1}(\alpha,\beta).$$

To show the asymptotic performance of the sequential sign detector, we now evaluate its ARE with respect to other detectors. Denote the sequential sign detector by D_1 and the FSS linear detector by D_2. Assume that the symmetric noise density $f(x)$ obeys the central limit theorem. As in Section IV, the sample size needed by an FSS linear detector to achieve (α,β) is given by

$$N \cong \left[\frac{\Phi^{-1}(1-\alpha) + \Phi^{-1}(\beta)}{2(S/\sigma_0)} \right]^2, \tag{7}$$

where $\sigma_0^2 = \int x^2 f(x)\, dx$, and where the approximation is asymptotically exact as the signal-to-noise ratio S/σ_0 approaches zero. The expected sample sizes, from Section III, for D_1 are

$$E(N|H) = \frac{P_{eH}a - (1-P_{eH})b}{q-p} \cong \frac{\alpha a - (1-\alpha)b}{q-p}$$

and

$$E(N|K) = \frac{(1-P_{eK})a - P_{eK}b}{p-q} \cong \frac{\beta a - (1-\beta)b}{p-q},$$

where the approximation errors approach zero as α approaches zero and β approaches unity. Therefore we have, given that H is true,

$$\text{ARE}_{D_2,D_1}(\alpha,\beta) \cong \lim_{S/\sigma_0\to 0} 4\frac{(S/\sigma_0)^2}{q-p} \cdot \frac{\alpha a - (1-\alpha)b}{(\Phi^{-1}(1-\alpha)+\Phi^{-1}(\beta))^2}.$$

Now, $a \cong \ln(\beta/\alpha)/\ln(p/q)$ and $b \cong \ln((1-\alpha)/(1-\beta))/\ln(p/q)$ for small S/σ_0; we obtain

$$\text{ARE}_{D_2,D_1} = \left\{ 4 \lim_{S/\sigma_0\to 0} \frac{-(S/\sigma_0)^2}{(p-q)\ln(p/q)} \right\} \left\{ \lim_{\substack{\alpha\to 0\\ \alpha=c^*(1-\beta)}} \frac{\alpha \ln\left(\frac{\beta}{\alpha}\right) - (1-\alpha)\ln\left(\frac{1-\alpha}{1-\beta}\right)}{(\Phi^{-1}(1-\alpha)+\Phi^{-1}(\beta))^2} \right\}.$$

With $\alpha \cong (-1/(\Phi^{-1}(\alpha)\sqrt{2\pi}))\exp(-(\Phi^{-1}(\alpha))^2/2)$ [10, p. 175] for small α, it is straightforward, although lengthy, to show that

$$\lim_{\alpha=c^*(1-\beta)\to 0} \frac{\alpha \ln\left(\frac{\beta}{\alpha}\right) - (1-\alpha)\ln\left(\frac{1-\alpha}{1-\beta}\right)}{(\Phi^{-1}(1-\alpha)+\Phi^{-1}(\beta))^2} = -\frac{1}{8}$$

so that we have

$$\text{ARE}_{D_2,D_1} = \frac{1}{2} \lim_{S/\sigma_0\to 0} \frac{(S/\sigma_0)^2}{(p-q)\ln(p/q)} \tag{8}$$

given that H is true. The same result is obtained when K is true.

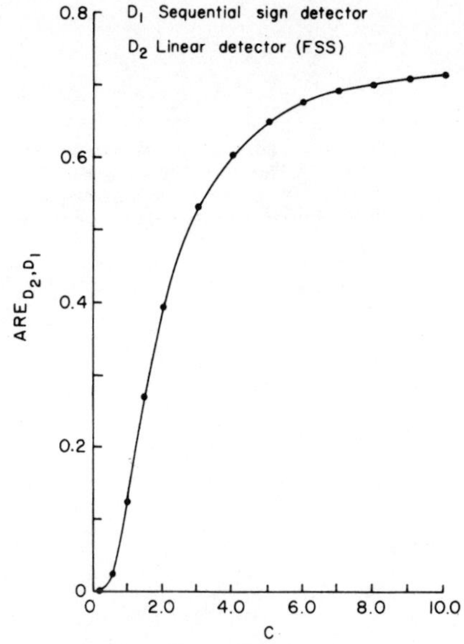

Fig. 10. ARE_{D_2,D_1} for generalized Gaussian noise.

We now find the numerical values of (8) for the generalized Gaussian noise density [9] given by

$$f(x) = \frac{c\eta(\sigma_0,c)}{2\Gamma(1/c)} \exp\{-[\eta(\sigma_0,c)|x|]^c\},$$

where $\eta(\sigma,c) = [\Gamma(3/c)/\Gamma(1/c)]^{1/2}/\sigma$ and c is a positive parameter. The values of q and p are given by

$$q = \int_{-\infty}^{0} \frac{c\eta(\sigma_0,c)}{2\Gamma(1/c)} e^{-(\eta(\sigma_0,c)|x-S|)^c}\, dx$$

$$= \frac{1}{2} - \int_0^S \frac{c\eta(\sigma_0,c)}{2\Gamma(1/c)} e^{-\eta^c(\sigma_0,c)|x-S|^c}\, dx$$

and

$$p = \frac{1}{2} + \int_0^S \frac{c\eta(\sigma_0,c)}{2\Gamma(1/c)} e^{-\eta^c(\sigma_0,c)|x-S|^c}\, dx.$$

Since $\ln((\frac{1}{2}+\Delta)/(\frac{1}{2}-\Delta)) \approx \ln(1+4\Delta) \approx 4\Delta$ for small Δ, we then have, for small S,

$$\ln(p/q) = 4\frac{c\eta(\sigma_0,c)}{2\Gamma(1/c)} \int_0^S e^{-\eta^c(\sigma_0,c)|x-S|^c}\, dx$$

$$\cong 2\frac{c}{\Gamma(1/c)}\eta(\sigma_0,c)S$$

$$= 2c(S/\sigma_0)\Gamma^{1/2}(3/c)/\Gamma^{3/2}(1/c)$$

and

$$p - q \cong c(S/\sigma_0)\Gamma^{1/2}(3/c)/\Gamma^{3/2}(1/c)$$

so that (8) becomes

$$ARE_{D_2,D_1} = \frac{1}{4c^2}\Gamma^3(1/c)/\Gamma(3/c).$$

This last expression is plotted in Fig. 10. In particular, when $c = 1$, which is the Laplace noise case, $ARE_{D_2,D_1} = \frac{1}{8}$ and when $c = 2$, which is the Gaussian noise case, $ARE_{D_2,D_1} = \pi/8$. As c becomes large, ARE_{D_2,D_1} approaches $\frac{3}{4}$.

The ARE of the sequential sign detector with respect to detector D_3 can be evaluated similarly, provided that the test statistic of D_3 has finite first and second moments. As an example, take D_3 as the FSS amplifier-limiter which is the Neyman–Pearson optimal detector in Laplace noise. The test for D_3 is

$$\sum_{i=1}^{N} g(x_i) \underset{H}{\overset{K}{\gtrless}} T,$$

where

$$g(x) = \begin{cases} 1, & x \ge S \\ -1, & x \le -S \\ x, & \text{otherwise.} \end{cases}$$

The first moment and the variance are

$$m_1 = E(g(x)|K) = \int_{-\infty}^{\infty} g(x)(\gamma/2)e^{-\gamma|x-S|}\, dx = -E(g(x)|H)$$

and

$$\sigma^2 = \int_{-\infty}^{\infty} g^2(x)(\gamma/2)e^{-\gamma|x-S|}\, dx - m_1^2.$$

With m_1/σ in place of S/σ_0 in (7), we have

$$ARE_{D_3,D_1} = \frac{1}{2}\lim_{S\gamma\to 0} \frac{(m_1/\sigma)^2}{(p-q)\ln(p/q)}, \tag{9}$$

where $p = \frac{1}{2} + \int_0^S(\gamma/2)e^{-\gamma x}\, dx \cong \frac{1}{2} + \gamma S/2$ for small S and $q \cong \frac{1}{2} - \gamma S/2$. Evaluating m_1 and σ and substituting into (9), we have $ARE_{D_3,D_1} = \frac{1}{4}$.

VI. Truncation of Test

Although the expected sample size of the sequential sign detector is considerably smaller than the sample size of FSS detectors, the sample size of a particular experiment in a sequential test can be undesirably large. One way to avoid this situation is to truncate the test at some prespecified number, say N_T. The value of N_T should be chosen such that $P(N \ge N_T)$ is small enough to keep P_{eH} and P_{eK} virtually unchanged.

We now examine the possibility of truncating the sequential sign test. We have given in Section III the expressions for probabilities of terminating the test at the kth sample. Some examples of these probabilities are plotted in Fig. 11. Fortunately, the tail of the distribution becomes relatively small for reasonable values of k. To find the sum of the probabilities at the tail, assume that the truncation is at N_T. We then have

$$P(N \ge N_T|H) = P(N \ge N_T \text{ and terminate at } a|H) + P(N \ge N_T \text{ and terminate at } -b|H)$$

with

$$P(N \ge N_T \text{ and terminate at } a|H) = Q(p,q,a)$$

and

$$P(N \ge N_T \text{ and terminate at } -b|H) = Q(q,p,b),$$

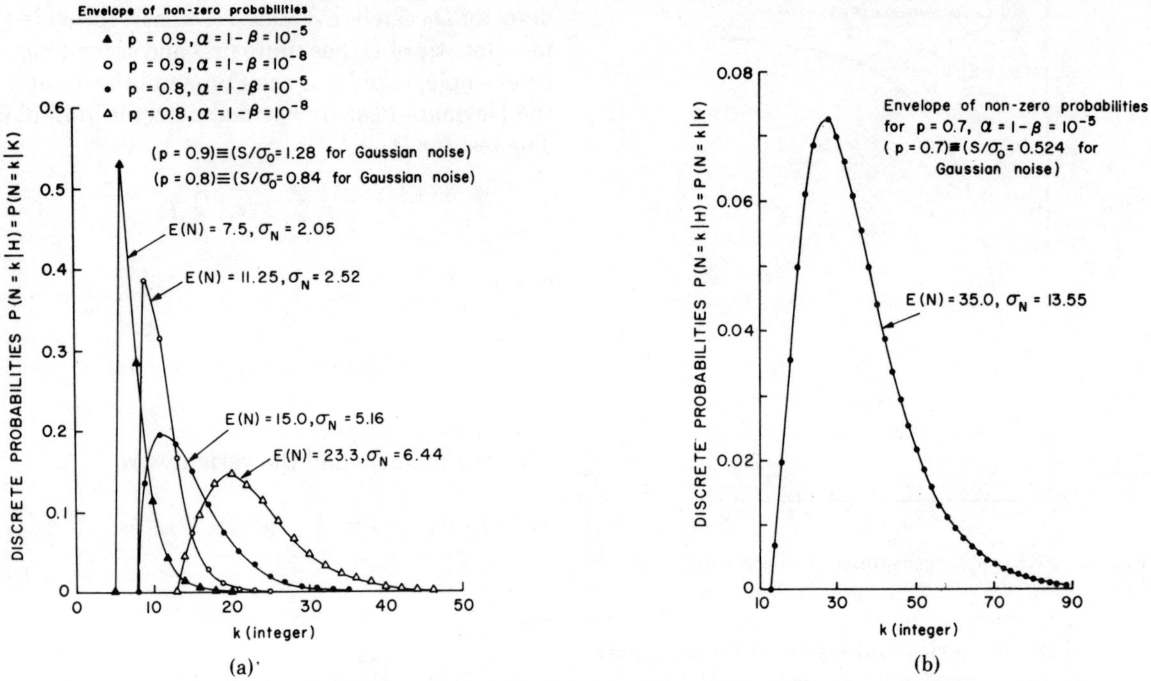

Fig. 11. Probability of terminating at kth sample.

TABLE III
WORST CASE TRUNCATION FOR $\alpha = 10^{-8}$ AND $1 - \beta = 10^{-5}$
$E_{\theta=0}(N)$ AND $\sigma_{N|\theta=0}$ ARE GIVEN IN TABLE II

Designed p \ $N_T = E_{\theta=0}(N) +$	$P(N \geq N_T \mid \theta=0)$: $\sigma_{N\|\theta=0}$	$2\sigma_{N\|\theta=0}$	$3\sigma_{N\|\theta=0}$	$4\sigma_{N\|\theta=0}$	$5\sigma_{N\|\theta=0}$	$6\sigma_{N\|\theta=0}$
0.93	.1376	.0476	.0175	.0060	.0022	.00077
0.84	.1310	.0474	.0167	.0060	.0021	.00077
0.69	.1320	.0473	.0168	.0060	.0021	.00076
0.54	.1321	.0471	.0168	.0060	.0021	.00076

where

$$Q(u,v,c) = \begin{cases} \dfrac{2^{N_T+1}}{a+b}\,(u^{N_T-c}v^{N_T+c})^{1/2} \displaystyle\sum_{i=1}^{M} \sin\left(\frac{\pi i}{a+b}\right) \sin\left(\frac{\pi c i}{a+b}\right) \frac{\cos^{N_T-1}\left(\dfrac{\pi i}{a+b}\right)}{1 - 4uv\cos^2\left(\dfrac{\pi i}{a+b}\right)}, & N_T - c \text{ is even} \\[2ex] \dfrac{2^{N_T+2}}{a+b}\,(u^{N_T-c+1}v^{N_T+c+1})^{1/2} \displaystyle\sum_{i=1}^{M} \sin\left(\frac{\pi i}{a+b}\right) \sin\left(\frac{\pi c i}{a+b}\right) \frac{\cos^{N_T}\left(\dfrac{\pi i}{a+b}\right)}{1 - 4uv\cos^2\left(\dfrac{\pi i}{a+b}\right)}, & N_T - c \text{ is odd.} \end{cases}$$

Similar expressions hold for the alternative K. Some numerical results for $P(N \geq N_T|H)$ and $P(N \geq N_T|K)$ are given in Table II. If N_T is chosen as $E(N|H) + 4\sigma_{N|H}$ (or $E(N|K) + 4\sigma_{N|K}$) and larger, then the probability under H (or K) of terminating at N_T is small so that the truncation has little effect on P_{eH} and P_{eK}.

The truncation may also be considered on the basis of the worst possible case, i.e., when $u = 0.5$ or $\theta = 0$. The probability of terminating the test at the truncation point for this case is

$$P(N \geq N_T|\theta = 0) = Q(0.5,0.5,a) + Q(0.5,0.5,b).$$

Some numerical values are shown in Table III which suggests that truncation of sampling at $N_T = E_{\theta=0}(N) + 4\sigma_{N|\theta=0}$ or larger results in a small probability of terminating the test at sample sizes N_T or larger. With N_T so chosen, the performance of the detector is practically unaffected.

VII. Conclusion

We have set up a sequential sign test for the detection of antipodal signals imbedded in noise with a symmetric density. The only knowledge needed to set up the test is the probability of a positive observation under either the hypothesis or the alternative. The resulting test gives a considerable saving in the number of samples needed for termination as compared to the FSS Linear detector. The asymptotic relative efficiency was investigated. We observed that, even with the simple sign detector, a sequential test can show a considerable saving in the number of samples over the optimum FSS test. Finally, to avoid excessively large sampling number in a particular experiment, the possibility of truncating the test was considered and shown to be practical.

Appendix A

Proof of (3) and (4)

Since $r^a(r^b-1)/(r^{a+b}-1) \geq \beta > 0$ and $(r^{a+b}-1)/(r^b-1) \geq 1/\alpha > 0$, we have $r^a \geq \beta/\alpha$ which is condition (3). Also $(r^b-1) \leq \alpha(r^{a+b}-1)$ implies $r^a(\alpha r^b) \geq (r^b-1+\alpha) > 0$ and $(r^{a+b}-r^a) \geq \beta(r^{a+b}-1)$ implies $r^a(1+\beta r^b - r^b) \leq \beta$, we then have $(1+\beta r^b - r^b)/(\alpha r^b) \leq \beta/(r^b-1+\alpha)$ which implies $(r^b-1)(\beta r^b - r^b + 1 - \alpha) \leq 0$. Since $(r^b-1) > 0$, this reduces to $(\beta r^b - r^b + 1 - \alpha) \leq 0$ and condition (4) follows.

Appendix B

Proof of (5) and (6)

To show (5), we note that $r^b \leq (1-\alpha)/(1-k\beta)$ implies $(r^b-1)/(k\beta r^b - \alpha) \leq 1$. Now $P_{eH} = (r^b-1)/(r^{a+b}-1) = (r^b-1)/(r^b(k\beta/\alpha)-1) = \alpha(r^b-1)/(r^b k\beta - \alpha) \leq \alpha$. Also $r^b \geq (1-\alpha/k)/(1-\beta)$ implies $(r^b-1)/(\beta r^b - \alpha/k) \geq 1$, so that $1-P_{eK} = r^a(r^b-1)/(r^{a+b}-1) = (k\beta/\alpha)(r^b-1)/(r^b k\beta/\alpha - 1) = \beta(r^b-1)/(r^b\beta - \alpha/k) \geq \beta$. For inequality (6), we have $P_{eH} = (r^b-1)/(r^{a+b}-1) = \alpha(r^b-1)/(r^b k\beta - \alpha) \leq \alpha(r^b-1)/(r^b-\alpha) < \alpha(r^b-1)/(r^b-1) = \alpha$, and $1-P_{eK} \geq \beta$ can be shown in the same way as for (5).

References

[1] A. Wald, *Sequential Analysis*. New York: Wiley, 1947.

[2] A. Wald and J. Wolfowitz, "Optimum character of the sequential probability ratio test," *Ann. Math. Stat.*, vol. 19, pp. 326–339, 1948.

[3] A. J. Viterbi, "The effect of sequential decision feedback on communication over the Gaussian channel," *Information and Control*, vol. 8, pp. 80–92, Feb. 1965.

[4] J. J. Bussgang, "Sequential methods in radar detection," *Proc. IEEE*, vol. 58, pp. 731–743, May 1970.

[5] S. A. Kassam and J. B. Thomas, "Asymptotically robust detection of a known signal in contaminated non-Gaussian noise," *IEEE Trans. Inform. Theory*, vol. IT-22, pp. 22–26, Jan. 1976.

[6] J. B. Thomas, "Nonparametric detection," *Proc. IEEE*, vol. 58, pp. 623–631, May 1970.

[7] S. W. Abel and J. R. Betten, "Application of the nonparametric sign test to sequential detection," *IEEE Trans. Inform. Theory*, vol. IT-16, p. 107, Jan. 1970.

[8] R. A. Worley, "Multilevel quantization of data for sequential testing," Navy Electronics Lab. Rept. 1218, Apr. 1964.

[9] J. H. Miller and J. B. Thomas, "Detectors for discrete-time signals in non-Gaussian noise," *IEEE Trans. Inform. Theory*, vol. IT-18, pp. 241–250, March 1972.

[10] W. Feller, *An Introduction to Probability Theory and Its Applications Vol. I*, 3rd Edition. New York: Wiley, 1968.

[11] J. Miller and J. B. Thomas, "Small sample performance of detectors in non-Gaussian noise," *Proceeding of the Sixth Asilomar Conference on Circuits and Systems*, pp. 438–443, 1972.

[12] E. Paulson, "A note on the efficiency of the Wald sequential test," *Ann. Math. Stat.*, vol. 18, pp. 447–450, 1947.

[13] R. Bechhofer "A note on the limiting relative efficiency of the Wald sequential probability ratio test," *J. Amer. Stat. Assoc.*, vol. 55, pp. 660–663, Dec. 1960.

Part III

NONPARAMETRIC DETECTION BASED ON RANKS

Editors' Comments on Papers 9 Through 15

Although tests based on signs only of observations are easy to implement and can result in nonparametric detectors, it is clear that in most cases hard-limiting of the observations results in a considerable loss of information. This is reflected in the generally poor performance of detectors based on signs relative to detectors that are optimum for any specific assumptions on the underlying statistical distributions.

By using further information in the form of the relative ranks of the received data, it is generally possible to obtain much more efficient tests. The reduction of data to relative rank information still implies an irreversible transformation, which in many specific situations of interest gives rise to rank data whose distribution under the null hypothesis is independent of the specific underlying probability distribution. As an example, consider the detection of a constant signal θ in additive noise with a density function $f(x)$ symmetric about the origin. Assume that n independent observations $X_1, X_2, \ldots, X_n$ are available. A polarity-based detector operates only on the vector of signs $S = [sgn\,(X_1), sgn\,(X_2), \ldots, sgn\,(X_n)]$. Suppose that we now also use the vector R^+ of ranks, $R^+ = (R_1^+, R_2^+, \ldots, R_n^+)$, R_i^+ being the rank of $|X_i|$ from among the set $\{|X_1|, |X_2|, \ldots, |X_n|\}$. Now it can be shown that the vectors S and R^+ are independent under the null hypothesis (noise only) (Hajek and Sidak).[1] Under the null hypothesis it is also true that any combination of polarities is equally likely for S and any permutation of the integers 1 through n equally likely for R^+. Thus a nonparametric test may be based on the use of both S and R^+; a simple test for this example is the Wilcoxon test.

The papers in this part are concerned with the analysis of some useful types of rank-based detectors. Capon's paper (Paper 9) represents one of the earliest considerations of the application of nonparametric statistical inference methods in signal detection. He considers the detection of a constant signal in additive noise, under the assumption that a reference noise sample is available. For this two-sample detection problem, the Mann-Whitney nonparametric rank test is proposed and analyzed on the basis of ARE. Capon's paper also contains a discussion of the efficacy for one- and two-sample detectors and the relationship of efficacy to ARE.

A considerable amount of statistical theory of rank tests was already available in 1959, when Paper 9 was presented. Most of the papers in this part are concerned with an application and adaptation as well as extension of these results for signal detection problems.

Paper 10 by Daly and Rushforth is also often referenced by later authors. It considers the detection of signals that are not constant but are either completely known or have a random phase. The observed sample is in the form of $N = nm$ independent observations under either noise-only or signal-plus-noise conditions, there being n repeated observations of each of m components of the possibly present known signal vector. The rank test considered is a locally optimum rank test for Gaussian alternatives and is an extension for non-constant signals of the c_1 test (Fraser)[2] of earlier statistical theory. The test is based on the relative ranks of all of the N observations and depends on the assumption that the signal is not constant. It is compared to the optimum detector for Gaussian noise (in both the coherent and incoherent cases) using the ARE criterion. An interesting conclusion is that, for this nonparametric rank test compared to the optimum test for Gaussian noise, the ARE is lower bounded by unity, making it asymptotically superior to the optimum correlator.

Hansen in Papers 11 and 12 looks at the performance of some one-sample nonparametric rank tests and treats specific applications in constant false-alarm rate (CFAR) radar. In Paper 11 tests are considered for the presence of a trend in data due to the presence of a signal in additive noise. Some examples of the so-called simple linear rank statistics, specifically the simple Spearman-rho test and the normal-scores test, are considered, and finite-sample, finite-signal detection performance results are obtained for Gaussian noise through extensive Monte Carlo simulations. The cases analyzed include the situation where successive inputs are phase incoherent so that the data are the output of square-law envelope detectors. The Kendall-tau statistic is also similarly analyzed, and a specific radar detector based on the use of two Kendall-tau tests is proposed and analyzed for performance. This scheme utilizes the fact that successive return pulses arriving at the radar detector have a characteristic trend imposed by the beam pattern of the scanning antenna. In Paper 12 Hansen analyzed another nonparametric scheme for radar applications. The Wilcoxon detector modified to be applicable for the case of a semicoherent pulse train in noise (Helstrom[3] and Paper 1) is studied through Monte Carlo simulations.

In Paper 13, Dillard and Rickard propose a different rank-based detector for nonparametric (CFAR) radar. In addition to a random overall phase for the train of radar pulse returns, a Doppler frequency offset is admitted, leading to an unknown phase parameter in the in-phase and quadrature samples. Their interesting scheme is based on the availability of a reference-noise sample and also yields an estimate of the Doppler frequency if signal is detected, the test statistic having been chosen with this specific practical requirement in mind.

Woinsky in Paper 14 considers the use of a rank test for detection of signals based on spectral characteristics of the observed time series. The data are reduced to blocks of the energy distribution over time-frequency cells, such that for each block a separate range of frequencies is covered, over which the noise has a constant spectral density. A signal appears as a larger measured output at a specific frequency (or frequencies), and the detection procedure is based on the multiple use of a two-sample rank test applied to each frequency column, with the rest of the data in the block pooled to form the second sample. Simplified rank procedures and computer simulation performance results (in addition to AREs) are also discussed.

The reason for the widespread use of the ARE in evaluating nonparametric tests, and specifically rank tests, is that it has been difficult to obtain an analytical measure of relative performance that is applicable in the more practical case of finite signals and finite sample sizes. Recently Papantoni-Kazakos, in work reported here as Paper 15, has obtained some useful analytical performance comparisons for one-sample rank tests for shift alternatives, under the assumption of Gaussian noise. In Paper 15 it is shown that for small sample sizes and for signal parameter (location shift) tending to infinity, the relative efficiency of the optimum rank test compared to the unconstrained optimum test is lower bounded by 3/4 for Gaussian statistics. Similarly, for the Wilcoxon rank test this lower bound is $1/\sqrt{2}$. These results are the first theoretical results on small-sample efficiencies of rank tests.

REFERENCES

1. J. Hajek and Z. Sidak. *Theory of Rank Tests*. Academic Press, 1967, p. 40.
2. D. A. S. Fraser. *Nonparametric Methods in Statistics*. John Wiley, 1957, p. 195.
3. C. W. Helstrom. *Statistical Theory of Signal Detection*. Pergamon Press, 1968, p. 325.

9

Reprinted with permission from pages 92–103 of the *1959 IRE WESCON Conv. Rec.*, Pt. 4, 201pp.

A NONPARAMETRIC TECHNIQUE FOR THE DETECTION OF A CONSTANT SIGNAL IN ADDITIVE NOISE*

Jack Capon**

Sperry Gyroscope Company

Abstract

In this investigation we consider several methods for the detection of a constant signal in additive noise. One such optimum detection system, known as the Neyman-Pearson observer, has been studied in the past. This detector bases its decisions on the likelihood ratio, a quantity which depends on the functional form of the probability density function (pdf) of the noise. Since the optimum detector is designed for a specific pdf, and since this pdf can be completely specified by a finite number of parameters, it has been termed a parametric detector.

Ordinarily the pdf of the noise is assumed to be normal, and the optimum detector obtained is a summing device. However, the summing device is optimum only for the normal pdf, and may perform badly for non-normal pdf's.

If we assume that a pure noise waveform is available at the detector, then this waveform can be compared with the observed waveform to determine whether there is a signal present. This comparison is performed by means of the Mann-Whitney test. The detection scheme is called nonparametric, since it remains valid for a large class of pdf's, which cannot be completely specified by a finite number of parameters.

A concept known as asymptotic relative efficiency is used to compare the Mann-Whitney detector to the summing device. It is shown that the Mann-Whitney detector can be much more efficient than the summing device for non-normal pdf's, and that the efficiency is never less than 108/125. Since the use of the Mann-Whitney detector does not entail a serious loss of efficiency, it is proposed as an alternative to the summing device for detecting constant signals in additive noise, when the pdf of the noise is unknown.

Introduction

The reception of signals in the presence of noise is a problem of considerable importance in communication theory. The two major subdivisions of this problem are known as detection and extraction. The detection problem is concerned with the design of systems which determine only the presence or absence of a signal in noise. Extraction deals with the actual reconstruction of the signal from the noise background. This investigation will treat only the detection problem.

The detection problem is encountered quite frequently in practice. An example of considerable importance is afforded by the radar detection problem, in which it is desired to determine the presence of a target by detecting the presence of a radar return signal in noise. Another example is obtained from a particular communication link known as scatter propagation. In this case the detection of a signal in noise is an indication of whether or not the signal has been successfully transmitted over the horizon.

Detectors which examine a sample time function of a random process with the object of supplying a yes or no answer to the question of whether a signal is present in noise have been studied in the past. Early approaches to the detection problem were based on the signal-to-noise ratio criterion due to North[1], and discussed by Zadeh and Ragazzini[2]. In this analysis it is assumed that one wishes to maximize the signal-to-noise ratio at the input of the detector. This may not be a valid assumption if detection is the main objective. However, this approach does possess the advantage of requiring relatively little information concerning the underlying distributions of the noise, and the mixture of signal and noise.

In some detection problems the information required by the detectors formulated in the past may be available. However, in other situations there may be much less information. In this investigation we consider the problem of detecting a constant signal in additive noise, when all that is known about the distribution of the noise is that it is continuous. Hence, it is known that the underlying distributions are continuous, and differ in location, but nothing else. In such cases the detectors studied in the past may be inappropriate.

Thus there is a need for a theory of detection systems which require much less a priori

---

* This work was supported by Air Force Cambridge Research Center, under Contract No. AF 19(604)-4140.

** Formerly with the Department of Electrical Engineering, Columbia University.

information than the detectors studied previously. It is the object of this investigation to contribute to the development of such a theory based on the methods of nonparametric statistical inference. These nonparametric detectors may be more appropriate in situations in which little a priori information is available, than the more conventional detectors in use at the present time.

The Neyman-Pearson Observer

The amplitude of the constant signal which we desire to detect will be denoted by θ, and without loss of generality we assume that θ is positive. The noise voltage will be denoted by $N(t)$, and is considered to be a sample function of a continuous parameter stochastic process $\{N(t)\}$ [3]. We let $Z(t)$ denote the observed waveform and we assume that $Z(t)$ is a sample function of the continuous parameter stochastic process $\{Z(t)\}$. We have

$$Z(t) = N(t) + \theta,$$

when signal is present; and

$$Z(t) = N(t),$$

when signal is absent.

A detector is defined as a device whose input is $Z(t)$, and whose output is a yes-no answer, indicating the detector's guess as to whether or not the signal is present (Fig. 1).

The errors committed by the detector can be of the following two exhaustive and mutually exclusive types: (a) the detector says there is a signal present, when in reality the signal is absent; the probability of such an error is denoted by α, and is known as the false alarm probability; (b) the detector says there is no signal present, when in reality there is a signal present; the probability of this type of error is denoted by β, and is known as the false dismissal probability.

The information upon which the detector must base its decision is contained in the observed waveform $Z(t)$. We shall assume that the detector samples $Z(t)$ at times t_i, $i=1,\ldots.,n$, to obtain the samples $Z_1,\ldots..,Z_n$ $(Z_i = Z(t_i)$, $i=1, \ldots.., n)$, and that the detector's decision is based on these n samples.

In general, $Z_1,\ldots.,Z_n$ is a set of random variables with a certain joint probability distribution function which depends on whether or not the signal is present. We shall let $P(z_1,\ldots..,z_n, \theta = 0)$ be the joint probability distribution function of $Z_1,\ldots.,Z_n$ when signal is absent and let $P(z_1,\ldots.,z_n, \theta = \theta_1)$ be the joint probability distribution function of $Z_1,\ldots.., Z_n$ when signal is present, where θ_1 is the particular value of the amplitude of the constant signal when signal is present; i.e.,

> $P(z_1,\ldots.,z_n, \theta=0)$ is the joint probability that $Z_i \leq z_i$, $i=1,\ldots., n$, when signal is absent,

and

> $P(z_1,\ldots., z_n, \theta=\theta_1)$ is the joint probability that $Z_i \leq z_i$, $i=1, \ldots., n$, when signal is present.

The detection decision procedure is equivalent to testing the null hypothesis H_o that the joint probability distribution function of $Z_1,\ldots.,Z_n$ is $P(z_1,\ldots.., z_n, \theta=0)$ (signal is absent), against the alternative hypothesis H_1 that the joint probability distribution function of $Z_1,\ldots.,Z_n$ is $P(z_1,\ldots.,z_n, \theta=\theta_1)$ (signal is present). The Neyman-Pearson observer is defined as that detector, which for fixed α and n, formulates its decisions concerning the validity of either H_o (signal is absent) or H_1 (signal is present) in such a way that β is minimized. Various aspects of the Neyman-Pearson detection scheme have been considered by Reich and Swerling[4], and Middleton[5].

In order to design the Neyman-Pearson detector the information concerning the functional form of the underlying distributions is required. Usually these distributions are known to have some simple functional form, such as that of the normal distribution, and are completely specified by one, two, or at most a countable number of real parameters that serve as indices, or labels for the probability distributions. It is for this reason that the Neyman-Pearson observer has been termed a parametric detector.

At this point we might pose the problem of what happens to the performance of the Neyman-Pearson observer when the underlying distributions are not those for which the Neyman-Pearson observer is optimum. In particular, we might fix both α and β, and examine the effect on the sample size n of changing the underlying distributions.

In this investigation we shall formulate a detection criterion which is valid for a large number of pairs of underlying distributions. In fact, the number of such distributions is so large that they cannot be completely specified by a finite number of parameters. Hence, these detectors will be known as nonparametric detectors. We shall compare (in a sense yet to be defined) the nonparametric detector with the Neyman-Pearson observer, as the functional forms of the underlying distributions change.

Nonparametric Formulation of the Detection Problem

In order to state the detection criterion on which the nonparametric detectors are based we shall introduce some additional assumptions and notation. It will be assumed that $\{Z(t)\}$ is stationary, and that the data elements $Z_1, \ldots, Z_n$ are independent. Since $\{Z(t)\}$ is identical to $\{N(t)\}$ when signal is absent, it follows that $\{N(t)\}$ is also stationary. The cumulative distribution function (cdf) of any one of the data elements when signal is absent is the same as that of any other data element since $\{N(t)\}$ is stationary. This cdf is denoted by $F(z)$, and is assumed to be continuous. The corresponding probability density function (pdf) is denoted by $f(z)$ $(= \frac{d}{dz} F(z))$. Since the samples $Z_1, \ldots, Z_n$ are independent it follows that

$\prod_{i=1}^{n} F(z_i)$ is the joint probability that $Z_i \leqslant z_i$ when signal is absent, $i=1, \ldots, n$.

The presence of the constant signal in additive noise changes the cdf of each of the data elements to $F(z - \theta_1)$. Hence

$\prod_{i=1}^{n} F(z_i - \theta_1)$ is the joint probability that $Z_i \leqslant z_i$ when signal is present, $i = 1, \ldots, n$.

In this notation the parametric detection criterion becomes one of testing

H_o : cdf of Z_i is $F(z)$, $i=1, \ldots, n$, signal is absent,

against

H_1 : cdf of Z_i is $F(z - \theta_1)$ $i=1,\ldots,n$, signal is present.

Thus, the parametric detection criterion is equivalent to a determination of whether the cdf of the Z_i's is $F(z)$ (signal is absent), or $F(z - \theta_1)$ (signal is present).

We now introduce the key assumptions on which the nonparametric detection criterion is based:

(1) There is available a sample function $N'(t)$ from the stationary continuous parameter stochastic process $\{N(t)\}$,

(2) It is possible to obtain m independents samples $Z_{n+1}, \ldots, Z_{n+m}$ from the sample function $N'(t)$.

On the basis of the two samples $Z_1, \ldots, Z_n$ and $Z_{n+1}, \ldots Z_{n+m}$, a decision procedure for detecting signals can be formulated by testing

H'_o : cdf of Z_i is $F(z)$, $i=1,\ldots,n+m$, signal is absent,

against

H'_1 : cdf of Z_i is $F(z - \theta_1)$, $i=1, \ldots, n$, and the cdf of Z_{n+j} is $F(z)$, $j=1, \ldots, m$, signal is present.

In essence this detection criterion is a statement of the following obvious facts: if the signal is absent, then the cdf of the Z_i's is $F(z)$ and must be the same as the cdf of the Z_{n+j}'s, since both sets of observations were obtained from sample functions of the same continuous parameter stochastic process $\{N(t)\}$: if the signal is present, then the cdf of the Z_i's is $F(z - \theta_1)$, which is not the same as $F(z)$, the cdf of the Z_{n+j}'s.

The nonparametric statistical tests that can be used to test H'_o against H'_1 have the property that the false alarm probability is independent of the cdf $F(z)$, and depends only on m and n. However, the false dismissal probability of these tests does depend on the cdf $F(z)$ and θ_1 in some manner, in addition to depending on m and n. Usually this dependence is given in terms of some functional associated with $F(z)$ and θ_1, such as $\int F(z)dF(z - \theta_1)$; i.e., β is known if the quantity Δ is known. It is to be noted that for fixed θ_1, there are many cdf's $F(z)$, for which Δ is the same. Thus, since α is independent of the cdf $F(z)$, and β is dependent upon only Δ, it follows that for fixed θ_1, the test remains valid for all those cdf's $F(z)$ for which Δ is the same. Hence we have formulated a detection criterion for which α and β remain the same for a large (usually infinite) class of cdf's. If the quantity Δ is the only information concerning $F(z)$ and θ_1 which is available in the formulation of the detection problem, then we have at our disposal a procedure for designing a detector on the basis of only this limited amount of information. The detectors based on this detection procedure will be referred to as nonparametric detectors, since the underlying distribution cannot be completely specified by a finite number of parameters; i.e., the parameter space cannot be simply represented as a subset of a real space of a finite number of dimensions. The basic form of such a detector is shown in Fig. 2.

Asymptotic Relative Efficiency of Detectors

The nonparametric detection procedure bases its decisions on a test statistic $U_{mn}(z_1,\ldots,z_{n+m})$ a function which is defined on the observations

$Z_1, \ldots, Z_{n+m}$. Since these observations are random variables it follows that U_{mn} is also a random variable (assuming that U_{mn} is a Borel-measurable function[6] of $Z_1, \ldots, Z_{n+m}$). The introduction of U_{mn} serves the useful purpose of mapping the outcome of our measurements on $Z(t)$ and $N'(t)$ from the $(n+m)$-dimensional sample space of $Z_1, \ldots, Z_{n+m}$ to the uni-dimensional sample space of U_{mn}; i.e., the real line.

A statistical test may be thought of as equivalent to a dichotomy of the sample space of the observations. Thus the nonparametric decision procedure is tantamount to a dichotomy of the $(n+m)$-dimensional sample space into a critical region and an acceptance region. If the observed sample $Z_1, \ldots, Z_{n+m}$ lies in the acceptance region, H'_o is accepted, otherwise H'_1 is accepted.

If the function U_{mn} is to serve as a useful mapping of our measurements, it must map the critical and acceptance regions of this $(n+m)$-dimensional sample space into two non-overlapping regions on the real line. In this event the dichotomy of the $(m+n)$-dimensional sample space is reduced to the considerably simpler problem of dichotomizing the unidimensional sample space of U_{mn}. Thus, the real line is dichotomized into non-overlapping regions -- a critical region and an acceptance region. If the test statistic U_{mn}, as calculated for a particular set of observations, lies in the acceptance region H'_o (signal is absent) is accepted, otherwise H'_1 (signal is present) is accepted.

In this investigation we shall consider only the following two possibilities for the form of the critical and acceptance regions:

(P.1) The critical region is an interval of the form $(U_{\alpha,mn}, \infty)$, and the acceptance region is an interval of the form $(-\infty, U_{\alpha,mn})$:

(P.2) The critical region is an interval of the form $(-\infty, U_{\alpha,mn})$, and the acceptance region is an interval of the form $(U_{\alpha,mn}, \infty)$,

where $U_{\alpha,mn}$ is a constant chosen to make the false alarm probability of the detector equal to α. These two possibilities cover most cases of interest.

The Neyman-Pearson detector bases its decisions on a statistic $U_n(z_1, \ldots, z_n)$, a function which is defined on the observations $Z_1, \ldots, Z_n$. Hence, in a similar manner we have that the Neyman-Pearson observer accepts H_o if U_n lies in the acceptance region, and accepts H_1 otherwise. The two possibilities for the form of the critical and acceptance regions are as in (P.1) and (P.2), except that $U_{\alpha,mn}$ is replaced by $U_{\alpha,n}$.

We will restrict our level of generality by considering only the detection of weak constant signals in additive noise. The weak signal assumption implies that the amplitude of the constant signal θ is very close to zero. This is done for two important reasons. First and foremost, this assumption leads to a considerable simplification of the analytical expressions encountered. In fact, many of our computations would be virtually impossible from the analytical standpoint without this assumption. The second reason is that we are usually interested primarily in the behavior of a detector when the signal strength is small. It can be argued from the physical viewpoint that almost any detector will operate satisfactorily when the signal strength is large.

We shall now introduce a criterion for comparing a nonparametric detector with a Neyman-Pearson detector, when the signal strength is very small. This criterion is an adaptation of the concept of asymptotic relative efficiency (A.R.E.) due originally to Pitman[7]. An exposition of Pitman's work has been given by Noether[8] and Fraser[9].

In order to define the A.R.E. of detection systems it is necessary to define what is meant by a sequence of detectors. Let $\{m_i\}$, $\{n_i\}$ be two increasing sequences of integers, and let the sequence of statistics $\{U_{m_i n_i}\}$ determine a detection system for each m_i and n_i, with false alarm and false dismissal probabilities α, $\beta_{m_i n_i}$, respectively. Also let $\{\theta_i\}$ be a sequence of signal strengths, such that

$$\lim_{i \to \infty} \theta_i = 0 \qquad (1)$$

In a similar manner let $\{n_i^*\}$ be an increasing sequence of integers, and let the sequence of statistics $\{U^*_{n_i^*}\}$ determine a detector for each n_i^*, with false alarm and false dismissal probabilities α, $\beta^*_{n_i^*}$, respectively.

If at the i-th stage the $\{U_{m_i n_i}\}$-detector and the $\{U^*_{n_i^*}\}$-detector always detect the same signal, with signal strength θ_i, and if

$$\lim_{i \to \infty} \beta_{m_i n_i} = \lim_{i \to \infty} \beta^*_{n_i^*} \qquad (2)$$

then, the A.R.E. of the $\{U_{m_i n_i}\}$-detector with respect to the $\{U^*_{n_i^*}\}$-detector is defined as

$$E_{U,U^*} = \lim_{i\to\infty} \frac{n_i^*}{n_i} \qquad (3)$$

if this limit exists the same for all sequences $\{m_i\}$, $\{n_i\}$, $\{n_i^*\}$ satisfying Eq. (2).

One of the important considerations in any given detection problem is the length of time required to detect the signal with a certain accuracy α, β. In most cases the m observations on $N'(t)$ can be obtained by the nonparametric detector before the n observations are obtained from $Z(t)$, and can be stored in the nonparametric detector. In such an event, the only time consumed is that used in obtaining the n^* samples from $Z(t)$. If periodic sampling is used, then n and n^* are proportional, respectively, to the amount of time required by the nonparametric and parametric detectors to detect the same weak signal with the same error probabilities α, β. Thus, the justification for the criterion of A.R.E. given in Eq. (3) is, that for periodic sampling, it gives an indication of how much faster (or slower) the nonparametric detector is than the Neyman-Pearson detector, in detecting the same weak signal, for prescribed error probabilities, α, β.

It should be stressed that the A.R.E. defined in Eq. (3) is dependent upon the pdf $f(z)$. In order to point out this dependence the A.R.E. should be written as $E_{U,U^*}(f)$, instead of E_{U,U^*}. However, for simplicity we shall use the latter notation, and reserve the former for those occasions when we wish to stress the dependence of the A.R.E. on $f(z)$.

We shall now show that subject to certain regularity conditions for U_{mn}, and $U^*_{n^*}$, there is a simple expression for the A.R.E. defined in Eq. (3).

Regularity Conditions for U_n

(A) $\dfrac{(U_n - E_\theta(U_n))}{\sigma_\theta(U_n)}$ is asymptotically normal with mean zero, and variance one*, uniformly for θ in the closed interval $[0, a]$, $a > 0$,

(B) $E'_\theta(U_n) = \dfrac{d}{d\theta} E_\theta(U_n)$ exists for all θ in $(0, a)$, and is continuous at $\theta=0$,

(C) $\lim_{n\to\infty} \dfrac{1}{n}\left[\dfrac{E'_\theta(U_n)}{\sigma_o(U_n)}\right]^2_{\theta=0} = k$, where k is a positive constant,

(D) there exists a sequence $\{\theta_n\}$ such that $\lim_{n\to\infty} \theta_n = 0$,

(E) $\lim_{n\to\infty} \dfrac{\sigma_{\theta_n}(U_n)}{\sigma_o(U_n)} = 1.$

Regularity Conditions for U_{mn}

(A') $\dfrac{U_{mn} - E_\theta(U_{mn})}{\sigma_\theta(U_{mn})}$ is asymptotically normal with mean zero and variance one, uniformly for θ in the closed interval $[0, a]$, $a > 0$,

(B') $E'_\theta(U_{mn})$ exists for all θ in $(0, a)$, and is continuous at $\theta=0$,

(C') $\lim_{\substack{m\to\infty\\ n\to\infty}} \dfrac{m+n}{mn}\left[\dfrac{E'_\theta(U_{mn})}{\sigma_o(U_{mn})}\right]^2_{\theta=0} = k'$, where k' is a positive constant,

(D') there exists a sequence $\{\theta_{mn}\}$ such that $\lim_{\substack{m\to\infty\\ n\to\infty}} \theta_{mn} = 0$,

(E') $\lim_{\substack{m\to\infty\\ n\to\infty}} \dfrac{\sigma_{\theta_{mn}}(U_{mn})}{\sigma_o(U_{mn})} = 1.$

We now prove two preliminary lemmas which will enable us to derive a simple expression for the A.R.E. defined in Eq. (3). In these theorems we shall use the notation that a size-α detector is one whose false alarm probability is α.

Lemma 1 (Pitman (7))

If U_{mn} satisfies the regularity conditions (A')-(E'), then the size-α detector which rejects H'_o when $U_{mn} > U_{\alpha,mn}$ has a limiting false dismissal probability given by

$$\lim_{\substack{m\to\infty\\ n\to\infty}} \beta_{mn} = \Phi\left(x_\alpha - \lim_{\substack{m\to\infty\\ n\to\infty}} \theta_{mn}\left(\frac{k'mn}{m+n}\right)^{\frac{1}{2}}\right)$$

* If $X_1, X_2, \ldots$ is a sequence of random variables, we say that X_n is asymptotically normal with mean zero and variance one, if

$$\lim_{n\to\infty} \mathrm{Prob}\{X_n \leq x\} = \frac{1}{\sqrt{2\pi}}\int_{-\infty}^{x} \exp(-\tfrac{1}{2}y^2)dy$$

where

$$\Phi(x) = (2\pi)^{-\frac{1}{2}} \int_{-\infty}^{x} \exp(-\tfrac{1}{2}y^2)dy \quad (4)$$

and where x_α is defined as

$$\Phi(x_\alpha) = 1 - \alpha \quad (5)$$

Proof: Letting $P_o(\cdot)$ represent the probability of an event under H_o', we have

$$\alpha = P_o\left(\frac{U_{mn}-E_o(U_{mn})}{\sigma_o(U_{mn})} > \frac{U_{\alpha,mn}-E_o(U_{mn})}{\sigma_o(U_{mn})}\right) \quad (6)$$

Hence, as a consequence of condition (A') we obtain

$$x_\alpha = \lim_{\substack{m\to\infty \\ n\to\infty}} \frac{U_{\alpha,mn}-E_o(U_{mn})}{\sigma_o(U_{mn})} \quad (7)$$

Letting $P_{\theta_{mn}}(\cdot)$ represent the probability of an event under H_1', we have

$$\beta_{mn} = P_{\theta_{mn}}\left(\frac{U_{mn}-E_{\theta_{mn}}(U_{mn})}{\sigma_{\theta_{mn}}(U_{mn})} < \frac{U_{\alpha,mn}-E_{\theta_{mn}}(U_{mn})}{\sigma_{\theta_{mn}}(U_{mn})}\right) \quad (8)$$

As a consequence of condition (B'), we may use the mean value theorem[10] to rewrite the second term in the brackets in Eq. (8) as

$$\frac{\sigma_o(U_{mn})}{\sigma_{\theta_{mn}}(U_{mn})}\left\{\frac{U_{\alpha,mn}-E_o(U_{mn})}{\sigma_o(U_{mn})} - \theta_{mn}\frac{E_\theta'(U_{mn})/\theta=\hat{\theta}}{\sigma_o(U_{mn})}\right\} \quad (9)$$

where $0 < \hat{\theta} < \theta_{mn}$. If we let m and n approach infinity, then we may use conditions (C'), (E'), and Eq. (7), in conjunction with the fact that $E_\theta'(U_{mn})$ is continuous at zero, to rewrite Eq. (9) in the form

$$x_\alpha - \lim_{\substack{m\to\infty \\ n\to\infty}} \theta_{mn}\left(\frac{k'mn}{m+n}\right)^{\frac{1}{2}} \quad (10)$$

We now make use of condition (A') to find that the limiting false dismissal probability is

$$\lim_{\substack{m\to\infty \\ n\to\infty}} \beta_{mn} = \Phi\left(x_\alpha - \lim_{\substack{m\to\infty \\ n\to\infty}} \theta_{mn}\left(\frac{k'mn}{m+n}\right)^{\frac{1}{2}}\right)$$

which completes the proof of the lemma.

An analogous lemma and proof can be given for the size-α detector which rejects H_o' when $U_{mn} < U'_{\alpha,mn}$; in this case we have

$$\lim_{\substack{m\to\infty \\ n\to\infty}} \beta_{mn} = 1 - \Phi\left(\lim_{\substack{m\to\infty \\ n\to\infty}} \theta_{mn}\left(\frac{k'mn}{m+n}\right)^{\frac{1}{2}} - x_\alpha\right)$$

Lemma 2 (Pitman (7))

If U_n satisfies the regularity conditions (A)-(E), then the size-α detector which rejects H_o when $U_n > U_{\alpha,n}$ has a limiting false dismissal probability given by

$$\lim_{n\to\infty} \beta_n = \Phi\left(x_\alpha - \lim_{n\to\infty} \theta_n(kn)^{\frac{1}{2}}\right)$$

Proof: The proof is analogous to that of Lemma 1.

An analogous lemma and proof can be given for the size-α detector which rejects H_o when $U_n < U'_{\alpha,n}$; in this case we have

$$\lim_{n\to\infty} \beta_n = 1 - \Phi\left(\lim_{n\to\infty} \theta_n(kn)^{\frac{1}{2}} - x_\alpha\right)$$

The constants defined in conditions (C) and (C') are of such importance in the study of the detection of weak signals in noise, that they are given a special name; they are known as the efficacy of U_n, and U_{mn}, respectively, and are defined as

$$e(U_n) = \lim_{n\to\infty} \left(\frac{1}{n}\right)\left[\frac{E_\theta'(U_n)}{\sigma_o(U_n)}\right]^2_{\theta=0} \quad (11)$$

$$e(U_{mn}) = \lim_{\substack{m\to\infty \\ n\to\infty}} \left(\frac{m+n}{mn}\right)\left[\frac{E_\theta'(U_n)}{\sigma_o(U_n)}\right]^2_{\theta=0} \quad (12)$$

We now prove a theorem which gives a simple expression for the A.R.E. in terms of the efficacies defined in Eqs. (11) and (12), when the regularity conditions are satisfied.

Theorem 1

If two detectors are based on sequences of statistics $\{U_{mn}\}$, $\{U^*_{n*}\}$ that satisfy the regularity conditions, then the A.R.E. of the $\{U_{mn}\}$-detector with respect to the $\{U^*_{n*}\}$-detector is

$$E_{U,U^*} = \frac{e(U_{mn})}{e(U_n)} \frac{1}{1+r}$$

where $r = \lim_{\substack{m\to\infty \\ n\to\infty}} n/m$.

Proof: If the two detectors have the same limiting false dismissal probability, and detect the same signal strength, then we have from Lemmas 1 and 2 that

$$\lim_{m,n,n^*\to\infty} n^*/n = (k/k^*)\frac{1}{1+r}$$

Therefore, as a consequence of the definition of efficacies in Eqs. (11) and (12), we obtain

$$E_{U,U^*} = \lim_{m,n,n^*\to\infty} (n^*/n) = \frac{e(U_{mn})}{e(U_n)}\,\frac{1}{1+r}$$

which completes the proof of the theorem.

Since $e(U_n)$ and $e(U_{mn})$ are constants which are independent of m and n, it follows that the efficiency of the nonparametric detector is as large as possible when r is chosen to be much less than one. In this case the A.R.E. can be written as

$$E_{U,U^*} \doteq \frac{e(U_{mn})}{e(U_n)}, \quad r \ll 1 \tag{13}$$

We shall find that Eq. (13) is of the utmost importance in comparing various detection systems when the signal strength is very small, and the sample sizes are very large.

The Optimum Neyman-Pearson Detector for Detecting a Constant Signal in Additive Noise

It is well known[5] that the optimum Neyman-Pearson detector bases its decisions on the likelihood ratio, which for the problem of detecting constant signals in noise can be written as

$$L_n(z_1;\ldots\ldots;z_n;\theta) = \prod_{i=1}^{n} \frac{f(z_i-\theta)}{f(z_i)} \tag{14}$$

At this point we assume that $f(u)$ is differentiable in $(z-\theta, z)$, uniformly in z, and $0 \leqslant \theta \leqslant a$; then, by the mean value theorem[10] we have

$$f(z-\theta) = f(z) - \theta \left.\frac{df(u)}{du}\right/_{u=\hat{u}} \tag{15}$$

where $z > \hat{u} > z-\theta$. Substituting Eq. (15) into (14) we find

$$L_n(z_1;\ldots;z_n;\theta) = 1 - \theta \sum_{i=1}^{n} \frac{\left.\frac{df(u)}{du}\right/_{u=\hat{u}_i}}{f(z)} + O(\theta^2) \tag{16}$$

where $O(\theta^2)$ indicates a quantity which when divided by θ^2 remains bounded as θ approaches zero, uniformly in $z_1,\ldots,z_n$, except possibly for sets of probability zero, and where $z_i-\theta < \hat{u}_i < z_i$. If θ is sufficiently small we may neglect the term which is $O(\theta^2)$ in Eq.(16); if in addition we can assume that $f'(u)$ is continuous at $u=z$, then we find that for sufficiently small θ, an equivalent statistic for L_n is L_n^* defined as

$$L_n^*(z_1;\ldots\ldots;z_n) = \frac{1}{n} \sum_{i=1}^{n} \frac{f'(z_i)}{f(z_i)} \tag{17}$$

since θ can be considered as a constant for any particular hypothesis testing problem.

In order to find the statistic L_n^* it is necessary to know the functional form of the pdf $f(z)$. If we assume that $f(z)$ is the normal pdf $\phi(z)$, defined as

$$\phi(z) = (2\pi\sigma_f^2)^{-\frac{1}{2}} \exp(-\tfrac{1}{2})\left(\frac{z-u_f}{\sigma_f}\right)^2 \tag{18}$$

then, the statistic L_n^* is

$$L_n^*(z_1;\ldots\ldots;z_n) = -\frac{1}{n} \sum_{i=1}^{n} \frac{z_i-u_f}{\sigma_f} \tag{19}$$

An equivalent statistic for L_n^* is t_n, defined as

$$t_n(z_1;\ldots\ldots;z_n) = \frac{1}{n} \sum_{i=1}^{n} z_i \tag{20}$$

The optimum detector in this case is a summing device -- a well known result. However, this detector is optimum only when $f(z)$ is the normal pdf $\phi(z)$. Indeed, we shall see subsequently that this detector performs poorly with respect to other detectors when $f(z)$ is allowed to take on functional forms other than $\phi(z)$.

We now calculate the efficacy of the test statistic t_n when the $\{t_n\}$-detector is used for the more general problem of detecting a constant signal in additive noise with mean and variance u_f, and σ_f, respectively, where the noise is not necessarily a sample function of a normal random process.

Since t_n is equal to a sum of independent and identically distributed random variables, we have from the central limit theorem, that t_n is asymptotically normal, provided that σ_f is finite. The mean and variance of this limiting distribution are given by

$$E_\theta(t_n) = \theta + u_f \tag{21}$$

$$\sigma_\theta^2(t_n) = \sigma_o^2(t_n) = \sigma_f^2/n \tag{22}$$

The asymptotic normality of t_n is true uniformly in θ, and hence condition (A) is satisfied. Conditions (D) and (E) are usually satisfied by sequences of statistics, and are not ordinarily checked. Conditions (B) and (C) are checked by calculating the efficacy, and noting whether it exists. This, in addition to checking condition (A) is ordinarily all that is required to check the regularity conditions.

We have from Eq. (21) that

$$E'_\theta(t_n) = 1 \tag{23}$$

Thus, the efficacy of t_n is

$$e(t_n) = \underset{n\to\infty}{\text{limit}} \left(\frac{1}{n}\right) \frac{1}{(\sigma_f^2/n)} = 1/\sigma_f^2 \tag{24}$$

The Mann-Whitney (Wilcoxon) Test

We shall now study a nonparametric test which can be used by a nonparametric detector to determine the presence or absence of a constant signal in additive noise. The Mann-Whitney (Wilcoxon) test was introduced by Wilcoxon[11] and by Mann and Whitney[12], and is based on the statistic $V_{mn}(z_1;\ldots;z_{n+m})$ defined as

$$V_{mn}(z_1;\ldots;z_{n+m}) = (1/mn) \sum_{i=1}^{n} \sum_{j=1}^{m} c(z_i - z_{n+j}) \tag{25}$$

where

$$c(x) = 1, \quad \text{if } x > 0$$
$$= 0, \quad \text{if } x \leqslant 0.$$

In essence, the statistic V_{mn} counts the number of times that the magnitude of an observation Z_i exceeds the magnitude of an observation Z_{n+j}. As an example, let us consider that the Z_i's are 5, 2, 9, 7 and that the Z_{n+j}'s are 6, 3, 1, 10, 12. The statistic V_{mn} for this example is equal to 0.45.

Mann and Whitney have shown (see reference 12) that V_{mn} has asymptotically a normal distribution when H'_o is true, if $F(z)$ is continuous, and limit of m/n exists as m,n approach infinity. Lehmann[13] has shown that V_{mn} has asymptotically a normal distribution when H'_1 is true, if the cdf $F(z)$ is continuous, and if the limit of m/n exists as m and n approach infinity. The limit m/n is said to exist, if it tends $m,n\to\infty$ to a definite value as both m and n approach infinity. The mean and variance of the limiting distribution of V_{mn} are given by (see reference 12)*

$$E_\theta(V_{mn}) = \int F(z)\,dF(z-\theta) = 1-\int F(z-\theta)\,dF(z) \tag{26}$$

$$mn\,\sigma_\theta^2(V_{mn}) = (m+n+1)/12 + (m-1)(\lambda - \epsilon_1) + (n-1)(\lambda - \epsilon_2) - \lambda^2(m+n-1) \tag{27}$$

where

$$\lambda = 1/2 - \int F(z)\,dF(z-\theta)$$
$$\epsilon_1 = 1/3 - \int F^2(z)\,dF(z-\theta)$$
$$\epsilon_2 = 1/3 - \int (1-F(z-\theta))^2\,dF(z)$$

In order to obtain the mean and variance of V_{mn} under H'_o we let θ equal zero in Eqs. (26) and (27), respectively. Upon doing this we find that

$$E_o(V_{mn}) = 1/2 \tag{28}$$

$$\sigma_o^2(V_{mn}) = (m+n+1)/12mn \tag{29}$$

which are independent of $F(z)$. Thus, the limiting distribution of V_{mn} under H'_o is independent of the cdf $F(z)$, and depends only on m and n. Therefore, if m and n are fixed, and sufficiently large, it follows that the false alarm probability of the $\{V_{mn}\}$-detector is independent of the cdf $F(z)$. We note in passing that this result is true even when m and n are not large (see reference (13)).

The asymptotic normality of V_{mn} is true uniformly in θ, so that condition (A') is satisfied. Since $f(z)$, $F(z-\theta)$, and $\frac{d}{d\theta}F(z-\theta)$ have been assumed to be continuous in the region $-\infty < z < \infty$, $0 < \theta < a$, it follows that we can interchange differentiation and integration[14] to obtain from Eq. (26) that

$$E'_\theta(V_{mn})/_{\theta=0} = \int f^2(z)\,dz \tag{30}$$

The efficacy of V_{mn} is obtained from Eqs. (29) and (30) to be

$$e(V_{mn}) = \underset{\substack{m\to\infty\\ n\to\infty}}{\text{limit}} \left(\frac{m+n}{mn}\right) \frac{\left(\int f^2(z)\,dz\right)^2}{(m+n+1)/12mn} = 12\left(\int f^2(z)\,dz\right)^2 \tag{31}$$

Asymptotic Relative Efficiency with Respect to the $\{t_n\}$-Detector

We are now in a position to compute the A.R.E. of the nonparametric detector which is based on the statistics $\{V_{mn}\}$, with respect to the $\{t_n\}$-detector, for the problem of detecting constant signals in additive noise. This expression is obtained from Eqs. (13), (24) and (31) as

*Hereafter, all limits of integration are minus and plus infinity, unless otherwise stated.

$$E_{V,t}(f) \doteq 12\sigma_f^2(\int f^2(z)dz)^2, \quad r \ll 1 \qquad (32)$$

The notation $E_{V,t}(f)$ is used to point out that the A.R.E. is a function of the pdf $f(z)$. A result which is analogous to that given in Eq.(32) has been obtained by Pitman[7] and Andrews[15].

The A.R.E. for the problem of detecting a constant signal in additive normal noise is obtained from Eq. (32) by letting $f(z)$ equal $\phi(z)$. Thus

$$E_{V,t}(\phi) \doteq 12(\int (2\pi)^{-1}\exp(-z^2)dz)^2$$

$$\doteq 3/\pi \doteq 0.955, \quad r \ll 1 \qquad (33)$$

An analogous result has been obtained by Pitman[7] and Mood[16]. We see that $E_{V,t}(f)$ is quite high for this detection problem, for which the $\{t_n\}$-detector is known to be the optimum detector. In view of this we might ask the question as to how large, or small, a value $E_{V,t}(f)$ can have. In order to consider the problem of how large $E_{V,t}(f)$ can be we shall substitute various pdf's in Eq. (32). The question of how small $E_{V,t}(f)$ can be will be answered subsequently.

Using results which are due to Pitman[7] we have that when

$$f(z) = 1, \quad 0 \leq z \leq 1$$
$$= 0, \quad \text{otherwise},$$

then,

$$E_{V,t}(f) \doteq 1, \quad r \ll 1$$

and when

$$f(z) = (z^2/2)\exp(-z), \quad z \geq 0$$
$$= 0, \quad \text{otherwise},$$

then

$$E_{V,t}(f) \doteq 81/64 \doteq 1.265, \quad r \ll 1.$$

All of these values are surprisingly high. We shall now construct a sequence of pdf's $\{f_i(z)\}$ which lead to very large values of $E_{V,t}(f)$. In order to do this we define

$$f_i(z) = (1/i!)\exp(-z^{1/i}), \quad z \geq 0, \; i=1,2,\ldots$$
$$= 0, \quad \text{otherwsie}. \qquad (34)$$

It is easily found that

$$\sigma_{f_i}^2 = (3i-1)!/(i-1)! - ((2i-1)!/(i-1)!)^2 \qquad (35)$$

and that

$$\int f_i^2(z)dz = 2^{-i}/i! \qquad (36)$$

Hence, substituting Eqs. (35) and (36) in (32) we obtain

$$E_{V,t}(f_i) \doteq \frac{(12)(2^{-2i})}{((i!)(i-1)!)^2}((3_i-1)!(i-1)!$$
$$- ((2i-1)!)^2), \quad r \ll 1 \qquad (37)$$

for which some typical values are

$$E_{V,t}(f_1) \doteq 3, \quad r \ll 1$$

$$E_{V,t}(f_2) \doteq 63/4 \doteq 15.75, \quad r \ll 1$$

$$E_{V,t}(f_3) \doteq 345/4 \doteq 86.25, \quad r \ll 1$$

Thus we conclude that $E_{V,t}(f)$ can be very large.

In order to answer the question as to how small $E_{V,t}(f)$ can be, it is shown in the Appendix that the minimum possible value of $E_{V,t}(f)$ is $108/125 = 0.865$, and occurs for the density $f_m(y)$ given by

$$f_m(y) = (3(5)^{\frac{1}{2}}/100)(5-y^2), \quad y^2 \leq 5 \qquad (38)$$
$$= 0, \quad \text{otherwise}.$$

This density function is shown in Fig. 3. An analogous result has been obtained by Hodges and Lehmann (17). Thus, the use of the Mann-Whitney test instead of the t_n-test for the problem of detecting constant signals in additive noise cannot entail a serious loss of efficiency. As a consequence we may conclude that the nonparametric detector which uses the Mann-Whitney statistic is quite useful for this detection problem. Thus, we have an alternative method for detecting constant signals in additive noise, which can be used to replace the classical detector for such a problem; the classical detector is of course a summing device.

A discussion of some other nonparametric tests which would be useful for detecting constant signals in additive noise has been given elsewhere[18]. In particular, it is shown in reference 18 that the so-called c_1-test is an excellent test for such detection problems.

Appendix

Minimum Asymptotic Relative Efficiency of the Mann-Whitney Test with Respect to the t_n-Test

We have shown that the A.R.E. of the Mann-Whitney test with respect to the t_n-test for the problem of detecting a constant signal in additive noise is given by (see Eq. (32))

$$E_{V,t}(f) \doteq 12\sigma_f^2\left(\int f^2(z)dz\right)^2, \quad r \ll 1 \qquad \text{(A.1)}$$

where σ^2 is the variance of the random variable whose pdf[f] is $f(z)$. We shall now prove a theorem concerning the minimum possible value that $E_{V,t}(f)$ can attain. An analogous result has been obtained by Hodges and Lehmann (see reference (17)).

<u>Theorem A.1</u>

The minimum value of $E_{V,t}(f)$ is $108/125 = 0.864$, and occurs for the density $f_m(z)$ given by

$$f_m(z) = (3(5)^{\frac{1}{2}}/100)(5-z^2), \quad z^2 \leqslant 5$$
$$= 0, \quad \text{otherwise.}$$

<u>Proof</u>: Since $E_{V,t}(f)$ is invariant under a change of location or scale we may take

$$\int zf(z)dz = 0 \qquad \text{(A.2)}$$

$$\int z^2f(z)dz = 1 \qquad \text{(A.3)}$$

Hence, σ_f^2 is equal to one, and $E_{V,t}(f)$ can be written as

$$E_{V,t}(f) \doteq 12\left(\int f^2(z)dz\right)^2, \quad r \ll 1 \qquad \text{(A.4)}$$

Therefore, the problem is equivalent to minimizing

$$\int f^2(z)dz \qquad \text{(A.5)}$$

subject to the conditions that

(1) $\int zf(z)dz = 0$
(2) $\int f(z)dz = 1$
(3) $\int z^2f(z)dz = 1$
(4) $\int f(z) \geqslant 0$, for all z.

This problem can be solved by applying the Euler equation of the calculus of variations (19). Thus, letting a_0, a_1, and a_2 denote Lagrangian multipliers, we have

$$\frac{d}{df(z)}\left(f^2(z) + (a_0 + a_1z + a_2z^2)f(z)\right) = 0 \qquad \text{(A.6)}$$

from which we see that

$$f_m(z) = -\tfrac{1}{2}(a_0 + a_1z + a_2z^2), \quad \text{for those values of } z \text{ for which } (a_0 + a_1z + a_2z^2) \leqslant 0,$$
$$= 0, \quad \text{otherwise.} \qquad \text{(A.7)}$$

It is easily seen from Eq. (A.7), that in order to satisfy the integral restraint in condition (1) the Lagrangian multiplier a_1 must equal zero. Hence, we may rewrite Eq. (A.7) as

$$f_m(z) = -\tfrac{1}{2}(a_0 + a_2z^2), \quad z^2 \leqslant -(a_0/a_2)$$
$$= 0, \quad \text{otherwise.} \qquad \text{(A.8)}$$

In order to satisfy the integral restraints in conditions (2) and (3) the Lagrangian multipliers a_0 and a_2 must satisfy the following pair of equations:

$$-a_0(-a_0/a_2)^{\frac{1}{2}} - \frac{a_2}{3}(-a_0/a_2)^{3/2} = 1 \qquad \text{(A.9)}$$

$$-\frac{a_0}{3}(-a_0/a_2)^{3/2} - \frac{a_2}{5}(-a_0/a_2)^{5/2} = 1 \qquad \text{(A.10)}$$

The solutions for a_0 and a_2 are

$$a_0 = -3(5)^{\frac{1}{2}}/10$$
$$a_2 = 3(5)^{\frac{1}{2}}/50$$

Therefore, we obtain that

$$f_m(z) = (3(5)^{\frac{1}{2}}/100)(5-z^2), \quad z^2 \leqslant 5$$
$$= 0, \quad \text{otherwise.} \qquad \text{(A.11)}$$

The A.R.E. for the pdf $f_m(z)$ is obtained from Eqs. (A.4) and (A.11) as

$$E_{V,t}(f_m) \doteq 12\left((45)(10^{-4})\left(\int_{-\sqrt{+5}}^{\sqrt{5}} (5-z^2)^2dz\right)\right)^2$$
$$\doteq 12(9/125) \doteq 108/125, \quad r \ll 1$$

which completes the proof of the theorem.

<u>References</u>

1. North, D.O., "Analysis of the Factors which Determine Signal/Noise Discrimination in Radar," Report PTR-6C RCA Lab., (1943).

2. Zadeh, L.A., and J.R. Ragazzini, "Optimum Filters for the Detection of Signals in Noise," Proc. I.R.E., vol. 40, No. 10, p. 1223, (1952).

3. Doob, J.L., <u>Stochastic Processes</u>, John Wiley and Sons, Inc., New York, N.Y., 1953.

4. Reich, E., and P. Swerling, "Detection of a Sine Wave in Gaussian Noise," J. Appl. Phys., <u>24</u>, p. 289, (1953).

5. Middleton, D., "Statistical Criteria for the Detection of Pulsed Carriers in Noise; I, II," J. Appl. Phys., 24, p. 371, p.379, (1953).

6. Cramer, H., Mathematical Methods of Statistics, Princeton University Press, p. 162, Princeton, N.J., (1946).

7. Pitman, E.J.G., Lecture Notes on Nonparametric Statistical Inference, Columbia University, Spring, 1948.

8. Noether, G.E., "On a Theorem of Pitman," Ann.Math.Stat., 26, p. 64, (1955).

9. Fraser, D.A.S., Nonparametric Methods in Statistics, John Wiley and Sons, Inc., p. 270, New York, N.Y., (1957).

10. Rudin, W., Principles of Mathematical Analysis, McGraw-Hill Book Co., Inc., p. 80, New York, N.Y., (1953).

11. Wilcoxon, F., "Individual Comparisons by Ranking Methods," Biometric Bull., vol. 1, p. 80, (1945).

12. Mann, H.B., and D.R. Whitney, "On a Test of Whether One of Two Random Variables is Stochastically Larger than the Other," Ann. Math. Stat., 18, p. 50, (1947).

13. Lehmann, E.L., "Consistency and Unbiasedness of Certain Nonparametric Tests," Ann.Math. Stat., 22, p. 165, (1951).

14. Titchmarsh, E.C., The Theory of Functions, Oxford University Press, p. 59, London, (1939).

15. Andrews, F.C., "Asymptotic Behavior of Some Rank Tests for Analysis of Variance," Ann. Math.Stat., 25, p. 724, (1954).

16. Mood, A.M., "On the Asymptotic Efficiency of Certain Nonparametric Two-Sample Tests," Ann.Math.Stat., 25, p. 514, (1954).

17. Hodges, J.L., Jr., and E.L. Lehmann, "The Efficiency of Some Nonparametric Competitors of the t-Test," Ann.Math.Stat., 27, p.324, (1956).

18. Capon, J., "Nonparametric Methods for the Detection of Signals in Noise," Department of Electrical Engineering Tech.Report No. T-1/N, Columbia University, March 12, 1959.

19. Hildebrand, F.B., Methods of Applied Mathematics, Prentice-Hall Inc., p. 143, Englewood Cliffs, N.J., 1952.

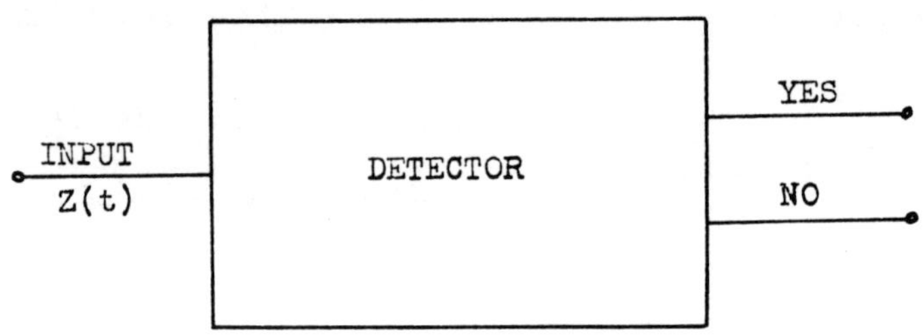

Basic Form of a Detector

Figure 1

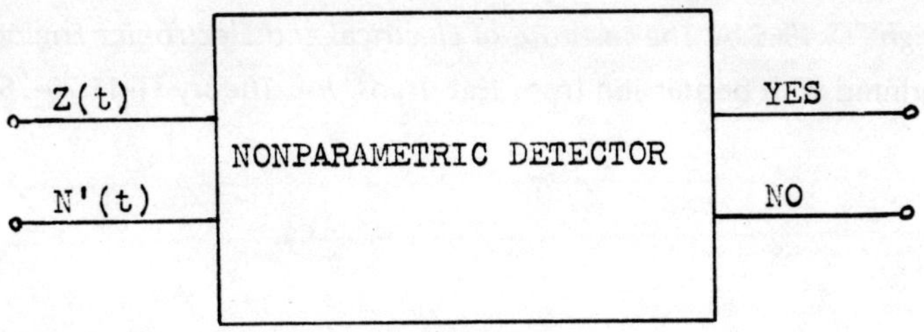

Basic Form of a Nonparametric Detector

Figure 2

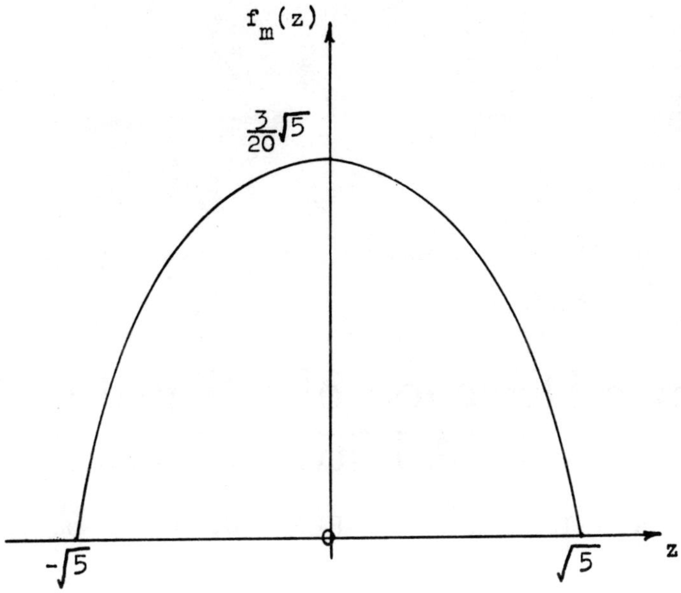

The Density Function for which $E_{V,t}(f)$ Assumes its Minimum Value

Figure 3

Reprinted with permission from *IEEE Trans. Inf. Theory* **IT-11**:70–76 (1965)

Nonparametric Detection of a Signal of Known Form in Additive Noise

R. F. DALY AND C. K. RUSHFORTH, MEMBER, IEEE

Abstract—This paper treats the problem of detecting a signal of known form in the presence of additive noise whose distribution function may not be known. Two detection schemes are considered, one for a signal which is known exactly and one for a signal with unknown phase. Using the criterion of asymptotic relative efficiency (ARE), the performance of each detector is compared with that of the corresponding detector which is optimum for Gaussian noise. It is shown that for detecting a known signal, the ARE of the nonparametric detector relative to the correlation detector is equal to one if the noise is Gaussian, and is strictly greater than one for any continuous non-Gaussian noise distribution. An identical result holds for the ARE of the nonparametric detector for a signal with unknown phase relative to the standard envelope detector.

Manuscript received September 3, 1963, revised April 16, 1964. Most of this work was performed while both authors were with Stanford Research Institute. Some of the results were obtained by C. K. Rushforth at Utah State University where he was supported by Funds from the Engineering Experiment Station.
R. F. Daly is with the Communication Lab., Stanford Research Institute, Menlo Park, Calif.
C. K. Rushforth is with Utah State University, Logan, Utah.

I. Introduction

THE DETECTION of signals in the presence of noise is one of the most basic and important problems encountered by communications engineers. Although a great deal of work has been done on this problem, most of this work has been based on the assumption that the signal and noise statistics are known [1]. This is a reasonable assumption for many important problems, but it is invalid for many others. In this paper, the detection problem is considered under more general assumptions using the techniques of nonparametric statistics.

The application of nonparametric methods to the problem of signal detection has been considered previously. Capon [2], [3] discusses the detection of a constant signal in the presence of additive noise by nonparametric methods. He concludes that for this problem, certain nonparametric detection procedures are much more flexible than the standard parametric procedures in the sense that they work well in a wide variety of statistical environments. Similar conclusions are reached by Wolff, et al. [4] for a somewhat different problem.

In the remainder of this paper, two nonparametric detection procedures are considered—one for detecting a signal which is exactly known at the receiver, and the second for detecting a signal with known form but unknown phase. When the signal is exactly known, the procedure used is a vector version of the c_1 test of nonparametric hypothesis—testing theory [5]. The results obtained for this detector represent a generalization of the work of Capon [3] in that the signal is no longer required to be constant. A modification of the vector c_1 test yields a nonparametric device for detecting a signal with unknown phase. Using the criterion of asymptotic relative efficiency (ARE),[1] which is closely connected with the information rate which can be achieved with a given error probability, it is shown that both nonparametric procedures considered here are considerably more versatile than the corresponding parametric detectors.

II. Detection of a Known Signal

A. Statement of the Problem

In this section, the detection of a signal which is exactly known at the receiver is considered. Specifically, the problem we consider is the following: A detector is presented with n independent observations of an m-dimensional vector of the form $\mathbf{X} = \xi\mathbf{C} + \mathbf{Y}$, where $\mathbf{Y}$ is a vector of noise samples, $\mathbf{C}$ is a known vector formed by sampling the signal waveform, and ξ is an unknown non-negative constant gain factor. This can be accomplished with n separate receivers, or by transmitting $\mathbf{C}n$ times and using a single receiver. The components of $\mathbf{Y}$ are assumed to be independent and identically- distributed samples from a population with absolutely continuous cumulative distribution function $F(y)$. The ith coordinate of $\mathbf{X}$ then has distribution function $F(x - \xi C_i)$ where C_i is the ith coordinate of $\mathbf{C}$. The function of the detector is to decide whether $\xi = 0$ (noise alone) or $\xi > 0$ (signal plus noise).

At this point, it is convenient to introduce some additional notation. The coordinates of the n vector observations may be grouped into m classes corresponding to the m coordinates of an individual vector, yielding the sample vector

$$\mathbf{S} = [X_1^{(1)}, \cdots, X_n^{(1)}, \cdots, X_1^{(m)}, \cdots, X_n^{(m)}] \quad (1)$$

where $X_j^{(i)}$ is the ith coordinate of the jth vector observation. In this notation we have $X_j^{(i)} = \xi C_i + Y_j^{(i)}$.

[1] Fraser [5], sec. 7.3.

B. Optimum Parametric Detector

If we assume that the noise samples have a Gaussian distribution with zero mean and known variance σ^2, the above detection problem reduces to the familiar problem of detecting a known signal in additive white Gaussian noise. The optimum detection procedure for this problem is the well-known correlation detector or matched filter [1]. In the notation of the present paper, this optimum detector calculates the quantity

$$t_n(\mathbf{S}) = \frac{1}{n} \sum_{i=1}^{m} C_i \sum_{k=1}^{n} X_k^{(i)} \quad (2)$$

That is, the optimum detector correlates the known signal with the sample mean of the received vectors.

C. Nonparametric Detector

There are several ways in which we could formulate a nonparametric version of the signal-detection problem stated in Section II-A. The particular formulation we choose is motivated by a desire to obtain a detector which functions as well or nearly as well as the optimum parametric detector in Gaussian noise but which, at the same time, is useful in a wider variety of statistical environments. We obtain such a detector as the solution to the following hypothesis-testing problem:

Hypothesis:

$$X^{(i)} = Y^{(i)},\ Y^{(i)} \sim F,\ i = 1, \cdots, m$$

Alternative: (3)

$$X^{(i)} = \xi C_i + Y^{(i)},\ Y^{(i)} \sim N(\mu, \sigma^2),\ i = 1, \cdots, m$$

where F is an arbitrary absolutely continuous distribution function, $\mu \in (-\infty, \infty)$ is unknown, and $\sigma^2 \in (0, \infty)$ is unknown. $Y^{(i)} \sim F$ means that $Y^{(i)}$ has distribution function F. Thus we seek an optimum or good solution to a hypothesis-testing problem consisting of a very general hypothesis and a Gaussian alternative. As always, the ultimate justification of an approach such as this lies in the evaluation of the detector so obtained. We shall see that this particular compromise leads to a detection scheme which has some very useful and important properties.

Unfortunately, there does not exist a detector which is optimum for this problem if we consider all possible detection procedures and all values of the gain ξ. If we restrict the class of detectors which we consider, and simultaneously restrict our attention to small SNR's, we can find an optimum detector within this smaller class.

This problem has been considered in detail for the case of a constant signal by Fraser.[2] Upon restricting his attention to small SNR's and to the class of detection procedures based on the relative ordering rather than the magnitudes of the received signal samples, he obtains

[2] *Ibid.*, sec. 5.3.

the so-called c_1 test of nonparametric hypothesis-testing theory.

The development[2] can be easily extended to the more general problem being treated here. This extension leads to a detector based on the statistic

$$C_{1n}(\mathbf{S}) = \frac{1}{n} \sum_{i=1}^{m} C_i \sum_{k=1}^{N} \beta_{Nk}^{(i)} E[Z_{(k)}] \tag{4}$$

where $N = nm$, $E[Z_{(k)}]$ is the expected value of the kth largest of N samples from a normal population with mean zero and variance one, and

$$\beta_{Nk}^{(i)} = \begin{cases} 1 \text{ if } k\text{th largest sample in } \mathbf{S} \text{ is an } i\text{th coordinate;} \\ \text{i.e., if } S_{(k)} = X_j^{(i)} \text{ for some } j = 1, \cdots, n \\ 0 \text{ otherwise} \end{cases}$$

It is assumed here that $\bar{C} = 1/m \sum_{i=1}^{m} C_i = 0$. It should be pointed out that for a given sample size N, the $E[Z_{(k)}]$ are known constants and are available in tables [6].

Although (4) looks rather complicated, it is actually relatively easy to implement. The main task which must be performed by the detector is that of ordering the N received samples in an increasing sequence. This yields the order-statistic vector $\mathbf{S}^0 = [S_{(1)}, \cdots, S_{(N)}]$, where

$$S_{(1)} \leq S_{(2)} \leq, \cdots, \leq S_{(N)}$$

These order statistics are divided into m classes of n samples each, corresponding to the m coordinates of the signal vector. This determines a corresponding partition of the constants $E[Z_{(1)}], \cdots, E[Z_{(N)}]$ which are stored at the detector. If $S_{(k)} = X_j^{(i)}$ for some $j = 1, \cdots, n$, $E[Z_{(k)}]$ is included in the ith class. The n quantities $E[Z_{(k)}]$ in the ith class are summed and multiplied by C_i, the ith coordinate of the signal vector $\mathbf{C}$. The m quantities formed in this way are then summed, and the result is compared with a threshold. If this final sum is greater than the threshold, the detector decides that signal is present.

Intuitively, the nonparametric detector described here is effective for essentially the same reasons that the correlation detector is effective. In the presence of signal, those sample coordinates corresponding to large C_i will tend to be large, and hence will tend to have large rank. The corresponding $E[Z_{(k)}]$ will therefore tend to be large, as will the sum of all the $E[Z_{(k)}]$ in the ith class. This sum is weighted heavily by multiplying it by C_i. Similarly, sums which correspond to small C_i tend to be smaller, and they are weighted less heavily by multiplying them by the corresponding small C_i. Thus, in the presence of signal, the detector output tends to be large. In the absence of signal, these correspondences are absent, and the detector output tends to be smaller. Thus, we see that the nonparametric detector has much in common with the correlation detector. The primary difference is that the nonparametric detector does not use absolute amplitude information, but only information pertaining to the relative ordering or rank of various samples.

III. Detection of a Signal with Unknown Phase

A. Statement of the Problem

In many important detection problems, the form of the signal is known but its phase is not. In this section, we consider modifications of the above detection procedures which are appropriate for this problem.

It is assumed that the signal is given by

$$s(t) = e(t) \cos [\omega_0 t - \psi(t) - \theta] \tag{6}$$

where $e(t)$ is the envelope and $\psi(t)$ is the supplementary phase, both assumed to vary much less rapidly than the carrier. The unknown phase θ is assumed to be uniformly distributed on $[0, 2\pi]$. The signal vector is again $\mathbf{C} = (C_1, C_2, \cdots, C_m)$ where

$$C_i = e_i \cos (\omega_0 t_i - \psi_i - \theta), i = 1, \cdots, m \tag{7}$$

As before,

$$\bar{C} = \frac{1}{m} \sum_{i=1}^{m} C_i = 0$$

Furthermore, we assume this holds for all values of θ. The remaining assumptions made previously are retained throughout this section.

B. Optimum Parametric Detector

The optimum parametric device for detecting a signal with unknown phase in the presence of additive white Gaussian noise is the well-known envelope detector [7] based on the statistic

$$\mathcal{E}_p = \sqrt{x^2 + y^2} \tag{8}$$

where

$$x = \sum_{i=1}^{m} e_i \cos (\omega_0 t_i - \psi_i) U_i \tag{9}$$

and

$$y = \sum_{i=1}^{m} e_i \sin (\omega_0 t_i - \psi_i) U_i \tag{10}$$

Here

$$U_i = \frac{1}{n} \sum_{k=1}^{n} X_k^{(i)} \tag{11}$$

is the sample mean of the ith coordinate.

According to the central limit theorem, the U_i are asymptotically Gaussian for the distributions considered here [5]. Hence, the probability density function of $\mathcal{E}_p$ in the presence of signal, which will be needed later, is [7]

$$f(\mathcal{E}_p) = \frac{\mathcal{E}_p}{\beta^2} \exp \left[- \left(\frac{\mathcal{E}_p^2 + M^2}{2\beta^2} \right) \right] I_0 \left(\frac{\mathcal{E}_p M}{\beta^2} \right) \tag{12}$$

where M is the mean and β^2 the variance of

$$t_n(\mathbf{S}) = \sum_{i=1}^{m} C_i U_i \tag{13}$$

C. Nonparametric Detector

From (4), it is seen that the nonparametric detector for a known signal based on the vector c_1 test calculates the quantity

$$C_{1n}(\mathbf{S}) = \sum_{i=1}^{m} C_i T_n^{(i)} \tag{14}$$

where

$$T_n^{(i)} = \frac{1}{n} \sum_{k=1}^{N} \beta_{Nk}^{(i)} E[Z_{(k)}] \tag{15}$$

Since (14) has the same form as (13), this detector can be modified in the same way as was the optimum parametric detector to facilitate the detection of a signal with unknown phase. Although it is intuitively reasonable that such a procedure should give satisfactory results, the ultimate justification must be based on quantitative evaluation. This evaluation is carried out in Section V.

These considerations lead to a nonparametric procedure for detecting a signal of unknown phase based on the quantity

$$\mathcal{E}_N = \sqrt{u^2 + v^2} \tag{16}$$

where

$$u = \sum_{i=1}^{m} e_i \cos(\omega_0 t_i - \psi_i) T_n^{(i)} \tag{17}$$

and

$$v = \sum_{i=1}^{m} e_i \sin(\omega_0 t_i - \psi_i) T_n^{(i)} \tag{18}$$

Just as in the known-signal case, this detector depends only on the relative ordering or rank of the received signal samples, and is nonparametric in its false-alarm rate. It has the same form as the parametric envelope detector with U_i replaced by $T_n^{(i)}$.

We now consider the problem of finding the probability density function of $\mathcal{E}_N$ in the presence of signal. By extending some results of Chernoff and Savage [8] (see Appendix I below), it can be shown that the quantities $T_n^{(i)}$, and hence u and v, are asymptotically Gaussian. The means of u and v are

$$E(u) = a = \sum_{i=1}^{m} e_i \cos(\omega_0 t_i - \psi_i) \mu^{(i)} \tag{19}$$

$$E(v) = b = \sum_{i=1}^{m} e_i \sin(\omega_0 t_i - \psi_i) \mu^{(i)} \tag{20}$$

where

$$\mu^{(i)} = E[T_n^{(i)}] \tag{21}$$

If we neglect sums of rapidly–oscillating terms [7], we obtain

$$\text{var}(u) = \text{var}(v) = \text{var}[C_{1n}(\mathbf{S})] \triangleq \sigma^2 \tag{22}$$

$$\text{cov}(u, v) = 0 \tag{23}$$

Hence, u and v are independent Gaussian random variables, and

$$f(\mathcal{E}_N) = \frac{\mathcal{E}_N}{\sigma^2} \exp\left[-\left(\frac{\mathcal{E}_N^2 + L^2}{2\sigma^2}\right)\right] I_0\left(\frac{\mathcal{E}_N L}{\sigma^2}\right) \tag{24}$$

where $L^2 = a^2 + b^2$. Thus, we see that the distribution of the nonparametric detector output (24) has the same form as that of the parametric detector output (12). This fact will be used in evaluating the relative performances of these two detectors.

IV. Asymptotic Relative Efficiency

Having obtained several parametric and nonparametric detection schemes, we now wish to compare their performances in both Gaussian and non-Gaussian environments. The criterion we use for this comparison is Pitman's ARE. The ARE of one detection procedure relative to another is the limiting ratio of the sample sizes required by the two detectors to yield the same error probabilities for a decreasing sequence of gains $\{\xi_i\}$. This criterion has been fully discussed and defined elsewhere [9].[1]

If the outputs of the two detectors to be compared satisfy certain reasonable regularity conditions and have distributions which are asymptotically of the same form, the ARE of one detector relative to the other can be obtained quite easily [10]. Let T_n and T_n^* be test statistics satisfying these conditions and having means $E_\xi(T_n)$ and $E_\xi(T_n^*)$ and variances $\sigma_\xi^2(T_n)$ and $\sigma_\xi^2(T_n^*)$, respectively, where ξ is the value of the gain parameter. We distinguish two cases.

Case 1:

$$\frac{d}{d\xi} E_\xi(T_n)\bigg|_{\xi=0} > 0, \quad \frac{d}{d\xi} E_\xi(T_n^*)\bigg|_{\xi=0} > 0$$

In this case, the ARE of $\{T_n\}$ relative to $\{T_n^*\}$ is

$$e_{T,T^*} = \lim_{n\to\infty} \left[\frac{\frac{d}{d\xi} E_\xi(T_n)\Big|_{\xi=0}}{\frac{d}{d\xi} E_\xi(T_n^*)\Big|_{\xi=0}}\right]^2 \frac{\sigma_0^2(T_n^*)}{\sigma_0^2(T_n)} \tag{25}$$

Case 2:

$$\frac{d}{d\xi} E_\xi(T_n)\bigg|_{\xi=0} = \frac{d}{d\xi} E_\xi(T_n^*)\bigg|_{\xi=0} = 0$$

$$\frac{d^2}{d\xi^2} E_\xi(T_n)\bigg|_{\xi=0} > 0, \quad \frac{d^2}{d\xi^2} E_\xi(T_n^*)\bigg|_{\xi=0} > 0$$

In this case, the ARE is

$$e_{T,T^*} = \lim_{n\to\infty} \left[\frac{\frac{d^2}{d\xi^2} E_\xi(T_n)\Big|_{\xi=0}}{\frac{d^2}{d\xi^2} E_\xi(T_n^*)\Big|_{\xi=0}}\right] \frac{\sigma_0(T_n^*)}{\sigma_0(T_n)} \tag{26}$$

We shall use (25) to evaluate the detection procedures for a known signal, and (26) when the phase is unknown.

V. ARE of Nonparametric Detectors Relative to Corresponding Parametric Detectors

A. Detectors for a Known Signal

Chernoff and Savage [8] have treated the problem of finding the asymptotic distributions and efficiencies of a class of test statistics closely related to the vector c_1 test (14) considered above. Their work is concerned with the problem of drawing and comparing samples from two populations to determine whether or not these two populations are identical, and is directly applicable to the detection of a constant signal. This detection problem has been considered by Capon [3].

The essential difference between the work of Chernoff and Savage and that of the present paper is in the number of populations from which samples are drawn. Whereas Chernoff and Savage consider two populations, one of which is noise alone, we have m populations (some but not all of which may be identical) corresponding to the m coordinates of the signal vector $\mathbf{C}$, with no specific requirement that one of these must be noise alone. The techniques and results of Chernoff and Savage may be extended to the present problem by taking care of details involving the differences between sampling from two classes and sampling from m classes. An outline of the important steps in this extension is contained in Appendix I, and the details are contained in [10].

It is shown in Appendix I that the ARE of the nonparametric vector c_1 test relative to the correlation detector for noise with distribution function F and finite second moment is

$$e_{N,P} = I^2(F)\sigma_F^2 \tag{27}$$

where

$$I(F) = \int_{-\infty}^{\infty} J'[F(x)]f^2(x)\,dx \tag{28}$$

$$J(F) = \Phi^{-1}(F) \tag{29}$$

σ_F^2 is the variance of the noise, f is the density function of the noise, and Φ is the unit normal distribution function.

This is identical with the result obtained by Chernoff and Savage. Hence, their variational argument and conclusions apply, and we have the following theorem.

Theorem

Let F be an absolutely continuous distribution function with finite second moment. The ARE of the nonparametric detector based on the vector c_1 test relative to the correlation detector for detecting a known signal in the presence of additive white noise with distribution function F is equal to or greater than one, with equality if and only if the noise is Gaussian.

This is an extremely important result. It tells us that the nonparametric detector described here performs as well as the optimum parametric detector (in terms of ARE) when the noise is Gaussian, and performs strictly better (again in terms of ARE) for any other continuous noise distribution. Thus, in this sense, the c_1 detector is more useful and versatile than the correlation detector for detecting the presence of an arbitrary known signal in the presence of additive noise with unknown statistics.

B. Detectors for a Signal with Unknown Phase

Using the work of Rice [11],[3] the moments of the distributions (12) and (24) can be evaluated. We have

$$\overline{\mathcal{E}_p^2} = 2\beta^2 + M^2 \tag{30}$$

$$\overline{\mathcal{E}_p^4} - (\overline{\mathcal{E}_p^2})^2 = 4\beta^2(\beta^2 + M^2) \tag{31}$$

$$\overline{\mathcal{E}_N^2} = 2\sigma^2 + L^2 \tag{32}$$

$$\overline{\mathcal{E}_N^4} - (\overline{\mathcal{E}_n^2})^2 = 4\sigma^2(\sigma^2 + L^2) \tag{33}$$

These quantities are the means and variances of the squares of the two detector outputs. Hence, it is convenient to use these squares rather than the output variables themselves to evaluate the ARE.

Using (26), the ARE of the nonparametric detector based on (16) relative to the standard envelope detector is evaluated in Appendix II. The result is

$$e_{N,P} = I^2(F)\sigma_F^2 \tag{34}$$

This expression is identical to that obtained for the ARE of the vector c_1 test relative to the correlation detector for detecting a known signal. Hence, the same conclusions apply; i.e., the ARE of the nonparametric "envelope" detector relative to the standard envelope detector is never less than one for noise with an absolutely continuous distribution function, and is equal to one if and only if the noise is Gaussian. Thus, the nonparametric procedure described here is more flexible than the standard envelope detector and can be used to detect the presence of a signal with unknown phase in a wide variety of statistical environments.

VI. Summary and Conclusions

In this paper, two nonparametric detection schemes have been considered, one for detecting a known signal in the presence of additive noise, and the other for detecting a signal with unknown phase. Using the criterion of ARE, it was shown that these detectors are more flexible than the corresponding detectors designed to be optimum for Gaussian noise. Hence, in situations where the noise may not be Gaussian, these nonparametric detectors should be considered as potential alternatives to the correlation detector or the envelope detector.

Appendix I

ARE of Known-Signal Detectors

Denote the (absolutely continuous) distribution function of the ith coordinate of the vector $\mathbf{X} = (X^{(1)}, \cdots, X^{(m)})$

[3] See also Wainstein and Zubakov [7], p. 198.

by $F^{(i)}(x)=F(x-\xi C_i)$. Thus $X_j^{(i)} \sim F^{(i)}$ for $j=1, \cdots, n$. The sample distribution function of the ith class is

$$F_n^{(i)}(x) = \frac{\text{number of samples from class } i \le x}{n} \tag{35}$$

The combined distribution function for the entire sample is

$$H(x) = \frac{1}{m} \sum_{i=1}^{m} F^{(i)}(x) \tag{36}$$

and the combined sample distribution function is

$$H_n(x) = \frac{1}{m} \sum_{i=1}^{m} F_n^{(i)}(x) \tag{37}$$

It is also convenient to define combined distribution functions for that part of the sample not in class i. Accordingly, we define

$$G^{(i)}(x) = \frac{1}{m-1} \sum_{j \ne i} F^{(j)}(x) \tag{38}$$

$$G_n^{(i)}(x) = \frac{1}{m-1} \sum_{j \ne i} F_n^{(j)}(x) \tag{39}$$

Following Chernoff and Savage [8], we express $T_n^{(i)}$ from (15) in the integral form

$$T_n^{(i)} = \int_{-\infty}^{\infty} J_N[H_n(x)]dF_n^{(i)}(x) \tag{40}$$

where

$$J_N(j/N) = E[Z_{(j)}] \tag{41}$$

Observe that $H_n[S_{(j)}] = j/N$ where $S_{(j)}$ is the jth largest sample from the vector **S** in (1).

By identifying the quantities $F^{(i)}(x)$ and $G^{(i)}(x)$ from above with $F(x)$ and $G(x)$, respectively, in Chernoff and Savage, and by noting from (36) to (39) that

$$H(x) = \frac{1}{m} F^{(i)}(x) + \left(1 - \frac{1}{m}\right) G^{(i)}(x) \tag{42}$$

and

$$H_n(x) = \frac{1}{m} F_n^{(i)}(x) + \left(1 - \frac{1}{m}\right) G_n^{(i)}(x) \tag{43}$$

we can apply their Theorem 1 immediately. From this theorem, we conclude that the quantities $T_n^{(i)}$ are asymptotically normal with means

$$\mu^{(i)} = \int_{-\infty}^{\infty} J[H(x)]dF^{(i)}(x) \tag{44}$$

and variances given by equation (4.3) of Chernoff and Savage [8]. Here $J(H) = \phi^{-1}(H)$ as in (29).

To calculate the ARE of the vector c_1 detector relative to the correlation detector from (25), we must obtain expressions for

$$\frac{d}{d\xi} E_\xi(C_{1n}) \bigg|_{\xi=0}$$

and the asymptotic variance $\sigma_0^2(C_{1n})$. Differentiating (44) and using the result in (14) yields

$$\frac{d}{d\xi} E_\xi(C_{1n}) \bigg|_{\xi=0} = I(F) \sum_{i=1}^{m} C_i^2 \tag{45}$$

where $I(F)$ is given by (28).

The calculation of the asymptotic variance is complicated by the fact that the $T_n^{(i)}$ are correlated. Using (35)–(40) above and following Chernoff and Savage, it can be shown that

$$\sigma_0^2(C_{1n}) = \frac{1}{n} \sum_{i=1}^{m} C_i^2 \tag{46}$$

The quantities (45) and (46) are used in (25), together with the corresponding and easily-calculated quantities for the correlation detector, to obtain (27).

Appendix II

ARE of Detectors for Signal with Unknown Phase

From (30) and (31), using obvious expressions for M and β^2, we obtain for the parametric detector

$$E_\xi(\mathcal{E}_P^2) = \frac{2\sigma_F^2}{n} \sum_{i=1}^{m} C_i^2 + \xi^2 \left(\sum_{i=1}^{m} C_i^2\right)^2 \tag{47}$$

$$\frac{d}{d\xi} E_\xi(\mathcal{E}_P^2) \bigg|_{\xi=0} = 0 \tag{48}$$

$$\frac{d^2}{d\xi^2} E_\xi(\mathcal{E}_P^2) \bigg|_{\xi=0} = 2 \left(\sum_{i=1}^{m} C_i^2\right)^2 \tag{49}$$

$$\sigma_0^2(\mathcal{E}_P^2) = \text{var}\,(\mathcal{E}_P^2) \bigg|_{\xi=0} = \frac{4\sigma_F^2}{n^2} \left(\sum_{i=1}^{m} C_i^2\right)^2 \tag{50}$$

For the nonparametric detector, we have

$$\frac{d}{d\xi} E_\xi(\mathcal{E}_N^2) \bigg|_{\xi=0} = 2L \frac{dL}{d\xi} \bigg|_{\xi=0} = 0 \tag{51}$$

since for $\xi = 0$,

$$\mu^{(i)} \equiv \mu = \int_{-\infty}^{\infty} J[F(x)]f(x)\, dx \tag{52}$$

$$a \bigg|_{\xi=0} = \mu \sum_{i=1}^{m} e_i \cos(\omega_0 t_i - \psi_i) = 0 \tag{53}$$

$$b \bigg|_{\xi=0} = \mu \sum_{i=1}^{m} e_i \sin(\omega_0 t_i - \psi_i) = 0 \tag{54}$$

$$L|_{\xi=0} = \sqrt{a^2 + b^2}|_{\xi=0} = 0 \tag{55}$$

Using (19), (20), and (44) and neglecting sums of rapidly-oscillating terms, we obtain

$$\frac{d^2}{d\xi^2} E_\xi(\mathcal{E}_N^2)|_{\xi=0} = 2I^2(F) \left(\sum_{i=1}^{m} C_i^2\right)^2 \tag{56}$$

Finally, from (33) and (46), we have

$$\sigma_0^2(\mathcal{E}_N^2) = \text{var}\,(\mathcal{E}_N^2)|_{\xi=0} = \frac{4}{n^2}\left(\sum_{i=1}^{m} C_i^2\right)^2 \qquad (57)$$

Equations (49), (50), (56), and (57) are used in (26) to obtain (34).

References

[1] Middleton, D., *An introduction to statistical communication theory*, McGraw-Hill Book Co., Inc., New York, N. Y., pt. 4, 1960.

[2] Capon, J., A nonparametric technique for the detection of a constant signal in additive noise, *1959 Wescon Convention Record*, pt. 4, pp. 92–103.

[3] Capon, J., Nonparametric methods for the detection of signals in noise, Tech. Rept. No. T-1/N, Columbia University, New York, N. Y., Mar 1959.

[4] Wolff, S. S., J. B. Thomas, and T. R. Williams, The polarity-coincidence correlator: a nonparametric detection device, *IRE Trans. on Information Theory*, vol. IT-8, pp. 5–9, Jan 1962.

[5] Fraser, D. A. S., *Nonparametric methods in statistics*, John Wiley & Sons, Inc., New York, N. Y., 1957.

[6] Fisher, R. A., and F. Yates, *Statistical tables for biological, agricultural, and medical research*, Hafner Publishing Co., Inc., New York, N. Y., 3rd ed., 1949.

[7] Wainstein, L. A., and V. D. Zubakov, *Extraction of signals from noise*, Prentice-Hall, Inc., Englewood Cliffs, N. J., ch. 5, 1962.

[8] Chernoff, H., and I. R. Savage, Asymptotic normality and efficiency of certain nonparametric test statistics, *Ann. Math. Stat.*, vol. 29, pp. 972–994, 1958.

[9] Capon, J., On the asymptotic efficiency of locally optimum detectors, *IRE Trans. on Information Theory*, vol. IT-7, pp. 67–71, Apr 1961.

[10] Rushforth, C. K., and R. F. Daly, Nonparametric detection of a known signal in additive noise, Research Memo., Stanford Research Institute, Menlo Park, Calif., Feb 1964.

[11] Rice, S. O., Mathematical analysis of random noise, in *Selected papers on noise and stochastic processes*, N. Wax, Ed., Dover Publications, New York, N. Y., sect. 3.10, 1954.

11

Detection Performance of Some Nonparametric Rank Tests and an Application to Radar

VILHELM GREGERS HANSEN, MEMBER, IEEE

Abstract—**The detection performance of some one-sample nonparametric rank tests for signals in Gaussian noise exhibiting a characteristic trend is presented in this paper for a finite number of observations and asymptotically. The tests considered are the Kendall tau, the Spearman rho, and a version of the c_1 normal-scores test; both coherent and incoherent signal processing are considered.**

The asymptotic performance is determined through a simple extension of existing analytical results, whereas for a finite number of observations a direct Monte Carlo simulation is used. The latter requires a preliminary determination of false-alarm probability versus decision threshold. For the Spearman rho and the c_1 tests an importance-sampling technique is used in a Monte Carlo simulation to determine the required results.

Finally an application of the Kendall tau test to conventional radar is considered, in which the characteristic trend due to the amplitude modulation of the received train of echoes by the scanning of the antenna is used for detecting the presence of a target.

I. Introduction

THE MAJOR reason behind the increasing engineering interest in nonparametric signal-detection schemes is their ability to maintain a constant false-alarm rate (CFAR) at the detector output for a broad class of noise inputs. This is an important requirement for radar or sonar systems that have to operate in an environment of heavy external interference. In addition, the detection performance of such nonparametric detectors for signals in non-Gaussian noise may often surpass that of detectors optimized against Gaussian noise.

Even though non-Gaussian noise may occur frequently enough to warrant serious consideration in the system design, Gaussian noise will usually be the prevailing type of interference. Therefore, a very important characteristic of a nonparametric detector is its detection performance against Gaussian noise as compared with the optimum parametric detector.

Manuscript received June 23, 1969; revised November 3, 1969.
The author was with the Laboratory for Communication Theory, Technical University of Denmark, Lyngby, Denmark. He is now with the Raytheon Company, Wayland, Mass. 01778.

This comparison, in engineering terminology, would normally be expressed as the decibel difference between the signal-to-noise ratios required by the two detectors in order to obtain specified values of false-alarm probability P_f and detection probability P_d for a given number of observations. The practical procedure for obtaining such results is, however, often complicated, and Monte Carlo simulation on a digital computer is frequently the only feasible approach.

A more universal and mathematically convenient criterion used by statisticians is the asymptotic efficiency of a detector. This criterion relates to the detection performance as the number of observations of approach infinity, and a corresponding loss figure expressed as a degradation in signal-to-noise ratio for the nonparametric detector relative to the optimum detector can usually be defined. In many cases this loss will be independent of the assumed false-alarm and detection probabilities. A major disadvantage of the asymptotic efficiency as a performance criterion is, however, that results obtained only will apply reliably for practical cases when the number of observations is very large.

In this paper both of the above criteria will be used to evaluate some nonparametric detectors based on ranks against Gaussian noise. Thus, the results obtained will provide some clarification of the important question of how many observations are required in order for an asymptotic analysis to provide a reasonably accurate indication of expected performance.

The nonparametric rank tests to be considered are based on a single set of observations that is due to noise only (hypothesis H_0) or to signal and noise (hypothesis H_1). For noise only, it is assumed that all observations are statistically independent and have the same continuous cumulative distribution function (cdf). For signal and noise, the observations are influenced by a signal that has different amplitudes for the individual observations so that a trend—and therefore a most probable rank order—will be present in the received set of observations. The form of the cdf's is unknown under both hypotheses.

The rank tests that will be considered here are those that lend themselves most easily to digital hardware implementation. These are the so-called simple linear rank tests [1] as well as the nonlinear Kendall tau (τ) test, originally proposed for testing the hypothesis of independence between two sets of observations [2]. The tests are evaluated for detection of narrow-band signals in Gaussian noise, and both coherent and incoherent processing are considered.

Finally, a nonparametric detector based on the Kendall tau test is considered for use in a conventional pulsed search radar. The modulation of the train of echoes due to the pattern of the scanning antenna will cause a characteristic rank order to appear when a target is present. The nonparametric detector is designed on this basis, and the performance of the resulting system is determined.

II. Detection Model and Rank Tests

The nonparametric-detection situation considered in this paper can be specified as follows.

Denote the received set of observations by $\mathbf{x} = (x_1, \cdots, x_M)$. With noise only present, all observations are assumed independent and identically distributed with unknown continuous cdf $F_n(x_i)$. Thus the joint probability density function (pdf) of the received set of observations is of the form

$$f(\mathbf{x}) = \prod_{i=1}^{M} f_n(x_i) \qquad (H_0) \tag{1}$$

where $f_n(x)$ is the pdf corresponding to $F_n(x)$.

In the signal-and-noise case a signal with the set of amplitudes $\mathbf{a} = (a_1, \cdots, a_M)$, which are all different, is present. The resulting observations are again independent and have unknown continuous cdf's $F_s(x_i \mid a_i)$, and the joint pdf of the set of observations is

$$f(\mathbf{x}) = \prod_{i=1}^{M} f_s(x_i \mid a_i) \qquad (H_1). \tag{2}$$

The signal amplitudes are assumed (without loss of generality) to be arranged in an increasing order of the indices (i.e., $a_1 < a_2 < \cdots < a_M$), and the continuous cdf's $F_s(x_i \mid a_i)$ are restricted to the class for which

$$a_j > a_i \Rightarrow \Pr\{x_j > x_i\} > \tfrac{1}{2} \tag{3}$$

so that the presence of a signal is accompanied by an upward trend from x_1 to x_M in the set of observations. The requirement of (3) can be expressed as

$$\int_{-\infty}^{\infty} f_s(x \mid a_i)\{F_s(x \mid a_i) - F_s(x \mid a_j)\}\, dx > 0. \tag{4}$$

This is shown by writing down the expression for $\Pr\{x_j > x_i\} > \frac{1}{2}$ and using

$$\int_{-\infty}^{\infty} f_s(x \mid a)F_s(x \mid a)\, dx = \tfrac{1}{2}. \tag{5}$$

With the above assumptions, the problem of detecting the presence of a signal is equivalent to deciding whether or not a trend is present in the received set of observations. Various nonparametric tests based on ranks may be used for testing such trend alternatives, and several of these will be considered in this paper.

The rank information in a set of observations $\mathbf{x}$ is normally specified by the vector of ranks $\mathbf{R} = (R_1, \cdots, R_M)$, where R_i is the rank of the observation x_i in the complete set.[1] For example with $\mathbf{x} = (3, 5, 9, 1)$ we have $\mathbf{R} = (2, 3, 4, 1)$.

An alternative way of specifying the rank information, which will be used later, is the following. The position of the largest observation is denoted l_1 as counted from the left. The position of the second largest observation among the remaining $M - 1$ observations is denoted l_2, etc., down to the smallest observation, which necessarily must have the position $l_M = 1$. Thus in the example above, we have $\boldsymbol{l} = (l_1, l_2, \cdots, l_M) = (3, 2, 1, 1)$. In the following we shall call $\boldsymbol{l}$ the relative rank order of the observations.

In a general rank test, the test statistic T is some function of the vector of ranks $\boldsymbol{R}$. In the following, we shall consider some examples of so-called simple linear rank tests,[2] which are of the form[3]

$$T = t(\boldsymbol{R}) = \sum_{i=1}^{M} c_i \cdot g(R_i). \tag{6}$$

With $c_i = i$ and $g(R_i) = R_i$, we obtain the Spearman rho test with the test statistic

$$T_\rho = \sum_{i=1}^{M} i \cdot R_i. \tag{7}$$

Of all simple linear rank tests, this appears to be the one best suited for digital implementation (integer arithmetic), and it will therefore be considered in some detail.

For the case of a signal in additive Gaussian noise ($x_i = a_i + n_i$), we consider the c_1 test for which

$$T_c = \sum_{i=1}^{M} a_i \cdot g_N(R_i). \tag{8}$$

In this expression the $g_N(i)$ are the expected values of normal order statistics. If M independent random variables ξ are drawn from a standardized Gaussian distribution (zero-mean, unit variance) and ordered according to their magnitude, so that

$$\xi^1 < \xi^2 < \cdots < \xi^M, \tag{9}$$

then $g_N(i)$ is equal to the expected value of ξ^i, i.e.,

$$g_N(i) = E\{\xi^i\}. \tag{10}$$

The test of (8) is of the normal-scores type [1], [3] and has the property that for Gaussian alternatives and with the signal amplitudes chosen so that their mean is zero

[1] With the assumption of continuous CDF's the probability of ties (equality between two observations) is zero and all observations will therefore have unambiguous ranks.

[2] See ([1], p. 61).

[3] In statistics such tests are used for regression alternatives, in which context the constants c_i are the regression constants and the function $g(R_i)$ is the score associated with the rank R_i.

(i.e., $\sum a_i/M = 0$), its asymptotic efficiency is 1 [3]. Thus, the test is asymptotically as good as the optimum likelihood-ratio test.

In addition to the above-mentioned simple linear rank tests, the nonlinear Kendall tau test will be considered. The Kendall tau test can be written in several equivalent forms [2]. According to one of these, each observation x_i $(i = 1, 2, \cdots, M)$ is compared with all following x_j $(j = i + 1, i + 2, \cdots, M)$ and the total number of cases where $x_j > x_i$ are summed. This leads to the test statistic T_τ, which is written

$$T_\tau = \sum_{i=1}^{M-1} \sum_{j=i+1}^{M} u(x_j - x_i) \tag{11}$$

where $u(x)$ is the unit step function. The same test statistic can also be expressed in terms of the vector of ranks[4] $\boldsymbol{R}$ or in terms of the relative rank order $\boldsymbol{l}$ defined previously. In the latter case we have

$$T_\tau = \sum_{i=1}^{M} (M + 1 - i - l_i) = \frac{M(M+1)}{2} - \sum_{i=1}^{M} l_i. \tag{12}$$

The Kendall tau test is well suited for digital implementation and in this respect is quite similar to the Spearman rho test. An advantage of the Kendall tau test is, however, that the effect of adding one further observation x_{M+1} can be taken into account by the recursive expression [2]

$$T_\tau' = T_\tau + \sum_{i=1}^{M} u(x_{M+1} - x_i) \tag{13}$$

where T_τ is given by (11) or (12). Similarly, the effect of removing the observation x_1 can be taken into account by calculating

$$T_\tau'' = T_\tau - \sum_{i=2}^{M} u(x_i - x_1). \tag{14}$$

In the radar application of the Kendall tau test described in Section V, this will be seen to be a useful property.

Finally, it should be noted that the Kendall tau test is asymptotically equivalent to the Spearman rho test.[5] This fact is used for determining the asymptotic efficiency of the Kendall tau test.

III. Cases Considered

The rank tests described in the previous section will be considered for both coherent and incoherent detection of signals in Gaussian noise.

In both cases a narrow-band signal in Gaussian noise is considered. For coherent detection exact knowledge of the signal phase is assumed for all observations, and the noise-only and signal-and-noise hypotheses can be represented as

$$\left.\begin{aligned} x_i &= n_i \qquad (H_0) \\ x_i &= a_i + n_i \qquad (H_1) \end{aligned}\right\} \quad i = 1, 2, \cdots, M \tag{15}$$

[4] See ([1], p. 115).
[5] See ([1], p. 61).

where a_i is the signal amplitude and n_i is a normalized Gaussian random variable $n_i \sim N\{0, 1\}$.[6] For the pdf's in (1) and (2) we therefore have

$$f_n(x_i) = (2\pi)^{-1/2} \exp(-\tfrac{1}{2}x_i^2) \tag{16}$$

$$f_s(x_i \mid a_i) = (2\pi)^{-1/2} \exp[-\tfrac{1}{2}(x_i - a_i)^2]. \tag{17}$$

The signal-to-noise ratio parameter in this case is defined as

$$S = \sum_{i=1}^{M} a_i^2. \tag{18}$$

If the observations (15) are obtained by matched filter processing of each of m signals in additive white noise, we would have $S = 2E/N_0$, where E is the total signal energy and N_0 is the noise spectral density. For the signal, the amplitudes used are

$$a_i = a_0\{i - \tfrac{1}{2}(M + 1)\}, \tag{19}$$

which, as discussed in Section IV, leads to the best possible asymptotic performance for the three rank tests considered.

In the incoherent case, the phase is unknown for each observation, and a square-law envelope detector is assumed. Thus, only the envelope information of the individual observations is used by the receiver, and we have

$$\left.\begin{aligned} x_i &= \tfrac{1}{2}\{n_{i1}^2 + n_{i2}^2\} \qquad (H_0) \\ x_i &= \tfrac{1}{2}\{(a_i + n_{i1})^2 + n_{i2}^2\} \qquad (H_1) \end{aligned}\right\} \quad i = 1, 2, \cdots, M \tag{20}$$

where n_{i1} and n_{i2} are independent normalized Gaussian random variables such that $n_{i1}, n_{i2} \sim N\{0, 1\}$. The pdf's in (1) and (2) in this case are

$$f_n(x_i) = \begin{cases} \exp(-x_i) & x_i \geq 0 \\ 0 & \text{otherwise} \end{cases} \tag{21}$$

$$f_s(x_i \mid a_i) = \begin{cases} \exp(-x_i)\exp(-\tfrac{1}{2}a_i^2) I_0(a_i\sqrt{2x_i}) & x_i \geq 0 \\ 0 & \text{otherwise.} \end{cases} \tag{22}$$

For small signals ($a_i \ll 1$), the latter expression can be approximated by

$$f_s(x_i \mid a_i) \simeq \begin{cases} \exp(-\tfrac{1}{2}a_i^2)\exp(-x_i \exp(-\tfrac{1}{2}a_i^2)) & x_i \geq 0 \\ 0 & \text{otherwise} \end{cases} \tag{23}$$

by introducing the small-signal approximations for the exponential and Bessel functions.

The signal-to-noise ratio parameter in the incoherent case is defined as[7]

[6] The notation $n_i \sim N\{\mu, \sigma^2\}$ signifies that the random variable n_i is Gaussian with mean μ and variance σ^2.
[7] Except in the application to radar where in accordance with standard practice we define $S = \tfrac{1}{2}a^2$ max.

$$S = \sum_{i=1}^{M} \tfrac{1}{2}a_i^2/M. \tag{24}$$

If the observations (20) are obtained by match filter processing and square-law envelope detection of each of m signals, we would have $S = E/N_0$, where E now is the average signal energy per signal pulse. Because of the incoherent signal processing, only positive (or only negative) amplitudes can be used for the signal without violating the condition (3).[8] In the Monte Carlo analysis of detection performance, we use a set of amplitudes given by

$$a_i = a_0 \cdot i \qquad i = 1, 2, \cdots, M, \tag{25}$$

which as shown in Section IV gives an asymptotic performance close to the best possible for the class of signals,

$$a_i = a_0 \cdot i^{\nu} \qquad 0 < \nu < \infty. \tag{26}$$

The radar application described in Section VI also belongs to the incoherent case, but is based on a different set of amplitudes as determined by the antenna pattern.

IV. Asymptotic Performance

In [1] the asymptotic efficiency of a test is defined by a comparison of its asymptotic power (detection performance) with that of the asymptotically most powerful test. The following definition is taken from [1].

> Assume that the asymptotically most powerful test for H_0 against H_1 may be based on a test statistic T_{opt}, which is asymptotically normal $N(0, \sigma_0^2)$ under H_0 and which is asymptotically normal $N(\mu_0, \sigma_0^2)$ under H_1. A different test is based on the statistic T_* which is asymptotically normal $N(0, \sigma_*^2)$ under H_0 and asymptotically normal $N(\mu_*, \sigma_*^2)$ under H_1. Then the number
>
> $$e = \{\mu_* \sigma_0\}^2 \{\mu_0 \sigma_*\}^{-2} \tag{27}$$
>
> will be called the asymptotic efficiency of the test T_*.

The asymptotic efficiency as defined by (27) is easily shown to be equal to the asymptotic relative efficiency [5] for a test based on the statistic T_* and the optimum test based on T_{opt}.

The asymptotic performance of the rank tests described in Section II for the coherent and the incoherent cases can be determined by a slight modification of results presented in [1]. A brief summary of the pertinent theory is given in the Appendix, from which it is seen that the coherent and the incoherent cases, respectively, correspond to the pdf's (34) and (35).[9]

The asymptotic efficiency can be written (42)

$$e = \rho_1^2 \rho_2^2 \rho_3^2 \tag{28}$$

in which the values of the factors ρ_1, ρ_2, and ρ_3 are determined as explained in the Appendix.

[8] On the basis of the envelope information, positive and negative amplitudes of the same magnitude will produce outputs that are statistically equivalent.

[9] In the incoherent case this correspondence is only found when the signal-to-noise ratio is small (23). This is, however, the case for the asymptotic analysis.

TABLE I

	c_1-test	Spearman rho (=Kendall tau)
Coherent	$e = 1.0$ $L_\infty = 0.0$ dB	$e = 0.912$ $L_\infty = 0.4$ dB
Incoherent $= 1.0$		$e = 0.312$ $L_\infty = 2.54$ dB

In the coherent case, the signal amplitudes are chosen in accordance with (19). Since the average of these amplitudes is zero ($\sum a_i = 0$), we have $\rho_3 = 1.0$ for both the Spearman rho[10] and the c_1 tests.

For the c_1 test, any set of amplitudes[11] will result in $\rho_2 = 1.0$, whereas the specific choice of a uniform spacing also leads to $\rho_2 = 1.0$ for the Spearman rho test.

The factor ρ_1 is finally found from (37), (39), and (40). In [1, ch. 1, Table 1], the functions $\phi(u, f)$ and $\phi_1(u, f)$ can be found for various probability density functions. When $f_s(x \mid a)$ is Gaussian, $\phi(u, f_s) = \Phi^{-1}(u)$, where

$$\Phi(u) = \int_{-\infty}^{u} (2\pi)^{-1/2} \exp(-\tfrac{1}{2}t^2)\, dt$$

is the error function. For the c_1 test, we also find $\phi(u) = \Phi^{-1}(u)$, and therefore have $\rho_1 = 1.0$, whereas for the Spearman rho test $\phi(u) = u$, $0 < u < 1$, and thus $\rho_1 = 0.912$.

We therefore find the results shown in Table I for the coherent case when the amplitudes are given by (19).

In the incoherent case, the class of amplitudes given by (26) is considered. Both of the factors ρ_2 and ρ_3 will depend on ν, and the optimization of ν, therefore, is carried out with respect to the product $\rho_2\rho_3$. From the expressions (38) and (43) we find,

$$\rho_2\rho_3 = \frac{\nu(12\nu + 3)^{1/2}}{(\nu + 1)(2\nu + 1)}. \tag{29}$$

When $f_s(x \mid a)$ is of the exponential type as in (23), we find from [1, ch. 1],

$$\phi_1(u, f_s) = -\{1 + \ln(1 - u)\} \qquad 0 < u < 1, \tag{30}$$

so that for the Spearman rho test where $\rho(u) = u$, $0 < u < 1$, we find $\rho_1 = (\tfrac{3}{4})^{1/2} = 0.8660$. Using the relation (46) the minimum value of the asymptotic loss is found to be $L_\infty = 2.23$ dB for $\nu = 1.87$. For $\nu = 1.0$, the loss, however, is only increased by about 0.3 dB, and $\nu = 1.0$ was used in the Monte Carlo analysis of detection performance described in Section V, resulting in the asymptotic loss shown in Table I.

The c_1 test was not considered in the incoherent case

[10] Because the Kendall tau test is asymptotically equivalent to the Spearman rho test, all results obtained for the latter will in fact apply for both tests.

[11] Subject to certain restrictions as stated in ([1], ch. 7).

because its unique properties are only obtained when the noise is additive Gaussian. From the Appendix it is seen that incoherent detection by suitable choice of $g(i)$ and c_i will lead to a different asymptotically optimum detector for which $\rho_1 = \rho_2 = 1.0$. However, for a given signal the value of ρ_3 is determined by (43), which therefore will set an upper limit on the achievable asymptotic efficiency. For the type of signal considered in (26) and for $\nu = 1.0$, we find $\rho_3 = 0.667$, and thus the minimum asymptotic loss is $L_\infty = 1.76$ dB. This is only 0.8 dB below the value found above for the Spearman rho (or Kendall tau) test, and a further investigation of the performance of asymptotically optimum detectors for a finite sample size was therefore not carried out.

V. Detection Performance

From the start of this study it was expected that a Monte Carlo simulation would be the only feasible method of determining the detection performance of the various rank tests for a finite number of observations. It was also clear, however, that if the false-alarm probability and the corresponding decision thresholds would have to be determined by a direct simulation, this would severely limit the range of results that could be obtained. Emphasis was therefore placed on finding more effective methods of determining the required threshold settings.

Very few tables of the distribution of the Kendall tau test statistic under hypothesis H_0 are available in the statistical literature. A recursive procedure [2, sec. 5.4] could, however, be programmed easily to obtain the required results for values of M up to 64 and a false-alarm probability lower than $P_f = 10^{-6}$.

Several tables of the distribution of the Spearman rho test statistic are available under hypothesis H_0. For small values of M these have been obtained [2, p. 174, Table II] by exhaustive generation of all permutations in (7), a method that for larger M becomes computationally intractable. The most comprehensive tables are found in [4], where results for $M \leqq 30$ and $P_f \geqq 10^{-3}$ have been obtained by approximating the distributions by means of Gram–Charlier series. This same procedure might have been used to extend the tables toward lower P_f and larger M.

A more direct approach, however, did appear to be the use of an importance-sampling technique in a Monte Carlo simulation, in order to determine the required distributions. The fundamental principle of the technique of importance-sampling is to modify the probabilities that govern the outcome of the basic experiment of the simulation in such a way that the event of interest (the false alarm) occurs more frequently. This distortion is then compensated for by weighting each event by a factor that is determined as the ratio between the probability that this specific event would have occurred if the true probabilities had been used in the simulation and the probability of the same event with the distorted probabilities. By proper choice of the distortion the variance of the estimate obtained by the simulation can be significantly reduced,[12] whereas its expected value remains unchanged.

In the undistorted case, a completely random vector of ranks can be generated in terms of the relative rank order introduced in Section II, as follows.

The relative position l_1 of the largest observation is first determined as the outcome of an experiment in which the possible outcomes are the integers $i = 1, 2, \cdots, M$, all with equal probabilities $\Pr\{i\} = p_{1i} = 1/M$. The position l_2 is determined as the outcome of an experiment with outcomes $i = 1, 2, \cdots, M-1$ and probabilities $\Pr\{i\} = p_{2i} = 1/(M-1)$, and the position l_j as the outcome of an experiment with outcomes $i = 1, 2, \cdots, M+1-j$ and probabilities $\Pr\{i\} = p_{ji} = 1/(M+1-j)$. In this case, all possible vectors of ranks will have the same probability of occurrence, which is $P_0 = 1/M!$

With importance sampling, an almost identical procedure is used except that the probabilities p_{ji} for the individual experiments are modified in such a way that high values of l_i will have an increased probability of occurrence. This then leads to vectors of ranks for which large values of the test statistic are occurring more frequently, and thus in the Monte Carlo simulation will generate events on the upper tail of the distribution.

The probabilities p_{ji} used in the importance sampling were chosen as geometric progressions, with parameter Q such that

$$p_{ji} = \frac{1-Q}{1-Q^{M+1-j}} Q^{M+1-j-i} \qquad \begin{array}{l} i = 1, 2, \cdots, M+1-j \\ j = 1, 2, \cdots, M. \end{array} \tag{31}$$

The fraction preceding $Q^{M+1-j-i}$ is a normalization constant such that

$$\sum_{i=1}^{M+1-j} p_{ji} = 1 \qquad j = 1, 2, \cdots, M. \tag{32}$$

When the probabilities given by (31) are used in the simulation, the probability that a certain relative rank order $l_1, l_2, \cdots, l_M$ is generated will be

$$P' = p_{1l_1} p_{2l_2} \cdots p_{M-1, l_{M-1}}, \tag{33}$$

and the weight factor to be applied to the corresponding event is P'/P_0.

With this method used in the simulation of the test (7), a histogram showing the relative frequencies with which T_ρ occurs, but with each event weighted by P'/P_0, can be generated; and from this, the false-alarm probability versus threshold can be determined by integration. By varying the parameter Q any part of this curve can be generated. As an example of the results obtained, the case where $M = 32$ is shown in Fig. 1. Two different

[12] When the technique of importance sampling is used, it is important that the distortion of probabilities be chosen so that the events that are forced to occur are those which are the most probable when the true probabilities are used. Failure to achieve this may not lead to the desired reduction of variance.

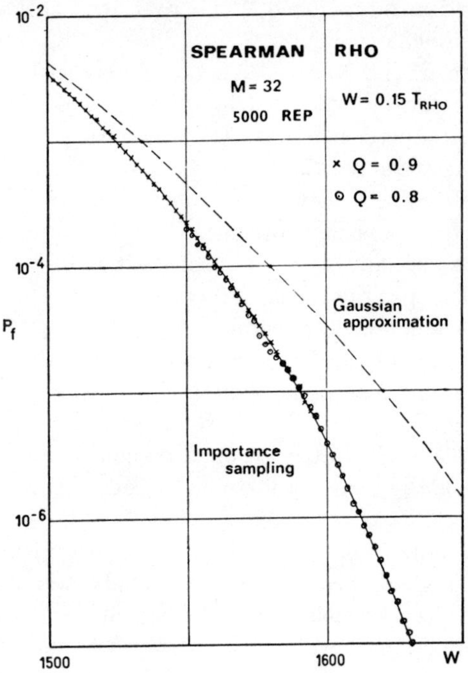

Fig. 1. False-alarm probability versus threshold for the Spearman rho test. The points on the curve were obtained by Monte Carlo simulation using an importance-sampling technique.

values of Q were used to generate the curve, and for each of these 5000 repetitions were made. For comparison, the curve resulting from a Gaussian approximation is also shown and is seen to differ substantially. Comparison of the results obtained by simulation with importance sampling and available tables showed excellent agreement [4].

No tables were discovered for the c_1 test in the form considered here, and curves of false-alarm probability versus threshold were determined by using the importance-sampling technique described above.

With the decision thresholds required to give specified values of P_f thus determined, detection curves for the rank tests could be determined by a straightforward Monte Carlo simulation.

For coherent detection, the observations were generated by means of pseudo-random-number generators according to (15) and by using the signal amplitudes given by (19). For each set of observations, the test statistics of interest were determined and compared with the decision thresholds. All detection events were counted, and after 500 repetitions the detection probability was estimated as the ratio between the number of detections and 500.

The results obtained for the Kendall tau test for $P_f = 10^{-6}$ are shown in Fig. 2. The reference curve in broken line is for the optimum parametric detector.

Fig. 3 shows results for the Spearman rho and the c_1 tests. The two sets of curves coincide and are shown in full line. A comparison with the dash–dotted curves for the Kendall tau test indicates that the difference in performance for all three detectors is small.

The results for the Kendall tau test are summarized in Fig. 4, which shows the loss in signal-to-noise ratio as a

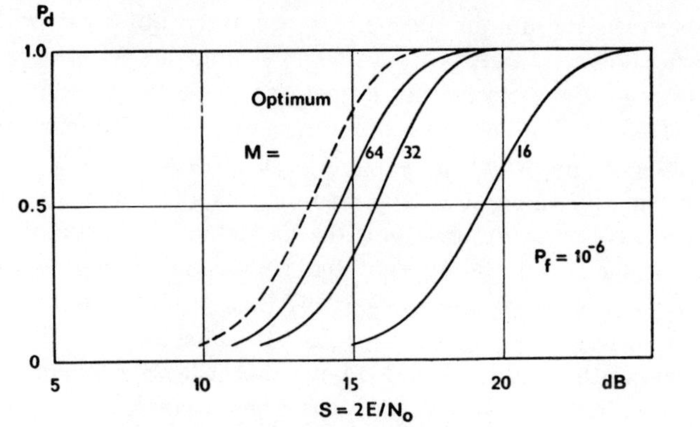

Fig. 2. Detection performance of the Kendall tau test in the coherent Gaussian case for $P_f = 10^{-6}$. The curve in broken line is for the optimum parametric detector.

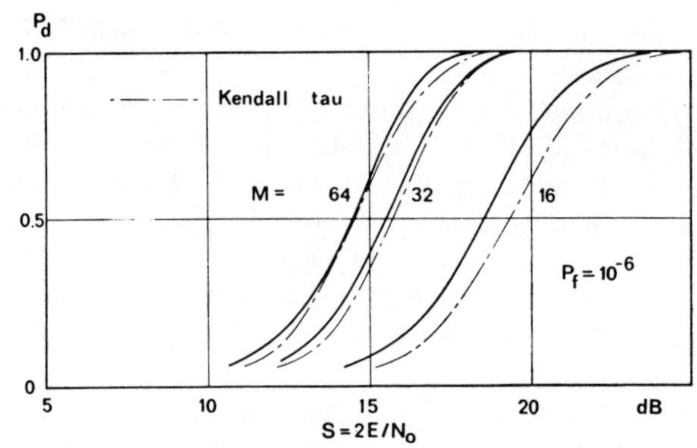

Fig. 3. Detection performance of the Spearman rho and c_1 tests in the coherent Gaussian case for $P_f = 10^{-6}$ shown in full line (the two sets of curves coincide). The dash–dotted curves are for the Kendall tau test.

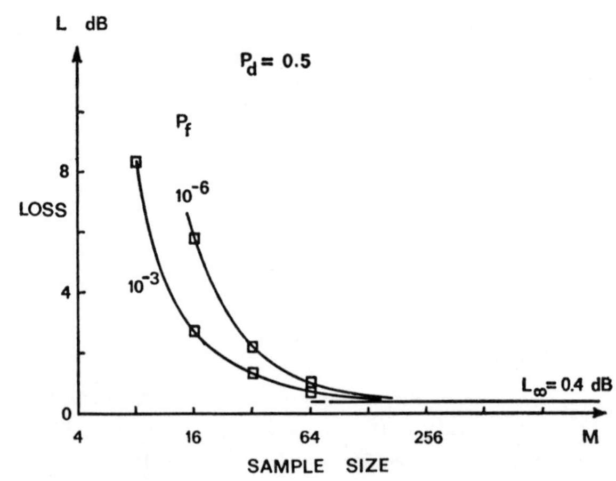

Fig. 4. Loss in signal-to-noise ratio for the Kendall tau test as a function of the number of observations in the coherent Gaussian case.

function of the number of observations. It is seen that the large loss for small values of M decreases rapidly for increasing M and becomes quite close to the asymptotic value $L_\infty = 0.4$ dB for $M = 64$.

In the incoherent Gaussian case, the simulation was performed according to (20), using the signal amplitudes given by (25). Using again 500 repetitions, the results obtained for the Kendall tau test are shown in Fig. 5 for

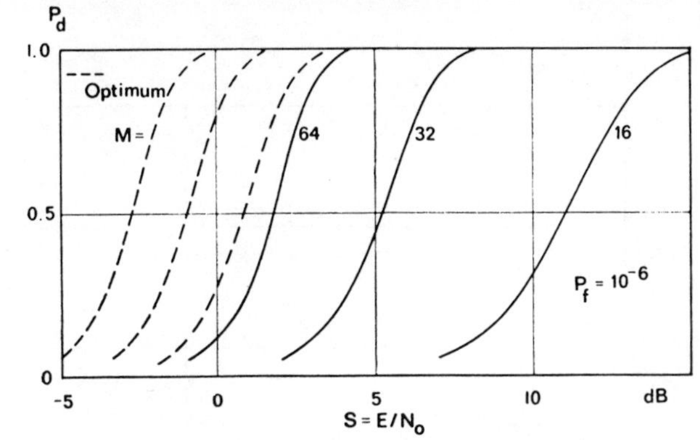

Fig. 5. Detection performance of the Kendall tau test in the incoherent Gaussian case for $P_f = 10^{-6}$. The reference curves in broken line are for the optimum parametric detector.

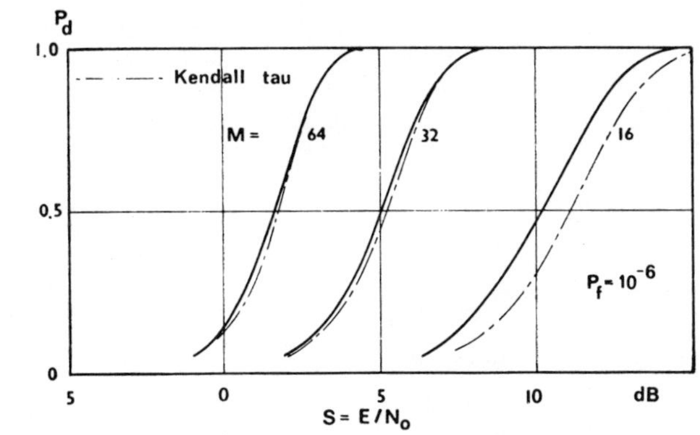

Fig. 6. Detection performance of the Spearman rho test in the incoherent Gaussian case for $P_f = 10^{-6}$. The dash–dotted curves are for the Kendall tau test.

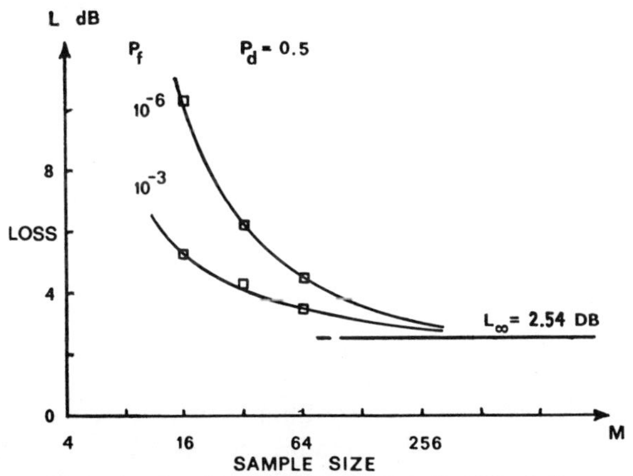

Fig. 7. Loss in signal-to-noise ratio for the Kendall tau test as a function of the number of observations in the incoherent Gaussian case.

$P_f = 10^{-6}$. The reference curves in broken line are for the optimum incoherent receiver, and were obtained by applying a correction of 1.3 dB to the optimum curves for m equal amplitude signals. The factor of 1.3 dB represents the asymptotic improvement when amplitudes (25) are used instead of equal amplitudes, for the same average signal-to-noise ratio (24).

In Fig. 6 results for the Spearman rho test are compared to the Kendall tau. As in the coherent case, the difference between the two detectors is small.

The results for the Kendall tau test in the incoherent case are summarized in Fig. 7. The curves show the same tendency as in Fig. 4 but are further from the asymptotic value for $M = 64$.

VI. Application to Radar

Previous adaptations of one-sample nonparametric tests to narrow-band signal detection have assumed a partially coherent case where phase coherence can be maintained between individual observations, but the overall signal phase is unknown. The nonparametric tests are then usually designed to operate independently on the quadrature components of the received data, and the two resulting statistics are subsequently combined [6], [7]. When applications to radar are considered, this approach is rendered difficult by the presence of unknown Doppler shifts, which makes a Doppler-filter bank mandatory.

To avoid this complication in a conventional scanning pulsed-radar system, use can be made of the fact that the received train of echoes is amplitude modulated owing to the radiation pattern of the scanning antenna. Thus, when a target is present, a characteristic rank order will appear in the received data; and it appears to be possible to use one of the nonparametric rank tests described previously for this detection situation. Because of the closeness in performance of these tests, we shall consider only the Kendall tau test, which in this particular application has a distinct practical advantage.

Such a nonparametric radar detector, as it would apply to a single range element, is shown in Fig. 8. A matched filter is used to maximize the signal-to-noise ratio against white noise, and after envelope detection the amplitudes are converted into digital numbers in an analog-to-digital (A/D) converter. For a given range bin, these observations enter a multibit shift register of length m, which is divided into two halves. A Kendall tau test is applied to the contents of each of these two shift registers and the smallest of the resulting outputs is selected in a type of AND circuit and compared with the decision threshold W in the comparator.[13] The two Kendall tau tests were chosen so that a target return centered on the shift register, and thus, consisting of a series of amplitudes decreasing towards the ends of the shift register, will maximize the two outputs.

With this detector, advantage is taken of the ease with which the Kendall tau test statistic is updated when a new observation enters the shift register and an old observation is shifted out.

The length M of the shift register can be related to the number of observations N received between the 3-dB points of the one-way antenna pattern, by the ratio $\gamma = M/N$. In the detector described above, $\gamma = 2.0$ was chosen

[13] The practical consequences of the digital implementation of the radar detector have not been considered, and the analysis of this section would only be strictly valid for an analog system. In particular, the digital representation of the amplitudes will make the earlier assumption of continuous cdf's invalid, and the possibility of ties between individual observations would have to be taken into account.

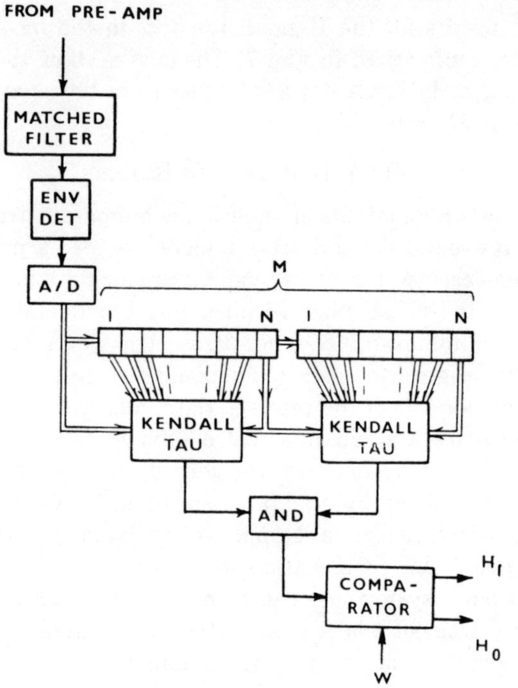

Fig. 8. Block diagram of nonparametric radar detector. The system is only shown as it would apply to a single range element.

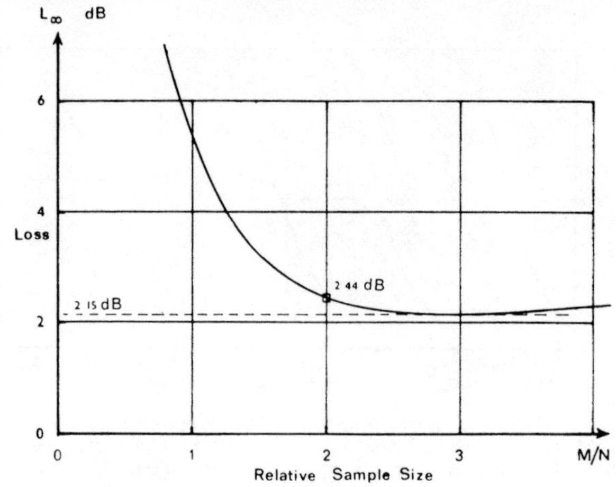

Fig. 9. Asymptotic optimization of nonparametric radar detector performance.

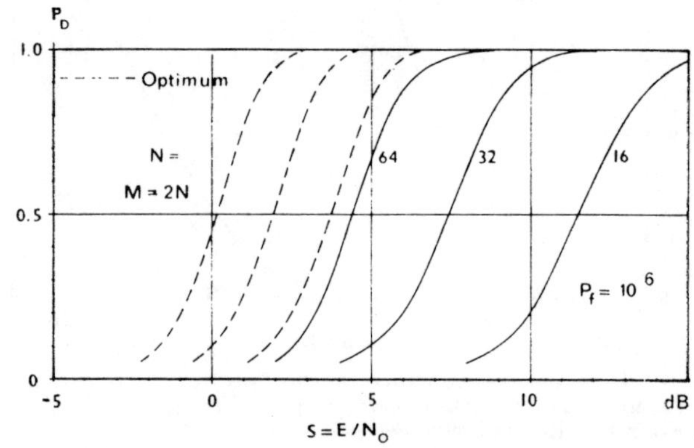

Fig. 10. Detection performance of nonparametric radar detector in the incoherent Gaussian case for $\gamma = 2.0$. The optimum curves are shown in broken line.

and thus the two halves of the shift register each have the length N. This choice is justified by its closeness to the asymptotically optimum value as shown below.

If instead of the AND logic a summation circuit had been used in Fig. 8, the asymptotic efficiency could be determined directly from the formulas in the Appendix. This is because the asymptotic efficiency of a test leading to an asymptotically normal test statistic is not changed by splitting the total number of observations into a finite number of statistically equivalent blocks (in our case two) and adding the individual test statistics. Thus, the asymptotic efficiency has been determined by assuming a Gaussian antenna radiation pattern and treating γ as a parameter. In Fig. 9 the asymptotic loss is shown as a function of γ. A minimum value of $L_\infty = 2.15$ dB is found for $\gamma = 2.8$, but the value $\gamma = 2.0$ only increases the loss by 0.3 dB.

A further loss, independent of the above optimization, has to be taken into account in the asymptotic analysis; it results from the use of the AND logic. Since the two inputs to this circuit are asymptotically normal, this loss can be calculated easily for given values of P_f and P_d. For $P_f = 10^{-3}$ and $P_d = 0.5$, a loss of 0.41 dB was found, and for $P_f = 10^{-6}$ and $P_d = 0.5$ a loss of 0.35 dB. In total, we therefore find an asymptotic loss for the radar detector shown in Fig. 8, which is typically $L_\infty = 2.15 + 0.3 + 0.35 = 2.8$ dB.

The detection performance for a finite number of observations was obtained by a Monte Carlo simulation in which for simplicity only the output from one of the Kendall tau computations is compared with the threshold W. If the false-alarm and detection probabilities for this case are P_f' and P_d', we find directly for the overall radar detector owing to the symmetry,

$$P_d = P_d'^2, \qquad P_f = P_f'^2.$$

Thus with the threshold set to yield $P_f' = 10^{-3}$ for the output from one Kendall tau test, the overall false-alarm probability will be $P_f = 10^{-6}$.

With the signal-to-noise ratio parameter defined as $S = \frac{1}{2}a_{\max}^2$, where $a_{\max}$ is the signal amplitude corresponding to the central part of the antenna pattern, the results obtained by the simulation are shown in Fig. 10 for $P_f = 10^{-6}$. For reference, the optimum curves for incoherent radar detection are shown in broken line. In [8] it has been shown that these curves, to a very good degree of accuracy, can be taken as the optimum result for N equal amplitude observations modified by a beam-shape factor of 1.6 dB, for $4 < N < \infty$ the maximum error is ± 0.2 dB. The loss for the nonparametric radar detector is seen to be 6–8 dB for $N = 16$ and 3–4 dB for $N = 64$.

A summary of the results for the radar detector is shown in Fig. 11. As in the earlier results for incoherent detection (Fig. 7), a large number of observations is required for the performance to approach the asymptotic value.

Another possibility for the nonparametric radar detector would be to use a single Kendall tau test for all M observations. Simulation results for this case showed an improve-

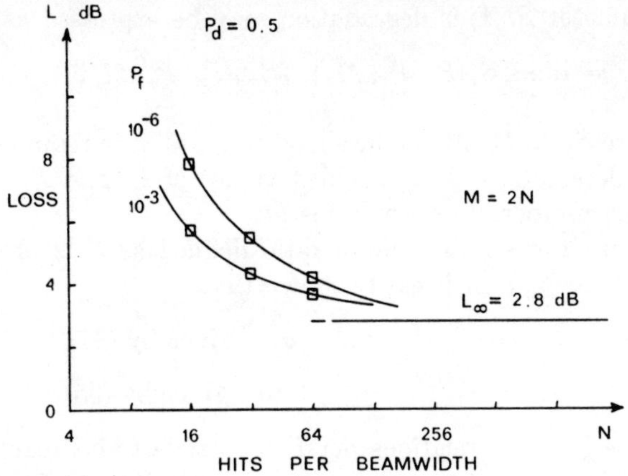

Fig. 11. Loss in signal-to-noise ratio for nonparametric radar detector as a function of the number of hits per beamwidth N and with $\gamma = 2.0$.

ment in performance of 0.5–1.0 dB, whereas asymptotically the difference would only amount to around 0.35 dB as shown above. This principle would not, however, allow a simple updating of the test statistic when a new observation enters the detector, and a more complex implementation would be required.

VII. Concluding Remarks

The detection performances of various nonparametric rank tests have been determined for signals in Gaussian noise exhibiting a characteristic trend. For a finite number of observations, results were obtained by Monte Carlo simulation, whereas analytical results could be obtained for the asymptotic case.

The resulting loss in signal-to-noise ratio depends considerably on the number of observations M. Taken as a whole, the results obtained indicate that in order for an asymptotic analysis to provide a reliable prediction of performance, the number of observations should exceed 50–100. Results not presented here, for the case where the signal amplitudes are Rayleigh fading, have shown that an even larger number may sometimes be required.

A comparison of the results obtained shows that for the situations considered, the detection performance of the Kendall tau, Spearman rho, and c_1 tests are very similar. A choice between these tests for practical applications can then be made on the basis of their hardware complexity. From this point of view the Kendall tau test has certain advantages.

A specific application of the Kendall tau test to the case of conventional radar has demonstrated the feasibility of exploiting the modulation of the target echo due to the beam shape of the antenna in a nonparametric radar detector. As one might expect, the resulting loss for a moderate number of observations is significant. To reduce this loss without introducing the need for phase coherence between individual observations (and consequently, requiring a Doppler-filter bank for radar use) one would probably have to resort to a two-sample test for which a reference set of noise observations has to be available.

Appendix

Asymptotic Efficiency of Rank Tests

The asymptotic efficiency of simple linear rank tests is treated in [1] for a model slightly different from the one considered in this paper. In the following, the main results from [1] will be reviewed and the necessary modification to obtain the results needed here is introduced.

In [1] the asymptotic efficiency is determined for a case where the hypothesis H_0 (noise only) is nonparametric and of the form stated in Section II in connection with (1). The hypothesis H_1 (signal and noise), however, assumes knowledge of the signal amplitudes $\boldsymbol{a}$ and the form of the pdf $f_s(x_i \mid a_i)$, but otherwise is as stated in connection with (2). The following two functional forms for this pdf are considered in [1]:

$$f(\boldsymbol{x}) = \prod_{i=1}^{M} f_s(x_i - a_i) \tag{34}$$

$$f(\boldsymbol{x}) = \prod_{i=1}^{M} \exp\left(-\tfrac{1}{2}a_i^2\right) \exp\left[-x_i \exp\left(-\tfrac{1}{2}a_i^2\right)\right]. \tag{35}$$

In the first of these density functions, the signal appears as a location parameter and in the second as a scale parameter.

For this hypothesis test, it is shown in [1] that the asymptotically optimum test is also a rank test and that the asymptotic efficiency of a general rank test is given by

$$e = \rho_1^2 \rho_2^2, \tag{36}$$

where

$$\rho_1 = \left\{\int_0^1 \phi(u, f_s)\phi(u)\, du\right\} \cdot \left\{\int_0^1 \phi^2(u, f_s)\, du \int_0^1 (\phi(u) - \bar{\phi})^2\, du\right\}^{-1/2}, \tag{37}$$

$$\rho_2 = \lim_{M\to\infty} \left\{\sum_{i=1}^{M} (c_i - \bar{c})(d_i - \bar{d})\right\} \cdot \left\{\sum_{i=1}^{M} (c_i - \bar{c})^2 \sum_{i=1}^{M} (d_i - \bar{d})^2\right\}^{-1/2}, \tag{38}$$

and $\phi(u)$ is a square-integrable function related to the scores $g(i)$ by

$$\lim_{M\to\infty} \int_0^1 \{g(1 + \operatorname{int}(uM)) - \phi(u)\}^2\, du = 0. \tag{39}$$

For the pdf given by (34)[14], $d_i = a_i$ and

$$\phi(u, f_s) = \{f_s'(F_s^{-1}(u))\} \cdot \{f_s(F_s^{-1}(u))\}^{-1} \qquad 0 \leqq u \leqq 1. \tag{40}$$

For the pdf as given by (35)[14], $d_i = \frac{1}{2}a_i^2$ and the function $\phi(u, f_s)$ is replaced by

$$\phi_1(u, f_s) = -1 - F_s^{-1}(u) \cdot f_s'(F_s^{-1}(u))\{f_s(F_s^{-1}(u))\}^{-1}. \tag{41}$$

In this paper a situation is considered in which both

[14] See Section III, which describes the cases considered.

hypotheses are parametric, and the pdf $f_n(x_i)$, therefore, is also of known form. This changes the asymptotic results for the optimum test, whereas the performance for the nonparametric rank tests is unchanged since they are invariant to changes in the noise pdf. The optimum test when both hypotheses are parametric can be found easily, and by taking into account the difference in performance between the optimum parametric test and the asymptotically optimum rank test, we find that the asymptotic efficiency has to be modified by including a factor ρ_3^2, so that it becomes

$$e = \rho_1^2 \rho_2^2 \rho_3^2 \tag{42}$$

where

$$\rho_3 = \lim_{M\to\infty} \left\{ \sum_{i=1}^{M} (d_i - \bar{d})^2 \right\}^{1/2} \left\{ \sum_{i=1}^{M} d_i^2 \right\}^{-1/2} \tag{43}$$

This factor reflects the fact that when the pdf of both $f_n(x_i)$ and $f_s(x_i \mid a_i)$ are known, the optimum detector can take advantage of an average difference in location or scale, a possiblity that does not exist when $f_n(x_i)$ is unknown.

From the above result (42), it is seen that the asymptotic efficiency is determined by three factors, each of which can be given a specific interpretation.

- ρ_1 depends on the relationship between the scores, through $\phi(u)$, and the pdf $f_s(x_i \mid a_i)$, through $\phi(u, f_s)$. By proper choice of the scores one can make $\phi(u) = \phi(u, f_s)+$ constant and thus ensure that $\rho_1 = 1$.
- ρ_2 is the correlation coefficient between the regression constants c_i and the parameters d_i. By choosing $c_i = d_i$ we can obtain $\rho_2 = 1$.
- ρ_3 is a measure of the effectiveness of the signal used. If $\bar{d} = \sum d_i/M = 0$, we have $\rho_3 = 1$. It may, however, not always be possible to choose the signal to satisfy this condition (e.g., the incoherent case).

Instead of the asymptotic efficiency as defined above, one may express the asymptotic performance in terms of the asymptotic degradation of the signal-to-noise ratio parameter S. This degradation may be expressed as

$$L_\infty = \lim_{M\to\infty} \{S_*(P_d, P_f, M)\} \cdot \{S_{\text{opt}}(P_d, P_f, M)\}^{-1} \tag{44}$$

where $S(P_d, P_f, M)$ is the signal-to-noise ratio required by the detector to yield specified values of P_d and P_f for a given number of observations M.

With the signal-to-noise ratio defined as S is proportional to $\sum a_i^2$ it is easily shown that

$$L_\infty = e^{-1} \qquad f_s(x_i \mid a_i) \text{ given by (34)} \tag{45}$$

$$L_\infty = e^{-1/2} \qquad f_s(x_i \mid a_i) \text{ given by (35).} \tag{46}$$

These last two relations make it possible to compare the asymptotic performance with the results obtained in Sections V and VI for a finite number of observations as shown in Figs. 1, 4, and 7.

Acknowledgment

The author is indebted to Dr. A. J. Zöttl for his careful criticism of the manuscript, and to Dr. H. Brøns for a helpful discussion. Computer time for this research was generously provided by the Northern Europe University Computing Centre (NEUCC) on their IBM 7094 installation.

References

[1] J. Hájek and Z. Sidak, *Theory of Rank Tests.* Prague: Academic, 1967.
[2] M. G. Kendall, *Rank Correlation Methods.* London: Griffin, 1948.
[3] D. A. S. Fraser, *Nonparametric Methods in Statistics.* New York: Wiley, 1957.
[4] G. J. Glasser and R. F. Winter, "Critical values of the coefficient of rank correlation for testing the hypothesis of independence," *Biometrika,* vol. 48, pp. 444–448, 1961.
[5] J. Capon, "Optimum coincidence procedures for detecting weak signals in noise," *1960 IRE Internatl. Conv. Rec.,* pt. 4, pp. 154–166.
[6] R. F. Daly and C. K. Rushforth, "Nonparametric detection of a signal of known form in additive noise," *IEEE Trans. Information Theory,* vol. IT-11, pp. 70–76, January 1965.
[7] J. W. Carlyle, "Nonparametric methods in detection theory," in A. V. Balakrishnan, *Communication Theory.* New York: McGraw-Hill, 1968.
[8] W. M. Hall, "Antenna beam-shape factor in scanning radars," *IEEE Trans. Aerospace and Electronic Systems,* vol. AES-4, pp. 402–409, May 1968.

Detection Performance of the Narrow-Band Wilcoxon Detector Against Gaussian Noise

V. GREGERS HANSEN

***Abstract*—Two adaptations of the nonparametric Wilcoxon test for use in detecting a coherent train of narrow-band signal pulses with unknown starting phase in a background of statistically undefined noise have been proposed in the literature. The performance of these detectors when the noise is stationary Gaussian is determined and compared. For a finite train of signal pulses, results were obtained by Monte Carlo simulation using the variance-reducing technique of importance sampling. For the best of the two narrow-band Wilcoxon detectors a comparison with the optimum parametric detector shows that for a train of 30 signal pulses and a false-alarm probability of $P_f = 10^{-6}$ the loss is about 2 dB, while for 50 signal pulses the loss is reduced to 1 dB. The theoretical asymptotic loss is 0.2 dB.**

Manuscript received November 9, 1971; revised February 13, 1972.
The author is with the Raytheon Company, Wayland, Mass. 01778.
[1] It is assumed that the coherent oscillator frequency is exactly equal to the signal frequency. A frequency offset would require phase shifting to be performed on the tapped delay line outputs.

I. Introduction

The familiar problem of detecting a signal consisting of a coherent train of M narrow-band pulses with unknown starting phase is considered. These M pulses may be because of an M-element coded signal, or the target return on M consecutive sweeps in a classical pulse radar. The carrier frequency of the signal is assumed known exactly, but a Doppler shift could, as usual, be taken into account by introduction of a Doppler filter bank or the equivalent.

After single-pulse filtering, quadrature detection,[1] and sampling of the output in a tapped delay line, the set of observations available to the detector under each of the two hypotheses can be written in complex vector notation as

$$\boldsymbol{x} = \boldsymbol{s} + \boldsymbol{n} \qquad (\mathrm{H}_1) \tag{1}$$

$$\boldsymbol{x} = \boldsymbol{n} \qquad (\mathrm{H}_0), \tag{2}$$

where $\boldsymbol{n} = (n_1,\cdots,n_M)$ represents the M complex noise observations, each of which is defined in terms of a pair of quadrature components, i.e., $n_i = n_{ci} + jn_{si}$, $i = 1,\cdots,M$. The signal vector $\boldsymbol{s} = (s_1,\cdots,s_M)$ has, in accordance with the assumptions, the components

$$s_i = A \exp(j\theta) = A \cos\theta + jA \sin\theta, \tag{3}$$

where A is the (constant) signal amplitude and θ the unknown signal phase.

If the statistical properties of the noise are uncertain, nonparametric detection procedures can be used to ensure a constant false-alarm rate in this detection situation. Under rather general conditions regarding the statistical properties of the noise, various one-sample nonparametric tests [1], [2] can be adapted to the above detection problem. The modification required is due to the unknown phase in the narrow-band signal as specified by (3). Several authors [3], [4] have shown how this adaptation can be accomplished for the Wilcoxon test, the sign test, or the normal-scores test.

Considerable difficulties often arise in the performance evaluation of such nonparametric detectors for a finite number of observations. Analytical results are in general difficult to obtain except when the number of observations is small [5], and Monte Carlo simulation on a digital computer is often the only feasible approach. Such a performance analysis for the narrow-band Wilcoxon detector is presented in the following. However, in order to avoid some of the usual limitations of a Monte Carlo simulation, the variance-reducing technique of importance sampling was adapted to the problem.

II. The Narrow-Band Wilcoxon Detector

The observation vector at the detector input is $\boldsymbol{x} = \{x_1, \cdots, x_M\}$ with complex components $x_i = x_{ci} + jx_{si}$. The implementation of the narrow-band Wilcoxon detector that we will primarily consider was proposed by Carlyle [3]. In this implementation the classical Wilcoxon statistic is first computed individually for the real (in-phase) and the imaginary (quadrature) components of the observations. For this the absolute values of the observations have to be ranked so that R_{ci}^+ is the rank of $|x_{ci}|$ in

$$|x_{c1}|, |x_{c2}|, \cdots, |x_{cM}| \tag{4}$$

and R_{si}^+ is the rank of x_{si} in

$$|x_{s1}|, |x_{s2}|, \cdots, |x_{sM}|. \tag{5}$$

From these the required Wilcoxon statistics are

$$t_{cW} = \sum_{i=1}^{M} R_{ci}^+ \cdot u(x_{ci}) \tag{6}$$

$$t_{sW} = \sum_{i=1}^{M} R_{si}^+ \cdot u(x_{si}), \tag{7}$$

where $u(x)$ is the unit step function. These statistics are therefore simply the sum of the ranks R^+ of the positive observations with the ranks calculated using absolute values. Then defining

$$t_{cW}' = t_{cW} - \tfrac{1}{4}M(M+1) \tag{8}$$

$$t_{sW}' = t_{sW} - \tfrac{1}{4}M(M+1) \tag{9}$$

it is known from the properties of the Wilcoxon statistic [1] that the statistics t_{cW}' and t_{sW}' have pdf's that are symmetric around zero. The test statistic of the narrow-band Wilcoxon detector can then be defined as [3]

$$T_W = \sqrt{t_{cW}'^2 + t_{sW}'^2}. \tag{10}$$

A slightly different implementation was proposed by Helstrom [4]. In this the Wilcoxon statistics given by (6) and (7) are also used but (8) and (9) are replaced by

$$t_{cW}' = \max\{t_{cW}, [\tfrac{1}{2}M(M+1) - t_{cW}]\} \tag{11}$$

$$t_{sW}' = \max\{t_{sW}, [\tfrac{1}{2}M(M+1) - t_{sW}]\} \tag{12}$$

and the detector output is again given by (10). Equivalent expressions for (11) and (12) are

$$t_{cW}' = |t_{cW} - \tfrac{1}{4}M(M+1)| + \tfrac{1}{4}M(M+1) \tag{13}$$

$$t_{sW}' = |t_{sW} - \tfrac{1}{4}M(M+1)| + \tfrac{1}{4}M(M+1). \tag{14}$$

As shown in the following sections this detector was found to be slightly inferior to that specified by (6)–(10) in the specific case of a Gaussian noise background.

For stationary white Gaussian noise the optimum parametric detector computes the statistic

$$T_{\text{opt}} = \sqrt{\left(\sum_{i=1}^{M} x_{ci}\right)^2 + \left(\sum_{i=1}^{M} x_{si}\right)^2}. \tag{15}$$

This detector was taken as reference in the following performance evaluation.

III. False-Alarm Threshold

In order to determine the detection performance of a detector it is necessary to know the threshold setting required to yield a specified false-alarm probability P_f.

To determine this relationship analytically for the narrow-band Wilcoxon detector appeared to be difficult and recourse was taken to a Monte Carlo simulation on a digital computer. However, for this to yield results for low values of the false-alarm probability with an acceptable computational effort, the technique of importance sampling [6] was used in the simulation.

A direct Monte Carlo simulation of the narrow-band Wilcoxon detector in the case where only noise is present can probably be implemented most simply by generating values of $u(x_{ci})$ and $u(x_{si})$ in (6) and (7) as either 0 or 1 with equal probabilities $p = q = \frac{1}{2}$ and using an arbitrary set of ranks, for example $R_{ci}^+ = R_{si}^+ = i$. It is then straightforward to generate sample values of the statistic T_W from the equations given.

The problem encountered when such a direct simulation is attempted is that the smallest probability that can be estimated is of the order of $(N_R)^{-1}$ (corresponding to one single event in N_R repetitions of the basic simulation). Thus, in order to estimate a low value of the false-alarm probability a prohibitively large number of repetitions N_R may be required. The technique of importance sampling [6] can be used to overcome this limitation. The fundamental principle of importance sampling is to distort the probabilities used in the simulation in such a way that the events of interest take place more frequently. This distortion of the probabilities is then accounted for by weighting each event by a factor that is equal to the likelihood of obtaining this specific event if the true probabilities had been used in the simulation, divided by the likelihood of obtaining this event with the distorted probabilities actually used.

For the narrow-band Wilcoxon detector the events of interest are large values of the detector statistic T_W, since such values determine the tail of the distribution and therefore the false alarm thresholds. A large value of T_W results when either $|t_{cW}'|$ or $|t_{sW}'|$ is large. This in turn is seen to correspond to situations where the Wilcoxon statistics t_{cW} or t_{sW} are either small or large. Therefore by changing the probabilities used in generating sample values of $u(x_{ci})$ and $u(x_{si})$ from $p = q = \frac{1}{2}$ to $p = \alpha$ and $q = 1 - \alpha$ this can be made to happen if either $\alpha > \frac{1}{2}$ or $\alpha < \frac{1}{2}$

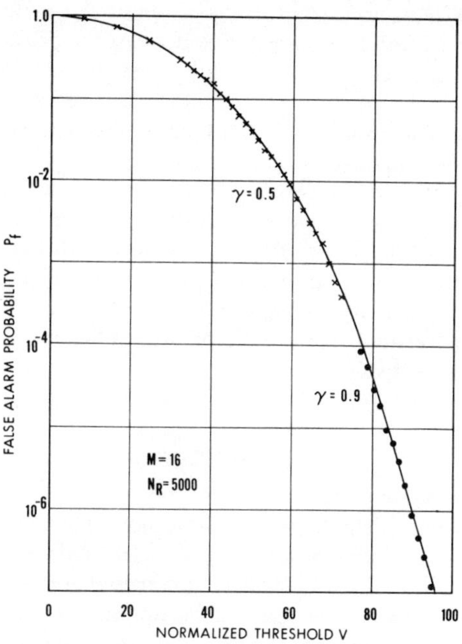

Fig. 1. False-alarm probability versus normalized threshold for the Carlyle narrow-band Wilcoxon detector as determined by Monte Carlo simulation using importance sampling.

TABLE I
THRESHOLDS FOR NARROW-BAND WILCOXON DETECTORS NORMALIZED TO A MAXIMUM VALUE OF 100.0

P_f	M: 8	16	32	64
10^{-1}	59.2 (78.2)	43.2 (70.2)	31.2 (64.4)	21.6 (60.2)
10^{-2}	76.8 (88.0)	58.4 (78.0)	43.2 (70.2)	30.4 (64.4)
10^{-3}	88.8 (94.2)	69.2 (83.8)	51.2 (74.4)	37.2 (67.6)
10^{-4}		76.8 (88.0)	58.4 (78.2)	43.2 (70.4)
10^{-5}		83.6 (91.6)	64.0 (81.2)	47.6 (72.8)
10^{-6}		89.6 (94.6)	68.8 (83.8)	51.2 (75.0)

The first value given is for the Carlyle version of the detector, which has a maximum output of $M(M+1)/2\sqrt{2}$. The second threshold value (in parentheses) is for the Helstrom version with maximum value $M(M+1)/\sqrt{2}$.

is used. In order for the importance sampling to result in the desired variance reduction it is required that the likelihood of all events of interest is increased to a similar extent. This can be achieved in the present situation by selecting the value of α for use in each successive repetition in a probabilistic manner.

The following procedure was therefore used to implement importance sampling in the Monte Carlo simulation. First, select a fixed probability $\gamma > \frac{1}{2}$. This is the importance sampling parameter of the simulation. On each repetition of the basic simulation decide with probability $\frac{1}{2}$ whether to use $\alpha = \gamma$ or $\alpha = 1 - \gamma$. Then generate sets of values of $u(x_{ci})$ and $u(x_{si})$ as described previously using the resulting value of α. Finally compute the sample value of the statistic T_W.

The weight w to be attached to the event that a specific value of T_W has occurred is determined as follows. If the true probabilities ($p = q = \frac{1}{2}$) are used in the simulation, all (discrete) outcomes are equally likely and each has a probability of $(\frac{1}{2})^{2M}$. In the simulation using importance sampling denote the total number of ones that occur among the $2M$ values of $u(x_{ci})$ and $u(x_{si})$ as L. The likelihood of the corresponding event T_W is then $0.5\gamma^L(1-\gamma)^{2M-L} + 0.5(1-\gamma)^L\gamma^{2M-L}$ where the two terms represent the equally likely cases of $\alpha = \gamma$ and $\alpha = 1 - \gamma$, respectively. The value of the weight is consequently

$$w = \frac{(0.5)^{2M}}{0.5[\gamma^L(1-\gamma)^{2M-L} + (1-\gamma)^L\gamma^{2M-L}]}. \tag{16}$$

After N_R repetitions, a histogram of events, each weighted according to (16), is constructed. By numerical integration of this histogram a curve of false-alarm probability versus threshold can be determined. An example of such a curve obtained for the Carlyle version of the detector is shown in Fig. 1 for $M = 16$, two values of γ, and $N_R = 5000$ repetitions for each value of γ. The crosses obtained for $\gamma = 0.5$ represent a direct simulation and the lower limit is $P_f \cong N_R^{-1}$ as stated previously. The solid points were obtained using the importance sampling technique for $\gamma = 0.9$. The abscissa in Fig. 1 is normalized to a maximum value of T_W of 100.0, which, in the case of Carlyle's version of the narrow-band Wilcoxon detector, corresponds to the value $M(M+1)/2\sqrt{2}$. These results, as well as those obtained for other values of M, are summarized in terms of the false-alarm thresholds as shown in Table I. The thresholds are again normalized to a maximum value of 100.0 as described previously. No randomization of the thresholds was found necessary as the largest deviation of P_f, observed in the worst case of $M = 8$ and $P_f = 10^{-3}$, was only 20 percent.

Similar results for the Helstrom version of the detector were also obtained; the thresholds are shown between parentheses in Table I. The maximum was again normalized to 100.0, which, for the Helstrom narrow-band Wilcoxon, corresponds to $M(M+1)/\sqrt{2}$.

IV. DETECTION PERFORMANCE

With the false-alarm thresholds determined as described in the previous section, the detection performance against stationary Gaussian noise was determined by a direct Monte Carlo simulation. In this simulation the detector input specified by (1) is generated on each successive repetition by adding independent Gaussian noise vectors to a complex signal with random phase as specified by (3). If the quadrature noise power is normalized to unity, the overall (integrated) signal-to-noise ratio is defined as

$$S = M \cdot \frac{A^2}{2},$$

where A is the signal amplitude. For each value of S a total of $N_R = 500$ repetitions was carried out in order to estimate the probability of detection. The results for the Carlyle version of the narrow-band Wilcoxon detector are shown in Figs. 2 and 3 for $P_f = 10^{-3}$ and 10^{-6}, respectively. With $N_R = 500$ repetitions it is estimated that these results are accurate to within ± 0.2 dB. The performance of the optimum parametric detector is shown in the broken-line curve in these figures. The equivalent losses in signal-to-noise ratio for $P_d = 0.5$ as read from Figs. 2 and 3 are shown in Fig. 4 as a function of the number of observations M, with the false-alarm probability as parameter. Also shown in Fig. 4 are results for $P_f = 10^{-1}$ and 10^{-2} and the asymptotic loss of $L_\infty = 0.2$ dB. This value of the loss follows directly from the fact that the asymptotic relative efficiency of the usual Wilcoxon detector against Gaussian noise is $E = 3/\pi$ and that we have $L_\infty = E^{-1} = 1.0472 \sim 0.201$ dB.

The results obtained for the implementation of the narrow-band Wilcoxon detector proposed by Helstrom are very similar in shape to Figs. 2 and 3 and are summarized in Fig. 4 as the broken-line curves. The loss is seen to be slightly larger for this detector.

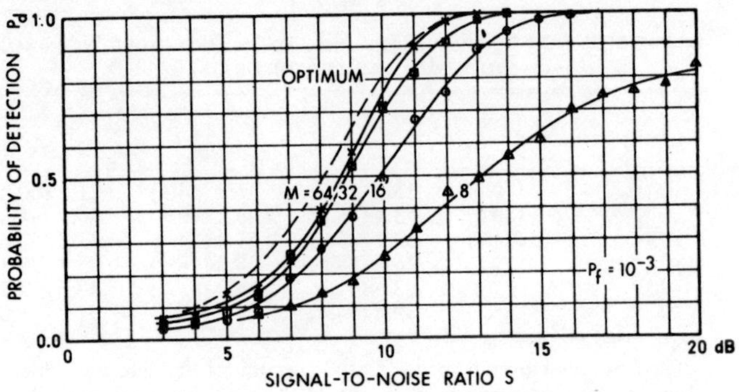

Fig. 2. Detection performance of Carlyle narrow-band Wilcoxon detector for $P_f = 10^{-3}$ with M as parameter. Broken-line curve—optimum parametric detector.

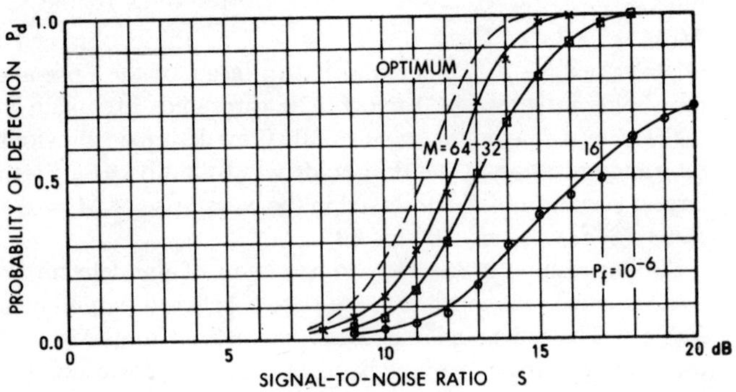

Fig. 3. Detection performance of Carlyle narrow-band Wilcoxon detector for $P_f = 10^{-6}$ with M as parameter. Broken-line curve—optimum parametric detector.

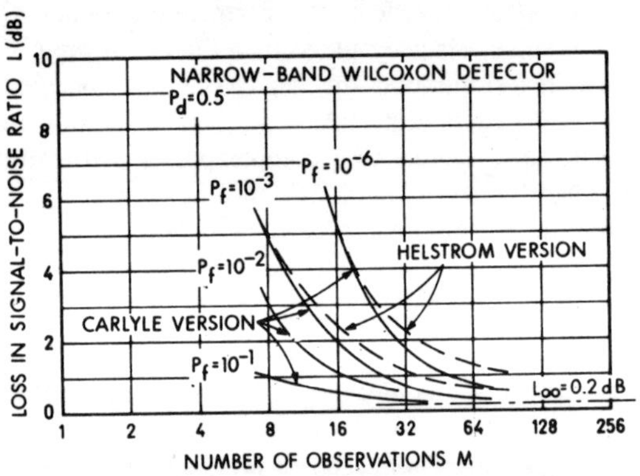

Fig. 4. Solid line—equivalent loss in signal-to-noise ratio for $P_d = 0.5$ from Figs. 2 and 3. Dash-dotted line—asymptotic loss of 0.2 dB. Broken line—corresponding results for the Helstrom version of the detector. Additional results for $P_f = 10^{-1}$ and 10^{-2} are also shown for the Carlyle version of the detector.

The asymptotic loss in this case depends on the particular values of P_d and P_f assumed. For large M the detector given by (13), (14), and (10) essentially adds a bias to the absolute value of each of the components t_{cW}' and t_{sW}' before squaring and summing. The resulting effect will be similar to the loss resulting from approximating an envelope detector by the sum of the absolute values of the two quadrature components. This loss has been found in recent work by Nathanson and Luke [7] to be about 0.5 dB for $P_f = 10^{-3}$ and about 0.7 dB for $P_f = 10^{-6}$. Adding the asymptotic loss of 0.2 dB for the Wilcoxon detector to these values, the asymptote of the curves in Fig. 4 for the Helstrom version of the detector are estimated to be 0.7 and 0.9 dB, respectively.

V. Conclusions

The detection performance for a coherent train of narrow-band signal pulses in stationary Gaussian noise has been determined for two adaptations of the nonparametric Wilcoxon test. It has been shown that the equivalent loss in signal-to-noise ratio for the best of the two detectors is asymptotically within 0.2 dB of the optimum. When the number of observations is $M = 50$, the loss for $P_d = 0.5$ and $P_f = 10^{-6}$ is about 1 dB, and it increases rapidly as M decreases further. Thus, these results demonstrate the inadequacy of asymptotic results in predicting detector performance for a small number of observations.

The major complication in a practical implementation of the Wilcoxon detector is the requirement for a ranking of the absolute values of the quadrature observations. This requirement is avoided in the narrow-band adaptation of the sign detector at the cost of an additional loss of $1\frac{1}{2} - 2$ dB.[2]

[2] These detailed results will be presented in a subsequent paper.

References

[1] J. Hájek, *Nonparametric Statistics*. San Francisco: Holden-Day, 1969.
[2] G. E. Noether, *Elements of Nonparametric Statistics*. New York: Wiley, 1967.
[3] J. W. Carlyle, "Nonparametric methods in detection theory," in *Communication Theory*, A. V. Balakrishnan, Ed. New York: McGraw-Hill, 1968, ch. 8.
[4] C. W. Helstrom, *Statistical Theory of Signal Detection*. London: Pergamon, 1969.
[5] J. Klotz, "Small sample power and efficiency for the one sample Wilcoxon and normal scores tests," *Ann. Math. Statist.*, vol. 34, pp. 624–632, 1963.
[6] J. M. Hammersley and D. C. Handscomb, *Monte Carlo Methods*. London: Methuen, 1964.
[7] F. E. Nathanson and P. J. Luke, "Loss from approximations to square-law detectors in quadrature systems with postdetection integration," *IEEE Trans. Aerosp. Electron. Syst.*, vol. AES-8, pp. 75–77, Jan. 1972.

Reprinted with permission from *IEEE Trans. Aerosp. Electr. Syst.* **AES-10**:479–486 (1974)

A Distribution-Free Doppler Processor

GEORGE M. DILLARD, Senior Member, IEEE
JOHN T. RICKARD, Student Member, IEEE
Naval Electronics Laboratory Center
San Diego, Calif. 92152

Abstract

A distribution-free Doppler processor (DFDP) is described. This procedure is applicable to the detection of signals with unknown phase or signals which undergo a change in phase from one observation to the next. The procedure has the characteristic that the false-alarm probability is a constant, independent of the probability distribution of the received data when no signal is present. Also, an estimate of the Doppler frequency, when applied to a Doppler radar processor, is provided.

Introduction

Distribution-free methods of detection are described extensively in the literature [1]-[3].[1] In radar detection systems, these procedures are usually applied to the video output or the IF output, under the assumption that the signal phase is known. Some procedures, such as the Wilcoxon test [4], can be applied when the phase is unknown; Helstrom [5, p. 325] describes this application of the Wilcoxon test. This procedure has the drawback that if the signal phase changes significantly during the observation period, the test is likely to fail; that is, the signal will not be detected.

[1]These are only a few of the many references on the subject of distribution-free detection. They are given because they also contain extensive bibliographies.

This paper describes a distribution-free Doppler processor (DFDP) that is applicable to the detection of signals with unknown phase. It is assumed that the received signals are Doppler shifted such that the signal phase may change from one observation to the next. That is, the phase at the jth observation is $jw_dT + \phi$, where ϕ is a random variable, $w_d/2\pi$ is the Doppler frequency, and T is the period between observations.

Several other assumptions are made and are discussed in more detail later. The most important assumption is that there is available a set of observations, called the known-noise sample (KNS), which will be used as a reference against which the possible-signal sample (PSS) is "compared." A practical method of obtaining the KNS is also described.

The distribution-free procedure described here has the characteristic that the false-alarm probability is a constant, independent of the probability distribution of the received data when no signal is present. The procedure also provides an estimate of the Doppler frequency (and, thus, the target velocity) when applied to a Doppler radar processor. The procedure is easily implemented and requires only slightly more equipment than a conventional digital Doppler processor that utilizes the discrete Fourier transform (DFT).

The Detection Problem

The distribution-free Doppler processor (DFDP) is applied to the outputs from a two-path detector, which is shown in Fig. 1. The received echo signal, if it is present, is prefiltered so that the input to the detector is of the form $A \cos [2\pi(f_c + f_d)t + \phi]$, where f_c and f_d are the carrier and Doppler frequencies, respectively, and ϕ is a random phase angle. The local oscillator operates at the zero-Doppler frequency f_c, or differs from f_c by a known amount.

The outputs $I(t)$ and $Q(t)$ are observed at times $0, T, \cdots, (M-1)T$, to obtain $x_0, x_1, \cdots x_{M-1}; y_0, y_1, \cdots, y_{M-1}$, where x_j and y_j are values obtained from the I and Q

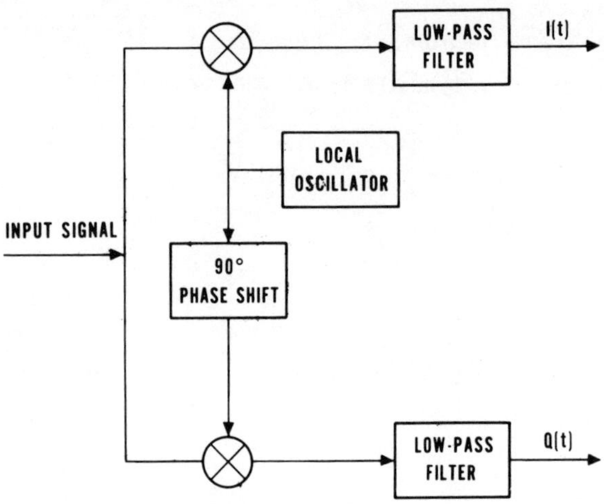

Fig. 1. A two-path detection system.

channels, respectively, at time jT.[2] The set of M pairs (x_j, y_j), $j = 0, 1, \cdots, M-1$, of observations is the possible-signal sample (PSS). Also assume that, for each j, observations $u_{j1}, u_{j2}, \cdots, u_{j,N-1}; v_{j1}, v_{j2}, \cdots, v_{j,N-1}$, are made such that, when no signal is present in the PSS, the random variables $\{X_j, Y_j, U_{jk}, V_{jk}\}$, $j = 0, 1, \cdots, M-1, k = 1, 2, \cdots, N-1$, are independent and identically distributed (IID) random variables. The set of M vectors $(u_{j1}, \cdots, u_{j,N-1}; v_{j1}, \cdots, v_{j,N-1})$, $j = 0, 1, \cdots, M-1$, is called the known-noise sample (KNS). Methods for obtaining the KNS are discussed later. The assumptions mentioned above, and further assumptions, are formalized in the statement of the hypotheses below.

H_0(No Signal): The random variables X_j, Y_j, U_{jk}, V_{jk}, $j = 0, 1, \cdots, M-1$; $k = 1, 2, \cdots, N-1$, are IID, with unknown density function $f_0(x)$ such that $f_0(x) = f_0(-x)$.

H_1 (Signal): The random variables $U_{jk}, V_{jk}, j = 0, 1, \cdots, M-1$; $k = 1, 2, \cdots, N-1$, are IID with density function $f_0(x)$. Also, X_j has density $f_0[x - A\cos(f\phi + \phi)]$ and Y_j has density $f_0[x - A\sin(j\theta + \phi)]$, $j = 0, 1, \cdots, M-1$, where $A \geqslant 0$ and $\theta = 2\pi f_d T$ are unknown (nonrandom) parameters, and ϕ is a random variable that is uniformly distributed over the interval $[0, 2\pi]$.[3]

In testing the above hypotheses, it is desirable that the false-alarm probability α (the probability that H_0 is rejected when it is true) be fixed and independent of the probability distribution of the data. This is the distribution-free nature of the procedure. Additionally, whenever H_1 is true, it is desirable that the procedure not only accepts H_1, but also produces an estimate of the parameter θ; that is, in radar terminology, the procedure estimates the Doppler frequency and, hence, the velocity of the target. A special case is the situation where θ is assumed known and, without loss of generality, can be assumed zero by properly offsetting the local oscillator frequency shown in Fig. 1. This corresponds to the detection of targets for which the velocity is known, and the procedure in this case only accepts or rejects H_0; that is, there is no estimate of θ.

Before describing the distribution-free procedure to be used for testing the above hypotheses, some observations are made that help to justify the selection procedure. First, further assumptions could be made about the probability distribution of the data so that a "best test" could be applied (e.g., Neyman-Pearson). But making such assumptions would be counter to the intent of distribution-free procedures; that is, the false-alarm probability is likely to be severely affected by differences between the assumed and true distributions. Second, since it is unwise to make enough assumptions to uniquely define a "best test," whatever procedure that is selected is, of necessity, ad hoc. Finally, the procedure that is selected must be subject to analysis and/or simulation and, to be practical, must be easily implemented.

The Distribution-Free Doppler Processor (DFDP)

The procedure to be described for testing the above hypotheses is based on the ranks of the samples. For each j, $j = 0, 1, \cdots, M-1$, the rank R_j of x_j with respect to $(x_j; u_{j1}, \cdots, u_{j,N-1})$ and the rank S_j of y_j with respect to $(y_j; v_{j1}, \cdots, v_{j,N-1})$ are determined and the decision to accept or reject H_0 is based on the M pairs of ranks $(R_0, S_0), (R_1, S_1), \cdots, (R_{M-1}, S_{M-1})$. The procedure for deciding between H_0 and H_1 is developed in the following.

Under H_0, the random variables $X_j; U_{j1}, \cdots, U_{j,N-1}$, are IID. Therefore,

$$P_0[R_j = a] = 1/N \tag{1}$$

for $a = 1, 2, \cdots, N$.[4] Similarly, it follows that

$$P_0[S_j = b] = 1/N \tag{2}$$

for $b = 1, 2, \cdots, N$. Note that (1) and (2) are independent of the density $f_0(x)$.[5] (In fact, it is only necessary that the random variables $X_j; U_{j1}, \cdots, U_{j,N-1}$ be IID with a continuous distribution function for (1) and (2) to be true.) Since the probability distribution of the ranks R_j and S_j is independent of the probability distribution of the data when H_0 is true, the false-alarm probability for any procedure based on (R_j, S_j), $j = 0, 1, \cdots, M-1$, is also independent of the probability distribution of the data.

[2]This sampling procedure describes the method by which observations are taken in a particular range bin of a range-gated pulse-Doppler radar. All other range bins of interest are treated in the same manner.

[3]Note than ϕ is independent of j; that is, the random part of the signal phase does not change during the observation period. Also note that $\theta = 2\pi f_d T$, where f_d is the Doppler frequency; therefore, $\theta/2\pi$ is the Doppler frequency normalized to the radar PRF and can be restricted to the interval $[0, 1]$.

[4]The notation $P_0[\cdot]$ and $P_1[\cdot]$ is used to denote the probability of the parenthetical expression when H_0 and H_1 are true, respectively.

[5]Equations (1) and (2) can be obtained from (3) and (4) with $A = 0$, and with $A = 0$, (3) and (4) are independent of $f_0(x)$.

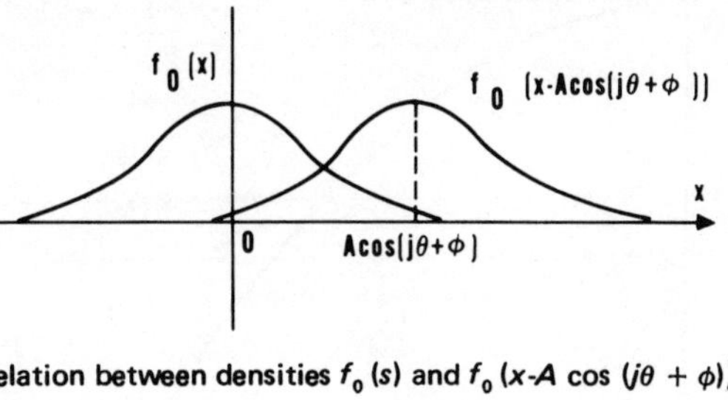

Fig. 2. Relation between densities $f_0(s)$ and $f_0(x\text{-}A\cos(j\theta+\phi))$.

When H_1 is true, the probability distributions of R_j and S_j depend on j, except for the case where $\theta = 0$. Also, the distributions of R_j and S_j depend on the specific choice for $f_0(x)$. The probability distributions of R_j and S_j under H_1 can be found from [1, p. 78]:

$$P_1[R_j = a] = \binom{N-1}{a-1} \int_{-\infty}^{\infty} [F_0(x)]^{a-1}[1-F_0(x)]^{N-a} \cdot f_0[x - A\cos(j\theta+\phi)]\,dx \quad (3)$$

$$P_1[S_j = b] = \binom{N-1}{b-1} \int_{-\infty}^{\infty} [F_0(x)]^{b-1}[1-F_0(x)]^{N-b} \cdot f_0[x - A\sin(j\theta+\phi)]\,dx \quad (4)$$

for a, $b = 1, 2, \cdots, N$, and where $f_0(x) = dF_0(x)/dx$. Note that (3) and (4) involve the unknown (at least the *assumed* unknown) density $f_0(x)$, and, to obtain results, such as detection probabilities, this density must be specified. Specifying $f_0(x)$ for purposes of computing detection probabilities, or for purposes of obtaining detection probabilities by Monte Carlo simulation, does not affect the fact that the procedure is distribution-free, since the false-alarm probability is still independent of the distribution $F_0(x)$.

From hypothesis H_1, when the phase is such that $\cos(j\theta + \phi) > 0$, the presence of a signal causes the density $f_0(x)$ to shift to the right, as indicated in Fig. 2. In this case, (3) indicates that the rank R_j tends to take on larger values than when no signal is present. Conversely, when $\cos(j\theta + \phi) < 0$, the density $f_0(x)$ is shifted left by $A\cos(j\theta + \phi)$, and the rank R_j tends to take on smaller values than when no signal is present.[6] Similar comments apply to the rank S_j relative to the sign of $\sin(j\theta + \phi)$. If a signal having Doppler frequency f_d is present, the density function $f_0(x)$, as a function of j, is shifted by an amount that varies as the cosine of the Doppler frequency in the I channel, and is shifted by an amount that varies as the sine of the Doppler frequency in the Q channel.

Define

$$U_j = R_j - (N+1)/2 \quad (5)$$

[6]When no signal is present, all rank values are equally likely, and the expected value of R_j and S_j is $(N + 1)/2$. When a signal is present and $\cos(j\theta + \phi) > 0$, $E[R_j] > (N + 1)/2$, and when $\cos(j\theta + \phi) < 0$, $E[R_j] < (N + 1)/2$. Similarly, $E[S_j] > (N + 1)/2$ for $\sin(j\theta + \phi) > 0$ and $E[S_j] < (N + 1)/2$ for $\sin(j\theta + \phi) < 0$.

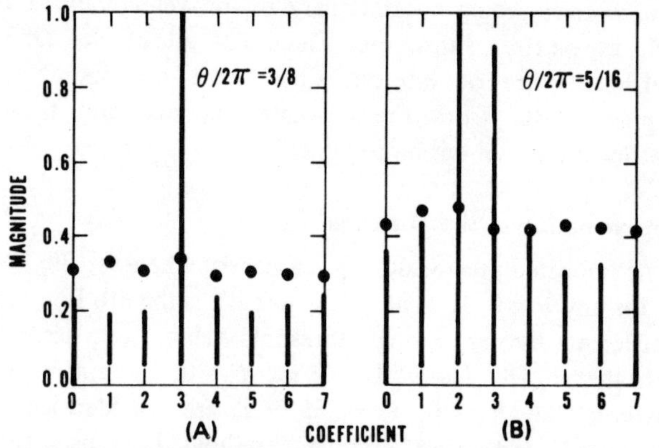

Fig. 3. Average Fourier coefficient magnitudes for $N = 10$, $M = 8$, and SNR = 3 dB. Circles are for noise only.

$$V_j = S_j - (N+1)/2 \quad (6)$$

(The "centered ranks" U_j and V_j are not to be confused with the random variables U_{jk} and V_{jk} of the known noise sample.) The above discussion then indicates that when a signal is present with Doppler frequency f_d, the vector defined by the point (U_j, V_j) "rotates" at the Doppler frequency. That is, the most probable value for $\tan^{-1}(V_j/U_j)$ is approximately $j\theta + \phi$. (The approximation occurs owing to the discreteness of U_j and V_j.) This suggests the following detection procedure. The transformed ranks U_j and V_j are treated as the complex number $Z_j = U_j + iV_j$, $i = \sqrt{-1}$, and the discrete Fourier transform (DFT)

$$W_h = \sum_{j=0}^{M-1} Z_j \exp(-2\pi ihj/M), \qquad h = 0, 1, \cdots, M-1 \quad (7)$$

is computed. The magnitude of each Fourier coefficient W_h, $h = 0, 1, \cdots, M-1$, is compared with a threshold T_0, and if $|W_h| \geqslant T_0$ for any h, H_1 is accepted and the doppler frequency, or target velocity, is estimated from the indices of the coefficients that exceed T_0. Since hypothesis H_1 tacitly assumes that only one signal is present, the estimate of θ is obtained from the index of the coefficient having the maximum magnitude. If $|W_h| < T_0$ for all h, then hypothesis H_0 is accepted.

Fig. 3 provides an indication of the relationship between the magnitudes of the Fourier coefficients under H_0 and H_1 for the indicated set of parameters. These data are averages obtained by Monte Carlo simulation, and for this simulation it is assumed that $f_0(x)$ is a zero-mean, unit-variance, Gaussian density. The coefficient magnitudes are normalized to have maximum value 1.0. In Fig. 3(A), the normalized Doppler frequency $\theta/2\pi$ is such that a "response" occurs only for the Fourier coefficient W_3; thus, the peak value is approximately five times larger than the remaining values. In Fig. 3(B), $\theta/2\pi$ is such that a response occurs for both coefficients W_2 and W_3; also, the peak values are only approximately 2.5 times larger than the remaining values.

In most cases, the purpose of frequency-domain processing is to eliminate returns from unwanted signals. These are usually the signals with Doppler frequencies near zero, corresponding to stationary or low-velocity targets, such as clutter, ships, etc. Thus, not all of the DFT coefficients are of interest, since some correspond to Doppler shifts from the unwanted targets, and these coefficients would not be computed.

Determination of the Threshold T_0

As indicated previously, H_1 is accepted whenever $|W_h| \geqslant T_0$ for any $h = 0, 1, \cdots, M-1$, where W_h is the hth Fourier coefficient. Ideally, a single threshold value T_0 is desired such that $P_0[|W_h| \geqslant T_0] = \alpha$ for $h = 0, 1, \cdots, M-1$. However, $|W_h|, h = 0, 1, \cdots, M-1$, are not identically distributed, and using a single threshold T_0 causes the false-alarm probability to vary with the index h of the Fourier coefficients. However, as shown later, this variation is expected to be small.

To simplify the problem, the value of T_0 is chosen so that $P_0[|W_0| \geqslant T_0] = \alpha$; that is, T_0 is determined from the distribution of $|W_0|$, and the same T_0 is used for $h = 0, 1, \cdots, M-1$. It is recognized that the 0th coefficient is normally of little interest, since it corresponds to zero-Doppler (i.e., stationary) targets. However, this procedure for determining T_0 is used, since the distribution of $|W_0|$ can be easily calculated. The calculation, and justification for its use in specifying false-alarm probabilities, is described below.

The real and imaginary parts of W_h are given by

$$X_h \triangleq \mathrm{Re}\,(W_h) = \sum_{j=0}^{M-1} U_j \cos(2\pi jh/M) + V_j \sin(2\pi jh/M) \tag{8}$$

$$Y_h \triangleq \mathrm{Im}\,(W_h) = \sum_{j=0}^{M-1} V_j \cos(2\pi jh/M) - U_j \sin(2\pi jh/M) \tag{9}$$

where U_j and V_j are defined in (5) and (6). Under H_0, U_j and V_j are IID with mean zero and variance $(N+1)(N-1)/12$. Thus, X_0 and Y_0 are IID random variables and are simply the sums of the transformed ranks U_j and V_j; that is,

$$X_0 = \mathrm{Re}\,(W_0) = \sum_{j=0}^{M-1} U_j \tag{10}$$

$$Y_0 = \mathrm{Im}\,(W_0) = \sum_{j=0}^{M-1} V_j \tag{11}$$

The distribution of X_0 (and, thus, Y_0) is found by an M-fold convolution of the discrete, uniform distribution of R_j, followed by the transformation

$$X_0 = \sum_{j=0}^{M-1} R_j - M(N+1)/2. \tag{12}$$

Given the distribution of X_0 and Y_0, the distribution of $|W_0|$ is found by a simple enumeration. This distribution has been tabulated for several choices of N and M. From the tabulated distributions the threshold T_0 that gives (at least approximately) the desired value of α can be found. By the central limit theorem, the distribution of the random variables $X_0/\sqrt{M}$ and $Y_0/\sqrt{M}$ approaches a Gaussian distribution with mean zero and variance $(N+1)(N-1)/12$, for large values of M. Therefore, the distribution of $|W_0|/\sqrt{M}$ approaches a Rayleigh distribution with parameter $(N+1)(N-1)/12$. Using the Rayleigh approximation for the distribution of $|W_0|$, the quantity $P_0[|W_0| > T]$ is plotted in Fig. 4 for $N = 10$ and $M = 32$, along with several points which have been calculated from the precise, tabulated distribution of $|W_0|$. This illustrates the accuracy of the Rayleigh approximation for this case.

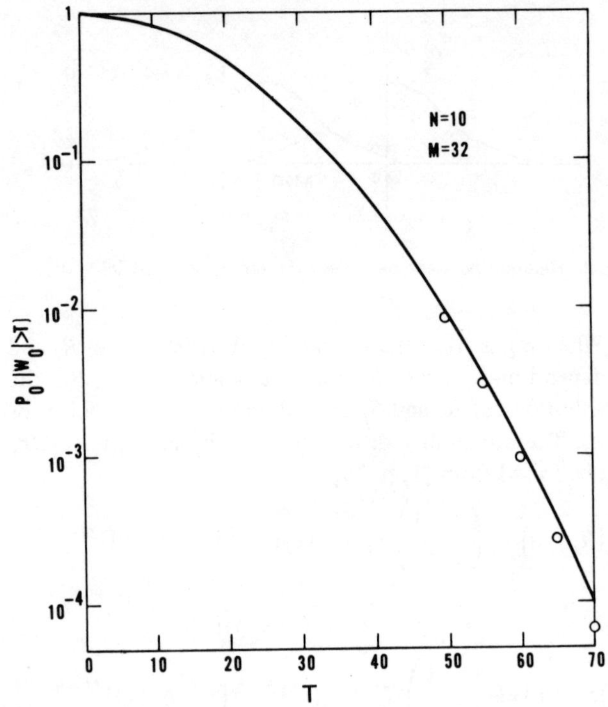

Fig. 4. Rayleigh approximation to $P_0[|W_0| > T]$. Circles indicate exact values.

For all values of h, the random variables X_h and Y_h are seen to be orthogonal, with mean zero and variance $M(N+1)(N-1)/12$. It is further shown in the Appendix that, for large M, the distributions of $X_h/\sqrt{M}$ and $Y_h/\sqrt{M}$ approach a Gaussian distribution with mean zero and variance $(N+1)(N-1)/12$. Hence, regardless of the value of h, $|W_h|/\sqrt{M}$ is approximately Rayleigh-distributed, with parameter $(N+1)(N-1)/12$, for large M. This lends some credibility to the use of a single threshold T_0.

Estimation of Detection Probability

The probability of detection is estimated by using Monte Carlo simulation. This simulation involves performing a "large" number of tests and recording the proportion of tests for which a detection occurs. The detection threshold T_0 is calculated as indicated in the previous section, and a detection occurs whenever the magnitude of any of the Fourier coefficients exceeds the threshold.

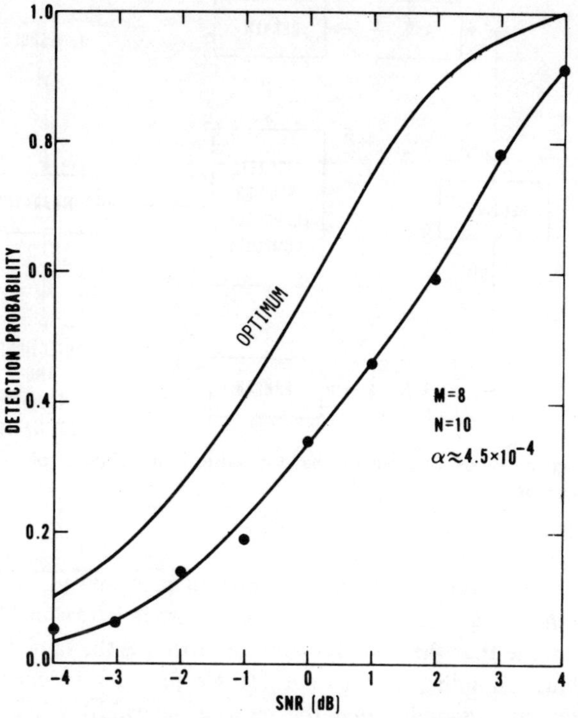

Fig. 5. Detection probability as a function of SNR = 10 $\log_{10}$ $A^2/2$.

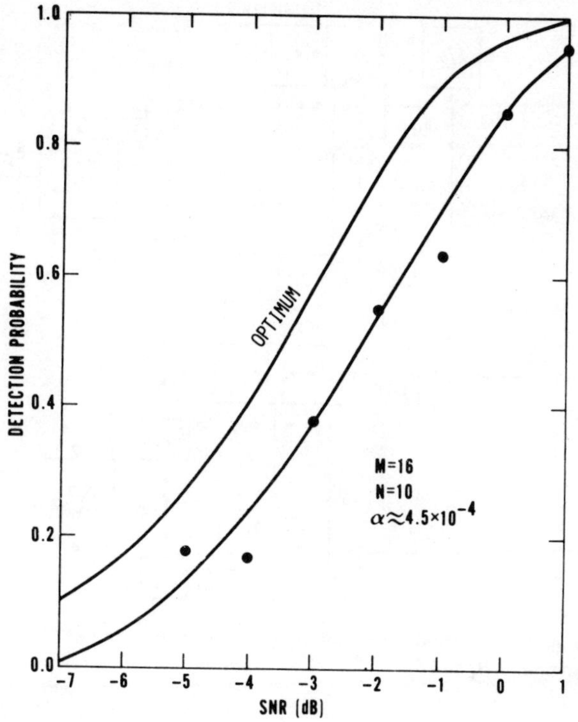

Fig. 6. Detection probability as a function of SNR = 10 $\log_{10}$ $A^2/2$.

Figs. 5 and 6 show the detection probability as a function of SNR = 10 $\log_{10} A^2/2$ for the indicated parameters. Each point represents the result of 100 simulated tests for which the underlying distributions are Gaussian. Also shown are the curves for optimum processing of the input, assuming the Doppler frequency is known but the phase ϕ is unknown, and assuming that $f(x)$ is a zero-mean, unit-variance, Gaussian density.

From the data of Figs. 5 and 6 it is seen that the DFDP procedure requires approximately 1.5 dB more SNR than the optimum procedure. However, this is not a fair comparison inasmuch as the "optimum" procedure is only optimum under very restrictive assumptions. For example, if the distribution of the data varies from that which is assumed, the false-alarm probability might become intolerable or the detection probability might become worse than for the DFDP [7]. Even though the DFDP may have a low detection probability under some conditions, the false-alarm probability is guaranteed to remain constant, provided only that the limited assumptions are satisfied.

Estimation of the Doppler Frequency

Whenever the Doppler frequency is such that $\theta/2\pi = m/M$, where m is an integer between 0 and M, then the mth Fourier coefficient should be the one, if any, that exceeds the threshold (provided that a false alarm does not occur). This is indicated in Fig. 3(A), where it is seen that, for $\theta/2\pi = 3/M$, the third coefficient has an average magnitude that is significantly larger than any of the remaining coefficients. This means that Doppler frequencies corresponding to values of θ such that $\theta/2\pi = m/M$, can be accurately estimated from the index of the coefficient having the maximum magnitude. The estimate is $f_d = m/MT$ Hz for $0 < m < M/2$, and is $f_d = -(M - m)/MT$ Hz for $M/2 \leq m < M$. Since the unambiguous Doppler frequencies are those in the interval $(-1/2T, 1/2T)$, the estimate will also be ambiguous unless a priori information provides for resolving the ambiguity.

Whenever the Doppler frequency is such that $(m-1)/M < \theta/2\pi < m/M$, either the $(m-1)$th or the mth coefficient (or both) should be the one, if any, that exceeds the threshold. If only one exceeds the threshold, the velocity estimate is as indicated in the previous paragraph. However, if both exceed the threshold, the velocity estimate is the average of the estimates corresponding to each coefficient.

Obtaining the Known-Noise Sample (KNS)

The KNS can be obtained by sampling in the time domain, the frequency domain, or in combinations of both. The most important objective to be met is to ensure that X_j, $Y_{jk}U_{jk}$, V_{jk}, $j = 0, 1, \cdots, M-1$; $k = 1, 2, \cdots, N-1$ are IID random variables when no signal is present, as required by hypothesis H_0. The method of obtaining the KNS that will be described here is time-domain sampling; this procedure is chosen because it is directly applicable to a pulsed-radar receiver.

In the following, assume that a pulsed radar transmits pulses at PRF pulses per second, and that each pulse is τ seconds duration. Thus, there will be approximately $1/\tau$ PRF range bins, and, as indicated previously, all range bins are treated in the same manner. The KNS and the ranks R_j

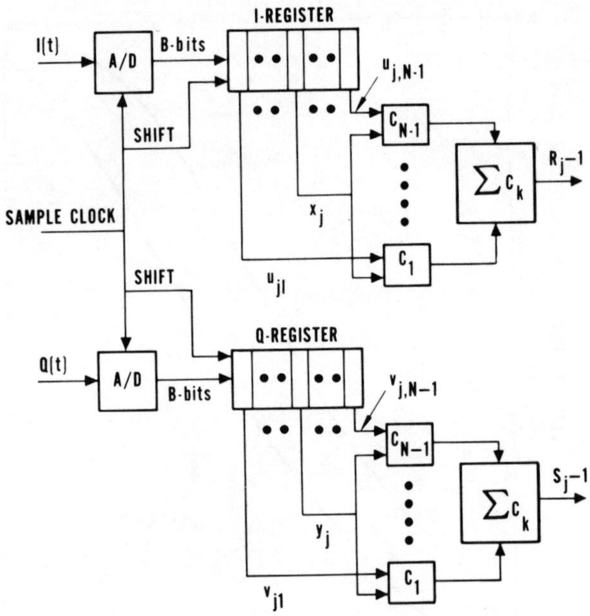

Fig. 7. A method for obtaining the known-noise sample (KNS) and the ranks R_j and S_j.

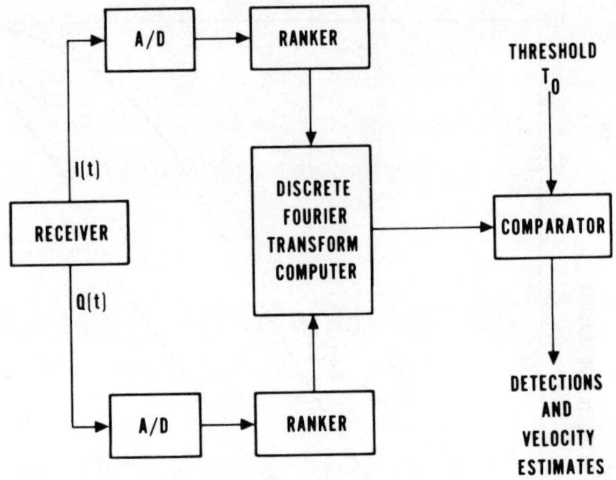

Fig. 8. Block diagram of the distribution-free Doppler processing system.

and S_j are obtained simultaneously as follows [1, p. 118]. (See Fig. 7.) The outputs $I(t)$ and $Q(t)$ from the two-path detector of Fig. 1 are sampled and converted to B-bit binary numbers by the A/D converters.[7] These sampled outputs are shifted into the I and Q registers at the clock rate $1/\tau$ Hz. The binary numbers x_j and y_j at the center of the registers are compared (in comparators $C_1, C_2, \cdots, C_N$) with the remaining samples, as indicated in Fig. 7. (The index that would correspond to the range bin number is suppressed, since all range bins are treated in the same manner.) The output from each comparator is a 1 if x_j (or y_j) is the larger of the two inputs, and is 0 otherwise. The comparator outputs are summed as indicated to produce $R_j - 1$ and $S_j - 1$.

Let $t = 0$ be the time at which the jth pulse is transmitted. At any time t', $\tau < t' < 1/\text{PRF}$, the outputs x_j and y_j comprise the PSS for the range bin corresponding to t', and the remaining contents of the I and Q registers are the KNS. Thus, if a single echo signal is propagating through the registers, the KNS is true noise only when that signal occupies the center stage (i.e., x_j and y_j) of the registers. As it "slides" through the other $N-1$ stages, it corrupts one noise sample in the KNS; however, this causes, at most, an error of ± 1 in the outputs of the rankers. A similar situation occurs with "mean-level detection," whereby the data from each range bin is compared with the mean obtained from adjacent bins. Problems associated with this method of obtaining the KNS are discussed in Dillard and Antoniak [7].

[7]This sampling transforms a continuous random variable into a discrete random variable for which no density function exists. However, we will assume that the number of bits in the A/D converters is sufficiently large that the outputs can be assumed continuous, so that ties occur with zero probability.

Whenever a point target is present in x_j and y_j, with no other targets within $\pm N/2$ range bins, it is reasonable to assume that the underlying noise process is the same for all bins. Sampling at the rate $1/\tau$ Hz provides independent samples, assuming that the receiver bandwidth is matched to the transmitted pulse. Thus, in this case it is reasonable to assume that the KNS does indeed contain IID random variables. If one or more additional target echoes are within $\pm N/2$ bins of a given target, the KNS is corrupted by these signals, and the probability of detection can be less than when no other target echoes are present. This situation is discussed in [7].

Fig. 8 is a block diagram of the entire distribution-free Doppler processor as applied to a digital Doppler radar system. In a conventional system, the A/D converter outputs would go directly to the DFT computer (or to some other set of digital filters), and the filter outputs (i.e., the DFT coefficients) would be compared with a threshold. The conventional system would also contain circuitry to compute the threshold adaptively, or to provide an adaptive gain control in order to maintain a constant false-alarm rate. Thus, the distribution-free processor only requires the insertion of the rankers of Fig. 7 after the A/D converters, and no longer requires an adaptive threshold computation.

Note that the storage and computation required in the DFT is reduced considerably by the insertion of the rankers. (The blocks labelled "ranker" are the devices shown in Fig. 7, exclusive of the A/D converters.) In the conventional system, the DFT operates directly upon the A/D converter outputs, which are likely to be at least 10 bits plus sign. However, in the distribution-free system, if N (the number of outputs being ranked) is less than 32, each ranker output requires at most 5 bits each, resulting in a large saving of data storage and computation time.

Conclusions and Discussions

The results presented here indicate that the DFDP can be applied effectively to a doppler radar without seriously degrading radar performance. The apparent loss of about 1.5 dB in sensitivity indicated in Figs. 5 and 6 is not necessarily indicative of the DFDP's performance in practice, since the "optimum" processor would not know the average noise power and thus would have to estimate it based on past history. This not only affects the detection sensitivity, but also can cause unwanted changes in the false-alarm rate, as discussed in [7]. Also, the optimum processor would suffer a further loss in sensitivity, and perhaps a change in false-alarm rate, if the distributions of the data change from those which were assumed. Thus, the DFDP performance is not necessarily inferior to the standard procedures applied in practice.

Implementing the DFDP is easily accomplished, especially when applied to a digital Doppler processor. This was indicated in the discussion of Fig. 8 in the previous section. The removal of the necessity for adaptive thresholding circuitry and the decrease in the number of bits required in the DFT (or digital filters) processor offsets the requirement of additional circuitry for performing the ranking indicated in Fig. 7.

Further analysis is required to determine if there is an optimum value for the size of the KNS. Based on results presented in [8], it is anticipated that increasing N will significantly improve the sensitivity (i.e., the detection probability for a given SNR will increase) up to a value of $N = 20$, with little improvement for larger values of N. Values of N less than 32 are significant in that the ranks can be represented in fewer than 6 bits, leading to the equipment savings discussed above.

Appendix

This appendix demonstrates the distributional convergence of $X_h/\sqrt{M}$ and $Y_h/\sqrt{M}$, (8) and (9), to a Gaussian distribution as $M \to \infty$. Although this result is clearly an extension of the basic central limit theorem, no appropriate version of this theorem has been found for the case described here, and so a proof is included for completeness.

The proof is accomplished by the characteristic function (CF) method. Under H_0, U_j and V_j are symmetrically, identically distributed random variables; hence, X_h and Y_h are identically distributed, and only X_h is considered. The probability mass function of U_j is discrete, uniform, and symmetric with probability $1/N$ at the points

$$K = (-N+1)/2, \cdots, (N-1)/2.$$

Let

$$C_{jh} \triangleq \cos[(2\pi jh)/M] \quad S_{jh} \triangleq \sin[(2\pi jh)/M].$$

Then the CF of $(U_j/\sqrt{M})\cos[(2\pi jh)/M]$ is

$$\phi_U(t) = (1/N) \cdot \left\{\sin[(tN/(2\sqrt{M}))C_{jh}]/\sin[(t/(2\sqrt{M}))C_{jh}]\right\}$$

Similarly, the CF of $(V_j/\sqrt{M})\sin[(2\pi jh)/M$ is

$$\phi_V(t) = (1/N) \cdot \left\{\sin[(tN/(2\sqrt{M}))S_{jh}]/\sin[(t/(2\sqrt{M}))S_{jh}]\right\}.$$

Since U_j and V_j are independent, the CF of

$$A_j \triangleq (U_j/\sqrt{M})\cos[(2\pi jh)/M] + (V_j/\sqrt{M})\sin[(2\pi jh)/M]$$

is given by

$$\begin{aligned}\phi_A(t) &= \phi_U(t)\phi_V(t)\\ &= (1/N^2)\left\{[\sin[(tN/(2\sqrt{M}))C_{jh}] \cdot \sin[(tN/(2\sqrt{M}))S_{jh}]]/ [\sin[(t/(2\sqrt{M}))C_{jh}]\sin[(t/(2\sqrt{M}))S_{jh}]]\right\}.\end{aligned}$$

It is easily shown that the nth moment M_n, of A_j is finite for all n, and that

$$\lim_{n\to\infty}(1/n)[M_n]^{1/n} < \infty.$$

Therefore, $\phi_A(t)$ can be expanded in a Maclaurin series [9]. This is most easily done by expanding the sine terms in the numerator and denominator of $\phi_A(t)$; viz.,

$$\begin{aligned}\phi_A(t) &\approx (1/N^2)\left\{[(t^2N^2)/4M]C_{jh}S_{jh}[1-((t^2N^2)/(24M)) \cdot C_{jh}^2][1-((t^2N^2)/(24M))S_{jh}^2]\right\}/\left\{[t^2/(4M)] \cdot C_{jh}S_{jh}[1-(t^2/(24M))C_{jh}^2][1-(t^2/(24M)) \cdot S_{jh}^2]\right\}\\ &\approx \left\{1-[(t^2N^2)/(24M)]\right\}/\left\{1-[t^2/(24M)]\right\}\end{aligned}$$

where only terms decreasing by the first order in M are retained. Since the A_j are independent of all j, the CF of

$$X_n/\sqrt{M} = \sum_{j=0}^{M-1} A_j$$

is

$$\begin{aligned}\phi_X^{(M)}(t) &= \prod_{j=1}^{M-1}\phi_A(t)\\ &= \left\{[1-[(t^2N^2)/(24M)]]/[1-[t^2/(24M)]]\right\}^M.\end{aligned}$$

Hence,

$$\begin{aligned}\lim_{M\to\infty}\phi_X^{(M)}(t) &= \exp[-(t^2N^2)/24]\exp[-(t^2/24)]\\ &= \exp[-(t^2/2)(N^2-1)/12].\end{aligned}$$

This is recognized as the CF of a Gaussian random variable with mean zero and variance $(N+1)(N-1)/12$.

References

[1] G.M. Dillard, "Distribution-free and nonparametric sequential signal detection," Ph.D. Thesis, University of California at San Diego, June 1971.

[2] J.B. Thomas, "Nonparametric detection," *Proc. IEEE, vol. 58, pp. 623-631, May 1970.*

[3] J.W. Carlyle and J.B. Thomas, "On nonparametric signal detectors," *IEEE Trans. Information Theory,* vol. IT-10, pp. 146-152, April 1964.

[4] F. Wilcoxon, "Individual comparisons by ranking methods," *Biometrics,* vol. 1, pp. 80-83, 1945.

[5] C.W. Helstrom, *Statistical Theory of Signal Detection,* 2nd Ed. New York: Pergamon, 1968.

[6] M.G. Kendall and W.R. Buckland, *A Dictionary of Statistical Terms.* Edinburgh: Oliver and Boyd, 1971.

[7] G.M. Dillard and C.E. Antoniak, "A practical distribution-free detection procedure for multiple-range-bin radars," *"IEEE Trans. Aerospace and Electronic Systems,* vol. AES-6, pp. 629-635, September 1970.

[8] G.M. Dillard, "Distribution-free detection of communication signals," Naval Electronics Lab. Center, San Diego, Calif., Rept. TR 1776, September 2, 1971.

[9] W. Feller, *An Introduction to Probability Theory and Its Applications,* vol. 2. New York: Wiley, 1966.

14

Reprinted with permission from *IEEE Trans. Inf. Theory* **IT-18**:110–118 (1972)

Nonparametric Detection Using Spectral Data

MELVIN N. WOINSKY, MEMBER, IEEE

***Abstract*—A detection system is considered that analyzes the spectrum of the time-series output from a sensing element. The spectral data consist of a matrix of estimates of the energy in many small time-frequency cells. A decision procedure is formulated that is based on the multiple use of a two-sample statistic operating on the columns of the matrix. If the input noise is Gaussian with unknown power, the asymptotically optimum statistic t is a ratio of two sample means. Since in certain applications the Gaussian input assumption may be unreliable, nonparametric techniques based on the Mann–Whitney U and Savage T statistics are studied.**

Asymptotic relative efficiency (ARE) is computed for general positive spectral noise data and a scale alternative. This alternative is appropriate since it includes, for SNR → 0, a Gaussian input with either a sinusoidal or Gaussian target. For a Gaussian input $\mathrm{ARE}_{U/t} \geq \frac{3}{4}$ and $\mathrm{ARE}_{T/t} \geq 0.816$. Non-Gaussian examples indicate that U and T can be much better than t. It is shown that, subject to a reasonable restriction on the noise cumulative distribution function (cdf), $\mathrm{ARE}_{U/t} \geq \frac{27}{64}$. The results obtained here for noncoherent detection, though not quite as strong, are analogous to the known bounds on ARE for linear coherent detection (a translation alternative).

I. Introduction and Summary

NONPARAMETRIC decision procedures have been previously applied to communication [2] and other detection problems [3]. This paper considers a different formulation of a detection problem that arises in some practical systems. Statistical techniques are applied after spectrum analysis of a time series. The data consist of a two-dimensional display (time and frequency) or matrix of spectral estimates. Such a data format leads quite naturally to a decision procedure based on the multiple use of a two-sample statistic.

Manuscript received January 27, 1970; revised October 9, 1970, and July 6, 1971.

The author is with Bell Telephone Laboratories, Inc., Whippany, N.J. 07981.

The system considered operates in the following manner. A sensing element, or a combination of such elements, receives energy from a background noise field with the possible addition of energy emitted from sources (targets) near the element. The output of the sensing element goes to a spectrum analyzer, which over a large band of frequencies estimates the energy in many small time-frequency cells of time duration T_f s and width 2Δ Hz. Over a longer decision interval a two-dimensional display of the data is formed, as shown in Fig. 1. It is assumed that targets appear as narrow lines in frequency. It is necessary to process the spectral data so as to detect possible targets and determine their frequency components. The postdetection statistical analyzer accomplishes this by deciding which frequency slots (vertical columns), if any, have target energy present (hits). The model is applicable to some radar and sonar systems where the time and frequency axis would represent delay and Doppler shift. A similar model is used in detection using a target's scattering function [4]. Applications also exist in radio astronomy and countermeasures. In the latter situation it may be necessary to detect electromagnetic jammers and identify the frequencies used. In speech analysis using sonograms, the techniques can be used to identify the formant frequencies. The procedures can also be adapted for use in an M-ary frequency-shift keying (FSK) communication system. In this case determining a target's frequency structure corresponds to decoding a transmitted word. The choice of the analyzer parameters T_f and 2Δ depends upon the particular application and desired goal, such as the degree of range and velocity resolution [7].

The block of spectral data considered (Fig. 1) represents the largest band for which it can be assumed that the noise

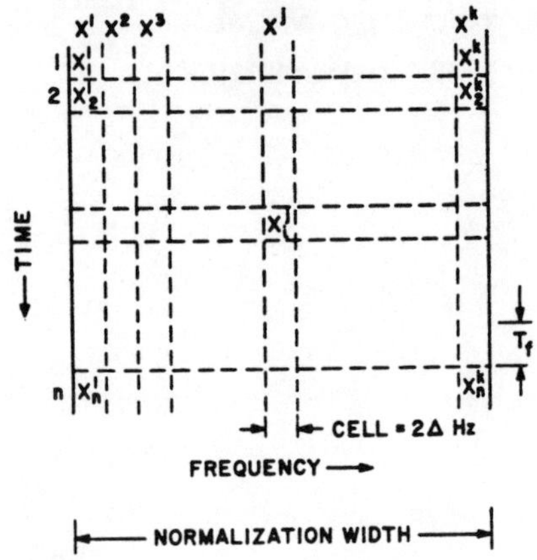

Fig. 1. Processing model.

spectrum is approximately flat. The entire frequency spectrum of interest is analyzed by forming several contiguous blocks covering the spectrum. The overall system performance can be obtained approximately by combining the results for each block, assuming each block to be independent. It is assumed that the noise spectrum level is unknown and may vary with frequency and possibly slowly with time. Therefore, breaking the band up in such a manner is necessary in order to specify a false-alarm rate. Each spectral estimate X_i^j is a random variable with the time index $i = 1,2,\cdots,n$ and the frequency index $j = 1,2,\cdots,k$. It is assumed that the spectral data X_i^j are independent for all i and j and that all elements of $X^j = (X_1{}^j, X_2{}^j, \cdots, X_n{}^j)$ are identically distributed with cdf $F(x)$ if there is only noise in slot j, or with $F_{\theta j}(x)$ if there is target energy in slot j. As usual the independence assumption is necessary to make the analysis tractable. To find some justification for independence in frequency, we consider the situation where the input process is white Gaussian noise and the X_i^j are obtained by forming a periodogram and filtering the result with ideal bandpass filters. The correlation coefficient $\rho(X_i^j, X_i^{j+l})$ is a function of the time-bandwidth product $B = 2\Delta T_f$. It can be shown in a tedious but straightforward manner that for $l = 1$ (adjacent slots), $\rho \doteq 0.208$ if $B = 1$ and $\rho \doteq 0.109$ if $B = 2$. For $l = 2$ the correlation drops by about a factor of ten. Note also that

$$\lim_{B \to \infty} \rho = 0.$$

Therefore, with $B \geq 1$ all correlations are small and the independence assumption is not totally unreasonable.

Under the null hypothesis H of only noise in the normalization block, all the X_i^j are distributed as $F(x)$. There are several alternative hypotheses depending on the number of hits and their frequency locations. One approach would be to test H and upon rejection use a ranking and selection procedure.[1] However, the method of this paper employs a procedure that simultaneously tests H and upon rejection indicates which of the many alternatives is correct. This problem is considered in the mathematical literature under the heading of multiple-decision procedures, multiple-comparison procedures, or simultaneous statistical inference. Miller [9] discusses much of the work in the literature on simultaneous statistical inference and provides a good bibliography. Paulson [10] considered detection for a Gaussian null density and a translation alternative where there could only be a single hit in a block. He derived an optimum procedure in a sense analogous to Neyman–Pearson theory. For the same problem Doornbos *et al.* [11] proposed a procedure where all the pooled data are ranked and the rank sum formed for each column. The maximum rank sum is the test statistic. Paulson [11] also considered a similar solution based on using normal-score weighted rank sums.

The decision procedure proposed here forms a test statistic for each frequency slot and compares the value of each statistic with a threshold to determine the frequency locations of target energy. A two-sample test statistic is used for each frequency slot. The spectral data in the slot being tested form the first sample, and all the remaining pooled data form the second sample. The data shown in Fig. 1 are then reduced to k dependent test statistics. Such a procedure is not necessarily optimum. Because of the nature of the problem, little work, except for Paulson's [10], has been done on finding optimum procedures. It will be shown that the procedure is intuitively reasonable when there are few hits per block. The main concern of this work is to decide which two-sample statistic should be used as the basis of the procedure.

A distribution model for the spectral data is required in order to be able to select a two-sample statistic. In some cases it is reasonable to assume that the input to the spectrum analyzer consists of a wide-band Gaussian noise process with either deterministic sinusoidal targets or narrow-band Gaussian targets. With the spectrum analyzer for each frequency slot characterized by a narrow-band filter followed by a squaring and integrating device, the spectral data with noise at the input is gamma distributed. The scale parameter is the unknown noise power in each frequency cell, and the number of degrees of freedom is twice the time-bandwidth product $4\Delta T_f$. A deterministic target leads to a noncentral distribution, and a Gaussian target yields a gamma distribution, but with an increased scale parameter. It is shown that the asymptotically optimum two-sample statistic t for both these cases is the sample mean of the data in the slot tested divided by the remaining pooled sample mean.

Because in certain applications the Gaussian input assumption may be unreliable, nonparametric (NP) techniques based on the Mann–Whitney U [11], and Savage T [14] statistics are studied. For nonparametric comparison purposes a general continuous noise cdf $F(x)$ for the spectral data is considered with $F(0) = 0$. The class of scale-alternative distributions (target plus noise) are studied with cdf's $F_\theta(x) = F(x/(1 + \theta))$, where θ is positive

[1] Ranking procedures [5], [6] attempt to order the populations (frequency slots) and determine the best population. Best is usually defined in terms of the largest (or smallest) translation or scale parameter. Selection procedures [12] determine a subset of the k populations, which with a preassigned probability, contain the best population.

and proportional to the cell SNR (ratio of average signal power to average noise power in a frequency cell). This is the appropriate formulation for noncoherent detection. The spectral data are positive, and the presence of a target increases the energy level. Also, this alternative is correct for Gaussian input noise plus Gaussian target, and for sinusoidal target the scale alternative is accurate as $\theta \to 0$.

The Pitman asymptotic relative efficiency (ARE) between multiple decision procedures considered here is the same as the ARE for the two-sample statistics upon which they are based. Relations for $\text{ARE}_{U/t}$ and $\text{ARE}_{T/t}$ are obtained as a function of $F(x)$ for the "one-side scale slippage" alternative previously described. The ARE depends upon the time-bandwidth product $B = 2\Delta T_f$ used in spectral analysis. For a Gaussian input the ARE is numerically evaluated for all values of B. At $B = 1$ (employed in many systems), $\text{ARE}_{U/t} = \frac{3}{4}$, $\text{ARE}_{T/t} = 1$, and as $B \to \infty$, $\text{ARE}_{U/t} \to 3/\pi$, $\text{ARE}_{T/t} \to 0.816$. On the basis of ARE the loss with Gaussian processes is quite small. Non-Gaussian examples of ARE indicate that U and T can be much better than t. It is shown that subject to a reasonable restriction on the noise cdf, $\text{ARE}_{U/t} \geq \frac{27}{64}$. The new results obtained here for noncoherent detection, though not quite as strong, are analogous to the known bounds on ARE for linear coherent detection (a translation alternative).

II. Decision Procedure

The spectral data to be processed in a single normalization block are shown in Fig. 1. Under the null hypothesis H, X_i^j is distributed as $F(x)$, $i = 1,2,\cdots,n$, $j = 1,2,\cdots,k$. It is necessary to test each of the k frequency slots for target energy. The processing procedure is as follows. Let

$$X^j = (X_1^j, X_2^j, \cdots, X_n^j)$$

$$Y^{-j} = (X_1^1, X_2^1, \cdots, X_n^1, \cdots, X_1^{j-1}, \cdots, X_n^{j-1}, X_1^{j+1}, \cdots, X_n^{j+1}, \cdots, X_1^k, X_2^k, \cdots, X_n^k),$$

where X^j are the data to be tested and Y^{-j} are the remaining pooled data. Let $S = D(X^j, Y^{-j})$, where $D(\cdot,\cdot)$ is a two-sample statistic. The data are reduced to k statistics $(S_1, S_2, \cdots, S_k)$. The S_j are obviously dependent. The decision procedure is to indicate there is a hit in slot j if $S_j > \lambda_\alpha$.

In an intuitive justification of such a procedure it is tacitly assumed that the elements of Y^{-j} are approximately distributed as $F(x)$, the noise distribution in all cases. The remaining pooled data are used in each case as a reference replica of the noise data. If in testing slot j there are hits in some other slots, this will make Y^{-j} appear "larger" and the actual single-slot false-alarm probability α_s will be reduced. The calculation of α_s is then conservative. However, the probability of detection P_d is also reduced since the artificially large Y^{-j} makes large values in slot j seem less significant. If only one hit is possible per block, the procedure is perfectly valid. It can be argued that if the maximum number of hits r is much less than k and if no hit is extremely large, then the assumption about the elements of Y^{-j} is good and the procedures are reasonable. To see this, assume that there are r hits with corresponding cdf's $F_{\theta 1}(x), \cdots, F_{\theta r}(x)$ and that $F_{\theta j}(x)$ corresponds to the slot j that is being tested. Then Y^{-j} has cdf

$$F'(x) = F(x) - \frac{1}{k-1} \sum_{\substack{i=1 \\ i \neq j}}^{r} [F(x) - F_{\theta i}(x)]$$

and X^j has cdf $F_{\theta j}(x)$. Since slippage alternatives will be considered, $F_{\theta i}(x) \leq F(x)$ and $F'(x) \leq F(x)$, so that the statistical separation between X^j and Y^{-j} or the slot SNR has been reduced by the presence of other targets. However, if $r \ll k$ the effect is small. Modifications to the procedure can be used to improve the detection probability when there are many hits per block or some strong lines. A possibility is to use multistage testing [9].

One approach to picking the threshold λ_α is to decide that it is necessary to specify the expected number R of false hits per block. Then if λ_α is found from $\Pr_H[S_j > \lambda_\alpha] = \alpha_s$, it follows that $R = k\alpha_s$ under the null hypothesis H, even though the S_j are dependent, and from the previous discussion $R < k\alpha_s$ when there are some hits. The maximum value of R can then be controlled. In some applications it may be desirable to specify the overall block false-alarm probability α_B; α_B is the probability of indicating that there is any combination of hits in the entire block of data when actually the entire block represents noise. By using a result in [10] it follows directly that $\alpha_B \simeq k\alpha_s$ if $\alpha_B \ll 1$.

III. Distributional Model

The nature of the distributions involved must be studied before a suitable choice can be made for the two-sample statistic. First consider a parametric model. A reasonable approximation to various practical spectrum analyzers is obtained by considering the data in each frequency slot to be generated by applying the time-series input to a narrow-band filter of width 2Δ Hz followed by a squaring device and an integrator operating over a time interval T_f. With stationary white Gaussian noise at the input, the density [17] of the data is

$$f(x) = \bar{N}^{-1} g(x/\bar{N}), \tag{1}$$

where

$$g(x) = \begin{cases} (x^{B-1} e^{-x/2})/(2^B \Gamma(B)), & x \geq 0, \\ 0, & x < 0, \end{cases} \tag{2}$$

is the gamma density, $\bar{N}$ is the unknown average noise power in a cell, and $\Gamma(\cdot)$ is the gamma function. The parameter $B = 2\Delta T_f$ is the time bandwidth product [17]. B is usually referred to as the shape parameter. Also for $2B$ a positive integer, the density of (2) is χ^2 and $2B$ is called the number of degrees of freedom.

If a sinusoidal target is added, the density is [17]

$$f_\theta(x) = \bar{N}^{-1} g_\theta(x/\bar{N}), \tag{3}$$

where

$$g_\theta(x) = \tfrac{1}{2}(x/\theta)^{1/2(B-1)} \exp\left[-\tfrac{1}{2}(x + \theta)\right] I_{B-1}(\theta^{1/2} x^{1/2}) \tag{4}$$

is the noncentral gamma density, $I_{B-1}(\cdot)$ is the modified Bessel function of the first kind and order $B - 1$, and θ is the cell SNR. If a flat narrow-band Gaussian target is added to the noise, the density is [18]

$$f_\theta(x) = (1 + \theta)^{-1} f(x/(1 + \theta)), \tag{5}$$

where θ is the cell SNR and $f(x)$ is given by (1). In practice $B \geq 1$ and frequently the analyzer is designed with $B = 1$, so that the null density is exponential. This will be considered the nominal operating point.

From (5), for Gaussian target, $f_\theta(x)$ is related to $f(x)$ by a scale change. Expression (4) for sinusoidal target is more complicated. However, using the series form [20] of (4) yields

$$g_\theta(x) = \pi^{-1/2} 2^{-B} x^{B-1} \exp\left[-\tfrac{1}{2}(x + \theta)\right] \sum_{l=0}^{\infty} \frac{(\theta x)^l \Gamma(l + \frac{1}{2})}{(2l)!\,\Gamma(B + l)}$$

$$= g(x) e^{-(1/2)\theta}(1 + \tfrac{1}{4}B^{-1}\theta x) + O(\theta^2)$$

$$g_\theta(x) = (1 + \tfrac{1}{2}B^{-1}\theta)^{-1} g(x/(1 + \tfrac{1}{2}B^{-1}\theta)) + O(\theta^2). \tag{6}$$

Then the noise density in (1) and the family of densities given by (5), with $\theta \propto$ SNR, represent the desired parametric class with Gaussian inputs for both flat Gaussian or sinusoidal targets at small SNR. The result of (6) indicates that the alternatives of (3) and (5) are identical for ARE comparisons.

For the purpose of making a more general study of performance, the family of all continuous cdf's with $F(0) = 0$ is taken as the nonparametric class of noise distributions. Since the spectral data are nonnegative, this does not represent a restriction. The most general one-sided slippage alternative considers $F_\theta(x) \leq F(x)$ for all x. This implies the presence of target, increases the sample values on the average, and is pertinent to this problem. Unfortunately, the class is too general to obtain definitive results. A more restrictive, but still reasonable family can be deduced from the previous parametric results. The one-sided scale slippage alternative,

$$F_\theta(x) = F(x/(1 + \theta)) \tag{7}$$

with $F(0) = 0$ and $\theta \propto$ SNR such that $\theta > 0$ will be studied. This generalization includes the Gaussian model.

In linear coherent detection $F_\theta(x) = F(x - \theta^{1/2})$, a translation alternative and very strong ARE results are available [3]. The noncoherent problem is quite different owing to the nonlinear processing, which causes the presence of target to increase both the mean and variance of the distribution.

IV. Choice of a Two-Sample Statistic

In the parametric case a choice of the two-sample statistic, for use in the decision procedure described in Section II, can be made by considering the gamma density. For the two-sample problem with the elements of Y^{-j} distributed as in (1) and the elements of X^j distributed as in (5) with $\theta = 0$ under the null hypothesis H and $\theta > 0$ under the alternative K, a test equivalent to the generalized-likelihood ratio can be found in a lengthy, but straightforward manner. It is based on

$$S_j = D(X^j, Y^{-j}) = t = \langle X^j \rangle / \langle Y^{-j} \rangle, \tag{8}$$

where $\langle X^j \rangle$ and $\langle Y^{-j} \rangle$ are the sample means of the components of X^j and Y^{-j}, respectively. This statistic is used for the reference parametric procedure. From the properties of the likelihood ratio [21] it follows that t is asymptotically optimum[2] for all B as $n \to \infty$, and from [22] the test is uniformly most powerful (UMP) if $B = 1$ (exponential null density). The slot false-alarm probability α_s can be found from the null distribution of t. The numerator and denominator in (8) are independent and gamma distributed under H. The ratio t is F or Snedecor distributed with $2nB$ and $2nB(k - 1)$ degrees of freedom, when $2B$ is a positive integer [21].

Two NP statistics are considered. The Mann–Whitney–Wilcoxon statistic [11]

$$U = \frac{1}{n^2(k - 1)} \sum_{i=1}^{n} R_{ij} - \frac{n + 1}{2n(k - 1)} \tag{9}$$

and the Savage statistic [14]

$$T = \frac{1}{n} \sum_{i=1}^{n} \left(\sum_{l=nk+1-R_{ij}}^{nk} l^{-1} \right), \tag{10}$$

where R_{ij} is the rank of X_i^j in the entire pool of data. The Savage statistic is studied because it is the optimum rank statistic [14] for an exponential distribution and a scale alternative. It has been applied to reliability problems [22] where the exponential distribution is important. The null distribution of U and T are tabulated [8]. Thresholds for a given α_s can be approximately determined from the limiting standard model cdf of $2[3n(k - 1)/k]^{1/2}(U - \frac{1}{2})$ and $[nk/(k - 1)]^{1/2}(T - 1)$.

V. Simplified Procedures

The nonparametric procedures require that all data in the block be stored and also require considerable processing time. This can be greatly simplified by only using row ranking in (9) and (10). The data are processed as generated by the analyzer, and no storage is required. Refer to Fig. 1 for the data format. The modified statistics are

$$U' = n^{-1}(k - 1)^{-1} \sum_{i=1}^{n} R_{ij}' - (k - 1)^{-1}, \tag{11}$$

$$T' = n^{-1} \sum_{i=1}^{n} \left(\sum_{l=k-R_{ij}'+1}^{k} l^{-1} \right), \tag{12}$$

where R_{ij}' is the rank of X_i^j among the data in row i only.

Under H, R_{ij}' has a discrete uniform distribution over the integers $1, 2, \cdots, k$ and $E_H(U') = \frac{1}{2}$, $\mathrm{var}_H(U') = (k + 1)/[12n(k - 1)]$. Asymptotic normality of $n^{1/2}(U' - \frac{1}{2})$ follows directly from the central limit theorem, and can be used to obtain an approximate threshold. Similarly, it can

[2] It appears, although it has not been proved, that for the extreme population problem ($r = 1$ *a priori*) using

$$\max_j \{t\}$$

is an optimum procedure in the same sense as Paulson's [10] procedure.

be shown that $E_H(T') = 1$ and

$$n \operatorname{var}_H(T') = k^{-1} \sum_{i=1}^{k} \left(\sum_{l=k-i+1}^{k} l^{-1} \right)^2 - 1. \qquad (13)$$

Also $n^{1/2}(T' - 1)$ is asymptotically normal. Relation (13) can be readily evaluated on a digital computer for any given value of k.

The effect of reduced ranking on asymptotic efficiency is found in Appendix II to be

$$\mathrm{ARE}_{U/U'} = \lim_{n\to\infty} n(k+1)/(nk+1) = 1 + k^{-1}. \qquad (14)$$

The loss due to the simplification in the Mann–Whitney procedure is negligible for reasonable values of k. A similar conclusion appears to hold for the modified Savage statistic, but this could not be rigorously shown. The idea of the simplified statistics of (11) and (12) is the same as used by Feustel and Davisson in their mixed statistical tests [15]. Only the arrangement of the subblock to be ranked is different. However, the result of (14) is a bit surprising, for in our formulation R_{ij}' remains uniformly distributed for all k, and this contradicts the intuitive reason given in [15] for the small loss in efficiency for large k.

If a large band of frequencies is to be analyzed, it is preferable to use a "sliding" normalization block of width k across the spectrum rather than fixed contiguous blocks. This approach guards against edge effects due to a noise spectrum that may be sloping within each normalization block. The block may slide one slot at a time, reranking the row each time. Only the rank of the cell at the center of the block is saved and used in forming the test statistic. Such processing is relatively easy with the simplified procedures.

VI. ARE for the One-Sided Scale Slippage Alternative

The decision procedure is based on the multiple use of dependent two-sample statistics. It is not obvious how the overall relative performance for the k slots is related to the ARE for the two-sample problem. However, Sherman [23] has shown that in making multiple comparisons based on the use of two-sample statistics, the ARE is the same as in the two-sample case. Her interpretation of ARE for multiple comparisons is that introduced by Lehmann. If the relative sample sizes used by the two procedures are related by the ARE for the two-sample case, all asymptotic performance characteristics are identical for each procedure. The result was obtained for translation alternatives and a comparison of the Mann–Whitney procedure with one based on the two-sample Student's statistic. It is easy to show that Student's statistic and our t statistic are asymptotically equiefficient, and from the results of Puri [5], [6] it follows that Sherman's result is valid for the Savage test and a scale alternative. Therefore, the $\mathrm{ARE}_{U/t}$ and $\mathrm{ARE}_{T/t}$ for the two-sample problem will be used as the performance measure.

Assume that the noise density $f(x)$ for the spectral data has variance σ^2 and mean μ. It is shown in Appendix I that for the Mann–Whitney processor and the one-sided scale slippage alternative (7)

$$\mathrm{ARE}_{U/t} = 12 \left(\frac{\sigma}{\mu}\right)^2 \left[\int_0^\infty x f^2(x)\, dx \right]^2 \qquad (15)$$

and for the Savage processor,

$$\mathrm{ARE}_{T/t} = \left(\frac{\sigma}{\mu}\right)^2 \left[\int_0^\infty \frac{x f^2(x)}{1 - F(x)}\, dx \right]^2. \qquad (16)$$

Insight into the efficiencies of (15) and (16) can be obtained by considering specific examples for $f(x)$. The results are independent of a scale factor in $f(x)$, so that the noise density of interest for the parametric model is $g(x)$, the gamma density as given in (2). In this case, B represents the time-bandwidth product used in spectral analysis, and the nominal operating point is considered to be $B = 1$. Also considered is the density

$$f^*(x) = \begin{cases} \exp(-\tfrac{1}{2}x^{1/B})/[B2^B\Gamma(B)], & x \geq 0 \\ 0, & x < 0. \end{cases} \qquad (17)$$

This family is of interest since it represents a class with both slower ($B > 1$) and faster ($B < 1$) upper tails as compared to $g(x)$. For the densities considered,

$$\mathrm{ARE}_{U/t}(g) = 12\Gamma^2(2B)/[B2^{2B}\Gamma^4(B)],$$

$$\mathrm{ARE}_{T/t}(g) = I^2/[B2^{2B}\Gamma^4(B)],$$

$$\mathrm{ARE}_{U,T/t}(f^*) = (XT)\mathrm{ARE}_{U,T/t}(g),$$

where $XT = B^{-1}([\Gamma(B)\Gamma(3B)/\Gamma^2(2B)] - 1)$, and

$$I = \int_0^\infty \left[1 - \frac{\gamma(B,\frac{1}{2}x)}{\Gamma(B)}\right]^{-1} e^{-x}x^{2B-1}\, dx \qquad (18)$$

with $\gamma(B,\frac{1}{2}x)$ the incomplete gamma function. Values of $\mathrm{ARE}_{U/t}$, $\mathrm{ARE}_{T/t}$, and $\mathrm{ARE}_{T/U}$ are shown for both classes of densities in Figs. 2–4, respectively. The values of I (18) were obtained by numerical integration.

At the nominal operating point of $B = 1$, $\mathrm{ARE}_{U/t}(g) = \frac{3}{4}$ and $\mathrm{ARE}_{T/t}(g) = 1$. For $g(x)$, arbitrary B, the efficiency cannot be greater than 1 since t is asymptotically optimum in this class. The loss is not very significant, and there is no loss for the T-processor. Capon's [16] work on locally most powerful rank tests also implies the latter result. Using [19],

$$\lim_{B\to 0} B\Gamma(B) = 1$$

indicates

$$\lim_{B\to 0} \mathrm{ARE}_{U/t} = 0,$$

and by numerical methods

$$\lim_{B\to 0} \mathrm{ARE}_{T/t} = 0.$$

The density $g(x)$ blows up at the origin for $B < 1$. The rate at which this occurs increases as $B \to 0$. However, the density does not have much physical significance for $B < \frac{1}{2}$. This illustrates the point that the NP procedures may be relatively poor if the density has a great concentration at the origin. For $B = \frac{1}{2}$, $\mathrm{ARE}_{U/t}(g) = 6/\pi^2$ and

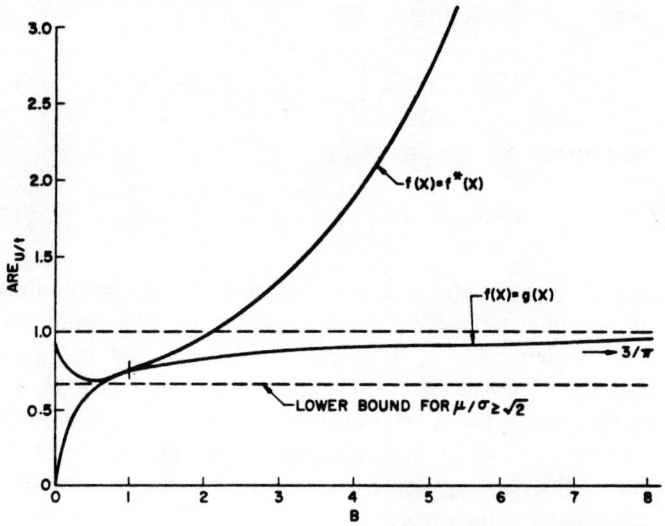

Fig. 2. Efficiency—Mann-Whitney procedure versus time-bandwidth product B.

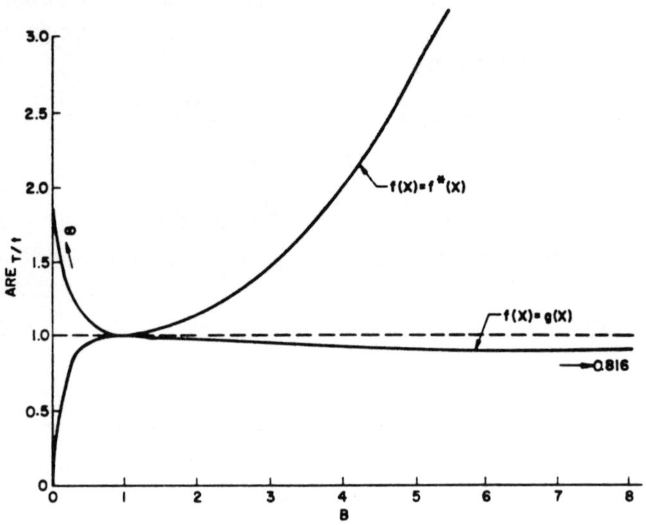

Fig. 3. Efficiency—Savage procedure versus time-bandwidth product B.

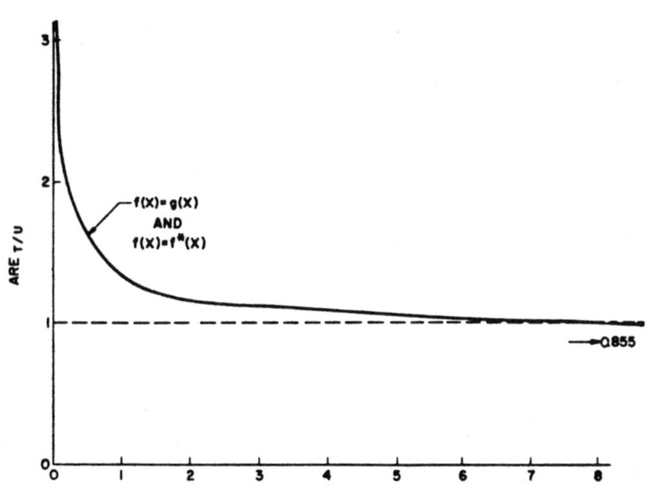

Fig. 4. Efficiency—Savage to Mann-Whitney versus time-bandwidth product B.

$\mathrm{ARE}_{T/t}(g) = 0.987$, so that even with an infinite discontinuity at the origin the loss is small.

As B increases, $g(x)$ shifts out and becomes more bell shaped, approaching the normal density. Actually the scale alternative approaches a translation alternative under these conditions. To see this, assume

$$\lim_{\mu\to\infty} \sigma f(\sigma x + \mu) = \hat{f}(x),$$

where the density $\hat{f}(x)$ is zero mean and has a variance of 1. Then from (15)

$$\lim_{\mu\to\infty} \mathrm{ARE}_{U/t} = 12\left[\int_{-\infty}^{\infty} \hat{f}^2(x)\,dx\right]^2 \tag{19}$$

if $\int_{-\infty}^{\infty} |x|\hat{f}^2(x)\,dx < \infty$, and similarly from (16)

$$\lim_{\mu\to\infty} \mathrm{ARE}_{T/t} = \left[\int_{-\infty}^{\infty} \frac{\hat{f}^2(x)}{1 - \hat{F}(x)}\right]^2 \tag{20}$$

if

$$\int_{-\infty}^{\infty} \frac{|x|\hat{f}^2(x)}{1 - \hat{F}(x)}\,dx < \infty.$$

Equations (19) and (20) are equivalent to translation results for the density $\hat{f}(x)$. If $f(x) = g(x)$ the gamma density, then $\hat{f}(x)$ and $\hat{F}(x)$ in (19) and (20) are the standard normal density and cdf, respectively. Making this substitution yields

$$\lim_{B\to\infty} \mathrm{ARE}_{U/t}(g) = 3/\pi$$

by direct integration and

$$\lim_{B\to\infty} \mathrm{ARE}_{T/t}(g) = 0.816$$

by numerical integration. From this result and Figs. 2 and 3 it follows that for the Gaussian input model, $\mathrm{ARE}_{U/t} \geq \frac{3}{4}$ and $\mathrm{ARE}_{T/t} \geq 0.816$ if consideration is restricted to time-bandwidth products not less than 1 ($B \geq 1$).

It is shown in [1] that for general $f(x)$,

$$\mathrm{ARE}_{U/t} \geq 0.864(1 - 0.458\sigma^2/\mu^2), \qquad \text{if } \mu/\sigma \geq \sqrt{2} \tag{21}$$

$$\geq (27/4)\frac{(\mu/\sigma)^6}{(1 + \mu^2/\sigma^2)^4}, \qquad \text{if } \mu/\sigma \leq \sqrt{2}. \tag{22}$$

The portion of the bound given in (22) was derived by Mallows. The bounds are monotonically increasing functions of μ/σ. Note that for $\mu/\sigma \geq \sqrt{2}$, $\mathrm{ARE}_{U/t} \geq \frac{2}{3}$ and that for $\mu/\sigma \geq 1$, $\mathrm{ARE}_{U/t} \geq \frac{27}{64}$. From the result for the gamma density it is clear that $\mathrm{ARE}_{U/t}$ can approach zero if $\mu^2/\sigma^2 = B$ is small. The bounds of (21) and (22) and their derivation indicate that this is true in general. If the noise density is such that there is a great concentration of mass at the origin as indicated by small values of μ/σ, $\mathrm{ARE}_{U/t}$ can be small. In [1] it is also shown how a composite test can be constructed that eliminates the cases of low efficiency. However, in our application, for physical reasons, densities with such a great concentration at the origin seem to be unlikely, and the lower bound for $\mu/\sigma = 1$ should be applicable. Therefore, modifying the tests is not necessary.

For the density $f^*(x)$, from Figs. 2 and 3, large improvements are possible for both the U and T procedures. The $\text{ARE}_{U/t}$ is high for densities with a heavy upper tail. The Savage procedure does well for densities with both slower and faster upper tails. It does particularly well when there is a sharp cutoff on this tail. For instance, for a triangular density (decreasing linearly from $x = 0$), $\text{ARE}_{U/t} = \frac{2}{3}$ while $\text{ARE}_{T/t} = 2$. The ARE results indicate that for reasonable densities the relative efficiency loss in using Mann–Whitney processing as compared to the parametric t-procedure is at most $\frac{27}{64}$, with an efficiency of $\frac{3}{4}$ at the nominal operating point. For a large class of densities great improvement is possible. Similar conclusions hold for the Savage processor except that at the nominal point there is no loss, and the degradation with densities concentrated at the origin falls off much slower than for the Mann–Whitney. From Fig. 4 the only place where the U-procedure performs better than the T-procedure is when the upper tail falls off very slowly or when the density is concentrated far away from the origin so that the alternative is similar to translation. Even in these cases the efficiency of the T-procedure to the U-procedure is not less than 0.855. For the problem being considered, one would expect the Savage processor to perform better than the Mann–Whitney processor. The previous discussion is based solely on ARE calculations.

VII. Statistical Simulation

A Monte Carlo simulation of the performance of the modified nonparametric statistics U' and T', and of the parametric statistic t, has been carried out using an analog spectrum analyzer. Sinusoids of controlled amplitude were added to pseudorandom white Gaussian noise, which was then spectrum analyzed. The thresholds were set empirically so that the cell false-alarm probability α_s was approximately 10^{-4}. The resulting empirical detection probability P_d is plotted against cell SNR (dB) in Fig. 5. The decibel scale has been arbitrarily translated to make the 0-dB point correspond to roughly 50 percent detection probability. The frequency window is $k = 64$ with a redundancy of 2, i.e., the frequency analyzing bands overlapped by 50 percent. The time length corresponds to $n = 416$ with a time redundancy of 4, i.e., the input data on each successive time scan overlapped by 75 percent. The time-bandwidth product is $B = 1$. Frequency and time redundancy were not considered in the theoretical analysis, but are frequently used in practical analyzers. The data block can be considered equivalent to a block of independent data smaller in size than 64 × 416. This point is not important for the relative comparison shown, since all thresholds were determined empirically.

Sufficient data were taken to give an accuracy for each curve alone of better than $\frac{1}{2}$ dB with 95 percent confidence. Since the same data are used for each procedure, the relative accuracy should be considerably better than $\frac{1}{2}$ dB.

To interpret the results, note that if θ_U and θ_t are the cell SNRs for the U' and t procedures, respectively, such

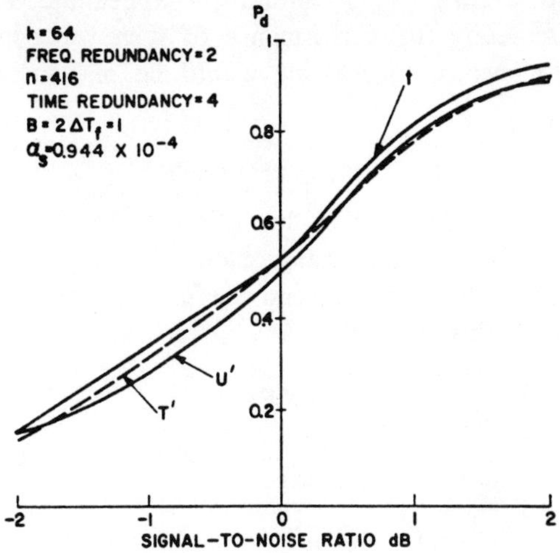

Fig. 5. Detection probability white Gaussian noise input versus SNR.

that both require the same sample size n to provide the same α_s and P_d, then it is readily shown that

$$\lim_{\substack{n \to \infty \\ \theta_{U'} \to 0 \\ \theta_t \to 0}} (\theta_t/\theta_{U'}) = (\text{ARE}_{U'/t})^{1/2},$$

for the scale slippage alternative. The ARE results then indicate that the Mann–Whitney procedure should be 0.625-dB poorer and the Savage procedure 0-dB poorer than the parametric procedure. This agrees quite well with the results shown in Fig. 5. The loss with the T' procedure is 0 dB at $P_d = 0.5$ and is never more than about 0.2 dB. The loss with the U' procedure is on the order of 0.3 dB. For the parameters used all procedures are essentially equivalent with white Gaussian noise at the input. Controlled experiments with non Gaussian noise have not been performed.

VIII. Conclusions

Nonparametric techniques have been applied to the processing of spectral data. The data format lends itself naturally to an NP solution based on multiple use of two-sample statistics. Comparisons were made using the "one-side scale slippage" alternative distribution since this generalization includes, at small SNR, a Gaussian input model with both deterministic and Gaussian target. It was shown that the performance of a procedure based on either the Mann–Whitney or the Savage statistic compares quite favorably with one based on the optimum parametric statistic. ARE results were obtained for noncoherent detection that, although not as strong, are analogous to the well-known results for coherent linear detection.

This paper has been mainly concerned with the choice of a two-sample statistic for a given multiple comparison procedure. An important factor, of course, is the choice of an optimum or "good" procedure. There is little work in the literature on optimum procedures except for Paulson's [10] on extreme populations. It appears that the procedure used here is good if there is a small number of hits in a normalization block. If this assumption is not valid, multistage testing [9] is intuitively appealing. Further analysis to study the performance of such procedures as applied to spectral processing would be of considerable interest.

IX. Acknowledgment

The author acknowledges the aid of Dr. M. I. Schwartz, who suggested studying the problem of "post detection" analysis and who made several useful suggestions during the course of this research.

Appendix I

$\text{ARE}_{U/t}$ and $\text{ART}_{T/t}$

Consider the two-sample problem,

$$H: \; X_1,X_2,\cdots,X_{n_1},Y_1,Y_2,\cdots,Y_{n_2} \sim F(x)$$

$$K: \; Y_1,Y_2,\cdots,Y_{n_2} \sim F(x)$$

$$X_1,X_2,\cdots,X_{n_1} \sim F_\theta(x),$$

where $F(0) = 0$ and the alternative cdf $F_\theta(x) = F(x/(1 + \theta))$ with $\theta > 0$. It is assumed that $F(x)$ is continuous with density $f(x)$ and $E(Y_i) = \mu$ and $\sigma^2(Y_i) = \sigma^2$. The statistics t, U, and T for this problem are given by (8)–(10), respectively, by dropping the index j and letting $n_1 = n$, $n_2 = n(k - 1)$ with $N = n_1 + n_2$. Let $E_\theta(W_i)$ and $\sigma_0^2(W_i) = \sigma_{\theta=0}^2(W_i)$ be the moments of statistics W_i representing t, U, or T. Subject to the usual regularity conditions,

$$\text{ARE}_{W_1/W_2} = \lim_{N\to\infty} \varepsilon(W_1)/\varepsilon(W_2)$$

where the efficacy of W_i is

$$\varepsilon(W_i) = \left[\left.\frac{dE_\theta(W_i)}{d\theta}\right|_{\theta=0}\right]^2 \Big/ \sigma_0^2(W_i). \tag{23}$$

Since $\ln(\cdot)$ is monotonic it follows that the test based on t and $z = \ln t$ are equivalent. If $Y \sim F(x)$, then $(1 + \theta)Y \sim F_\theta(x)$ and the distribution of z can be found from

$$z = \ln(1 + \theta) + \ln\left(n_1^{-1} \sum_{i=1}^{n_1} X_i\right) - \ln\left(n_2^{-1} \sum_{i=1}^{n_2} Y_i\right)$$

where X_i and $Y_i \sim F(x)$. From Lehmann [24] and the central limit theorem it follows that z and t are asymptotically normal with limiting moments, $E_\theta(z) = \ln(1 + \theta)$ and $\sigma_0^2(z) = \sigma^2(n_1^{-1} + n_2^{-1})/\mu^2$ so that, $\varepsilon(t) = \varepsilon(z) = n_1 n_2 N^{-1}\mu^2/\sigma^2$. From [2], $\sigma_0^2(U) = (N + 1)/12n_1n_2$ and $E_\theta(U) = \int_0^\infty [1 - F_\theta(x)]\, dF(x)$ so that $(dE_\theta(U))/d\theta|_{\theta=0} = \int_0^\infty xf^2(x)\, dx$ and from (23)

$$\varepsilon(U) = 12n_1n_2(N + 1)^{-1} \left[\int_0^\infty xf^2(x)\, dx\right]^2. \tag{24}$$

From Savage [13],

$$E_\theta(T) = \int_0^\infty J[n_1N^{-1}F_\theta(x) + n_2N^{-1}F(x)]\, dF_\theta(x), \tag{25}$$

where

$$J(x) = -\ln(1 - x), \qquad 0 < x < 1. \tag{26}$$

Equation (26) follows from [13, theorem 2]. Substituting $y = x/(1 + \theta)$, and (26) into (25) and differentiating yields

$$\left.\frac{dE_\theta(T)}{d\theta}\right|_{\theta=0} = n_2N^{-1} \int_0^\infty \frac{xf^2(x)\, dx}{1 - F(x)}.$$

From [13] using (26) and letting $\theta = 0$ gives $\sigma_0^2(T) = n_2/n_1N$ so that

$$\varepsilon(T) = n_1n_2N^{-1} \left[\int_0^\infty \frac{xf^2(x)}{1 - F(x)}\, dx\right]^2.$$

Appendix II

$\text{ARE}_{U/U'}$

The data format is shown in Fig. 1. Assume slot j is being tested. The Mann–Whitney U statistic is given in (9) of the test and from Appendix I, $\sigma_0^2(U) = (nk + 1)/[12n^2(k - 1)]$. The simplified statistic can be written as

$$U' = n^{-1} \sum_{i=1}^{n} \left((k - 1)^{-1} \sum_{\substack{l=1 \\ l\neq j}}^{k} u(X_i^j - X_i^l)\right)$$

$$= n^{-1} \sum_{i=1}^{n} U_i(1, k - 1), \tag{27}$$

where $u(\cdot)$ is the unit step function and $U_i(1, k - 1)$ is a Mann–Whitney statistic with one sample in the first group and $(k - 1)$ samples in the second group. The $U_i(1, k - 1)$ are independent for $i = 1,2,\cdots,n$, so that U' is asymptotically normal. It follows from (27) that $E_\theta(U') = E_\theta(U)$, so that from (23)

$$\text{ARE}_{U'U'} = \lim_{n\to\infty} \sigma_0^2(U')/\sigma_0^2(U)$$

$$= \lim_{n\to\infty} n(k + 1)/(nk + 1)$$

$$= 1 + k^{-1}. \tag{28}$$

References

[1] M. N. Woinsky, "A composite nonparametric test for a scale slippage alternative," *Ann. Math. Statist.*, to be published.
[2] M. N. Woinsky and L. Kurz, "An efficient nonparametric detector based on a three-sample classification model," *IEEE Trans. Inform. Theory*, vol. IT-14, Sept. 1968, pp. 744–751.
[3] J. W. Carlyle and J. B. Thomas, "On nonparametric signal detectors," *IEEE Trans. Inform. Theory*, vol. IT-10, Apr. 1964, pp. 146–152.
[4] J. V. Evans and T. Hagfors, *Radar Astronomy*. New York: McGraw-Hill, 1966, ch. 10.
[5] M. L. Puri and P. S. Puri, "Multiple decision procedures based on ranks for certain problems in analysis of variances," *Ann. Math. Statist.*, vol. 40, Apr. 1969, pp. 619–632.
[6] P. S. Puri and M. L. Puri, "Selection procedures based on ranks, scale parameter case," *Sankhya*, vol. 30, ser. A, 1968, pp. 291–302.
[7] R. B. Blackman and J. W. Tukey, *The Measurement of Power Spectra From the Point of View of Communications Engineering*. New York: Dover, 1959, p. 93.
[8] J. Hajek, *Nonparametric Statistics*. San Francisco: Holden-Day, 1969, p. 170.
[9] R. G. Miller, Jr., *Simultaneous Statistical Inference*. New York: McGraw-Hill, 1966, p. 226.
[10] E. Paulson, "An optimum solution to the k-sample slippage problem for the normal distribution," *Ann. Math. Statist.*, vol. 23, 1952, pp. 610–616.
[11] J. E. Walsh, *Handbook of Nonparametric Statistics*, vol. 2. Princeton, N.J.: Van Nostrand, 1965, pp. 543–545.
[12] S. Blumenthal and D. W. Patterson, "Rank order procedures for selecting a subset containing the population with the smallest scale parameter," *Sankhya*, vol. 31, ser. A, 1969, pp. 37–42.
[13] H. Chernoff and I. R. Savage, "Asymptotic normality and efficiency of certain nonparametric test statistics," *Ann. Math. Statist.*, vol. 29, 1958, pp. 972–994.
[14] I. R. Savage, "Contributions to the theory of rank order statistics—the two-sample case," *Ann. Math. Statist.*, vol. 27, 1956, pp. 590–615.
[15] E. A. Feustal and L. D. Davisson, "The asymptotic relative efficiency of mixed statistical tests," *IEEE Trans. Inform. Theory*, vol. IT-13, Apr. 1967, pp. 247–255.
[16] J. Capon, "Asymptotic efficiency of certain locally most powerful rank tests," *Ann. Math. Statist.*, vol. 32, 1961, pp. 88–100.
[17] H. Urkowitz, "Energy detection of unknown deterministic signals," *Proc. IEEE*, vol. 55, Apr. 1967, pp. 523–531.
[18] ——, "Energy detection of a random process in colored Gaussian noise," *IEEE Trans. Aerosp. Electron. Syst.*, vol. AES-5, Mar. 1969, pp. 156–162.

[19] B. O. Pierce, *A Short Table of Integrals*, 4th ed. New York: Ginn, 1956, pp. 132–136.
[20] B. W. Lindgren, *Statistical Theory*. New York: Macmillan, 1962, p. 289.
[21] S. S. Wilks, *Mathematical Statistics*. New York: Wiley, 1962, pp. 186, 422.
[22] K. Doksum, "Starshaped transformation and the power of rank tests," *Ann. Math. Statist.*, vol. 40, 1969, pp. 1167–1176.
[23] E. Sherman, "A note on multiple comparisons using rank sums," *Technometrics*, vol. 7, May 1965, pp. 255–256.
[24] E. L. Lehmann, *Testing Statistical Hypotheses*. New York: Wiley, 1964, p. 274.

15

Small-Sample Efficiencies of Rank Tests

P. PAPANTONI-KAZAKOS, MEMBER, IEEE

I. Introduction

THE EFFICIENCIES with respect to the optimum parametric test of several rank tests have been found in the limiting case of a large number of observations available and the two hypotheses approaching each other. An effort is given here towards the calculation of the efficiencies of these tests when the number of data is limited.

The problem in this case becomes tractable if the signal-to-noise ratio is large. In this paper we are concerned with rank tests based on a finite number N of independent identically distributed observations. The efficiencies with respect to the optimal parametric test of such tests is sought when the data are Gaussian distributed.

The key variable in every rank test is the rank vector Z. The vector Z has N components. The ith component z_i is equal to $+1$ if the ith absolutely smallest observation is positive and it is equal to -1 in the opposite case. In this section the probability $P(Z)$ of the vector Z will be found.

Before we proceed with the calculation of $P(Z)$, let us assume that the independent identically distributed observations are Gaussian with mean λ and, without loss of generality, variance 1.

Let us define the probability $P_1(Z_1)$ as the probability of the vector Z_1, which is defined such that its ith component z_{1i} is equal to $+1$ or -1 depending on the sign of the ith observation as received. Then, if Z_0 is an N-dimensional rank vector with ± 1 components, it is obviously true that

$$P(Z_0) = N!\, P_1(Z_0) \tag{1}$$

where $N!$ is N factorial. If $z_{0,i}$ is the ith component of the vector Z_0, then

$$P_1(Z_0) = \int \prod_{i=1}^{N} \phi(t_i - z_{0,i}\cdot\lambda)\, dt \tag{2}$$

where the integral is over the range

$$0 \le t_1 \le \cdots \le t_N \le \infty$$

and

$$dt = dt_1,\cdots,dt_N$$

$$\phi(t) = \frac{e^{-t^2/2}}{\sqrt{2\pi}}.$$

Thus, directly from (1) and (2), we derive

$$P(Z_0) = N! \int_{0\le t_1\le\cdots\le t_N\le\infty} \prod_{i=1}^{N} \phi(t_i - z_{0,i}\cdot\lambda)\, dt. \tag{3}$$

The integral in (3) cannot be generally calculated. However, if the vector

$$t_{\max} = \begin{bmatrix} t_{1\max} \\ \vdots \\ t_{N\max} \end{bmatrix}$$

that maximizes the integrand $\prod_{i=1}^{N} \phi(t_i - z_{0,i})$ can be found, then if one ignores the tail values for λ large, one can approximately break the integral in (3) into a product of lower dimensional integrals that can be calculated. Specifically, in this case we can write the approximation

$$\int_{0\le t_1\le\cdots\le t_N\le\infty} \prod_{i=1}^{N} \phi(t_i - z_{0,i}\lambda)\, dt = \prod_{i=1}^{n} \int_{0\le t_{i1}\le\cdots\le t_{im_i}\le\infty} \prod_{j=1}^{m_i} \phi(t_{ij} - z_{0,M_{i-1}+j}\cdot\lambda)\, dt_{i1}\cdots dt_{im_i} \tag{4}$$

where t_{ij}, $j = 1,\cdots,m_i$, correspond to the ith group of consecutive equally valued $t_{\max}$ components. The separation in groups goes from $t_{1\max}$ to $t_{N\max}$, where $t_{i\max}$ is the ith component of the vector $t_{\max}$. Also $M_i = \sum_{j=1}^{i} m_j$. It will be shown later that the integrals

$$P_i = \int_{0\le t_1\le\cdots\le t_{m_i}\le\infty} \prod_{j=1}^{m_i} \phi(t_{ij} - z_{0,M_{i-1}+j}\cdot\lambda)\, dt_{i1}\cdots dt_{im_i} \tag{5}$$

can be calculated. Apparently the vector $t_{\max}$ must be determined before any effort is given to the calculation of the integrals in (5). The vector $t_{\max}$ was found for the different cases of the rank vector Z_0 in Appendix I. In the same appendix, definitions of the parameters n, m_i, k_i, μ_i are given. These parameters will be used in the course of this paper.

After the vector $t_{\max}$ is explored and the separation of the integral in (3) into the integrals P_i is complete, the integrals P_i must be calculated. To begin with the calculation of P_i, we will examine the exponent in its integrand. The exponent is equal to

$$\sum_{j=1}^{m_i} (t_{ij} - z_{0,M_{i-1}+j}\cdot\lambda)^2$$

and it can obviously be expressed as follows:

$$\sum_{j=1}^{m_i} (t_{ij} - z_{0,M_{i-1}+j}\cdot\lambda)^2 = \sum_{j=1}^{m_i}\left[\left(t_{ij} - \frac{k_i-\mu_i}{k_i+\mu_i}\lambda\right) + \left(\frac{k_i-\mu_i}{k_i+\mu_i}\lambda - z_{0,M_{i-1}+j}\cdot\lambda\right)\right]^2 \tag{6}$$

From the definition of k_i,μ_i in Appendix I, it is obvious that

$$k_i + \mu_i = m_i \tag{7}$$

$$k_i - \mu_i = \sum_{k=1}^{m_i} z_{0,M_{i-1}+k}. \tag{8}$$

Substituting (7) and (8) in (6), we get

$$\sum_{j=1}^{m_i}(t_{ij} - z_{0,M_{i-1}+j})^2 = \sum_{j=1}^{m_i}\left[\left(t_{ij} - \frac{\sum_{k=1}^{m_i} z_{0,M_{i-1}+k}}{m_i}\lambda\right) + \left(\frac{\sum_{k=1}^{m_i} z_{0,M_{i-1}+k}}{m_i} - z_{0,M_{i-1}+j}\lambda\right)\right]^2. \tag{9}$$

Before anything further can be done, several cases must be considered.

A. $k_i > \mu_i$

In this case, (9) becomes

$$\begin{aligned}\sum_{j=1}^{m_i}(t_{ij} - z_{0,M_{i-1}+j})^2 &= \sum_{j=1}^{m_i}\Bigg[\left(t_{ij} - \frac{\sum_{k=1}^{m_i} z_{0,M_{i-1}+k}}{m_i}\lambda\right)^2 \\ &\quad + \lambda^2\left(\frac{\sum_{k=1}^{m_i} z_{0,M_{i-1}+k}}{m_i} - z_{0,M_{i-1}+j}\right)^2 \\ &\quad + 2\lambda\left(t_{ij} - \frac{\sum_{k=1}^{m_i} z_{0,M_{i-1}+k}}{m_i}\lambda\right) \\ &\quad \cdot\left(\frac{\sum_{k=1}^{m_i} z_{0,M_{i-1}+k}}{m_i} - z_{0,M_{i-1}+j}\right)\Bigg]. \\ &= \sum_{j=1}^{m_i}\left(t_{ij} - \frac{\sum_{k=1}^{m_i} z_{0,M_{i-1}+k}}{m_i}\lambda\right)^2 \\ &\quad + 2\lambda\sum_{j=1}^{m_i} t_{ij}\left(\frac{\sum_{k=1}^{m_i} z_{0,M_{i-1}+k}}{m_i} - z_{0,M_{i-1}+j}\right) \\ &\quad + \delta^2\sum_{j=1}^{m_i}\left(1 - \frac{\left[\sum_{k=1}^{m_i} z_{0,M_{i-1}+k}\right]^2}{m_i^2}\right).\end{aligned} \tag{10}$$

At this point we define a new parameter:

$$v_i = \frac{\sum_{k=1}^{m_i} z_{0,M_{i-1}+k}}{m_i}, \qquad -1 \le v_i \le 1. \tag{11}$$

If we substitute (11) in (10), we have the following simpler expression of the exponent in the P_i integrand

$$\begin{aligned}\sum_{j=1}^{m_i}(t_{ij} - z_{0,M_{i-1}+j}\cdot\lambda)^2 &= \sum_{j=1}^{m_i}(t_{ij} - v_i\lambda)^2 \\ &\quad + 2\lambda\sum_{j=1}^{m_i} t_{ij}(v_i - z_{0,M_{i-1}+j}) \\ &\quad + \lambda^2(1 - v_i^2)m_i.\end{aligned} \tag{12}$$

We notice that the last term in (12) does not include any operations with respect to j. We must also observe that since $k_i > \mu_i$ here, the parameter v_i is positive, and since λ is large, so is $v_i\lambda$.

If we now use (12) in (5), we get

$$\begin{aligned}P_i &= \exp\left(-\frac{m_i(1-v_i^2)}{2}\lambda^2\right) \\ &\quad\cdot\int_{0\le t_{i1}\le\cdots\le t_{im_i}\le\infty}\prod_{j=1}^{m_i}[\phi(t_{ij} - v_i\lambda) \\ &\quad\cdot\exp(-\lambda t_{ij}[v_i - z_{0,M_{i-1}+j}])]\,dt^i.\end{aligned} \tag{13}$$

For very large λ, the factor $\phi(t_{ij} - v_i\lambda)$ in the integrand in (13) is obviously the predominant one in the product $\phi(t_{ij} - v_i\lambda)\exp(-\lambda t_{ij}[v_i - z_{0,M_{i-1}+j}])$. Thus since $v_i\lambda$ is positive and very large, we can write, ignoring tail values

$$\begin{aligned}P_i &= \exp\left(-\frac{m_i(1-v_i^2)}{2}\lambda^2\right) \\ &\quad\cdot\int_{-\infty\le t_{i1}\le\cdots\le t_{im_i}\le\infty}\prod_{j=1}^{m_i}[\phi(t_{ij} - v_i\lambda) \\ &\quad\cdot\exp(-\lambda t_{ij}[v_i - z_{0,M_{i-1}+j}])]\,dt^i.\end{aligned} \tag{14}$$

Let us also note that for the negative t_{ij} we can change the parameters with no effect on the value of P_i (for large λ).

Continuing on (14), it can easily be shown that

$$\phi(t_{ij} - v_i\lambda)\cdot\exp(-\lambda t_{ij}[v_i - z_{0,M_{i-1}+j}]) = \phi(t_{ij} - b_{ij})\exp(-\lambda[t_{ij} - b_{ij}]a_j) \tag{15}$$

where

$$b_{ij} = \lambda[z_{0,M_{i-1}+j} + a_j] \tag{16}$$

$$a_i = \sqrt{1 - v_i^2}. \tag{17}$$

By substitution of (15) in (14) and changing of variables we achieve the following expression:

$$P_i = \exp\left(-\frac{m_i(1-v_i^2)}{2}\lambda^2\right)\cdot\int_{-\infty\le t_{i1}\le\cdots\le t_{im_i}\le\infty}\prod_{j=1}^{m_i}[\phi(t_{ij})\cdot\exp(-\lambda t_{ij}c_j)]\,dt^i \tag{18}$$

where

$$c_i = \begin{cases} a_i, & t_{ij} \ge 0 \\ -a_i, & t_{ij} < 0. \end{cases} \tag{19}$$

Here we may note that the integral

$$I = A^{m_i} \int_{-\infty \le t_{i1} \le \cdots \le t_{im_i} \le \infty} \prod_{j=1}^{m_i} [\phi(t_{ij}) \cdot \exp(-\lambda t_{ij} c_i)] \, dt^i$$

is actually the probability of receiving m_i independent data in such a way that the ith received is also the ith smallest. Furthermore, all these data have the same density $f(x) = A\phi(x) \exp(-\lambda t_{ij} c_i)$, where A is the normalizing constant. The probability L of receiving the same data in *any* order is then obviously equal to $(m_i!) \cdot I$.

The probability L is, of course, equal to

$$A^{m_i} \prod_{i=1}^{m_i} \left[\int_{-\infty}^{\infty} \phi(t) \exp(-\lambda t c_i) \, dt \right] \tag{20}$$

and the integral I is given by the following expression:

$$I = \frac{1}{m_i!} \prod_{j=1}^{m_i} \left[\int_{-\infty}^{\infty} \phi(t) \cdot \exp(-\lambda t c_i) \, dt \right]. \tag{21}$$

The P_i integral is then given by (21)

$$P_i = \exp\left(-\frac{m_i(1 - v_i^2)}{2} \lambda^2\right) \cdot \frac{1}{m_i!} \prod_{j=1}^{m_i} \left[\int_{-\infty}^{\infty} \phi(t) \exp(-\lambda t c_i) \, dt \right]. \tag{22}$$

In order to calculate P_i in the limit of $\lambda \to \infty$, we will bound it from both above and below and show that the two bounds tend to the same value for increasing λ.

Since $\lambda t c_i$ was chosen to be always positive, the product $\phi(t) \exp(-\lambda t c_i)$ is obviously bounded from above by $\phi(t)$ and consequently the upper bound to P_i is equal to

$$\frac{1}{m_i!} \exp\left(-\frac{m_i(1 - v_i^2)}{2} \lambda^2\right). \tag{23}$$

Finally, we can obviously write

$$I = \frac{1}{m_i!} \prod_{j=1}^{m_i} [E\{\exp(-\lambda t c_i)\}] \tag{24}$$

with $tG(0,1)$, and, using the convexity of the exponential function, we have

$$\frac{1}{m_i!} \prod_{j=1}^{m_i} [\exp(-\lambda E\{t c_i\})] \le I. \tag{25}$$

Thus, the lower bound to P_i is equal to

$$\frac{1}{m_i!} \exp\left(-\frac{m_i(1 - v_i^2)}{2} \lambda^2 - \lambda \sum_{j=1}^{m_i} E\{t c_i\}\right). \tag{26}$$

Since $v_i \ne 1$ in the limit $\lambda \to \infty$, the term $\{-[m_i(1 - v_i)]/2\}\lambda^2$ is the predominant one in the exponent in (26) and indeed the lower bound in (26) then approaches the upper bound in (23). Hence,

$$\lim_{\lambda \to \infty} P_i = \frac{\exp\left(-\frac{m_i(1 - v_i^2)}{2} \lambda^2\right)}{m_i!}. \tag{27}$$

With (27) the calculation of the integral P_i is completed for large λ, in the case where $k_i > \mu_i$. Two more cases remain: $k_1 = 0$, and $k_1 \le \mu_1$.

B. $k_1 = 0$

In Appendix I, it is shown that the only group of ± 1 components that can have the number of $+1$ smaller than the number of -1 is the first group. In the case we cover here, all the components in the first group are equal to -1. The integral P_1 is then equal to

$$P_1 = \int_{0 \le t_1 \le \cdots \le t_{m_1} \le \infty} \prod_{j=1}^{m_1} \phi(t_j + \lambda) \, dt^1. \tag{28}$$

Similarly to case A) we have the following equality:

$$P_1 = \frac{1}{m_1!} \left[\int_0^{\infty} \phi(t + \lambda) \, dt \right]^{m_1}. \tag{29}$$

We obviously have

$$\int_0^{\infty} \phi(t + \lambda) \, dt = \int_{-\infty}^{-\lambda} \phi(t) \, dt$$

and in the limit, the following equation is true:

$$\lim_{\lambda \to \infty} \left[\int_{-\infty}^{-\lambda} \phi(t) \, dt \right] = \frac{\phi(-\lambda)}{\lambda} = \frac{\exp(-\lambda^2/2)}{\lambda}. \tag{30}$$

Hence, directly from (29) and (30), the following equation is derived:

$$P_1 = \frac{1}{m_1! \, \lambda^{m_1}} \cdot \exp\left(-m_1 \frac{\lambda^2}{2}\right). \tag{31}$$

Finally, the case $k_1 \le \mu_1$ will be examined.

C. $k_1 \le \mu_1$

This is actually the case in which the Z components consist of only one group with k_1 components equal to $+1$ and μ_1 components equal to -1, as is shown in detail in Appendix I.

In this case we have an expression for P_1 which is similar to (13). That is,

$$P_1 = \exp\left(-\frac{m_1(1 - v_1^2)}{2} \lambda^2\right) \cdot \int_{0 \le t_1 \le \cdots \le t_{m_1} \le \infty} \prod_{j=1}^{m_1} [\phi(t_j - v_1 \lambda) \cdot \exp(-\lambda t_j [v_1 - z_{0j}])] \, dt^1. \tag{32}$$

The difference from (13) is that the parameter v_1 is negative here since $k_1 \le \mu_1$. The factor $\phi(t_j - v_1 \lambda)$ is still the predominant one in the product $\phi(t_j - v_1 \lambda) \cdot \exp(-\lambda t_j \cdot [v_1 - z_{0j}])$ when λ is large and we can write

$$\lim_{\lambda \to \infty} P_1 = \lim_{\lambda \to \infty} \cdot \exp\left(-\frac{m_1(1 - v_1^2)}{2} \lambda^2\right) \cdot \int_{0 \le t_1 \le \cdots \le t_{m_1} \le \infty} \prod_{j=1}^{m_1} \phi(t_j - v_1 \lambda) \, dt^1. \tag{33}$$

Furthermore, again using the concepts of ordered and unordered probabilities, we reach the following expression:

$$\lim_{\lambda \to \infty} P_1 = \frac{1}{m_1!} \exp\left(-\frac{m_1(1 - v_1^2)}{2} \lambda^2\right) \cdot \left(\int_0^{\infty} \phi(t - v_1 \lambda) \, dt \right)^{m_1}. \tag{34}$$

Proceeding as before, we finally obtain

$$\lim_{\lambda \to \infty} P_1 = \frac{1}{(v_1 \lambda)^{m_1}} \cdot \exp\left(-\frac{m_1}{2}\lambda^2\right). \quad (35)$$

In this case the parameter m_1 is equal, of course, to the number of observations N. At this point the calculation of the subintegrals P_i is complete.

Our objective is the calculation of the probability $P(Z_0) = \prod_{i=1}^{n} P_i$. From the results in cases A, B, and C, the following expressions for $P(Z_0)$ are derived:

$$\lim_{\lambda \to \infty} P(Z_0) = \begin{cases} k_{11}(Z_0) \cdot \exp\left(-\frac{\lambda^2}{2}\sum_{i=1}^{n} m_i(1 - v_i^2)\right), \\ \quad \text{when } k_i > \mu_i, \ \forall i \\ k_{12}(Z_0)\lambda^{-m_1} \exp\left(-\frac{\lambda^2}{2} m_1 - \frac{\lambda^2}{2}\sum_{i=2}^{n} m_i(1 - v_i^2)\right), \\ \quad \text{when } k_1 = 0, \ k_i > \mu_i, \ \forall i \ge 2 \\ k_{13}(Z_0)\lambda^{-N} \exp\left(-\frac{\lambda^2}{2} N\right), \\ \quad \text{when one only group with } k_1 \le \mu_1 \end{cases} \quad (36)$$

$k_{11}(Z_0), k_{12}(Z_0), k_{13}(Z_0)$ are just functions of Z_0. It is easy to develop a uniform formula covering all the cases in (36). Indeed, if we define

$$\rho_i = \begin{cases} v_i, & v_i > 0 \\ 0, & v_i \le 0 \end{cases} \quad (37)$$

$$s_1 = \begin{cases} m_1, & \mu_1 \ge k_1 \\ 0, & \mu_1 < k_1 \end{cases} \quad (38)$$

the following expression is true for the probability $P(Z_0)$:

$$\lim_{\lambda \to \infty} P(Z_0) = K(Z_0) \cdot \lambda^{-s_1} \exp\left(-\frac{\lambda^2}{2} N\right) \cdot \exp\left(+\frac{\lambda^2}{2}\sum_{i=1}^{n} m_i \rho_i^2\right). \quad (39)$$

Again, $k(Z_0)$ is only a function of the rank vector Z_0.

It is finally obvious that the probability of the rank vector $-Z_0$ is given by an expression similar to (39)

$$\lim_{\lambda \to \infty} P(-Z_0) = M(Z_0) \cdot \lambda^{-t_1} \exp\left(-\frac{\lambda^2}{2} N\right) \cdot \exp\left(+\frac{\lambda^2}{2}\sum_{i=1}^{m} n_i \phi_i^2\right) \quad (40)$$

where the following table of correspondence was used:

$$P(Z_0) \to P(-Z_0)$$
$$k(Z_0) \to M(Z_0)$$
$$s_1 \to t_1$$
$$m_i \to n_i$$
$$n \to m$$
$$\rho_i \to \phi_i.$$

Now that the probabilities $P(Z_0)$ and $P(-Z_0)$ are known, one can proceed towards finding the number to which the small sample efficiencies of several rank tests converge when the signal-to-noise ratio increases infinitely. Specifically, the limiting efficiencies with respect to the optimum parametric test of the optimum rank and the Wilcoxon nonparametric tests will be found.

II. Small-Sample Efficiency of the Optimum Rank Test

It is well known that the optimum rank test is the testing of a likelihood ratio against a fixed threshold. Let us assume that the two hypotheses to be tested, H_0 and H_1, are defined as follows.

H_1: The N independent observations are all from a symmetric (about the mean) density of variance 1 and mean λ.

H_0: The same observations are from a density as was shown except that the mean is equal to $-\lambda$.

If Z_0 is the rank vector defined in the Introduction, the likelihood ratio

$$\text{LR} = \frac{P_{H_1}(Z_0)}{P_{H_0}(Z_0)}$$

will be tested here against a threshold C.

If we define $P(Z_0) = P_{H_1}(Z_0)$, then obviously $P_{H_0}(Z_0) = P(-Z_0)$.

The objective of this section is the establishing of the efficiency of the optimum rank test under the Gaussian distribution. Using (39) and (40), we see that the likelihood ratio is given by

$$\text{LR} = \frac{K(Z_0)}{M(Z_0)} \lambda^{t_1 - s_1} \exp\left(\frac{\lambda^2}{2}\left[\sum_{i=1}^{n} m_i \rho_i^2 - \sum_{i=1}^{m} n_i \phi_i^2\right]\right). \quad (41)$$

At this point we will stop to elaborate on (41). Depending on the difference

$$\sum_{i=1}^{n} m_i \rho_i^2 - \sum_{i=1}^{m} n_i \phi_i^2$$

the decision is based on two different schemes. Namely,

a) if $\sum_{i=1}^{n} m_i \rho_i^2 = \sum_{i=1}^{m} n_i \phi_i^2$

the test is based on the difference $t_1 - s_1$, which is always nonzero. For large λ the threshold will be zero and the test is then

$$t_1 - s_1 \underset{H_0}{\overset{H_1}{\gtrless}} 0 \quad (42)$$

b) if $\sum_{i=1}^{n} m_i \rho_i^2 \ne \sum_{i=1}^{m} n_i \phi_i^2$

then for λ large, the exponential is the predominant factor in (41), and the test will be

$$\sum_{i=1}^{n} m_i \rho_i^2 \underset{H_0}{\overset{H_1}{\gtrless}} \sum_{i=1}^{m} n_i \phi_i^2. \quad (43)$$

We now study the probability of error. A first expression is

$$P_{r_e} = \sum_{Z_0:\, H_0 \text{ is decided}} \frac{PZ_0}{H_1}. \tag{44}$$

If the simplified test is the one in (42), then the terms in (44) are obviously all of the form $k\lambda^{-\theta}$ and from the expression of the probability of error P_{r_0} of the optimal parametric test, as it appears in Appendix II, it is clear that the efficiency of the optimum rank test then tends to zero. Hence, this case is of no interest to us.

If instead the simplified test is the one in (43), the terms in (44) are all of the form

$$G(Z_0,\lambda) \exp\left(-\frac{\lambda^2}{2} \sum_{i=1}^{n} m_i(1 - \rho_i^2)\right) \tag{45}$$

where $G(Z_0,\lambda)$ is a nonexponential function of λ. For large λ, the term of (45) that will dominate in (44) will be the one that has the sum $\sum_{i=1}^{n} m_i(1 - \rho_i^2)$ minimum among all Z_0 that lead to decision H_0. That is, since the sum $\sum_{i=1}^{n} m_i(1 - \rho_i^2)$ depends on the particular rank vector Z_0, if we call

$$M(Z_0) = \sum_{i=1}^{n} m_i(1 - \rho_i^2) \tag{46}$$

then the predominant term in (44) will be the one that has

$$M_0 = \sum_{i=1}^{n_0} m_{0i}(1 - \rho_{0i}^2) = \min_{Z_0:\, H_0 \text{ is decided}} M(Z_0). \tag{47}$$

The probability of error will in this case be equal to

$$P_{r_e} = G_0(Z_0,\lambda) \exp\left(-\frac{M_0}{2}\lambda^2\right) \tag{48}$$

where $G_0(Z_0,\lambda)$ is a nonexponential function of λ. The probability of error in (45) looks very much like the probability of error P_{r_0} in Appendix II. The parameter of interest that must be found is M_0. Let us go back to the definition of M_0

$$M_0 = \min_{Z:\, P(Z_0)/[P(-Z_0)] < 1} M(Z_0). \tag{49}$$

What (49) says is that the Z_0 vector that will actually achieve M_0 must be such that

$$M(Z_0) > M(-Z_0). \tag{50}$$

Hence, only the vectors that satisfy (50) will be candidates for achieving M_0.

Suppose now that an arbitrary constant N_0 is chosen. The question is, is there some Z_0 such that

$$M(Z_0) = N_1 \tag{51}$$

and at the same time (50) is satisfied and, if so, what is this Z_0? If the answer to this question exists, then by reducing N_1, the threshold M_0 after which (50) is not satisfied any more could eventually be found.

Going back to (51), if a Z_0 exists that satisfies both (50) and (51), that must be at least the Z_0 that minimizes $M(-Z_0)$ under the restriction $M(Z_0) = N_1$. Let us observe here that if we want to have $M(Z_0) > M(-Z_0)$, the constant $N_1 = M(Z_0)$ must necessarily be larger than $N/2$. On the other hand, $M(Z)$ is bounded from above by N. So, we first require that

$$\frac{N}{2} \leq N_1 \leq N. \tag{52}$$

The expression $M(-Z_0)$ is actually a sum $\sum_{i=1}^{n} m_i(1 - \rho_i^2)$, where $\sum_{i=1}^{n} m_i = N$, $0 \leq 1 - \rho_i^2 \leq 1$, $m_i > 0$, and $1 - \rho_i^2$ are not all equal to zero. In such a sum, the minimum is achieved if $m_1 = N$, $m_i = 0$, for all $i \geq 2$ with a_1 chosen accordingly.

From Appendix I it is clear that a vector $-Z_0$ realizing the preceding is the vector

$$-Z_0: \begin{cases} z_{01} = z_{02} = \cdots = z_{0k} = 1 \\ z_{0(k+1)} = \cdots = z_{0N} = -1. \end{cases} \tag{53}$$

Then we have

$$M(-Z_0) = N - \frac{(2k - N)^2}{N}$$

$$M(Z_0) = k. \tag{54}$$

So, if k was originally chosen to be equal to N_1, we have from (54)

$$\min M(-Z_0) = N - \frac{(2N_1 - N)^2}{N}. \tag{55}$$

We still have to restrict ourselves to the case where $M(Z_0) > M(-Z_0)$. Hence, the following inequality must be satisfied:

$$N_1 > N - \frac{(2N_1 - N)^2}{N}. \tag{56}$$

Directly from this we have

$$N_1 > \tfrac{3}{4}N \tag{57}$$

which leads to the conclusion

$$M_0 > \tfrac{3}{4}N. \tag{58}$$

Here if we ask that

$$\frac{P_{r_e}}{P_{r_0}} = 1 \tag{59}$$

where P_{r_e} is given in (48) and P_{r_0} in Appendix II, we end up wanting the following equation satisfied

$$\left.\frac{G_0(Z_0,\lambda)}{C_0(\lambda)} \exp\left(-\frac{\lambda^2}{2}(M_0 - N_0)\right)\right|_{\lambda\to\infty} = 1. \tag{60}$$

For large λ, (60) is actually true for $M_0 - N_0 \to 0$. Using (58), we directly get $N/N_0 \to \frac{4}{3}$. That means that the relative efficiency of the optimum rank test with respect to the optimal parametric test converges for the signal-to-noise ratio increasing to infinity to a number that is bounded from below by $\frac{3}{4}$. The efficiency calculated in this section is not, of course, nearly as good as the asymptotic relative efficiency of the same test which is 1. Considering, though, the fact that the number of samples here is small, we can surely say that efficiency $\frac{3}{4}$ is pretty good in this case.

III. The Small-Sample Efficiency of the Wilcoxon Nonparametric Test

In this section, the hypotheses of Section II are again tested. Also, the relative efficiency of the Wilcoxon nonparametric test is sought, again under the assumption of Gaussian data. The sum $\sum_{i=1}^{N} z_{0i} \cdot i$ is tested here against a threshold C, which in the limiting case of $\lambda \to \infty$ will be equal to zero. The components z_{0i} of the rank vector Z_0 were defined in the Introduction.

The probability of error here is given by the expression

$$P_{r_e} = \sum_{Z_0: \sum_{i=1}^{N} z_{0i} \cdot i < 0} P(Z_0). \tag{61}$$

Again the predominant term in (61) for λ large is the term

$$G(Z_0,\lambda) \exp\left(-\frac{\lambda^2}{2} \sum_{i=1}^{n} m_i(1 - \rho_i^2)\right)$$

that has the smallest exponent factor $\sum_{i=1}^{n} m_i(1 - \rho_i^2)$ among all the terms in P_{r_e}. Furthermore, the following equation is then satisfied:

$$\lim_{\lambda \to \infty} P_{r_e} = G(Z_0,\lambda) \cdot \exp\left(-\frac{\lambda^2}{2} M_0\right) \tag{62}$$

where

$$M_0 = \min_{Z_0: \sum_{i=1}^{N} z_{0i} \cdot i < 0} \sum_{i=1}^{n} m_i(1 - \rho_i^2). \tag{63}$$

Reasoning as in Section II, we notice that the rank vector Z_0 that will actually achieve M_0 is the one with the first l components equal to (-1) and the remaining $N - l$ components to $+1$. The number l will be found from the inequality

$$\sum_{i=1}^{N} z_{0i} \cdot i < 0. \tag{64}$$

Indeed, for a vector Z_0 such as the one mentioned, (64) is equivalent to

$$\frac{-l(l+1)}{2} + l(N - l) + \frac{(N - l)(N - l + 1)}{2} < 0. \tag{65}$$

From (65) one can easily establish the following inequality:

$$l > \frac{N}{\sqrt{2}}. \tag{66}$$

Finally, for an N-dimensional vector with its first l components equal to -1 and the rest equal to $+1$, it is easy to conclude that

$$M_0 = \sum_{i=1}^{n} m_i(1 - \rho_i^2) = l. \tag{67}$$

Relationships (66) and (67) obviously give

$$M_0 > \frac{N}{\sqrt{2}} \tag{68}$$

and one finally finds that the relative efficiency of the Wilcoxon nonparametric test with respect to the optimum parametric test is bounded from below by $1/\sqrt{2}$, when the number of data is small, the signal-to-noise ratio is infinitely large, and the observations are Gaussian.

IV. Conclusion

From Sections I–III of this paper it is clear that we have to limit ourselves to the case of large signal-to-noise ratio if we want to actually calculate small-sample relative efficiencies. One can find analogies between the computational difficulties here and the difficulties in finding asymptotic relative efficiencies. Indeed, one has to limit oneself to the case of hypothesis H_1 being very close to hypothesis H_0 and to the number of data being very large in order actually to calculate relative efficiencies in the latter case.

The small-sample relative efficiency of the optimum rank test was found to be, for large signal-to-noise ratios, at least as good as $\frac{3}{4}$ for Gaussian data. That result is comparable to the asymptotic relative efficiency (which is 1), since in this case we are dealing with a small number of data. The small-sample efficiency of the Wilcoxon nonparametric test was found to be $1/\sqrt{2}$, in contrast to $3/\pi$, which is the *asymptotic* relative efficiency of the same test.

As a final remark it will be pointed out here that since the nonparametric tests are very sensitive to dependence [16], their efficiency in the presence of dependent data is expected to be different than that found in this paper.

Acknowledgment

The author is grateful to Dr. J. Capon and the anonymous reviewers for the time they spent on this paper and their helpful suggestions, as well as to Prof. L. D. Davisson for his help.

Appendix I
Dynamic Programming Method for the Calculation of t_{max}

The objective here is the calculation of the vector t_{max} that maximizes the product

$$\prod_{i=1}^{N} \phi(t_i - z_{0i}\lambda) \tag{A1}$$

where the variables t_i are positive and ordered so that t_i is also the ith smallest. Also, z_{0i} are given constants equal to either $+1$ or -1, and they are taken to be the components of an N-dimensional vector Z_0. Finally,

$$\phi(t) = \frac{e^{-t^2/2}}{\sqrt{2}}.$$

The vector t_{max} will be found through dynamic programming methods, and its value will, of course, depend on the particular vector Z_0 used.

The problem breaks down to finding the minimum with respect to the t_i of the expression $\sum_{i=1}^{N} (t_i - z_{0i}\lambda)^2$, under the restriction of all t_i being positive and ordered. We proceed by stating the following lemma.

Lemma 1: If Z_0 is such that $z_{01} = \cdots = z_{0k} = -1, k \geq 1$, then t_{max} is such that its first k components $t_{1\,max}, \cdots, t_{k\,max}$ are all equal to zero.

Proof: Using dynamic programming, we fix the values of $t_2, \cdots, t_N$ and minimize the expression

$$\sum_{i=1}^{N} (t_i - z_{0i}\lambda)^2 = \sum_{i=1}^{N} (t_i + \lambda)^2 + \sum_{i=k+1}^{N} (t_i - z_{0i}\lambda)^2$$

with respect to t_1. That gives $t_1 = t_{1\,max} = 0$. Proceeding similarly, we get the result in the statement of the lemma.

Lemma 2: Let the k first components of the vector Z_0 be equal to $+1$, followed by μ components equal to -1. Then, t_{max} is such that

$$t_{1\,max} = t_{2\,max} = \cdots = t_{k+\mu,max} = 0, \quad \text{if } \mu \le k$$

$$t_{1\,max} = t_{2\,max} = \cdots = t_{k+\mu,max} = \frac{k-\mu}{k+\mu}\lambda, \quad \text{if } \mu < k \text{ and } t_{k+\mu+1,max} \le \frac{k-\mu}{k+\mu}\lambda$$

$$t_{1\,max} = t_{2\,max} = \cdots = t_{k+\mu,max} = t_{k+\mu+1,max}, \quad \text{if } \mu < k \text{ and } t_{k+\mu+1,max} < \frac{k-\mu}{k+\mu}\lambda.$$

Proof: In the case of a Z_0 vector as described in the statement of the lemma, the sum $\sum_{i=1}^{N} (t_i - z_{0i}\lambda)^2$ that we want to minimize with respect to the t_i is equal to

$$\sum_{i=1}^{k} (t_i - \lambda)^2 + \sum_{i=k+1}^{k+\mu} (t_i + \lambda)^2 + \sum_{i=k+\mu+1}^{N} (t_i - z_{0i}\lambda)^2. \quad \text{(A2)}$$

Using dynamic programming methods, we see that if $t_2, t_3, \cdots, t_N$ are kept constant, the $t_{1\,max}$ that minimizes the expression in (2) under the restriction of $t_1 \le t_2$ is then equal to t_2. We note here that to every t_i such that $\lambda < t_i \le 2\lambda$ corresponds a t_i' such that $0 \le t_i' \le \lambda$ and $(t_i - \lambda)^2 = (t_i' - \lambda)^2$. Hence we will assume here that the t_i that minimize the first partial sum in (A2) are all in the region $[0,\lambda]$.

Let us now suppose that the values $t_{1\,max}, t_{2\,max}$ up to $t_{k\,max}$ are determined. By the dynamic programming logic, they will all be equal to each other. The question is, what is the value $t_{k+1,max}$ that achieves the minimum of (A2) with respect to t_{k+1} under the restriction $t_{k+1,max} \ge t_{k,max}$? The answer is obviously $t_{k+1,max} = t_{k,max}$.

Finally, given the value $t_{k+1,max}$ and, since the terms $(t_i + \lambda)^2$ are minimum for the smallest possible positive t_i values, then under the restriction $t_{k+1,max} \le t_{k+2,max} \le \cdots \le t_{k+\mu,max}$, the minimum of the second partial sum in (A2) is assumed for $t_{k+1,max} = t_{k+2,max} = \cdots = t_{k+\mu,max} \le t_{k+\mu+1,max}$. From what has already been said in this proof, it is clear that the sum in (A2) becomes equal to

$$k(t_0 - \lambda)^2 + \mu(t_0 + \lambda)^2 + \sum_{i=k+\mu+1}^{N} (t_i - z_{0i}\lambda)^2 \quad \text{(A3)}$$

where t_0 is the common value that all the variables $t_1, \cdots, t_{k+\mu}$ assume in the minimum. To actually find this value t_0, we differentiate (A3) with respect to it and put the result equal to zero. The second derivative is always $k + \mu > 0$ so the extreme, if it is in the region $[0,\lambda]$, is the minimum.

Indeed, the first derivative gives $t_0 = \lambda(k - \mu)/(k + \mu)$. If $k - \mu \ge 0$, t_0 is positive, and it assumes the minimum under the restriction $t_1 \le \cdots \le t_N$. If $k - \mu < 0$, however, the value $\lambda(k - \mu)/(k + \mu)$ is outside the acceptable region. But for $\lambda t_0 > (k - \mu)/(k + \mu)$, (A3) increases monotonically, assuming the minimum, under the restriction $t_0 > 0$, at $t_0 = 0$.

By just extending the results in this lemma in a more-or-less straightforward manner, we reach the following lemma.

Lemma 3: Let the components of vector Z_0 be grouped in the following way.

The ith group consists of k_i consecutive Z_0 components of value equal to $+1$, followed by μ_i consecutive components of value equal to (-1). Then, the following results are true.

a) If $k_i > \mu_i$, and at the same time $k_i/\mu_i \le k_{i+1}/\mu_{i+1}$, for all i, then t_{max} is such that

$$t_{j,max} = \frac{k_1 - \mu_1}{k_1 + \mu_1}\lambda, \qquad j = 1, 2, \cdots, k_1 + \mu_1$$

$$t_{j,max} = \frac{k_i - \mu_i}{k_i + \mu_i}\lambda, \qquad j = \sum_{l=1}^{i-1} (k_l + \mu_l), \cdots, \sum_{l=1}^{i} (k_l + \mu_l).$$

b) If $k_i > \mu_i$, for all i, and, at the same time, $k_j/\mu_j \le k_{j+1}/\mu_{j+1}$, for $j = 1,2,\cdots, i-1$, while

$$\frac{k_i}{\mu_i} > \frac{k_{i+1}}{\mu_{i+1}}$$

then we find the smallest integer L such that

$$\frac{\sum_{m=0}^{L+1} k_{i+1-m}}{\sum_{m=0}^{L+1} \mu_{i+1-m}} \le \frac{k_{i-L-1}}{\mu_{i-L-1}}$$

and we have the same results as in case a) under the new groups $(k_1 + \mu_1),(k_2 + \mu_2),\cdots,(k_{i-L-1} + \mu_{i-L-1}),(\sum_{m=0}^{L-1} k_{i+1-m} + \sum_{m=0}^{L+1} \mu_{i+1-m})$.

c) If $k_i < \mu_i$, for some i, then again we find the smallest integer L such that

$$\frac{k_{i-L-1}}{\mu_{i-L-1}} \ge \frac{\sum_{m=0}^{L+1} k_{i-m}}{\sum_{m=0}^{L} \mu_{i-m}}$$

and proceed as in case b).

d) If $(\sum_{j=1}^{i} k_j)/(\sum_{j=1}^{i} \mu_j) \le 1$, for some i, then $t_{j\,max} = 0$, for $j = 1,2,\cdots,\sum_{l=1}^{i} (k_l + \mu_l)$.

e) If the last group of Z_0 components has all its members equal to $+1$, then the t_{max} components corresponding to this group are all equal to λ.

Summarizing the results in this appendix, we have

$$\max_{t_i, i=1,\cdots,N} \prod_{i=1}^{N} \phi(t_i - z_{0i}\lambda) = \prod_{i=1}^{n} \phi^{k_i}(\xi_i\lambda) \cdot \phi^{\mu_i}(\rho_i\lambda)$$

under the restriction $0 \le t_{1\,max} \le \cdots \le t_{N\,max}$, where

n	number of groups of Z_0 components
m_i	number of components in the ith group
k_i	number of $+1$ components in the ith group
μ_i	number of -1 components in the ith group

$$\xi_i = \begin{cases} 1, & \text{in case d)} \\ \dfrac{m_i - \sum_j z_{0j}}{m_i}, & \text{in cases a)–c)} \end{cases}$$

$$\rho_i = \begin{cases} 1, & \text{in case d) and Lemma 1} \\ \dfrac{m_i + \sum_j z_{0j}}{m_i}, & \text{in cases a)–c)} \end{cases}$$

$\sum_j z_{0j}$ summation of all Z_0 components in the ith group.

Appendix II
Probability of Error of the Optimum Parametric Test

Assuming the hypothesis

H_1: observations Gaussian with mean λ and unit variance
H_0: observations Gaussian with mean $-\lambda$ and unit variance

and, furthermore, assuming the N observations to be v_i independent of each other, the optimum parametric decision is given by the following relationship:

$$T(v) = \frac{\sum_{i=1}^{N} v_i}{N} \underset{H_0}{\overset{H_1}{\gtrless}} C \tag{B1}$$

C is, of course, the threshold.

The random variable $T(v)$ is Gaussian with variance $1/N$ and means λ or $-\lambda$ depending on the hypothesis under which it is examined.

For λ assuming large values, the threshold C becomes equal to zero. The probability of error under these assumptions is

$$\begin{aligned}
P_{r_0} &= P_r\left\{\text{decide } \frac{H_0}{H_1}\right\} + P_r\left\{\text{decide } \frac{H_1}{H_0}\right\} \\
&= \int_{-\infty}^{0} \frac{\exp\left(-\frac{1}{2}N(u-\lambda)^2\right)}{\sqrt{2\pi}\,(1/\sqrt{N})}\,du \\
&\quad + \int_{0}^{\infty} \frac{\exp\left(-\frac{1}{2}N(u+\lambda)^2\right)}{\sqrt{2\pi}\,(1/\sqrt{N})}\,du \\
&= 2\int_{0}^{\infty} \frac{\exp\left(-\frac{1}{2}N(u+\lambda)^2\right)}{\sqrt{2\pi}\,(1/\sqrt{N})}\,du \\
&= 2\int_{-\infty}^{-\lambda\sqrt{N}} \phi(u)\,du \xrightarrow[\lambda-\infty]{} \frac{\phi(-\lambda\sqrt{N})}{\lambda\sqrt{N}} = \frac{\exp\left(-(\lambda^2/2)N\right)}{\sqrt{2\pi N}\,\lambda} \\
&= C_0(\lambda)\exp\left(-(\lambda^2/2)N\right)
\end{aligned} \tag{B2}$$

where $C_0(\lambda)$ is a nonexponential function of λ.

References

[1] J. Neyman and E. S. Pearson, "On the problem of the most efficient tests of statistical hypotheses," *Phil. Trans. Roy. Soc. London*, vol. A231, p. 289, 1933.

[2] J. W. Carlyle and J. B. Thomas, "On nonparametric signal detectors," *IEEE Trans. Inform. Theory*, vol. IT-10, pp. 146–152, Apr. 1964.

[3] J. M. Wozencraft and I. M. Jacobs, *Principles of Communication Engineering*. New York: Wiley, 1965.

[4] E. A. Feustel and L. D. Davisson, "The asymptotic relative efficiency of mixed statistical tests," *IEEE Trans. Inform. Theory*, vol. IT-13, pp. 247–255. Apr. 1967.

[5] M. Kanefsky and J. B. Thomas, "On adaptive nonparametric detection systems using dependent samples," *IEEE Trans. Inform. Theory*, vol. IT-11, pp. 521–526, Oct. 1965.

[6] J. Capon, "Nonparametric technique for the detection of a constant signal in noise," *1959 IRE WESCON Conv. Rec.*, vol. 4, pp. 92–103, 1959.

[7] R. F. Daly and C. K. Rushforth, "Nonparametric detection of a signal of known form in additive noise," *IEEE Trans. Inform. Theory*, vol. IT-11, pp. 70–76, Jan. 1965.

[8] M. N. Woinsky and L. Kurz, "An efficient nonparametric detector based on a three-sample classification model," *IEEE Trans. Inform. Theory*, vol. IT-14, pp. 744–751, Sept. 1968.

[9] M. E. Terry, "Some rank-order tests which are most powerful against specific alternatives," *Ann. Math. Statist.*, vol. 23, pp. 346–366, 1952.

[10] D. A. S. Fraser, "Most powerful rank-type tests," *Ann. Math. Statist.*, vol. 28, pp. 1040–1043, 1957.

[11] J. Capon, "Asymptotic efficiency of certain locally most powerful rank tests," *Ann. Math. Statist.*, vol. 32, pp. 88–100, 1961.

[12] J. H. Klotz, "Small sample power and efficiency for the one-sample Wilcoxon and normal scores tests," *Ann. Math. Statist.*, vol. 34, pp. 335–337, 1963.

[13] J. B. Thomas, "Nonparametric detection," *Proc. IEEE*, vol. 58, pp. 623–632, May 1970.

[14] R. C. Milton, *Rank Order Probabilities*. New York: Wiley, 1970.

[15] U. V. Rao, I. R. Savage, and M. Sobel, "Contributions to the theory of rank order statistics: The two-sample censored case," *Ann. Math. Statist.*, vol. 31, pp. 415–426, 1960.

[16] J. L. Gastwirth and H. Rubin, "Effect of dependence on level of some one-sample tests," *J. Amer. Statist. Assoc.*, vol. 36, pp. 816–820, 1971.

Part IV

OTHER NONPARAMETRIC DETECTION METHODS AND RESULTS

Editors' Comments on Papers 16 Through 25

16 FEUSTEL and DAVISSON
The Asymptotic Relative Efficiency of Mixed Statistical Tests

17 CHING and KURZ
Nonparametric Detectors Based on m*-Interval Partitioning*

18 KASSAM and THOMAS
Dead-Zone Limiter: An Application of Conditional Tests in Nonparametric Detection

19 SHIN and KASSAM
Three-Level Coincidence-Correlator Detectors

20 GRIEVE
The Optimum Constant False Alarm Probability Detector for Relatively Coherent Multichannel Signals in Gaussian Noise of Unknown Power

21 MILLARD and KURZ
The Kolmogorov-Smirnov Tests in Signal Detection

22 HENRICHON and FU
Calamity Detection Using Nonparametric Statistics

23 ANDERSON and BENNING
A Distribution-Free Discrimination Procedure Based on Clustering

24 BEAKLEY and TUTEUR
Distribution-Free Pattern Verification Using Statistically Equivalent Blocks

25 SRINIVASAN
Distribution-Free Processors Based on the Theory of Runs

In addition to papers adapting, applying, and analyzing well-known nonparametric statistical procedures for signal detection systems, many contributors to the engineering literature have described new modifications and formulations of nonparametric detector structures that result in simpler hardware implementations and greater efficiencies, or that exploit nonparametric statistical procedures other than those based directly on signs or ranks of observations. In this part, we have collected ten such papers, most of them relatively recent.

A very interesting general technique for reducing the complexity of implementation of many rank tests was introduced by Feustel and Davisson in Paper 16. The basic idea is to divide the total number of n independent observations (from each sample, or channel) into p groups of m observations each. For each group of m observations, an intermediate rank statistic is defined, and the overall test statistic is the sum of the p intermediate statistics. The resulting statistic has been termed a *mixed statistic*. Since the number of operations for a ranking of n observations grows very quickly with n, it is clear that for reasonably small m this procedure can reduce substantially the processing time of a rank detector. Feustel and Davisson have shown that in many instances a small value of m (around 10) results in efficiencies that are almost as high as those of the unmodified rank tests where $m = n$ and $p = 1$. Both one- and two-sample tests have been considered, and performance comparisons are based on ARE computations.

Paper 17 by Ching and Kurz analyzes the performance of a nonparametric detector structure that uses partitioned or quantized data for the detection of a constant signal in noise based on one sample of independent observations. The idea developed here is that if a finite set of parameters of the class of noise distributions is known—specifically some quantiles in addition to the median—then nonparametric tests can be implemented that are more efficient than the sign test. The m-interval partitioning of data thus results in more efficient extensions of the sign test, for which m=2. Locally optimum test statistics based on partitioned data are derived, and asymptotic performance comparisons are made based on the usual definition of ARE (Pitman) as well as based on Bahadur and Hodges-Lehmann efficiencies. The paper also discusses briefly the use of adaptive systems for quantile estimation. Further details of these ideas have recently appeared in Kurz[1] and Kersten and Kurz.[2]

Kassam and Thomas have applied the concept of a conditional statistical test in Paper 18 to obtain a simple extension of the sign detector that has significantly better detection performance. The dead-zone limiter nonlinearity quantizes data symmetrically into three levels (+1, 0, −1); the detector studied uses this instead of the hard-limiter of the sign detector. Assuming only symmetry of the noise density function, the authors show that by using the limited data to set the threshold and randomization probability, nonparametric detection performance is obtained for any symmetric three-level quantization of the data. The paper investigates optimum quantization and contains both asymptotic performance comparisons, as well as finite-signal, finite-sample detection power comparisons. This basic idea can also be extended to higher-order quantization schemes (Kassam and Thomas),[3] which require only a symmetry assumption and conditional testing for nonparametric performance.

In Paper 19, Shin and Kassam have considered a similar extension of the polarity-coincidence correlator (Paper 4) based on conditional testing. The test statistic for the three-level coincidence-correlator detector is based on a symmetric partitioning into three regions of the two-dimensional observation space: once again the use of a conditional test allows nonparametric performance when only symmetry is assumed for the noise density function. An extension of this concept to higher-level quantization of the observation space can also be analyzed (Shin and Kassam).[4]

Paper 20 by Grieve is concerned with a multichannel detection problem for Gaussian noise with unknown variance. Although this is a parametric problem, a CFAR (constant false-alarm rate) processor is considered based on the theory of similar tests for a composite null hypothesis. The statistical theory summarized in this paper is useful in understanding the basic procedure applied in Papers 18 and 19, where only references are given to the theoretical basis of conditional tests.

Included in this part are a short correspondence and papers that are of interest because they represent attempts to adapt other statistical procedures (not based on signs or ranks) to detection applications. In Paper 21, Millard and Kurz study some procedures in the class of Kolmogorov-Smirnov, Cramer-von Mises tests, which are essentially tests based on the empirical distribution functions of the observations. Detection of a constant signal in additive noise is considered, assuming that a reference noise sample is available. The tests considered obtain the distance between the two empirical distribution functions and are compared for performance by Monte Carlo simulations. Paper 22 (Henrichon and Fu) applies the Kolmogorov-Smirnov test (and also the two-sample Mann-Whitney rank test) to detect a change in the normal operation of a linear dynamical system. It is assumed that a set of observations representing normal system operation is available, and the procedure compares a sample of observations from a possibly faulty system with this set. Monte Carlo simulation results are reported for the performance of the two nonparametric tests, and a specific control system example is considered for the rank test.

Papers 23 and 24 present an interesting application of nonparametric tolerance regions for distribution-free two-class discrimination and pattern-classification problems. Let $S(X)$ be a tolerance region based on a random sample $X = (X_1, X_2, \ldots X_n)$ of independent observations from some probability distribution, indexed by θ, in a possible collection Ω. Then S (X) is said to be a distribution-free β tolerance region for a proportion p if the probability measure $P_\theta\{S(X)\}$ of the region S (X), which is itself a random variable, is such that its probability of being at least as large as p is β, independent of θ; that is, $\text{Prob}_\theta\{P_\theta\{S(X)\} \geq p\} = \beta$. Thus a β tolerance region for a proportion p, constructed from a sample of n observations, will contain future outcomes with probability p, this statement having a confidence level β. A compact and readable discussion of nonparametric tolerance regions may be found in Fraser.[5]

In the situations considered in Papers 23 and 24, two training samples from each of two possible distributions are available, and an observation sample is to be classified as belonging to one of these. In Paper 23, Anderson and Benning describe the construction of a classification procedure in which a region of "reserve judgment" or "no decision" is allowed, with the conditional probabilities of wrong classification held below specified numbers within certain confidence levels. Their construction is based on the formation of nonparametric tolerance regions formed from a knowledge of the two training samples. Beakley and Tuteur in Paper 24 consider a more classical approach where the probability of one type of error is controlled (to within a specific confidence level), and a suboptimum decision function is constructed based on tolerance regions. They also describe results of an experiment for automatic speaker verification based on their procedures.

Finally, in Paper 25, Srinivasan proposes some nonparametric detection schemes based on the theory of runs. The detection problem studied is that of deciding if a possible signal sample and a reference noise sample have a common distribution (null hypothesis). Suppose that for the combined sample, the number of runs is determined—that is, the number of sequences of observations from one sample bounded by observations from the second sample, in the ordered data. Then under the null hypothesis this number tends to be larger; this fact and the fact that the distribution of the number of runs may be obtained without a knowledge of the distribution functions allows the formulation of nonparametric detection schemes. This paper does not, unfortunately, give any simulation or analytical results on performance against different alternatives.

REFERENCES

1. L. Kurz. "Nonparametric Detectors Based On Partition Tests." In *Nonparametric Methods in Communications*, P. Papantoni-Kazakos and D. Kazakos, eds. Marcel Dekker, 1977, pp. 73–144.
2. P. R. Kersten and L. Kurz. "Improved Operation of m-Interval Detectors by Optimum Signal Selection." *IEEE Trans. Inf. Theory* **IT-24**:477–481 (1978).

3. S. A. Kassam and J. B. Thomas. "Generalizations of the Sign Detector Based on Conditional Tests." *IEEE Trans. Commun.* **COM-24**:481–487 (1976).
4. J. G. Shin and S. A. Kassam. "Multilevel Coincidence Correlators for Random Signal Detection." *IEEE Trans. Inf. Theory* **IT-25**:47–53 (1979).
5. D. A. S. Fraser. *Nonparametric Methods in Statistics*. John Wiley, 1957, pp. 114–118, 147–158.

Reprinted with permission from *IEEE Trans. Inf. Theory* **IT-13**:247–255 (1967)

The Asymptotic Relative Efficiency of Mixed Statistical Tests

EDWARD A. FEUSTAL, MEMBER, IEEE, AND LEE D. DAVISSON, MEMBER, IEEE

I. Introduction

In the past ten years important results were obtained in the theory of distribution-free or nonparametric tests. Engineering interest was focused on the use of these statistics in detection theory, particularly in the sonar problem. The nonparametric detector is appropriate when insensitivity to change in input statistics, a fixed false alarm rate independent of input statistics, and sampled-data processing of input are desired.

In applications such as sonar where very large samples are available, the asymptotic relative efficiency (ARE) is the usual figure of merit for a detector. An additional consideration is the complexity of implementation. Simple detection tests such as the sign test and polarity coincidence correlator have very low efficiencies, but are easily implemented requiring $O(N)$ operations on the N observations. Complex detection tests which require ranking such as the Wilcoxon and Mann–Whitney have high efficiencies but require $O(N^2)$ operations.

A class of statistical tests is proposed in this paper that has relatively high efficiency with intermediate complexity. This class is called the class of "mixed statistical tests." A mixed statistic is a threshold test which is derived as follows. The N observations from each of k channels (samples) are divided into p groups of m observations. An intermediate statistic on each of the m observations appropriate for k channels reduces kN observations to p values. The p values are summed to form the final value which is compared with the threshold.

The advantage of this procedure is that the asymptotic relative efficiency as a function of m approaches its maximum value rapidly with increasing m and only requires $O(m^2p) = O(mN)$ operations. Hence recent advances in digital techniques allow real-time implementation of these more efficient statistical techniques.

In this paper the ARE of mixed statistics which use the Wilcoxon, Mann–Whitney, Kendall, and Fisher–Yates: Terry–Hoeffding procedures as intermediate statistics are investigated, and digital implementations are developed. These schemes allow increases of 30 percent to 200 percent in efficiency over sign tests and related techniques under the assumption of Gaussian noise input.

II. Asymptotic Relative Efficiency and Efficacy

The tests considered in this paper are based on the Neyman–Pearson criteria. Under these criteria, the probability of detection β is maximized while the false alarm rate is constrained to be less than a fixed level α. Thus the optimum test is generally uniquely determined if the probability distributions of signal and noise are known. Nonparametric tests are suboptimum tests which are applicable when only some general properties of the signal and noise are known and are not uniquely determined. Hence a further criterion is necessary to order the performance of such detectors. The selected criterion is Pitman's asymptotic relative efficiency.[1] It is particularly appropriate in the case of weak signal detection. If the hypothesis is taken to be noise alone and the alternative to be signal plus noise, the asymptotic relative efficiency applies when the alternative approaches the hypothesis—that is, the signal-to-noise ratio approaches zero. Specifically, it relates the number of samples required to maintain the false alarm rate below α while the probability of detection is β in one detector to the number required for a second in the limiting case:

$$\text{ARE}\,(d_1, d_2) = \lim_{\substack{K\to H \\ N_1, N_2 \to \infty}} \frac{N_1(\alpha, \beta, K)}{N_2(\alpha, \beta, K)} \qquad (1)$$

where

- α is the probability of false alarm not to be exceeded;
- β is the probability of detection to be maintained;
- N_i is the minimum number of observations required for detector d_i.

It is assumed that both d_1 and d_2 are consistent.

The tests of interest are threshold tests. A statistic S^N based on the N observations is compared with a threshold $C(N)$ which has been determined in an a priori fashion from the required value of α. In most previous electrical engineering work, functions on S^N, $C(N)$ were used to calculate the ARE in an extremely laborious fashion. Statisticians have used a concept known as efficacy.[2]

Manuscript received May 10, 1966; revised October 24, 1966. This work was partially supported under NASA Grant NGR-31-001-068.
E. A. Feustal is with the Division of Applied Science, California Institute of Technology, Pasadena, Calif. He was formerly with the School of Engineering and Applied Science, Princeton University.
L. D. Davisson is with the School of Engineering and Applied Science, Princeton University, Princeton, N. J.

It is a measure of efficiency for one test which may be directly compared to that of another to obtain the ARE:

$$\text{ARE}(d_1, d_2) = \frac{\mathcal{E}_1}{\mathcal{E}_2} \tag{2}$$

where $\mathcal{E}_i$ is the efficacy of the ith test. The advantage of efficacy is ease of calculation with complicated tests.

There are two forms of efficacy, one for the one-channel case and one for the two-channel case. The two-channel test calculates a statistic on the sets $\{x_i/i = 1, \cdots, m\}$ and $\{y_i/i = 1, \cdots, n\}$ where the first set is from the X channel and the second is from the Y channel. Depending on the detection problem, Y may be a reference channel or an independent signal channel.

The two-channel form follows:

$$\mathcal{E} = \lim_{N\to\infty} \left\{\frac{E_K(S^N) - E_H(S^N)}{\Delta_N \operatorname{Var}_H^{1/2}(S^N)}\right\}^2 \frac{1}{\gamma(1-\gamma)N} \tag{3}$$

where it is assumed that 1) the central limit theorem applies to the statistic S^N, 2) $\Delta_N = kN^{-\frac{1}{2}} \propto E_K(S^N) - E_H(S^N)$, and 3) $0 < \gamma < 1$ is the proportion of observations in the X channel (usually $\frac{1}{2}$), or alternatively:

$$\mathcal{E} = \lim_{\substack{N\to\infty \\ \Delta_N \to 0}} \left\{\frac{\frac{\partial E_K(S^N)}{\partial \Delta_N}}{\operatorname{Var}_{H_0}^{1/2}(S^N)}\right\}^2 \frac{1}{\gamma(1-\gamma)N}. \tag{4}$$

The one-channel form is very similar:

$$\mathcal{E}(S^N) = \lim_{N\to\infty} \left\{\frac{\frac{\partial E_K(S^N)}{\partial \Delta_N}}{\operatorname{Var}_{H_0}^{1/2} S^N}\right\}^2 \frac{1}{N} \tag{5}$$

where all but condition 3) apply. Alternatively, this latter form may be written:[1]

$$\mathcal{E}(S^N) = \lim_{N\to\infty} \frac{[\mu_{S^N}(kN^{-1/2}) - \mu_{S^N}(0)]^2}{k^2 \operatorname{Var}_H(S^N)}. \tag{6}$$

where $\mu_S{}^m(\theta)$ is the mean S^N for signal θ.

III. Mixed Statistical Tests

The sample mean test, a well-known parametric test which is optimum for independent Gaussian observations of known variance, requires about 22 000 samples for $\alpha = 0.001$, $\beta = 0.95$ and signal-to-noise ratio (SNR) $= -30$ dB. For this large number of observations, the central limit theorem should apply. It also shows the reason why only simple tests have been used in the past. The amount of computation for a sign test is $O(N)$ while the amount of computation for the Wilcoxon is $O(N^2)$. Yet the ARE of the sign test for normal alternatives is only about 0.63, whereas the ARE of the Wilcoxon for normal alternatives is about 0.95. If the Wilcoxon is used, a saving of about 30 percent in the sampling time results, but the processing time becomes so long as to preclude real-time application. Herein lies the justification for mixed statistical tests.

A mixed statistic is the sum of p observations of an intermediate statistic S_I^m operating on independent groups of m independent data points taken from each of k simultaneous channels or "samples." The subscript I denotes "intermediate" and subsequently stands for the exact intermediate statistic used (S for sign test, etc.). $N = mp$ is the total number of data points for each sample, and will be used to calculate the ARE with respect to any other k-sample test for fixed false alarm rate and probability of detection. The work involved is $O(m^2p)$. The question is, "What is the asymptotic relative efficiency of the mixed statistic as a function of m?" The answer is that the ARE rises sharply with increasing m, usually within the range $6 \le m \le 15$ to between 80 and 90 percent of the efficiency of the highly efficient rank statistic.

An intuitive motivation for this procedure is the idea that nonparametric techniques are transformations which convert arbitrary distributions into known ones, usually very close to Gaussian distributions for 20 or 30 observations. It is known that the optimum method of combining independent Gaussian samples for detection is to add them. The intermediate statistic serves to convert data from an arbitrary distribution to "almost" Gaussian data which are then summed and detected. The fact that the ARE of the mixed statistical test is sharply monotonic with m confirms this intuitive approach.

The discussion to follow first investigates the one-sample case and then the two-sample case. The k-sample case will be discussed at a later date. Three types of statistics will be investigated: the Wilcoxon (and subsequently the similar Mann–Whitney and Kruskal–Wallis), the Kendall τ, and the Fisher–Yates: Terry Hoeffding (FYTH) (and subsequently the k-sample expected rank-correlation test of Rushfort and Daly).

A simple, one-sample, distribution-free test that has been throughly investigated[3] is the sign test:

$$S_S^N = \sum_{i=1}^{N} \operatorname{sgn} x_i \tag{7}$$

or equivalently,

$$S_S^N = \sum_{i=1}^{N} u(x_i) \tag{8}$$

$$u(x_i) = \begin{cases} 1 & x_i \ge 0 \\ 0 & \text{otherwise.} \end{cases} \tag{9}$$

The ARE of this detector in location testing with respect to the student's t-test for Gaussian alternatives is $2/\pi$. The Wilcoxon, one-sample detector is a general extension of this detector and calculates the statistic:[4]

$$S_W^N = \sum_{i=1}^{N}\sum_{j=i}^{N} \operatorname{sgn}(x_i + x_j) \quad \text{or} \quad S_W^N = \sum_{i=1}^{N}\sum_{j=i}^{N} u(x_i + x_j). \tag{10}$$

[1] Suggested by a reviewer.

The ARE of this detector with respect to the student's t-test for Gaussian alternatives is $3/\pi$.

The following question arises: If p observations of the output of a Wilcoxon operating on m data points are summed and a decision is based on the statistic[2]

$$S_W^N = \sum_{i=0}^{p-1} S_W^m(i) \tag{11}$$

where

$$S_W^m(i) = \sum_{j=1}^{m} \sum_{k=j}^{m} u(x_{im+j} + x_{im+k}) \tag{12}$$

what is the ARE with respect to the optimum parameter test for Gaussian alternatives? First the efficacies for mixed statistics from one channel will be calculated. For sufficiently small θ:

$$\mu_{S^m}(a\theta) = a\mu_{S^m}(\theta). \tag{13}$$

For a sequence of alternatives $\theta = k'N^{-1/2}$, the efficacy of the mixed test based on

$$S^N = \sum_{i=0}^{p-1} S_I^m(i), \qquad N = mp \tag{14}$$

is

$$\mathcal{E}(S^N) = \lim_{N\to\infty} \frac{[\mu_{S^N}(k'N^{-1/2}) - \mu_{S^N}(0)]^2}{k'^2 \operatorname{Var}_H (S^N)} \tag{15}$$

$$= \lim_{N\to\infty} \frac{p^2[\mu_{S^m}(k'N^{-1/2}) - \mu_{S^m}(0)]^2}{pk'^2 \operatorname{Var}_H S^m} \tag{16}$$

$$= \lim_{N\to\infty} p\,\frac{k'^2 N^{-1}}{k^2 m^{-1}}\,\frac{[\mu_{S^m}(km^{-1/2}) - \mu_{S^m}(0)]^2}{k'^2 \operatorname{Var}_H (S^m)} \tag{17}$$

$$= \mathcal{E}(S_I^m). \tag{18}$$

In precisely the same manner, the efficacy of the two-channel mixed statistic is shown to be

$$\mathcal{E}(S^N) = \mathcal{E}(S_I^m). \tag{19}$$

Returning to the question, we find that

$$\begin{aligned} \text{ARE} &= \frac{24(m - 1 + \sqrt{2})^2}{4\pi(2m+1)(m+1)} \\ &= 2/\pi \qquad m = 1 \\ &\to 3/\pi \qquad m \to \infty \end{aligned} \tag{20}$$

Most interesting is the functional curve of the ARE versus m plotted in Fig. 1. Starting at 0.637 and rising to 0.955 at infinity, by $m = 6$ the ARE has reached a value of 0.863.

[2] $S_W{}^N$ has been used both for the mixed statistic and for the original to save subscripts. The meaning should be clear from the context.

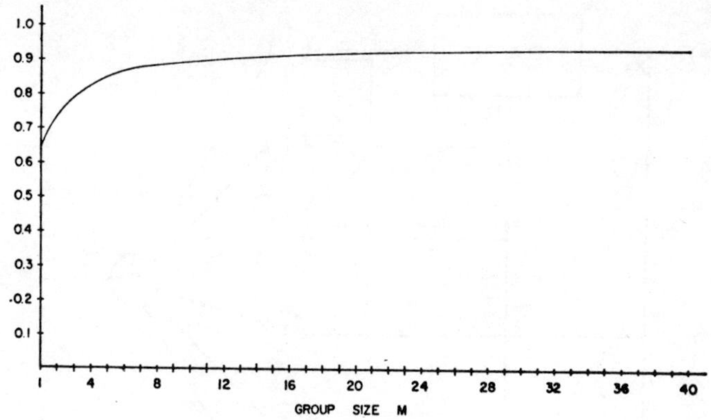

Fig. 1. Asymptotic relative efficiency versus group size for mixed Wilcoxon.

TABLE I

Type	ARE	N	Work
1) Sample Mean Test	1.000	22 400	$0(2.2\times10^4)$
2) Wilcoxon Test	0.955	23 500	$0(5.5\times10^8)$
3) 6-point Mixed Wilcoxon Test	0.863	26 000	$0(1.6\times10^5)$
4) Sign Test	0.637	35 000	$0(3.5\times10^4)$

Consider what this means in a practical situation. Let N be the number of data points for student's test required for α and β for:

$$\begin{aligned} H &: x_i \sim N(0, \sigma) \\ K &: x_i \sim N(\mu, \sigma) \\ N &= (\Phi^{-1}(1-\alpha) - \Phi^{-1}(1-\beta))^2 \frac{\sigma_x^2}{\mu_x^2} \end{aligned} \tag{21}$$

for $\alpha = 0.001$, $\beta = 0.95$, $A^2 = 22.4$

$$N = 22.4\,\frac{\text{NOISE POWER}}{\text{SIGNAL POWER}}. \tag{22}$$

Suppose the above ratio is 30 dB. Table I then summarizes the average number of samples required for four tests for Gaussian alternatives and indicates the order of magnitude of the work involved. Of the non-parametric tests, the 6-point mixed Wilcoxon is clearly superior to the sign test in efficiency and to the Wilcoxon test in the amount of work that is required.

When digital hardware is utilized to implement the mixed Wilcoxon, a real-time processor is possible for typical rates of 2 MHz. A single $(m-1)$-tap delay line and $\frac{1}{2}m(m+1)$ threshold devices plus a binary adder are sufficient. A detector for $m = 3$ is shown in Fig. 2. Since the output of the binary counter can be taken in parallel, no analog-to-digital conversion is necessary before transmission. In addition, the information rate for transmission to a central processing point is reduced in the same manner as in the sign detector, a desirable feature for sonobouy detection.

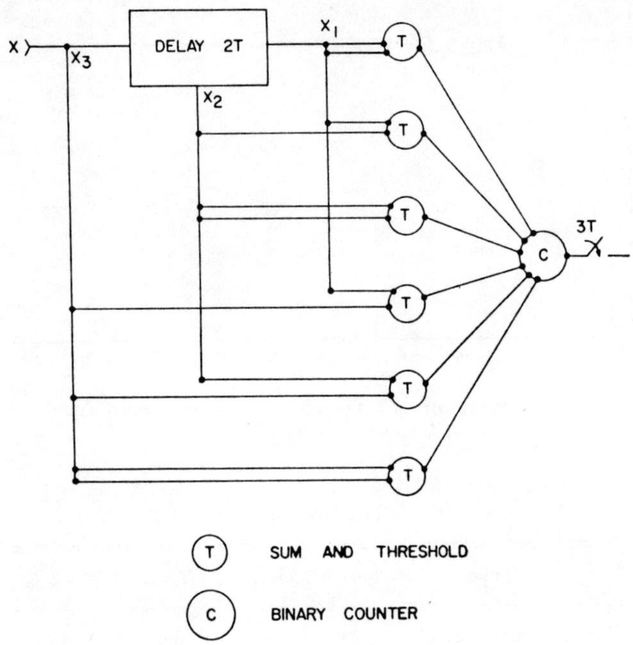

Fig. 2.

IV. Two-Sample Tests

While one-sample tests are very useful in giving information about mean location changes (or more generally when the signal is of any known form), they are not very useful when the signal is unknown. If two samples with independent noises are available, the presence of a common signal can be detected by testing for correlation or some other statistical test on dependence. A common method is polarity coincidence correlation (PCC).[5]–[8] The basis statistic for PCC is:

$$S_{\mathrm{PCC}}^{N} = \sum_{i=1}^{m} \operatorname{sgn} x_i \operatorname{sgn} y_i \tag{23}$$

or alternatively

$$S_{\mathrm{PCC}}^{N} = \frac{1}{2} \sum_{i=1}^{m} u(x_i y_i) \tag{24}$$

where the difference is a shift in the mean. Although such processing is extremely simple, the ARE is about $(2/\pi)^2$ with respect to the parametric detector which measures the sample correlation for Gaussian inputs, and the ARE with respect to the ideal Neyman–Pearson scheme is $\frac{1}{2}(2/\pi)^2$.[5] Several other statistics are of a similar type. The Kendall τ[9] is one such example:

$$S_{\tau}^{N} = \sum_{i=1}^{N} \sum_{j=i+1}^{N} \tfrac{1}{2}\{1 + \operatorname{sgn}(x_i - x_j) \operatorname{sgn}(y_i - y_j)\}. \tag{25}$$

The ARE for this test with respect to the correlator is $9/\pi^2$. Then the ARE of the Kendall detector with respect to the PCC is 9/4. This is a large improvement, but is accompanied by a good deal of additional work.

The success of the mixed method for the sign test gives rise to the hope for a similar result here. Calculations show that for a test using

$$S_{\tau}^{N} = \sum_{i=0}^{p-1} S_{\tau}^{m}(i) \tag{26}$$

where

$$S_{\tau}^{m}(i) = \sum_{j=1}^{m} \sum_{k=j+1}^{m} \tfrac{1}{2}\{1 + \operatorname{sgn}(x_{im+j} - x_{im+k}) \cdot \operatorname{sgn}(y_{im+j} - y_{im+k})\} \tag{27}$$

the ARE with respect to the correlator for Gaussian alternatives is:

$$\mathrm{ARE}\,(d_1, d_2) = \left[\frac{9(m-1)}{2(2m+5)}\right]\left(\frac{2}{\pi}\right)^2. \tag{28}$$

The functional curve of the ARE of PCC compared to the m-data point, summed Kendall test is interesting (see Fig. 3). Starting at a value of 1/2 for $m = 2$, it increases slowly, reaching 27/26 at $m = 4$ and 1.62 at $m = 10$. At $m = 29$, it is 2. For $m = 10$, only 5/8 as many samples are required for the Kendall mixed statistic as for the PCC detector. One possible implementation is shown in Fig. 4. $m(m-1)$ threshold devices are required and $m(m-1)/2$ exclusive-OR devices. All can be constructed from diodes and simple differential amplifier circuitry, which is very commonly microminaturized.

In some instances it may be desirable to utilize a test for randomness instead of one for independence. In this case the two-sample Mann–Whitney[10],[11] appears to be the "best" distribution-free test. It is well known that the Mann–Whitney has an ARE of $3/\pi$ with respect to the two-sample student's t test. Again we are interested in the sum of p observations of the Mann–Whitney operating on two samples of m data points:

$$S_{MW}^{m}(k) = \sum_{i=1}^{m} \sum_{j=1}^{m} \operatorname{sgn}(x_{km+i} - y_{km+j}) \tag{29}$$

$$S_{MW}^{N} = \sum_{k=0}^{p-1} S_{MW}^{m}(k) \tag{30}$$

The ARE for this case is: $3/\pi(2m/2m+1)$, $m \geq 1$ (see Fig. 5). By the time $m = 5$, the ARE is about 91 percent of the full value or 0.864 with respect to the Mann–Whitney. For any continuous distribution, the worst possible ARE for the Mann–Whitney or the Wilcoxon has been shown to be 0.864.[12] The preceding analysis remains valid for these distributions providing the factor $3/\pi$ is replaced by 0.864. This means that for any continuous distribution the ARE with respect to student's t test should not fall below $(0.864)^2$ for the five-data-point mixed statistic. An implementation for this statistic involves more threshold units than the others mentioned previously, requiring m^2 threshold gates in addition to the binary adder. An implementation for $m = 3$ is given in Fig. 6.

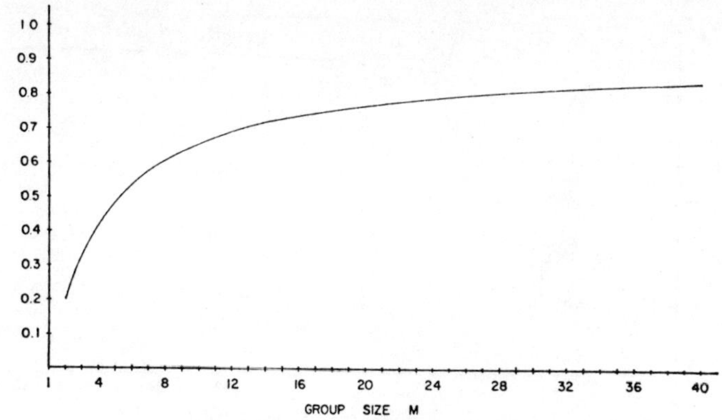

Fig. 3. Asymptotic relative efficiency versus group size for mixed Kendall.

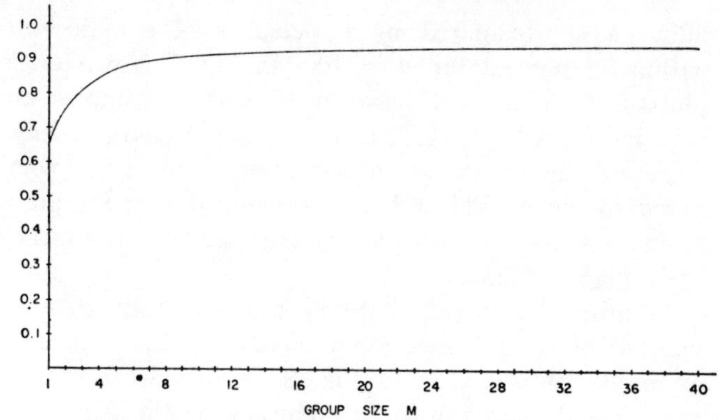

Fig. 5. Asymptotic relative efficiency versus group size for mixed Mann-Whitney.

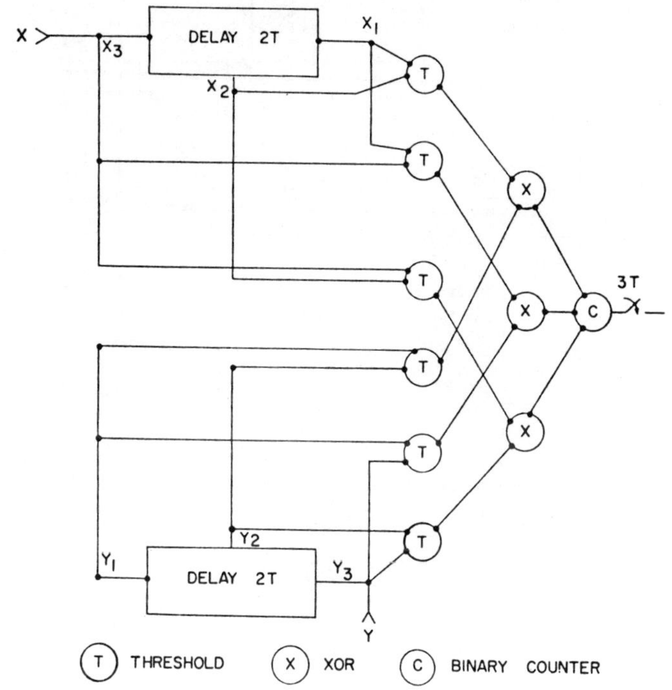

Fig. 4.

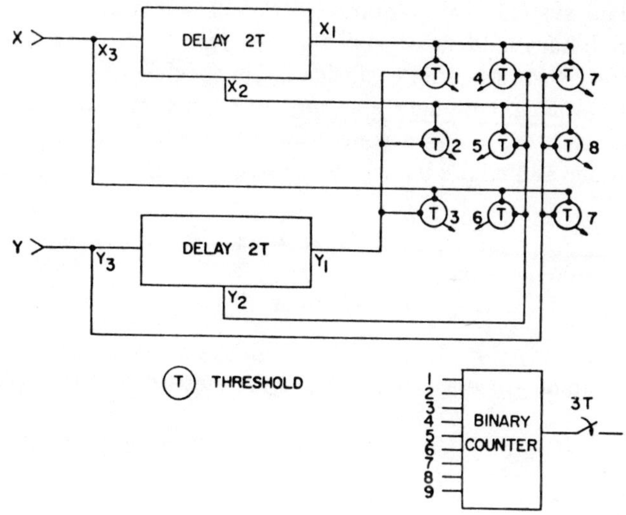

Fig. 6.

Analogous to the normal scores test for one sample is the Fisher–Yates: Terry–Hoeffding test (FYTH) for two samples.[13] Suggested by Fisher and Yates, it remained for Terry[14] to show the optimal properties. Chernoff and Savage[15] have shown that for hypotheses and alternatives with absolutely continuous densities the ARE of the FYTH against student's two-sample t must always be equal or greater than 1 with the equality holding for the case of normal alternatives. Capon[16] has shown that the test is the locally most powerful rank test for Gaussian alternatives. With all these advantages, one asks: "Why has it not been implemented?" A glance at the statistic is sufficient to show why. The FYTH statistic is:

$$S_{\text{FYTH}}^{N} = \frac{1}{N} \sum_{i=1}^{2N} a_i z_i \tag{31}$$

where a_i is the expected value of the ith smallest order statistic drawn from the absolute normal population of zero mean and unit variance. z_i, is the 1(0) function which has value 1 if the ith smallest observation is drawn from the first sample. It is immediately clear that for large N, the implementation is a horrendous problem; N coefficients must be provided, $2N$ items must be ranked, $O(4N^2)$, the proper coefficients must be selected, and then a sum must be computed.

As expected, the mixed statistic proves itself considerably easier to implement. The test will consist of the sum of p observations of the statistic operating on m data points of two samples. In Appendix I, the ARE of this mixed statistic is calculated with respect to the two-sample student's t test as:

$$\text{ARE}\,(d_1, d_2) = \frac{2m}{2m-1} \sum_{s=1}^{2m} \{E(s, 2m)\}^2 \tag{32}$$

where

$$E(s, 2m) = a_s.$$

Unfortunately, there is no simple interpretation of this expression. Ruben has shown[17] that the resulting value may be approximated by hyperspherical simplices to within 0.1 percent for $m = 100$. In Fig. 7, the ARE is plotted. Observe that between 15 and 20 samples are required for a high ARE. Lehmann and Hodges[18] have discussed the advantages of the Mann–Whitney as compared to the FYTH and have concluded that for probability distributions with heavy tails the Mann–Whitney has a higher efficacy.

Because the mixed statistic requires only a small number of comparisons for a relatively high efficiency, an implementation, as in Fig. 8, is considerably more complicated than the Mann–Whitney or the Kendall τ since pure digital hardware cannot be used, but the greater efficiency that it possesses may warrant its use. The system then requires $(2m - 1)m$ threshold gates, m binary adders of size $(2m - 1)m$, a diode selector matrix for selecting coefficients, $2m$ coefficient generators, an analog summer, and an A-D converter for transmission. By using a commutator and hold circuitry, one adder and a simpler selector matrix may be used.

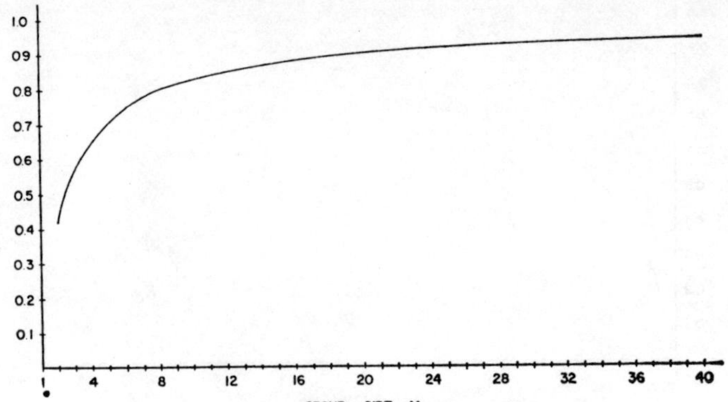

Fig. 7. Asymptotic relative efficiency versus group size for mixed Fisher-Yates: Terry-Hoeffding.

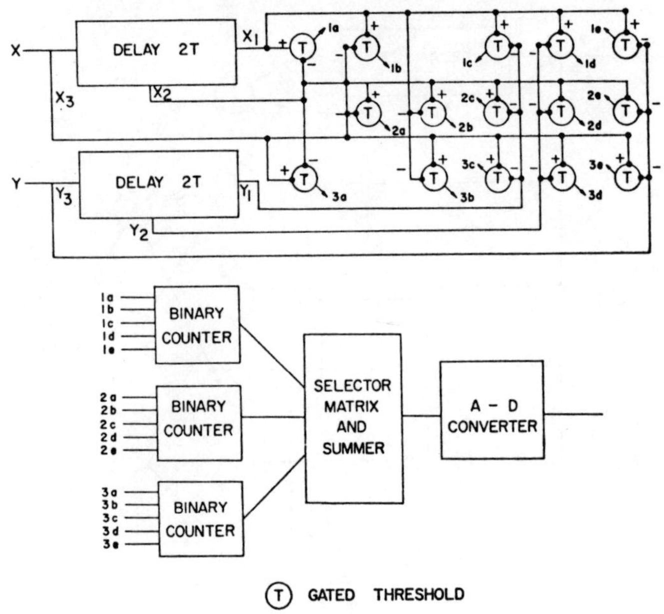

Fig. 8.

Appendix I

One-Sample Tests

Let x_i, $i = 1, \cdots, N$ be independent samples of a random process. To test the hypothesis-alternative pair:

$$\left.\begin{aligned} H &: x_i = n_i \\ \text{vs. } K &: x_i = \mu_0 + n_i \end{aligned}\right\} \begin{aligned} &n_i \text{ Gaussian, mean 0, variance } \sigma^2 \\ &\mu_0 \text{ is a positive constant.} \end{aligned}$$

The optimum Neyman–Pearson detector calculates

$$\sum_{i=1}^{N} x_i = S^N. \tag{33}$$

Let N_G be the number of samples required for false alarm rate α and probability of detection β. In the limit as $N_G \to \infty$, when $\mu_0 \to 0$ and since $\text{Var}_K(S^N) = \text{Var}_H(S^N)$,

$$\Phi^{-1}(1 - \alpha) - \Phi^{-1}(1 - \beta) = A = \frac{\sqrt{N_G}\mu_0}{\sigma} \tag{34}$$

$$N_G = \frac{\sigma^2}{\mu_0^2} A^2 \tag{35}$$

where Φ^{-1} is the inverse to the normal distribution function of zero mean and unit variance. For the nonparametric sign detector

$$S_S^N = \sum_{i=1}^{N} \text{sgn } x_i \tag{36}$$

under the same H and K:

$$E_K(S_S^N) = N_S(p - q) = N(1 - 2\Phi(-\mu_i/\sigma_i)) \tag{37}$$

where $p = Pr(x_i > 0)$ and $q = 1 - p$

$$\text{Var}_H(S_S^N) = N_S(\tfrac{1}{4}) \tag{38}$$

when

$$\mu_0 \to 0 \qquad E_k(S_S^N) \Rightarrow \frac{N}{\sqrt{2\pi}} \frac{\mu_0}{\sigma} \tag{39}$$

and

$$N_S = \frac{2}{\pi} \frac{\sigma^2}{\mu_0^2} A \tag{40}$$

$$\text{ARE}(d_S, d_G) = \frac{N_s}{N_G} = \frac{2}{\pi}. \tag{41}$$

For the Wilcoxon one-sample detector:

$$\sum_{i=1}^{N} \sum_{j=i}^{N} u(x_i + x_j) = S_W^N \tag{42}$$

$$E_K(S_W^N) = Np_2 + \frac{N(N - 1)}{2} p_1 \tag{43}$$

$$\text{Var}_H(S_W^N) = \frac{N(2N + 1)(N + 1)}{24} \tag{44}$$

where: $p_2 = Pr(x_i > 0)$ and $p_1 = Pr(x_i + x_j > 0)$.

Making a series expansion yields

$$E_K(S_W^N) - E_H(S_W^N) = \frac{N(N-1+\sqrt{2})\mu_0}{\sqrt{4\pi\sigma}} \quad \text{for } \mu_0/\sigma \text{ small.} \tag{45}$$

Then for the mixed statistic based on the Wilcoxon

$$S_W^N = \sum_{i=1}^{p} (S_W^m - E(S_W^m/\mu_0 = 0)) \tag{46}$$

and for $pm = N$,

$$E(S_W^N) = N \frac{(m-1+\sqrt{2})}{\sqrt{4\pi\sigma_i}} \mu_i \tag{47}$$

$$\text{Var}(S_W^N) = N \frac{(2m+1)(m+1)}{24} \tag{48}$$

$$\text{ARE}(d_G, d_W) = \frac{24(m-1+\sqrt{2})^2}{(2m+1)(m+1)4\pi} \tag{49}$$

which agrees as $m \to \infty$: ARE $(d_G, d_W) = 3/\pi$ and for $m = 1$: ARE $(d_G, d_W) = 2/\pi$; the result for the sign detector.

Appendix II

Two-Sample Tests

Here two parametric test results are required: the number of observations for the student's t, two-sample test, and the number required for the correlator. The hypothesis-alternative pair for the first detector, d_{st}, is

H : $\{x_i\}$, $\{y_i\}$ are statistically independent with bivariate normal distribution

$$F(x) = F(y)$$

K : $\{x_i\}$, $\{y_i\}$ are statistically independent with

$$F(x) = F(y - \theta), \quad \theta = \text{positive constant.}$$

The student's t statistic is[12]:

$$\frac{t^2}{2m-2} = \frac{(\bar{x}-\bar{y})^2 \dfrac{m_1 m_2}{2m}}{m_1 s_1^2 + m_2 s_2^2}. \tag{50}$$

This is the optimum parametric test for determining location shift with a Gaussian model of unknown variance where m_1 is the number of observations in the first sample, m_2 in the second, $m_1 + m_2 = m$ and s_1^2 and s_2^2 are the respective sample variances. In this case s_1 and s_2 are known and $m_1 = m_2 = m$ so

$$\frac{t^2}{2m-2} = \frac{1}{2}\frac{(\bar{x}-\bar{y})^2}{(s_1^2 + s_2^2)} \tag{51}$$

if $s_1^2 = s_2^2 = \sigma^2$

$$= \frac{2}{m^2\sigma^2}\left(\sum_{i=1}^{m}(x_i - y_i)\right)^2. \tag{52}$$

This tends to normality with increasing m with mean θ and variance $2\sigma^2/m$.

The second test d_c is a correlation test which for known variance and

H_2 : $\{x_i\}$, $\{y_i\}$ statistically independent Gaussian

$$\sigma_x = \sigma_y = \sigma_i, \quad \mu_x = \mu_y = 0, \quad \sigma_{xy} = 0$$

K_2 : $\{x_i\}$, $\{y_i\}$ correlated Gaussians

$$\sigma_x = \sigma_y = \sigma_i, \quad \mu_x = \mu_y = 0, \quad \sigma_{xy}^2 = \rho\sigma_x\sigma_y$$

is

$$S_c^m = \sum_{i=1}^{m} \frac{x_i y_i}{\sigma_i^2}. \tag{53}$$

Then

$$E_{K_2}(S_c^m) = \rho m \tag{54}$$

$$\text{Var}_{H_2}[S_c^m] = m \tag{55}$$

and the limiting distribution for $S_c^m/k \sim N(m\rho, m)$.

The Kendall τ statistic is:

$$S_\tau^m = \frac{2}{m(m-1)} \sum_{i<j}^{m}\sum h_{ij} \tag{56}$$

where

$$h_{ij} = \tfrac{1}{2}\{1 - \text{sgn}(x_i - x_j)\,\text{sgn}(y_i - y_j)\} \tag{57}$$

$$E_{K_2}(S_\tau^m) = \left[1 - \frac{2}{\pi}\sin^{-1}\rho\right] \tag{58}$$

and

$$|E_{K_2}(S_\tau^m) - E_{H_2}(S_\tau^m)| \simeq \frac{2}{\pi}\rho \quad \text{for small } \rho. \tag{59}$$

Kendall[9] gives the variance of S_τ^m as:

$$\text{Var}_{H_2} S_\tau^m = \frac{2(2m+5)}{9m(m-1)}. \tag{60}$$

Let $N = pm$. The statistic

$$S_\tau^N = \frac{1}{p}\sum_{i=1}^{p} S_\tau^m(i). \tag{61}$$

Then

$$E(S_\tau^N) \Rightarrow \frac{2\rho}{\pi} \quad \text{for small } \rho \tag{62}$$

and

$$\text{Var}(S_\tau^N) = \frac{2(2m+5)}{9N(m-1)}. \tag{63}$$

Then

$$\text{ARE}(d_c, d_\tau) = \frac{4\cdot 9(m-1)}{(\pi)^2(2m+5)} = \left(\frac{2}{\pi}\right)^2 \frac{9(m-1)}{2(2m+5)} \tag{64}$$

which for

$$m \to \infty \Rightarrow \frac{9}{\pi^2}.$$

The statistic for PCC is

$$S_{PCC}^N = \sum_{i=1}^{N} (u(x_i y_i) - \tfrac{1}{2}) \tag{65}$$

$$E_{K_2}(S_{PCC}^N) = \frac{N}{\pi} \sin^{-1} \rho \tag{66}$$

$$\mathrm{Var}_{H_2} (S_{PCC}^N) = \frac{N}{2}. \tag{67}$$

Then the ARE $(d_e, d_{PCC}) = (2/\pi)^2$. (68)

If, instead of a test for dependence, a test for homogeneity is desired, i.e., H_1 vs. K_1, the Mann-Whitney or Fisher-Yates: Terry-Hoeffding tests are used. Here it is expected that one channel will contain noise alone.

The Mann-Whitney test statistic for m samples in one channel and n in the other is:

$$S_{MW}^{m,n} = \sum_{i=1}^{m} \sum_{j=1}^{n} u(x_i - y_j) \tag{69}$$

has:

$$\mathrm{Var}\,(S_{MW}^{m,n}) = \frac{mn}{12} (m + n + 1) \tag{70}$$

and incremental expectation

$$E_{K_1}(S_{MW}^{m,n}) - E_{H_1}(S_{MW}^{m,n}) = \frac{nm}{2\sqrt{\pi}} \frac{\mu}{\sigma}. \tag{71}$$

The sum $N = pm$, and

$$S_{MW}^N = \sum_{i=1}^{p} S_{MW}^{m,n} \tag{72}$$

For $n = m$:

$$\frac{N_t}{N_{MW}} = \left\{ \frac{\dfrac{\sqrt{2}\sigma}{\mu}}{\dfrac{\sqrt{\pi}\sqrt{2m+1}\sqrt{m}}{\sqrt{3m}} \dfrac{\sigma}{\mu}} \right\}^2 = \frac{3}{\pi} \frac{2m}{2m+1} \tag{73}$$

$$\mathrm{ARE}\,(d_{t^2}, d_{MW}^m) = \left(\frac{3}{\pi}\right)\left(\frac{2m}{2m+1}\right). \tag{74}$$

Alternatively, the Fisher-Yates: Terry-Hoeffding statistic may be used. This statistic is for equal sample sizes:

$$C_1^m = \frac{1}{m} \sum_{i=1}^{m} E[x_i, 2m] z_i \tag{75}$$

where $E[x_i, 2m]$ is the expected value of the ith smallest observation of $2m$ drawn from a normal population of zero mean and unit variance and z_i is the indicator function having the value 1 if the ith smallest observation is from the first sample.

Capon[16] has shown that the efficacy of this detector using K_1 and H_1 is simply:

$$E^* = \inf G_{\varphi_0} \tag{76}$$

and that the variance is:

$$\mathrm{Var}\,(C_1^m) = \frac{(1-\gamma)}{\gamma M} \inf G_{\varphi_0} \tag{77}$$

where

$$\inf G_{\varphi_0} = E\left\{ \left(\frac{\partial}{\partial \theta} \ln g_\varphi(\theta) \Big|_{\theta=0} \right)^2 \right\} \tag{78}$$

and $\gamma M = m$, $N = mp$. Then let

$$e = \sum_{i=1}^{p} C_{1i}^m. \tag{79}$$

$$\mathrm{Var}\,(e) = p \,\mathrm{Var}\,(C_1^m) \tag{80}$$

$$E(e) = pE(C_1^m) \tag{81}$$

$$E_\varphi^* = \lim_{M\to\infty} \left\{ \frac{\dfrac{\partial}{\partial \Delta} E(e)}{\mathrm{Var}^{1/2}(e)} \right\}^2 \frac{1}{\gamma(1-\gamma)N} \tag{82}$$

$$\Delta = kM^{-1/2} \tag{83}$$

$$\chi = kN^{-1/2} \tag{84}$$

$$\Delta = p^{1/2}\chi \tag{85}$$

$$\frac{\partial}{\partial \Delta} = \frac{\partial}{\partial \chi} \cdot \frac{dx}{d\Delta} \tag{86}$$

$$= \frac{\partial}{\partial \chi} p^{-1/2} \tag{87}$$

$$= \lim_{M\to\infty} \left[\frac{p^{1/2} \dfrac{\partial}{\partial \chi} E(C_1^m)}{p^{1/2}\,\mathrm{Var}^{1/2}(C_1^m)} \right]^2 \frac{1}{\gamma(1-\gamma)N} \tag{88}$$

$$= \lim_{N\to\infty} \left[\frac{\dfrac{\partial}{\partial \chi} E(C_1^m)}{\mathrm{Var}^{1/2}(C_1^m)} \right]^2 \frac{1}{\gamma(1-\gamma)N} \tag{89}$$

$$= \inf G_{\varphi_0}. \tag{90}$$

From Kendall:[12]

$$\mathrm{Var}_{H_0}\,(C_1^m) = \frac{(1-\gamma)}{\gamma(M)(M-1)} \sum_{S=1}^{M} \{E(S, M)\}^2 \tag{91}$$

$$\inf G_{\varphi_0} = \frac{M}{M-1} \left\{ \frac{1}{M} \sum_{S=1}^{M} \{E(S, M)\}^2 \right\}. \tag{92}$$

From Ruben:[17]

$$\begin{aligned} \bar{G} &= \frac{1}{M} \sum_{i=1}^{M} [E(i, M)]^2 \\ &\simeq \frac{M-1}{M+1} \alpha_1 + \frac{(M-1)(M-2)(M-3)}{(M+1)(M+2)(M+3)} \alpha_3 \cdots \end{aligned} \tag{93}$$

$$E_e^* \simeq \frac{2m}{2m-1} \bar{G} \tag{94}$$

$$\mathrm{ARE}\,(d_{t^2}, d_{FYTH}) \simeq \frac{2m}{2m-1} \bar{G}. \tag{95}$$

References

[1] E. J. G. Pitman, Lecture notes on nonparameteric statistical inference, Columbia University, New York, N. Y., Spring 1948.

[2] G. E. Noether, "On a theorem of Pitman," *Ann. Math. Statistics*, vol. 26, pp. 64–68, 1955.

[3] J. E. Walsh, *Handbook of Nonparametric Statistics*, vols. I and II. Princeton, N. J.: Van Nostrand, 1962, 1965.

[4] J. W. Carlyle and J. B. Thomas, "On nonparametric signal detectors," *IEEE Trans. on Information Theory*, vol. IT-10, pp. 146–152, April 1964.

[5] S. S. Wolff, J. B. Thomas, and T. R. Williams, "The polarity-coincidence correlator; a nonparametric detection device," *IRE Trans. on Information Theory*, vol. IT-8, pp. 5–9, January 1962.

[6] M. Kanefsy and J. B. Thomas, "A comparison of some detectors for non-Gaussian inputs," Commun. Lab., Dept. of Elec. Engrg., Princeton University, Princeton, N. J. Tech. Rept. 3, February 1964.

[7] ——, "On detection by polarity coincidence correlation using dependent samples," Commun. Lab., Dept. of Elec. Engrg., Princeton University, Princeton, N. J., Tech. Rept. 6, May 1964.

[8] M. Kanefsky, "Detection of weak signals with polarity coincidence arrays," *Proc. 1966 Internat'l Conf. on Information Theory;* also published in *IEEE Trans. on Information Theory*, vol. IT-12, pp. 260–268, April 1966.

[9] M. G. Kendall, *Rank Correlation Methods*, 3rd ed. New York: Hafner, 1963.

[10] H. B. Mann and D. R. Whitney, "On a test of whether one or two random variables is stochastically larger than the other," *Ann. Math. Statistics*, vol. 18, p. 50, 1947.

[11] J. Capon, "Nonparametric methods for the detection of signals in noise," Columbia University, New York, N. Y., Tech. Rept. $T-1/N$, March 1959.

[12] M. G. Kendall and A. Stuart, *The Advanced Theory of Statistics*, vol. 2. New York: Hafner, 1961.

[13] D. A. S. Fraser, *Nonparameteric Methods in Statistics*. New York: Wiley, 1957.

[14] M. E. Terry, "Some rank-order tests which are most powerful against specific alternatives," *Ann. Math. Statistics*, vol. 23, pp. 346–366, 1952.

[15] H. Chernoff and I. R. Savage, "Asymptotic normality and efficiency of certain nonparametric test statistics," *Ann. Math. Statistics*, vol. 29, pp. 972–994, 1958.

[16] J. Capon, "Asymptotic efficiency of certain locally most powerful rank tests," *Ann. Math. Statistics*, vol. 32, pp. 88–100, 1961.

[17] H. Ruben, "On the sum of squares of normal scores," *Biometrics*, vol. 43, pp. 456–458, 1956.

[18] E. L. Lehman and J. L. Hodges, Jr., "Comparison of normal scores and Wilcoxon tests," *4th Berkeley Symp. on Mathematical Statistics and Probability*. Berkeley, Calif.: University of California Press, 1961, pp. 316–317.

17

Reprinted with permission from *IEEE Trans. Inf. Theory* **IT-18**:251–257 (1972)

Nonparametric Detectors Based on *m*-Interval Partitioning

YAU-CHAU CHING, MEMBER, IEEE, AND LUDWIK KURZ, MEMBER, IEEE

I. Introduction

Consider a binary signal detection problem where the signal can be represented by an n-dimensional vector and is corrupted by an additive noise vector whose elements are statistically independent and identically distributed. Mathematically, this problem is equivalent to that of separating the n-dimensional sample space into two regions in some optimal manner. Most often the functional form of the noise distribution is assumed to be known, with only a finite number of unknown parameters. Much research has been concentrated on these "parametric" detection problems and on the corresponding optimum and suboptimum partitioning of the sample space. However, in many practical digital communication systems, the functional form of the noise distribution is unknown, and the detection problem is then "nonparametric." Any detector designed without the knowledge of the functional form of the noise distribution is a nonparametric or distribution-free detector.[1]

The main advantage of the nonparametric detector is its robustness, i.e., the detector performs relatively well in a broad class of noise environments. The robustness of the nonparametric detectors is usually measured in terms of Pitman–Noether asymptotic relative efficiency (ARE) [7] by comparing them to a detector specifically designed for Gaussian noise. Most of the existing nonparametric detectors are rank detectors, which are attractive because of their efficiency [1]–[3]. For example, the ARE for the normal-scores detector is found to be 1 for Gaussian noise and strictly greater than 1 for any other noise [1]. This implies that the normal-scores detector is at least as efficient as the optimum detector for Gaussian noise in Pitman sense.

However, the rank detectors also have some severe shortcomings. For example, the distribution of the test statistic under the alternative hypothesis is very difficult to compute. This difficulty in computation not only hinders the exact analysis of the rank detectors under signal-plus-noise conditions, but also makes the most powerful rank detectors impossible to design, except for weak signals. The complexity of the rank detectors also grows rapidly as n increases. Furthermore, an adaptive rank detector is difficult to implement because of the slow covergence of the estimates of the pdf.

In this paper a new class of nonparametric detectors, called *m-interval detectors*, is proposed. These detectors are designed on the basis of a finite set of parameters of the noise distribution rather than its functional form. Their performance is readily analyzed because it involves only a one-dimensional cdf. Robustness of the detectors is obtained with little loss of efficiency. When the finite set of parameters associated with the noise distribution is completely unknown, an adaptive most powerful m-interval detector can be implemented with ease because only $(2m - 1)$ unknown parameters need to be estimated.

II. *m*-Interval Partitioning

There are many ways in which one can design a nonparametric detector. A common method is to design the detector in two stages. First, the n-dimensional sample space is partitioned into a finite number of regions, called *nonparametric partitions*, based only on a few general properties of the noise distribution. At the second design stage, a test statistic is defined as a set function of these nonparametric partitions, mainly on the basis of the signaling scheme. The nonparametric detectors can, therefore, be classified according to their nonparametric partitionings. For example, let

$$z = (z_1 z_2 \cdots z_n)$$

be an element of the n-dimensional Euclidean sample space. Then the test statistics of the rank detectors are defined on the nonparametric partitions generated by the set of hyperplanes

$$\{z_i = z_j, \qquad i \neq j\}$$

It should be noted that when the signal is absent, the nonparametric property of the nonparametric detectors is due entirely to the nonparametric partition of the sample space and not to the test statistics.

Because only a finite number of nonparametric partitions have been generated from the continuous sample space, the

Manuscript received May 3, 1971; revised July 29, 1971. This work was supported by the Natural Science Foundation through a graduate fellowship and under Grant GK-2947.

Y. C. Ching is with the Bell Telephone Laboratories, Inc., Holmdel, N.J. 07733.

L. Kurz is with the New York University, Bronx, N.Y. 10453.

[1] In the literature these terms are used interchangeably by some authors. Other authors refer to the nonparametric property just as the independence of the noise distribution of the type I error.

design as well as the implementation of the nonparametric detectors is usually much simpler than that of the parametric detectors. This simplicity is another advantage of the nonparametric detectors.

In the first part of this paper we shall consider the detection problem where the set of $(m-1)$ quantiles[2]

$$\{a_1,a_2,\cdots,a_{m-1}\}, \qquad a_i = F^{-1}(i/m)$$

of the noise cdf $F(z)$ is known. Assume further that $F(z)$ is strictly increasing in z, so that the quantiles are unique. On the basis of this initial information about the noise, the sample space is partitioned by the set of hyperplanes

$$\{z_k = a_i, \qquad i = 1,2,3,\cdots,m-1; \qquad k = 1,2,\cdots,n\}.$$

These hyperplanes partition the sample space into $N = m^n$ nonparametric partitions, called *m-interval partitions*, represented by the set

$$V = \{v_j, \qquad j = 1,2,\cdots,N\},$$

where

$$v_j = \{y_1^{(j)}y_2^{(j)}\cdots y_n^{(j)}\}$$

and $y_k^{(j)}$ takes on a value on

$$I = \{I_i, \qquad i = 1,2,\cdots,m\},$$

where the I_i denote the interval (a_{i-1},a_i) with $a_0 = -\infty$, $a_m = \infty$. When the signal is absent,

$$\Pr(z_k \in I_i \mid \Phi) = F\left[F^{-1}\left(\frac{i}{m}\right)\right] - F\left[F^{-1}\left(\frac{i-1}{m}\right)\right]$$

$$= 1/m, \qquad i = 1,2,\cdots,m, \quad k = 1,2,\cdots,n,$$

we obtain

$$\Pr(z \in v_j \mid \Phi) = \prod_{k=1}^{n} \Pr(z_k \in y_k^{(j)} \mid \Phi) = m^{-n}, \qquad j = 1,2,\cdots,N,$$

which is independent of the actual noise distribution. The detector defined on these m-interval partitions is, therefore, nonparametric when the signal is absent (the type I error probability is independent of the noise distribution), provided that the quantiles of the noise distribution are known, although the functional form of the distribution is actually unknown.

III. m-Interval Detection for a Known Signal

Consider binary signal detection when a dc signal

$$u = s(11\cdots 1)$$

is transmitted. The proposed test statistic is then of the form

$$T = \frac{1}{n}\sum_{k=1}^{n}\sum_{i=1}^{m} b_i \delta_{z_k I_i}, \tag{1}$$

where

$$\delta_{z_k I_i} = \begin{cases} 1, & z_k \in I_i \\ 0, & \text{otherwise} \end{cases}$$

and $\{b_1, i = 1,2,\cdots,m\}$ is a set of predetermined weighting constants. The test can be generalized to other signal forms by replacing $\{b_i,\ i = 1,2,\cdots,m\}$ with $\{b_{ik},\ i = 1,\cdots,m;\ k = 1,\cdots,n\}$. However, the analysis would be considerably more complicated.

Let

$$P_i = \Pr(z \in I_i \mid u), \qquad i = 1,2,\cdots,m. \tag{2}$$

Then

$$P_i = F(a_i) - F(a_{i-1}),$$

which can be readily computed for any given $F(z)$. The moments of the test statistic T are

$$E(T \mid \Phi) = \sum_{i=1}^{m} b_i/m$$

$$E(T \mid u) = \sum_{i=1}^{m} b_i P_i$$

$$\operatorname{var}(T \mid \Phi) = \frac{1}{n}\left[\sum_{i=1}^{m} b_i^2/m - E^2(T \mid \Phi)\right]$$

$$\operatorname{var}(T \mid u) = \frac{1}{n}\left[\sum_{i=1}^{m} b_i^2 P_i - E^2(T \mid u)\right].$$

Since T is the sum of n independent bounded random variables, it is asymtotically normal.

To find the exact distribution of T, we shall express T in terms of a multinomially distributed random variable. Let $n_i^{(j)}$ be the number of I_i in $v_j \in V$ and let the test statistic T be defined on v_j only through the set of numbers $\{n_i^{(j)}, i = 1,2,\cdots,m\}$. Specifically, T can be rewritten as

$$T(v_j) = \frac{1}{n}\sum_{i=1}^{m} b_i n_i^{(j)}.$$

Define a subset $V(n_1,n_2,\cdots,n_m) \subset V$ as

$$V(n_1,n_2,\cdots,n_m) = \{v_j : n_i^{(j)} = n_i, \qquad i = 1,\cdots,m\}.$$

For every m-interval partition $v_j \in V(n_1,n_2,\cdots,n_m)$,

$$T(v_j) = \frac{1}{n}\sum_{i=1}^{m} b_i n_i$$

$$\Pr(Z \in v_j \mid u) = \prod_{i=1}^{m} P_i^{n_i}.$$

Consider the random variable $(N_1N_2\cdots N_m)$ induced by the mapping

$$Z \in V(n_1,n_2,\cdots,n_m) \to [N_i = n_i, \qquad 1,2,\cdots,m].$$

Then $\sum_{l=1}^{m} N_l = n$ and it is apparent that $(N_1N_2\cdots N_m)$ is multinomially distributed, namely,

$$\Pr(N_i = n_i, i = 1,2,\cdots,m \mid u) = n!\left(\frac{P_1^{n_1}}{n_1!}\right)\cdots\left(\frac{P_m^{n_m}}{n_m!}\right)$$

$$\Pr(N_i = n_i, i = 1,2,\cdots,m \mid \Phi) = \frac{n!}{n_1!\cdots n_m!}\left(\frac{1}{m}\right)^m. \tag{3}$$

[2] Although in this paper nonparametric detectors based on the quantiles are discussed in detail, the same philosophy may be applied to other finite sets of parameters associated with the unknown distribution.

T can, therefore, be considered as a linear function of the multinomially distributed random variable.

The exact distribution of T depends on the set of weighting constants $\{b_i\}$. In general, it can be found with the help of a generating function as shown in Theorem 1 of the Appendix. Usually this distribution can be quite complicated because it is not of the lattice type and one would have to enumerate all possible values of T. Furthermore, this distribution may not give a clear indication of the concentration of T. To circumvent this difficulty, one should convert this discrete frequency function into a piecewise continuous cdf for more meaningful interpretation and analysis.

IV. Most Powerful m-Interval Detectors

Let the actual noise distribution approximate a known distribution, e.g., near-Gaussian distribution. One may then wish to find a detector that is robust yet optimum, in some sense, in the class of robust detectors. In other words, one may wish to find the most powerful nonparametric detector for a particular noise environment and particular nonparametric partitioning.

A most powerful m-interval detector can be formed on the basis of the set of conditional probabilities

$$\{\Pr(Z \in v_j \mid u), \Pr(Z \in v_j \mid \Phi), \qquad j = 1,2,\cdots,N\}.$$

Using the Neyman–Pearson fundamental lemma, this detector takes the form

$$T = \frac{1}{n} \sum_{k=1}^{n} \log [\Pr(z_k \in y_k \mid u)/\Pr(z_k \in y_k \mid \Phi)].$$

If we limit out discussion to the detection of a dc signal, the detector reduces to

$$T_m = \frac{1}{n} \sum_{i=1}^{m} n_i \log (mP_i),$$

which is of the same form as in (1) with $b_i = \log (mP_i)$.

Suppose the signal is weak. Then P_i can be approximated by

$$P_i = m^{-1} + s[f(a_{i-1}) - f(a_i)]$$

and

$$\log (mP_i) = sm[f(a_{i-1}) - f(a_i)].$$

The locally most powerful m-interval detector is then

$$T_{lm} = \frac{m}{n} \sum_{i=1}^{m} n_i[f(a_{i-1}) - f(a_i)]. \tag{4}$$

From the asymptotic normality of T, the performance comparison in terms of Pitman–Noether ARE is readily obtainable.

Consider the probabilities

$$P_i = F(a_i - s) - F(a_{i-1} - s). \tag{5}$$

Differentiating (8) with respect to s, we obtain

$$\left.\frac{dP_i}{ds}\right|_{s=0} = f(a_{i-1}) - f(a_i).$$

The efficacy of

$$T = \frac{1}{n} \sum_{i=1}^{m} b_i n_i$$

is then

$$\varepsilon(T) = \frac{\{\sum_{i=1}^m b_i[f(a_{i-1}) - f(a_i)]\}^2}{\sum_{i=1}^m b_i^2/m - E^2(T \mid \Phi)}.$$

For the locally most powerful m-interval detector as in (4), the efficacy reduces to

$$\varepsilon(T_{lm}) = m \sum_{i=1}^{m} [f(a_{i-1}) - f(a_i)]^2.$$

From Schwarz's inequality, it follows that

$$\varepsilon(T_{lm}) \geq \varepsilon(T)$$

and the equality holds if and only if

$$b_i = K[f(a_{i-1}) - f(a_i)],$$

where K is some arbitrary constant.

Having derived some general relationships for the m-interval detectors, a few specific examples will be now considered to give some additional insight into the operation of these detectors.

Example 1

Let $m = 2$, $a_1 = 0$ and $b_1 = -1$, $b_2 = +1$. In this case the m-interval detector reduces to the sign detector with

$$T = \frac{1}{n} \sum_{k=1}^{n} \operatorname{sgn} z_k.$$

Assuming

$$f(-\infty) = f(\infty) = 0,$$

the efficacy of the sign detector is

$$\varepsilon_{\text{sign}} = \frac{[(-1)[f(-\infty) - f(0)] + f(0) - f(\infty)]^2}{\frac{1}{2}[(-1)^2 + (1)^2] - 0} = 4f^2(0),$$

which agrees with the result obtained previously [1]. For Gaussian noise of unit variance, this efficacy is $2/\pi$.

Example 2

Suppose $m = 4$, and arbitrarily assign

$$b_1 = -2, \qquad b_2 = -1, \qquad b_3 = 1, \qquad b_4 = 2. \tag{6}$$

Then the efficacy is

$$\varepsilon(T) = \tfrac{2}{5}[f(a_1) + 2f(a_2) + f(a_3)]^2.$$

For Gaussian noise with unit variance,

$$f(a_1) = f(a_3) = 0.314 \tag{7}$$

$$f(a_2) = f(0) = 0.4 \tag{8}$$

and the efficacy is 0.814, which is an improvement over the sign detector.

For Laplacian noise of unit variance

$$f(x) = 2^{-1/2} \exp[-\sqrt{2}\,|x|]$$

and

$$f(a_1) = f(a_3) = 2^{-3/2}$$

$$f(a_2) = 2^{-1/2}$$

$$\varepsilon(T) = 1.8.$$

From (7) and (8) it is obvious that the weighting constants in (6) do not assume their optimum values for Gaussian noise.

To obtain the most powerful four-interval detector for the Gaussian noise, the weighting constants need to be selected as, from (4),

$$-b_1 = b_4 = 0.314$$

$$-b_2 = b_3 = 0.086. \tag{9}$$

From (9) we obtain the efficacy of this most powerful four-interval detector if the Gaussian noise is of unit variance, namely,

$$\varepsilon(T_{lm}) = 0.85. \tag{10}$$

Consider the locally most powerful detector of the form

$$T_{l0} = \frac{1}{n}\sum_{k=1}^{n} \log\left(\frac{f'(z_k)}{f(z_k)}\right).$$

The efficacy of this detector is found to be

$$\varepsilon(T_{l0}) = \int \left(\frac{f'(z)}{f(z)}\right)^2 f(z)\,dz = I(f), \tag{11}$$

where $I(f)$ is Fisher's information. For Gaussian noise of unit variance, Fisher's information is 1. Thus, from (10) and (11) the ARE for the most powerful four-interval detector relative to the locally most powerful detector is 0.85 for Gaussian noise. For some other noise environments, where the locally most powerful Gaussian noise detector performs poorly, the locally most powerful m-interval detector may be preferred. Since T_{l0} is the locally most powerful detector, its efficacy is the maximum among all the detectors in the same noise environment. Thus we obtain the inequality

$$\varepsilon(T,f) \le I(f)$$

for any detector T. In particular, for the m-interval detectors

$$\sum_{i=1}^{m} m[f(a_i) - f(a_{i-1})]^2 \le I(f). \tag{12}$$

It is interesting to observe that, as m increases, the left-hand side of (12) approaches $I(f)$. Thus the locally most powerful m-interval detectors have the efficacies asymptotically approaching the efficacy of the locally optimum detector. This is reasonable because our knowledge of F approaches the complete knowledge.

However, from the result of the four-interval detector in Gaussian noise one may conjecture that even for small m, the loss in efficiency is small for the locally most powerful m-interval detectors. On the other hand, since the sample space is partitioned only on the basis of a set of quantiles instead of the complete knowledge of the noise, the detector is relatively insensitive to the noise statistic. The robustness is therefore obtained with little loss of efficiency.

V. Performance of m-Interval Detectors Based on Bahadur and Hodges–Lehmann Efficiencies

The Pitman–Noether ARE is informative only for a sequence of signals that approach the null signal. In any practical situation the signal is fixed, and the ARE is not applicable. For a fixed signal strength, the asymptotic performance of the m-interval detectors can be examined using the Bahadur efficiency [4] and the Hodges–Lehmann efficiency [5].

Essentially, the Bahadur efficiency is obtained by observing the asymptotic behavior of the error of the first kind (type I error) for a fixed signal and a fixed error of the second kind (type II error). Since the detectors are consistent, the error of the first kind reduces to $\Pr[T > E(T \mid u) \mid \Phi]$ for which exponential bounds are available.

Let D_1 and D_2 be the detectors used to detect the presence of the dc signal

$$u = s(11\cdots1)$$

and let

$$\Pr[D_1 > E(D_1 \mid u) \mid \Phi] \le \exp[-nA_1(s)]$$

$$\Pr[D_2 > E(D_2 \mid u) \mid \Phi] \le \exp[-nA_2(s)],$$

where $A_1(s)$ and $A_2(s)$ are the tightest exponential bounds. Then the Bahadur efficiency BE of D_1 relative to D_2 is

$$\mathrm{BE}(D_1,D_2,s) = A_1(s)/A_2(s).$$

Since T is the sum of n independent random variables, the Chernoff bound is applicable. Let

$$g(h) = \sum_{i=1}^{m} e^{hb_i}/m$$

$$q(h) = \log[g(h)].$$

Then

$$\Pr(T > t \mid \Phi) \ge \exp[-n(ht - q(h))] \tag{13}$$

for all $h > 0$. Maximizing the exponent of (13) with respect to h, and substituting

$$t = E(T \mid u),$$

we obtain that the optimum h is such that

$$E(T \mid u) = \sum_{i=1}^{m} b_i e^{hb_i} \Big/ \sum_{i=1}^{m} e^{hb_i}.$$

Let T_m be the most powerful m-interval detector, i.e.,

$$T_m = \sum_{i=1}^{m} \log(P_i) n_i/n.$$

Then

$$\Pr[T_m > E(T_m \mid u) \mid \Phi] \le \exp[-nC(s)]$$

where

$$C(s) = \log m + \sum_{i=1}^{m} P_i \log P_i$$

is the capacity of a memoryless doubly uniform channel with the set of transition probabilities $\{P_1,P_2,\cdots,P_m\}$.

It has been shown that $A(s)$ can serve as a lower bound on the capacity of the memoryless doubly uniform channel with the preceding set of transition probabilities [8]. The fact that $A(s) = C(s)$ thus implies that $C(s)$ is the tightest exponential bound on the probability of error in (13).

For purpose of comparison, consider the error of the first kind for a sample-mean detector t in Gaussian noise. It is bounded by

$$\Pr\left[t > E(t \mid u) \mid \Phi\right] < \exp\left(-ns^2/2\right).$$

Hence, the Bahadur efficiency of the most powerful m-interval detector relative to the sample-mean detector in Gaussian noise is

$$\mathrm{BE}\left(T_m,t,s\right) = 2C(s)/s^2.$$

For the sign detector, this efficiency has been tabulated [4].

The Hodges–Lehmann efficiency is defined as the ratio of the logarithm of the bound on the error of the second kind while the error of the first kind is held constant. Let

$$\Pr\left[D_1 < E(D_1 \mid \Phi) \mid u\right] < \exp\left(-nB_1(s)\right)$$

$$\Pr\left[D_2 < E(D_2 \mid \Phi) \mid u\right] < \exp\left(-nB_2(s)\right).$$

Then the Hodges–Lehmann efficiency HLE of D_1 relative to D_2 is defined as

$$\mathrm{HLE}\left(D_1,D_2,s\right) = B_1(s)/B_2(s).$$

For the sign detector relative to the sample-mean detector, this efficiency has been tabulated [5].

The Bahadur efficiency and the Hodges–Lehmann efficiency for most powerful two-interval, four-interval, and eight-interval detectors relative to the sample-mean detector for Gaussian and Laplace noise are shown in Figs. 1–6. At low signal strength both efficiencies approach the Pitman efficiency. However, the capacity of the m-letter doubly uniform channel approaches $\log m$ as the signal strength approaches infinity. Thus the Bahadur efficiency of the most powerful m-interval detectors approaches zero for large signal power. On the other hand, the Hodges–Lehmann efficiency of the most powerful m-interval detectors relative to the sample-mean detector may approach some nonzero constant. For example, the Hodges–Lehmann efficiency of the sign detector relative to the sample-mean detector approaches one half when the signal power approaches infinity in Gaussian noise. Furthermore, it can be shown that the Hodges–Lehmann efficiency of the sign detector is uniformly higher than its Bahadur efficiency.

VI. Adaptive Systems

Thus far we have assumed the accurate measurements of a set of $(2m - 1)$ parameters

$$a_i = F^{-1}(m^{-1}), \qquad i = 1,2,\cdots,m-1$$

$$P_i = \Pr\left(z \in I_i \mid u\right), \qquad i = 1,2,\cdots,m$$

are available. Consider now the case where these parameters are estimated by a set of finite-sample-size observations.

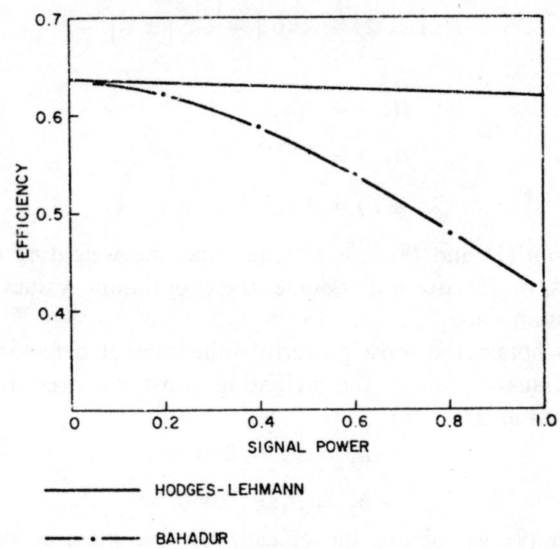

Fig. 1. Hodges–Lehmann efficiency and Bahadur efficiency of most powerful two-interval detector in Gaussian noise of unit variance.

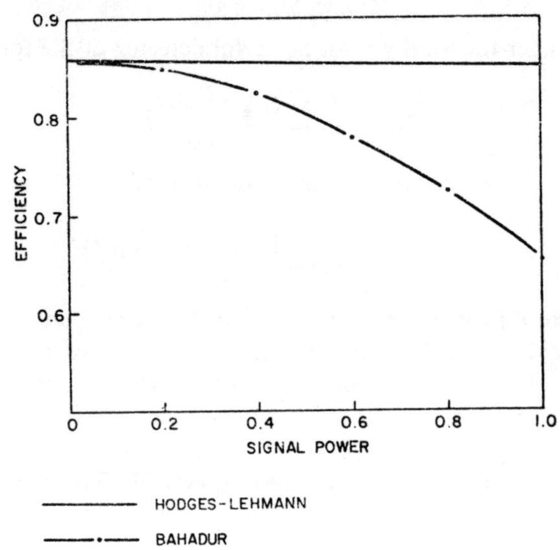

Fig. 2. Hodges–Lehmann efficiency and Bahadur efficiency of most powerful four-interval detector in Gaussian noise of unit variance.

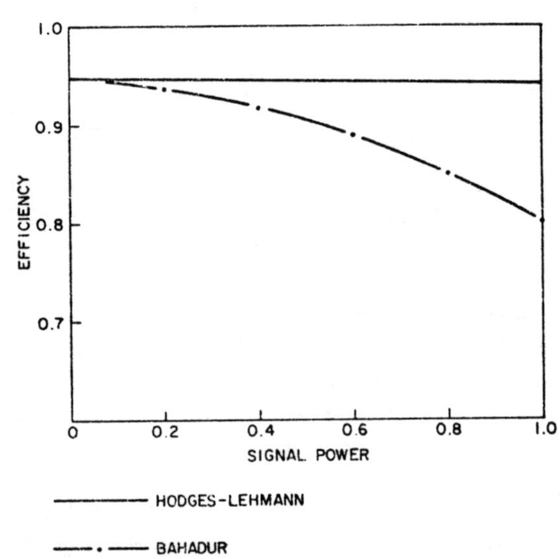

Fig. 3. Hodges-Lehmann efficiency and Bahadur efficiency of most powerful eight-interval detector in Gaussian noise of unit variance.

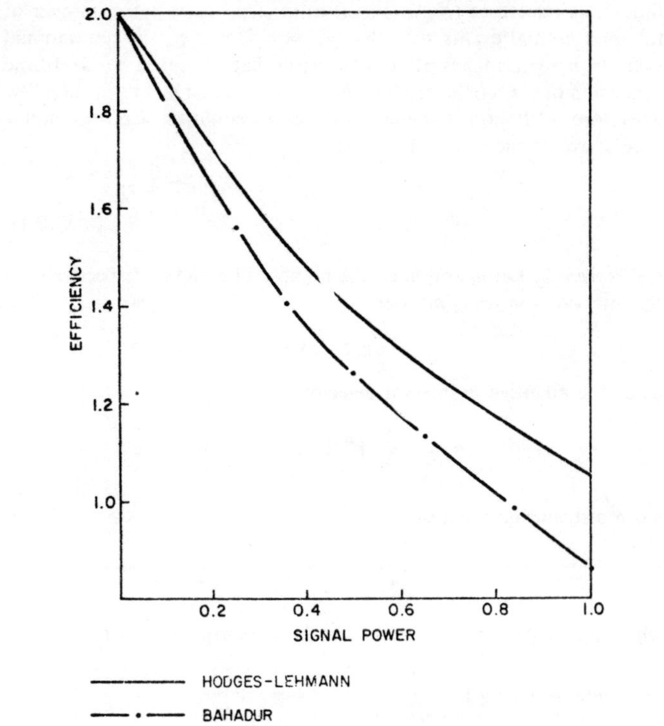

Fig. 4. Hodges–Lehmann efficiency and Bahadur efficiency of most powerful two-interval detector in Laplace noise of unit variance.

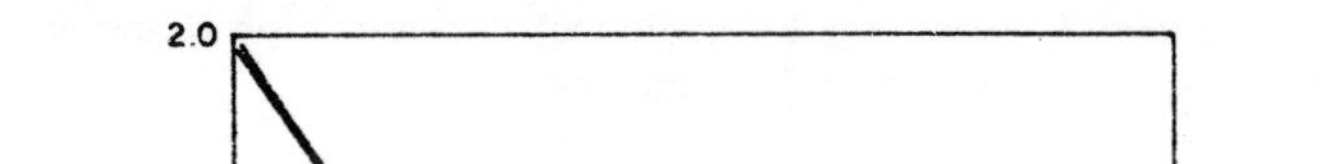

Fig. 5. Hodges–Lehmann efficiency and Bahadur efficiency of most powerful four-interval detector in Laplace noise of unit variance.

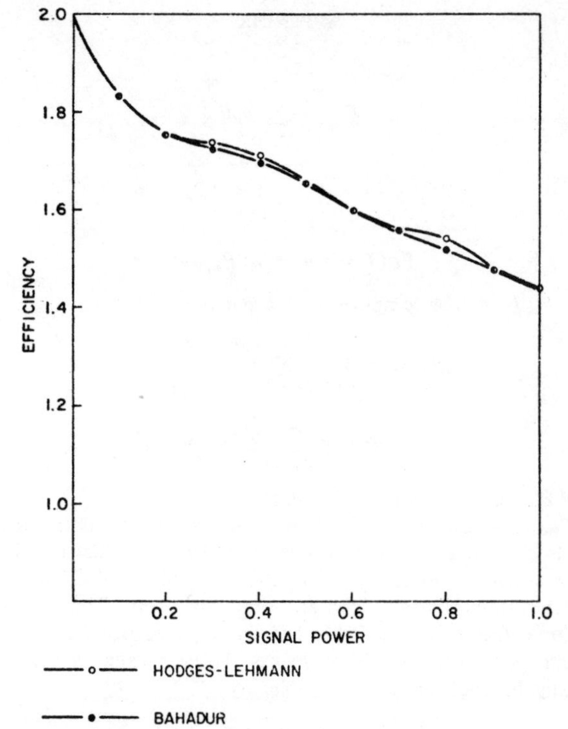

Fig. 6. Hodges–Lehmann efficiency and Bahadur efficiency of most powerful eight-interval detector in Laplace noise of unit variance.

Suppose that there is available at the receiver a set of observations from the pure noise sample

$$W = \{W_r, \qquad r = 1,2,\cdots,N(W)\}$$

and a set of observations from the signal-plus-noise sample

$$X = \{x_s, \qquad s = 1,2,\cdots,N(X)\}.$$

Let

$$Z = \{z_k, \qquad k = 1,2,\cdots,n\}$$

be the unknown sample being tested. For simplicity, let

$$N(W) = md - 1$$

for some integer d and let $w(j)$ be the jth-order statistic in W. Then $w^{(id)}$ is an estimate of a_i.

Let n_i and h_i be the numbers of z and x, respectively, in the interval $(w^{[id-1]},w^{(id)})$. Then an adaptive most powerful m-interval detector takes the form

$$T_a = \sum_{i=1}^{m} \left(\frac{n_i}{n}\right) \log \left(\frac{h_i}{N(X)}\right).$$

However, the distribution function of T_a is very complicated, and no closed form has been found. If $N(X)$ and $N(W)$ are large, then $w^{(id)}$ converges to a_i and $[h_i/N(X)]$ converges to P_i. In this case, T_a reduces to the most powerful m-interval detector as analyzed in Section IV. It should be noted that $w^{(id)}$ and $[h_i/N(X)]$ converge in any noise environments.

Suppose $N(W)$ is large, or $w^{(id)}$ can be replaced by a_i. T_a is then a function of two multinomially distributed random variables, and the distribution of T_a can be found by once again making use of the generating function in Theorem 2 of the Appendix. A more detailed study of adapative m-interval detectors will be the subject of a separate paper where methods for more efficient estimation of quantiles will be presented.

APPENDIX

Theorem 1: Let

$$T = \frac{1}{n} \sum_{i=1}^{m} b_i n_i \tag{14}$$

and

$$\Pr(z \in I_i) = P_i, \quad i = 1,2,\cdots,m.$$

Then

$$\Pr(T = t/n) = n!\, P_{m,n}(t),$$

where $P_{m,n}(t)$ can be computed from a generating function

$$g(x,y) = \prod_{i=1}^{m} \left[\sum_{k=0}^{n} \frac{(x^{b_i} y P_i)^k}{k!} \right]$$

$$= 1 + \sum_{r \in R} \sum_{s=1}^{n} P_{m,s}(r) x^r y^s$$

and R extends over all possible values of nT.

Proof: A generating function is in essence a counting device capable of finding systematically the number of product terms with certain specified properties [6]. In this case, the product terms are associated isomorphically with some random events. The generating function is therefore modified to find the total probabilities of all the random events with certain properties. To each value of the multinomially distributed random variable, say

$$(N_1 N_2 \cdots N_m) = (n_1 n_2 \cdots n_m)$$

corresponds a product term

$$(x^{b_1} y)^{n_1} (x^{b_2} y)^{n_2} \cdots (x^{b_m} y)^{n_m},$$

which appears once and only once in the expansion of

$$g^*(x,y) = \prod_{i=1}^{m} \left[\sum_{k=0}^{n} (x^{b_i} y)^k \right] \tag{15}$$

$$= 1 + \sum_{r \in R} \sum_{s=1}^{n} P^*_{m,s}(r) x^r y^s. \tag{16}$$

The term $x^r y^s$ in (16) is formed by all the product terms in (15) such that

$$r = \sum_{i=1}^{m} b_i n_i$$

$$s = \sum_{i=1}^{m} n_i.$$

Hence $P^*_{m,n}(t)$ is the total number of terms for which

$$nT = \sum_{i=1}^{m} b_i n_i = t$$

$$\sum_{i=1}^{m} n_i = n.$$

Since T is related to $(N_1 N_2 \cdots N_m)$ as in (1), $P_{m,n}(t)$ is the number of the multinomial events such that $nT = t$. However, the multinomial event $(n_1 n_2 \cdots n_m)$ takes place with the probability given by (3). In the expansion of $g(x,y)$, the product terms are modified by this probability. Therefore, $n!\, P_{m,n}(t)$ is the sum of the probabilities of all the multinomial events such that $nT = t$, or

$$n!\, P_{m,n}(t) = \Pr\{Z \in V(n_1, n_2, \cdots, n_m) : nT = t\}.$$

Q.E.D.

Theorem 2: Let n_i and h_i be the number of z and x, respectively, in the interval $I_i = (a_{i-1}, a_i)$. Let

$$\Pr(x \in I_i \mid u) = P_i.$$

Then the adaptive m-interval detector

$$T_a = \sum_{i=1}^{m} \left(\frac{n_i}{n}\right) \log \left(\frac{h_i}{N(X)}\right)$$

has a distribution function

$$\Pr\left(T_a = \frac{t}{n}\right) = n!\, N(X)!\, P_{m,n,N(X)}(t),$$

where $P_{m,n,N(X)}(t)$ is computed from the generating function

$$g(x,y,z) = \prod_{i=1}^{m} \left[\sum_{n_i=0}^{n} \sum_{h_i=0}^{N(X)} \frac{x^{n_i \log(h_i/N(X))} (yQ_i)^{n_i} (zP_i)^{h_i}}{n_i!\, h_i!} \right]$$

$$= 1 + \sum_{r \in R} \sum_{s} \sum_{u} P_{m,s,u}(r) x^r y^s z^u,$$

where R extends over all possible values of nT_a. This theorem is similar to Theorem 1, except for an additional dimension for accounting purposes. The proof is therefore omitted.

REFERENCES

[1] J. Capon, "A nonparametric technique for the detection of a constant signal in additive noise," in *1959 IRE Wescon Conv. Rec.*, pt. 4, pp. 92–103.

[2] E. Stoll and L. Kurz, "Most powerful rank detection—An extension and application of Hoeffding's theorem," in *Proc. 1st Princeton Conf. Information Sciences and System*, pp. 330–334, Mar. 1967.

[3] M. N. Woinsky and L. Kurz, "An efficient nonparametric detector based on a three-sample classification model," *IEEE Trans. Inform. Theory*, vol. IT-14, pp. 744–751, Sept. 1968.

[4] R. R. Bahadur, "Stochastic comparison of tests," *Ann. Math. Statist.*, vol. 31, pp. 276–295, 1960.

[5] J. L. Hodges and E. L. Lehmann, "The efficiency of some nonparametric competitors of the *t*-test," *Ann. Math. Statist.*, vol. 27, pp. 324–335, 1956.

[6] J. B. Millard and L. Kurz, "The Kolmogorov–Smirnov tests in signal detection," *IEEE Trans. Inform. Theory* (Corresp.), vol. IT-13, pp. 341–342, Apr. 1967.

[7] G. E. Noether, "On a theorem of Pitman," *Ann. Math. Statist.*, vol. 26, pp. 64–68, 1955.

[8] Y. C. Ching and L. Kurz, "Lower bounds on the capacity of permutation channels," *IEEE Trans. Inform. Theory*, vol. IT-17, pp. 134–140, Mar. 1971.

Reprinted from *Acoust. Soc. Am. J.* **60**:857–862 (1976)

Dead-zone limiter: An application of conditional tests in nonparametric detection*

Saleem A. Kassam

Department of Systems Engineering, Moore School of Electrical Engineering, University of Pennsylvania, Philadelphia, Pennsylvania 19174

John B. Thomas

Department of Electrical Engineering, Princeton University, Princeton, New Jersey 08540

INTRODUCTION

The theory of signal detection is fundamentally that of statistical hypothesis testing. If input statistics can be completely characterized, optimal decision rules can be implemented.[1] In cases where less than complete information is available about the input statistics, nonparametric techniques can produce detectors with a guaranteed level and reasonable power for large classes of input distributions, classes which may not be parameterizable by a finite number of parameters.[2]

For nonparametric hypothesis testing, sign and rank tests have been developed and applied successfully in many situations.[2,3] The general formulation of such permutation tests for nonparametric hypothesis testing, however, usually results in conditional tests.[4] Conditional tests also have a theoretical basis in the concepts of similarity, sufficiency, and completeness.[5] In this paper we describe a nonparametric detector based on such a test and show that it has desirable properties both in terms of performance (including a worst-case performance bound) and ease of implementation.

I. PRELIMINARIES

Let $\mathbf{X}_n$ be the vector of observed samples $X_1, X_2, \ldots, X_n$, with probability density function $p(\mathbf{x}_n)$. Let H and K denote any hypothesis and alternative, respectively, for $p(\mathbf{x}_n)$, and let D denote the detector which decides between H and K on the basis of the received data vector $\mathbf{X}_n$. The detector D is assumed to operate from a test function $\phi_n(\mathbf{x}_n)$ which takes on values in $[0, 1]$, $\phi_n(\cdot)$ being the probability with which the alternative K is accepted. The function $\phi_n(\cdot)$ is defined in general through a test statistic $T_n(\mathbf{x}_n)$

$$\phi_n(\mathbf{x}_n) = \begin{cases} 1, & T_n(\mathbf{x}_n) > \lambda_{n,\alpha}(\mathbf{x}_n) \\ a_{n,\alpha}(\mathbf{x}_n), & T_n(\mathbf{x}_n) = \lambda_{n,\alpha}(\mathbf{x}_n) \\ 0, & T_n(\mathbf{x}_n) < \lambda_{n,\alpha}(\mathbf{x}_n) , \end{cases} \quad (1)$$

where $a_{n,\alpha}(\mathbf{x}_n)$ and the thresholds $\lambda_{n,\alpha}(\mathbf{x}_n)$ are chosen so that

$$E_H[\phi_n(\mathbf{x}_n)] \leq \alpha \quad (2)$$

for some fixed size α in $(0, 1)$. (The size of a test is an upper bound on the probability of false alarm in testing H versus K.) The test described by Eq. (1) is a conditional test.[4] Otherwise, the functions $a_{n,\alpha}(\mathbf{x}_n)$ and $\lambda_{n,\alpha}(\mathbf{x}_n)$ are constants with respect to $\mathbf{x}_n$.

We now define a specific hypothesis and alternative pair (H_1, K_1) which we consider in this paper. Let (H_1, K_1) be given by

$$H_1 : p(\mathbf{x}_n) = \prod_{i=1}^{n} f(x_i) , \quad f(x) = f(-x) \quad (3)$$

$$K_1 : p(\mathbf{x}_n) = \prod_{i=1}^{n} f(x_i - s) , \quad f(x) = f(-x), \quad s > 0 \quad (4)$$

so that the test is for the location of a symmetric density function. The H_1 and K_1 of Eqs. (3) and (4) are appropriate for the detection of a constant positive signal s in additive noise with a symmetric density function.

The sign detector[2] is a well-known detector which can be used for testing H_1 versus K_1. This detector is based only on the signs of the observed data samples X_i. Its test statistic can be obtained by summing the output when data samples are passed through a hard limiter, a device with the nonlinear characteristic $l_0(\cdot)$ defined by

$$l_0(x) = \begin{cases} 1 , & x > 0 \\ 0 , & x = 0 \\ -1, & x < 0 \end{cases} . \quad (5)$$

The test function of the sign detector does not result in a conditional test; $a_{n,\alpha}(\mathbf{x}_n)$ and $\lambda_{n,\alpha}(\mathbf{x}_n)$ in Eq. (1) are not functions of $\mathbf{x}_n$.

The sign detector uses all available data samples to form the test statistic. If only part of the n samples are used, however, we now show that the resulting detector can have significantly better performance characteristics.

II. DEAD-ZONE LIMITER DETECTOR

We consider a detector based only on the signs of the samples X_i which satisfy $|X_i| > c$ for some fixed $c \geq 0$, where we assume that $F(c) < 1$, $F(\cdot)$ being the absolutely

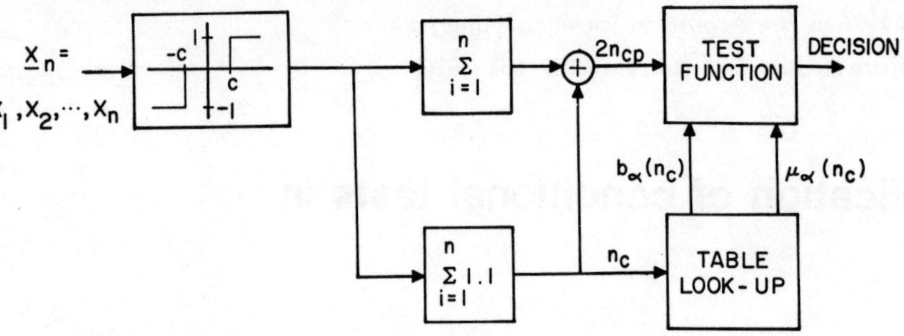

FIG. 1. Dead-zone limiter detector.

continuous distribution corresponding to $f(\cdot)$. The resulting detector is an extension of the sign detector, and employs the dead-zone limiter characteristic $l_c(\cdot)$ to operate on the data, where $l_c(\cdot)$ is given by

$$l_c(x) = \begin{cases} 1, & x > c \\ 0, & -c \leq x \leq c \\ -1, & x < -c \end{cases} \quad . \tag{6}$$

(For $c = 0$ we get the function $l_0(\circ)$ of Eq. (5), and the sign detector.)

The sign detector has a constant false-alarm probability α, i.e., it is *similar*[3] of size α, for the hypothesis H_1 because the distribution function of its test statistic under H_1 is independent of the particular density $f(\cdot)$. More specifically, its test statistics N_{0p}, which is the number of samples which are greater than (or equal to) zero, is distributed binomially under H_1 with parameters $\frac{1}{2}$ (the probability of success) and $N_0 = n$ (the number of trials). In the dead-zone limiter we assume that the data is operated on to obtain the number N_{cp} of samples X_i which satisfy $X_i > c$, and the number N_c of samples which satisfy $|X_i| > c$. Both N_{cp} and N_c are random variables, being functions of the observations, whereas for the sign detector ($c = 0$) the number N_0 is always equal to n. A test function based only on the statistic N_{cp} for $c > 0$ cannot therefore be the basis for a nonparametric detector for H_1 versus K_1. This is because the distribution of N_{cp} under H_1 depends on the density $f(\cdot)$. However, by going to a conditional test, a simple nonparametric detector can be constructed. Although the distribution of N_{cp} under H_1 depends on the specific $f(\cdot)$, the distribution of N_{cp} *conditioned* on N_c is again binomial, with parameters $\frac{1}{2}$ and N_c. Therefore a test function for which the functions $a_{n,\alpha}(\mathbf{x}_n)$ and $\lambda_{n,\alpha}(\mathbf{x}_n)$ depend on the observations through N_c can be made the basis for a nonparametric detector, with the test function described by

$$\phi(\mathbf{x}_n) = \begin{cases} 1, & n_{cp} > \mu_\alpha(n_c) \\ b_\alpha(n_c), & n_{cp} = \mu_\alpha(n_c) \\ 0, & n_{cp} < \mu_\alpha(n_c) \,. \end{cases} \tag{7}$$

The values of $b_\alpha(n_c)$ and the thresholds $\mu_\alpha(n_c)$ are such that for each realization n_c of the random variable N_c, we have $E_{H_1}[\phi(\mathbf{X}_n) | N_c] = \alpha$. This implies that $E_{H_1}[\phi(\mathbf{X}_n)] = \alpha$ and thus a constant false-alarm probability can be obtained.

In implementing this detector a table of values of $b_\alpha(n_c)$ and $\mu_\alpha(n_c)$ for $0 \leq n_c \leq n$ is required. This table can be generated and stored easily. It should be noted that the values of $b_\alpha(n_c)$ and $\mu_\alpha(n_c)$ do not depend on the particular value of c, which may be chosen to obtain good performance for a particular class of density functions. The values of $b_\alpha(n_c)$ and $\mu_\alpha(n_c)$ are also independent of the sample size n; therefore they may be considered to be fixed parameters of the detector.

Thus the dead-zone limiter detector remains nonparametric for any $c \geq 0$, and its false alarm probability remains invariant to odd nonlinear transformations of the data. Figure 1 shows the structure of the dead-zone limiter detector. In Fig. 2, the sign-detector structure is shown for comparison. The table look up adds to the complexity of the dead-zone limiter; however, as we will see in the next section, the performance of this detector may be decidedly superior to that of the sign detector. Furthermore, the table look up is not difficult to implement in a practical system.

III. PERFORMANCE OF THE DEAD-ZONE LIMITER DETECTOR

The performance of a nonparametric detector is usually evaluated in terms of its relative performance with respect to another detector, expressed as asymptotic relative efficiency (ARE).[5,6] The ARE is a ratio of sample sizes required by detectors operating at the same size α for a given hypothesis H to achieve the

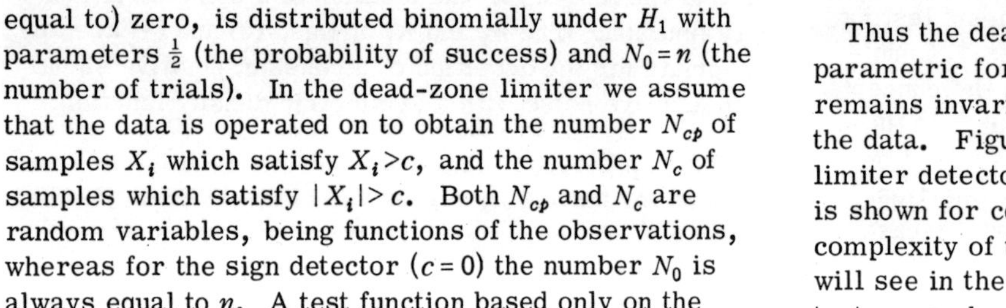

FIG. 2. Sign detector.

same detection power (i.e., the probability of detecting the alternative) for a given alternative, in the asymptotic case where signal strength is vanishingly small and sample sizes approach infinity. The ARE is defined in this way to allow analytic expressions to be derived. It is usually difficult to do this in the finite-sample and nonvanishing-signal case. For this latter case we will give computational results to show the performance of the dead-zone limiter detector.

Subject to certain regularity conditions (such as asymptotic normality of the test statistics), the ARE of two detectors can be expressed as a ratio of their efficacies.[6] The efficacy E of a detector is an indication of the performance of the detector for specific alternatives. In most cases of interest, the efficacy can be expressed as

$$E = \lim_{n\to\infty} \frac{\{(d/d\theta)E_\theta[T_n(\mathbf{X}_n)]|_{\theta=0}\}^2}{n\,\mathrm{var}_{\theta=0}[T_n(\mathbf{X}_n)]}, \qquad (8)$$

where $T_n(\cdot)$ is the test statistic of the detector and θ represents the "distance" of the particular alternative from the hypothesis ($\theta = 0$ represents a probability measure of the hypothesis). For testing H_1 versus K_1, the signal s has the role of the parameter θ.

The efficacy E_{DZ} of the dead-zone limiter detector can be obtained from Eq. (8), as is shown in Appendix A. We have the result, assuming $F(c) < 1$ and $f(c)$ is continuous at c, that

$$E_{DZ} = 2f^2(c)/[1 - F(c)]\,. \qquad (9)$$

For the sign detector, the efficacy E_S is given by

$$E_S = 4f^2(0)\,. \qquad (10)$$

This also follows from Eq. (9) with $c = 0$. Finally, for the linear detector, which uses as a test statistic the sum of the n observations, the efficacy E_L is

$$E_L = 1/\sigma^2, \qquad (11)$$

where σ^2 is the variance of the noise density $f(\cdot)$. The linear detector is not a nonparametric detector, but it is the locally optimum detector[6] for H_1 versus K_1 when $f(\cdot)$ is a Gaussian density function.

We can now obtain $\mathrm{ARE}_{DZ,L}$ and $\mathrm{ARE}_{DZ,S}$, the ARE's of the dead-zone limiter detector relative to the linear detector and sign detector, respectively, by forming the ratios E_{DZ}/E_L and E_{DZ}/E_S. We have already remarked that the dead-zone limiter detector remains nonparametric for any choice of $c > 0$; therefore, its asymptotic performance can be optimized for particular $f(\cdot)$ by choosing c optimally. For sufficiently smooth $f(\cdot)$ the equations for the optimum c can be obtained easily and can be readily solved numerically. We have performed this optimization for the class of density functions given by[7]

$$f_k(x) = \frac{k}{2\Gamma(1/k)A(k)} \exp\left\{-\left[\frac{|x|}{A(k)}\right]^k\right\}, \quad k \geq 1\,, \qquad (12)$$

where

$$A(k) = \left[\frac{\sigma^2\Gamma(1/k)}{\Gamma(3/k)}\right]^{1/2}, \qquad (13)$$

$\Gamma(\cdot)$ being the gamma function. This class allows the performance of the dead-zone limiter detector to be obtained not only for the case of Gaussian noise ($k = 2$) but also for specific densities which are close to the Gaussian density. Figure 3 shows $\mathrm{ARE}^*_{DZ,L}$ and $\mathrm{ARE}^*_{DZ,S}$ plotted against k, which parameterizes the class of generalized Gaussian densities of Eq. (12). The * denotes optimum values obtained by using the maximizing values of c for each k. Also plotted in Fig. 3 are the $\mathrm{ARE}_{DZ,L}$ and $\mathrm{ARE}_{DZ,S}$ for the fixed value of c which is the optimum for the case of Gaussian noise. (This value is 0.612σ). For comparison, $\mathrm{ARE}_{S,L}$ is also shown. It can be seen that, for Gaussian noise, the dead-zone limiter detector is significantly superior to the sign detector (in terms of asymptotic performance). Furthermore, significant improvements in performance are obtained for densities close to the Gaussian in the class of non-Gaussian densities defined by Eq. (12), even without using optimum values of c

To complete the asymptotic performance analysis we mention an interesting result on the worst-case performance of the dead-zone limiter detector relative to the linear detector. We can prove (Appendix B) that for $f(\cdot) \in \mathfrak{F}$, the class of symmetric density functions, the worst-case performance of the dead-zone limiter detector over $\mathfrak{F}$ [optimized by picking the appropriate c for each $f(\cdot) \in \mathfrak{F}$] results in an $\mathrm{ARE}_{DZ,L}$ of $\frac{2}{3}$. This can be compared with the corresponding value of $\frac{1}{3}$ for the sign detector and the class of unimodal and symmetric densities.[8] This result implies that the use of a dead-zone limiter detector instead of the linear detector in the case of known $f(x)$ results in a maximum loss of efficiency of $\frac{1}{3}$. Of course, in many cases an improvement in performance may be obtained. The advantage

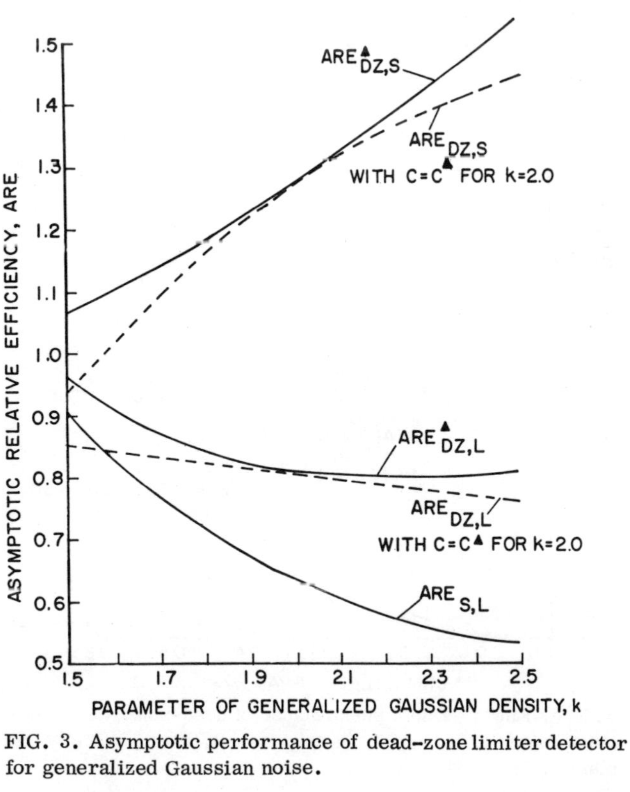

FIG. 3. Asymptotic performance of dead-zone limiter detector for generalized Gaussian noise.

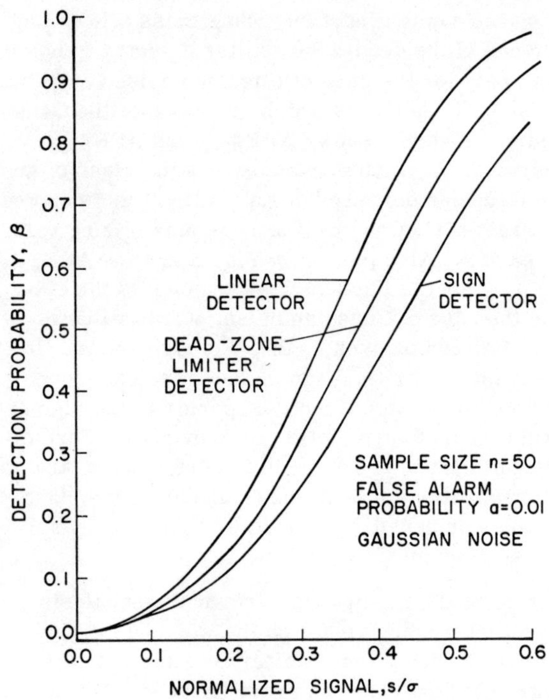

FIG. 4. Finite-sample performance of dead-zone limiter detector: $n=50$, $\alpha=0.01$, Guassian noise, asymptotically optimum c.

of the dead-zone limiter detector is that it is nonparametric, and the thresholds for any false-alarm probability do not depend on the particular $f(\cdot)$, even for finite sample sizes.

The finite-sample-size performance characteristics of the dead-zone limiter detector can be obtained by computing the exact detection probabilities for different

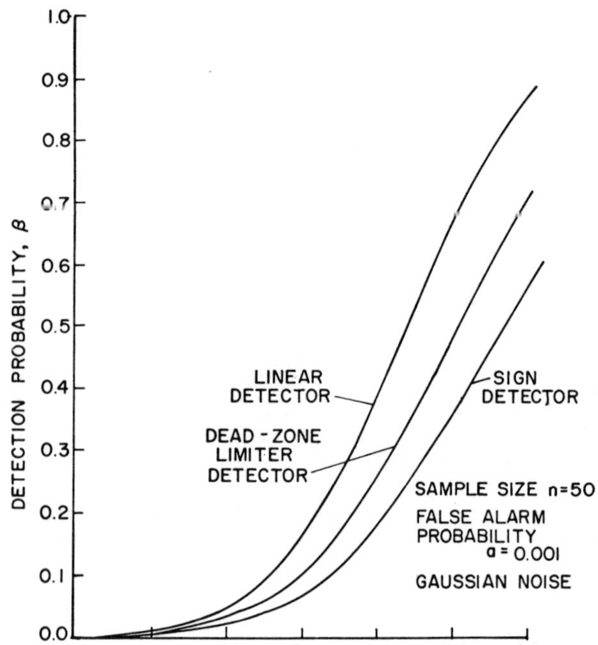

FIG. 5. Finite-sample performance of dead-zone limiter detector: $n=50$, $\alpha=0.001$, Gaussian noise, asymptotically optimum c.

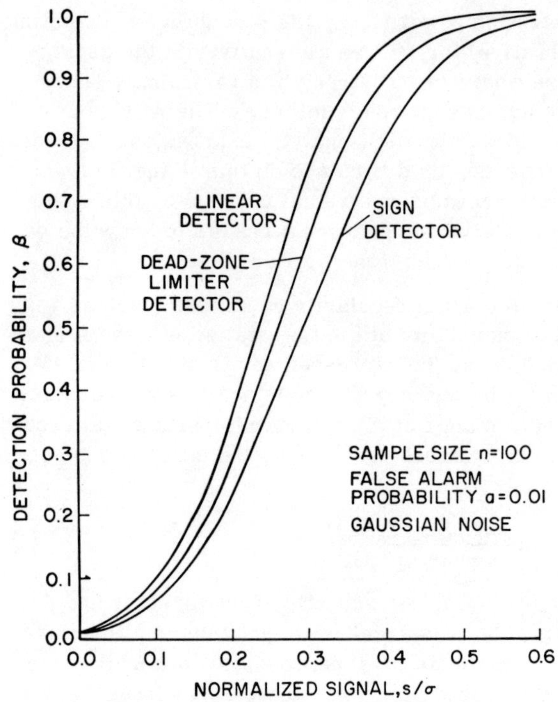

FIG. 6. Finite-sample performance of dead-zone limiter detector: $n=100$, $\alpha=0.01$, Gaussian noise, asymptotically optimum c.

signal strengths and sample sizes, with tables of $b_\alpha(n_c)$ and $\mu_\alpha(n_c)$ being used to achieve desired false-alarm probabilities. Since the binomial probability distribution function is involved in such computation and table generation, a digital computer can be used for the purpose. Figures 4–7 show the results obtained for some typical false-alarm probabilities. In the figures, detection probability β is plotted against normalized signal

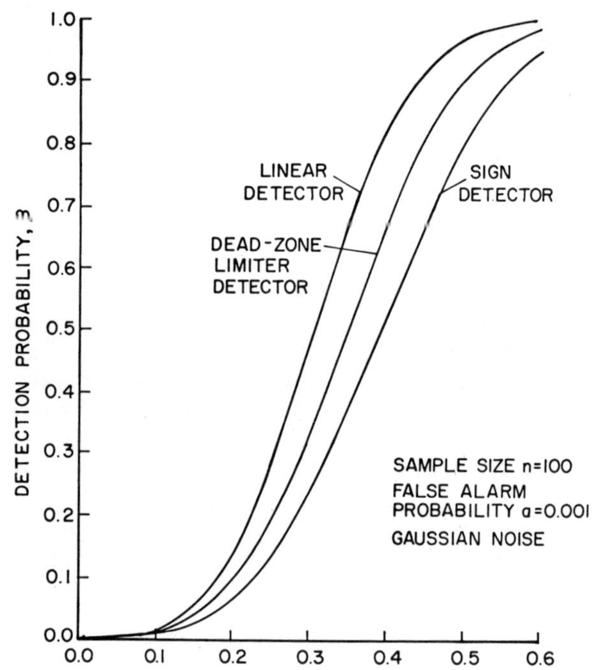

FIG. 7. Finite-sample performance of dead-zone limiter detector: $n=100$, $\alpha=0.001$, Gaussian noise, asymptotically optimum c.

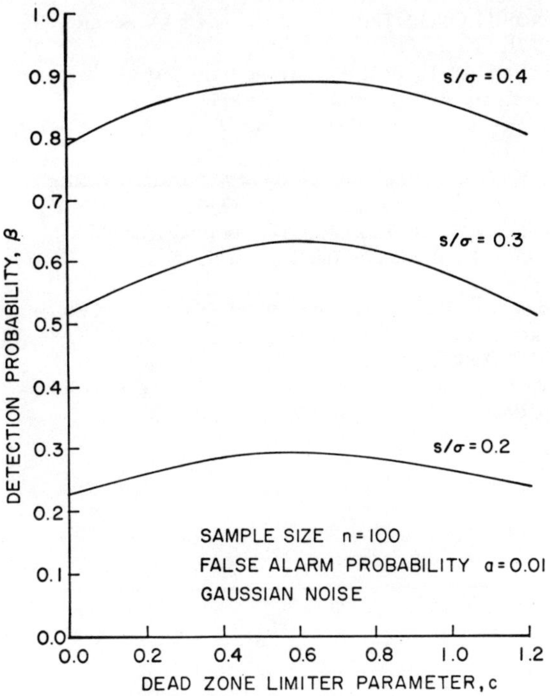

FIG. 8. Variation of finite-sample performance of dead-zone limiter detector with c: $n=100$, $\alpha=0.01$, Gaussian noise.

strength s/σ for different sample sizes for the linear, sign, and dead-zone limiter detectors. The noise is assumed to be Gaussian, and, in the latter detector, the value of c maximizing its asymptotic performance is used. From these results it is clear that, for Gaussian noise, the dead-zone limiter detector has a definitely better performance than the sign detector for finite samples and signals, as is true in the asymptotic case.

In Fig. 8 detection probability for the dead-zone limiter detector is plotted against c for fixed values of sample size, false-alarm probability, and signal strengths. The flatness of these curves shows the relative insensitivity of the performance for deviations of c from the optimum value and indicate that the selection of c is not very critical.

APPENDIX A

For the dead-zone limiter detector, the test statistic $T_n(\mathbf{X}_n)$ is the number N_{cp} of observations X_i satisfying $X_i > c$, and the test function is defined conditionally on the number N_c of observations such that $|X_i| > c$. The conditional distribution of N_{cp}, given N_c, is binomial, and we have

$$E_s\{N_{cp}|N_c\} = \frac{1-F(c-s)}{1-F(c-s)+F(-c-s)} N_c \tag{A1}$$

and

$$\sigma_s^2\{N_{cp}|N_c\} = E_s\{N_{cp}^2|N_c\} - E_s^2\{N_{cp}|N_c\}$$
$$= \frac{[1-F(c-s)][F(-c-s)]}{[1-F(c-s)+F(-c-s)]^2} N_c . \tag{A2}$$

From the strong law of large numbers,

$$N_c/n \to 1 - F(c-s) + F(-c-s) \text{ with probability 1 .} \tag{A3}$$

If $F(c)<1$, this implies that $N_c \to \infty$ with probability one as $n\to\infty$. From this and the DeMoivre–Laplace limit theorem, the conditional distribution of the (normalized) statistic $(N_{cp} - E_s\{N_{cp}|N_c\})/\sigma_s\{N_{cp}|N_c\}$ converges in probability to the normal distribution function with zero mean and unit variance. (For a detailed treatment of the asymptotic distribution theory of conditional tests see Refs. 4 and 5.) Using the asymptotic value of Eq. (A3) in Eqs. (A1) and (A2) we get the required results on the mean and variance as functions of s. Then it follows directly from Eq. (8) that

$$E_{DZ} = 2f^2(c)/[1-F(c)] , \tag{A4}$$

assuming that $f(c)$ is continuous at c.

APPENDIX B

From Eqs. (9) and (11), we have

$$\text{ARE}_{DZ,L} = 2\sigma^2 f^2(c)/[1-F(c)] . \tag{B1}$$

We want to obtain the worst-case (asymptotic) performance of the dead-zone detector relative to the linear detector, i.e., we want to obtain the least favorable symmetric density $f_L(\cdot)$ such that

$$\max_{c\geq 0} \frac{2f_L^2(c)}{1-F_L(c)} \leq \max_{c\geq 0} \frac{2f^2(c)}{1-F(c)} , \quad \text{for all } f(\cdot)\in \mathfrak{F}_1 , \tag{B2}$$

where $\mathfrak{F}_1$ is the class of all symmetric densities $f(\cdot)$ with unit variance, and with absolutely continuous distributions $F(\cdot)$. (We assume $\sigma^2 = 1$ in Eq. (B1) without loss of generality.)

Theorem: (i) The density $f_L(\cdot)$ which satisfies Eq. (B2) is given by

$$f_L(x) = \begin{cases} 6^{-1/2} - |x|/6 , & |x| \leq 6^{1/2} \\ 0 , & |x| > 6^{1/2} . \end{cases} \tag{B3}$$

(ii) With $f_L(\cdot)$ defined by Eq. (B3), the left-hand side of Eq. (B2) has the value $\frac{2}{3}$, with the maximum being achieved for any $c \in [0, 6^{1/2}]$.

Proof: It is simple to show that $\sigma^2 = 1$ for $f_L(x)$ of Eq. (B3), and that (ii) of the theorem is true.

Let $f(\cdot)$ be an arbitrary density function in $\mathfrak{F}_1$. We have $F(0) = F_L(0) = \frac{1}{2}$, and $\int_0^\infty x^2\,dF(x) = \int_0^{6^{1/2}} x^2\,dF_L(x) = \frac{1}{2}$. We first prove that there exists $c_f \in (0, 6^{1/2})$ for which

$$1 - F(c_f) < 1 - F_L(c_f) . \tag{B4}$$

Supposing this were not so; then

$$1 - F(x) \geq 1 - F_L(x) \quad \text{for all } x \in (0, 6^{1/2}) \tag{B5}$$

and[9]

$$\int_0^\infty x^2\,dF(x) = 2\int_0^\infty x[1-F(x)]\,dx$$
$$\geq 2\int_0^{6^{1/2}} x[1-F_L(x)]\,dx$$
$$= \int_0^{6^{1/2}} x^2\,dF_L(x)$$

with equality if and only if $F(x) = F_L(x)$, for all x. But this is a contradiction; therefore there exists

$c_f \in (0, 6^{1/2})$ satisfying Eq. (B4).

Now if $f(c_f) \geqslant f_L(c_f)$, then the theorem is proved, since (ii) is true. If $f(c_f) < f_L(c_f)$, then from Eq. (B4) and the continuity of $F(\cdot)$, there exists $c_{f0} \in (0, c_f)$ for which $f(c_{f0}) \geqslant f_L(c_{f0})$ and $1 - F(c_{f0}) \leqslant 1 - F_L(c_{f0})$. Thus, the theorem is proved.

*This research is supported by the National Science Foundation under Grant GK-24187 and by U.S. Army Research Office under Grant DAAG29-75-G-0192.

[1]C. W. Helstrom, *Statistical Theory of Signal Detection* (Pergamon, New York, 1968), 2nd ed.

[2]J. W. Carlyle and J. B. Thomas, "On Nonparametric Signal Detectors," IEEE Trans. Inf. Theory **IT-10**, 146–152 (1964).

[3]J. Hajek and Z. Sidak, *Theory of Rank Tests* (Academic, New York, 1967).

[4]M. L. Puri and P. K. Sen, *Nonparametric Methods in Multivariate Analysis* (Wiley, New York, 1971).

[5]D. A. S. Fraser, *Nonparametric Methods in Statistics* (Wiley, New York, 1957).

[6]J. Capon, "On the Asymptotic Efficiency of Locally Optimum Detectors," IEEE Trans. Inf. Theory **IT-7**, 67–71 (1961).

[7]M. Kanefsky and J. B. Thomas, "On Polarity Detection Schemes with Non-Gaussian Inputs," J. Franklin Inst. **280**, 120–138 (1965).

[8]J. L. Hodges and E. L. Lehmann, "The Efficiency of Some Nonparametric Competitors of the *t*-Test," Ann. Math. Stat. **27**, 324–335 (1956).

[9]W. Feller, *An Introduction to Probability Theory and its Applications* (Wiley, New York, 1971), Vol. 2, p. 150.

Reprinted from *Acoust. Soc. Am. J.* **63**:1389–1395 (1978)

Three-level coincidence-correlator detectors

Jung Gil Shin and Saleem A. Kassam

Department of Systems Engineering, Moore School of Electrical Engineering, University of Pennsylvania, Philadelphia, Pennsylvania 19104

INTRODUCTION

In many detection problems, such as in underwater-sound applications, the objective is to detect the presence of a weak zero-mean random signal against a background of additive zero-mean noise which is independent but statistically identical at each of two inputs. There are several systems which may be used for such problems, and in general the test statistic may be based on some combination of the magnitudes, the signs, and the ranks of the observed data.

The Neyman–Pearson (NP) detector for Gaussian statistics is obtained by applying the fundamental lemma of statistical hypothesis testing to the Gaussian model. It calculates the test statistic

$$T_{NP} = \sum_{i=1}^{n} (X_1^i + X_2^i)^2 , \tag{1}$$

where n is the number of samples in the observation vectors $\mathbf{X}_1 = (X_1^1, X_1^2, \ldots, X_1^n)$ and $\mathbf{X}_2 = (X_2^1, X_2^2, \ldots, X_2^n)$ at the two channels. In developing the rest of this paper it is assumed that each input sequence has statistically independent components which are identically distributed, and that the noise density function f is symmetric about the origin. For weak-signal detection it has been shown in Ref. 1 that a much simpler detection scheme outperforms the NP detector when the noise density function is assumed to be double exponential instead of Gaussian.

Another method for detecting weak random signals in two-input systems is based on the correlation of input data[2–6]; for example, the general correlation detector calculates the test statistic

$$T_C = \sum_{i=1}^{n} l(X_1^i) l(X_2^i) , \tag{2}$$

where usually

$$l(x) = -f'(x)/f(x) , \tag{3}$$

the locally optimum nonlinearity corresponding to the assumed noise density function f.

A useful class of detectors is the class of *coincidence detectors*[7]; in such detectors a threshold is set, and the test statistic is the number of two-dimensional samples for which some function of the samples exceeds the threshold. By retaining only the polarity of the samples in a correlation detector, i.e., with l in Eq. (2) a hard limiter, we obtain the polarity coincidence correlator (PCC).[4] It has been shown[1] that the PCC is useful as a nonparametric detection device, maintaining a fixed false-alarm probability for a large class of noise densities. However, the rather poor performance of the two-level PCC compared to the parametric detector for Gaussian noise has motivated the study of its extensions. One possibility is to use the idea of statistical conditional testing, and in Ref. 6 this was used to formulate the three-level DZC (dead-zone limiter coincidence) detector. The DZC detector is based on a test statistic

$$T_{DZ} = \sum_{i=1}^{n} l_c(X_1^i) l_c(X_2^i) , \tag{4}$$

where

$$l_c(x) = \begin{cases} 1 , & x \geq c \\ -1 , & x < -c \\ 0 & \text{otherwise} . \end{cases} \tag{5}$$

It maintains nonparametric operation by employing a threshold which depends on the observed data.[6,8] We will explain this concept briefly in Sec. II.

The DZC detector is a simple extension for two-input systems of the one-input dead-zone sign detector.[8] It can yield a considerable improvement in detector performance relative to the PCC, without losing the simplicity of implementation and nonparametric property of the PCC. The DZC detector may be viewed as a simple member of the class of *three-level coincidence-correlator* (CC) detectors. In general, a three-level CC detector partitions the two-dimensional data space into three regions P, N, and O, and weights of $+1$, -1, and zero are assigned to the observations which fall in P, N, and O, respectively. Several other specific three-level CC detectors may be listed which are intuitively comparable to the DZC detector; for example, the square dead-zone CC, the triangular dead-zone CC, the quarter-circle dead-zone CC, etc. We will describe, and examine the performance of, each of these CC detectors and compare them with the PCC and the DZC detectors.

Constraining our interest to the class of CC detectors,

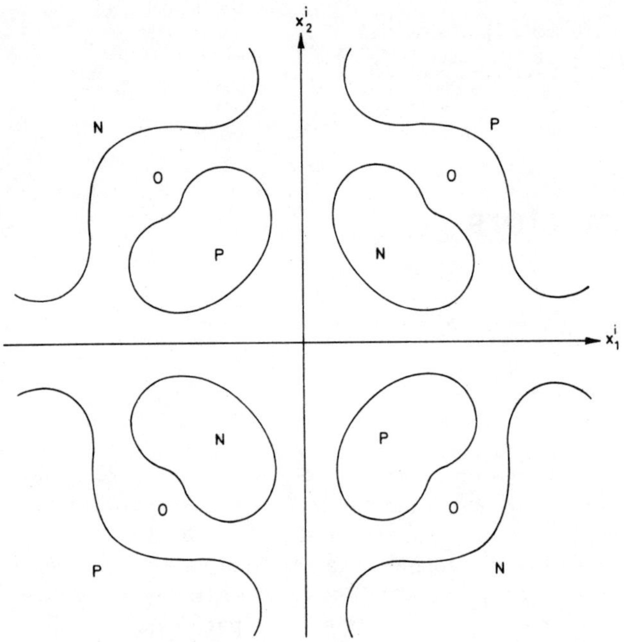

FIG. 1. Partitioning and weighting of input-data space for three-level CC detectors.

we obtain in Sec. II the structure of the optimum three-level CC detector for two-input systems. Our criterion of performance is detector efficacy which we define in the next section. In Sec. III a special class of three-level CC detectors is considered where the partitioning of the input-data space is defined by a single-valued function in each quadrant. Some numerical results are given in Sec. IV for the performance of the various detectors when noise is assumed to be Gaussian. Although the detectors considered are nonparametric, detector performance for Gaussian noise is usually required to be acceptable for a useful nonparametric detector. The results of this paper allow a comparison of the performances of different practical three-level extensions of the nonparametric PCC, and also give the best performance attainable in three-level CC detection using the optimum extension.

I. PRELIMINARIES

The problem of detecting random signals in two-input systems can be modeled as a hypothesis testing problem. For two random vectors $\mathbf{X}_1 = (X_1^1, X_1^2, \ldots, X_1^n)$ and $\mathbf{X}_2 = (X_2^1, X_2^2, \ldots, X_2^n)$ which represent the input data, we can formulate the problem as that of a comparison between a null hypothesis H_0 and an alternative H_1, where

$$\text{under } H_0\text{: } X_1^i = N_1^i, \; X_2^i = N_2^i, \quad i = 1, 2, \ldots, n, \tag{6}$$

$$\text{under } H_1\text{: } X_1^i = N_1^i + \theta S_i, \; X_2^i = N_2^i + \theta S_i, \quad i = 1, 2, \ldots, n. \tag{7}$$

The S_i, $i = 1, 2, \ldots, n$, representing the random signal are assumed to be of zero mean and unit variance, and the components N_1^i, N_2^i, $i = 1, 2, \ldots, n$, of the noise vectors are assumed to be independent and identically distributed with a density function f. This density function f is assumed to be symmetric about the origin, with the corresponding cumulative distribution function denoted by F. Let σ^2 be the noise variance. Then θ^2/σ^2 is the input signal-to-noise ratio (SNR). To test H_0 vs H_1, a test statistic $T(\mathbf{X}_1, \mathbf{X}_2)$ is compared to a threshold λ with a randomization probability ρ to give the desired false-alarm probability. In this paper we investigate a special class of detectors with CC-type test statistics.

The performance comparison of two detectors will be based on their asymptotic relative efficiency (ARE).[9,10] ARE is the limiting ratio of sample sizes required by the detectors to give the same performance, i.e., false-alarm probability and the power of the test, for SNR approaching zero. Under certain regularity conditions such as asymptotic normality of the sequence of the test statistics (which will be the case for our detectors), the ARE can be expressed as a ratio of the detector efficacies,[11] the efficacy $\mathcal{E}$ for a test statistic T being defined as

$$\mathcal{E} = \lim_{n\to\infty} \frac{[(d/d\theta^2)E_\theta\{T \mid H_1\}\mid_{\theta=0}]^2}{n \operatorname{Var}\{T \mid H_0\}} . \tag{8}$$

Note that in Eq. (8) the derivative of $E_\theta\{T \mid H_1\}$ is taken with respect to θ^2, which is the signal strength parameter. It is clear from Eq. (8) that $\mathcal{E}$ is a form of SNR for the test statistic. Although the ARE is an asymptotic relative performance measure, it has been shown[12] that for conditional tests similar to those considered here the ARE is generally consistent with finite-sample performance. Thus we will base our optimization and comparisons on this definition of the efficacy of a detector.

II. OPTIMUM THREE-LEVEL CC (COINCIDENCE-CORRELATOR) DETECTOR

Let us define a function h for sample pairs (X_1^i, X_2^i), $i = 1, 2, \ldots, n$, such that

$$h(x_1^i, x_2^i) = \begin{cases} 1, & (x_1^i, x_2^i) \in P \\ -1, & (x_1^i, x_2^i) \in N \\ 0, & (x_1^i, x_2^i) \in O, \end{cases} \tag{9}$$

where P, N, and O are three regions partitioning the two-dimensional data space. An example of a possible set of regions is shown in Fig. 1. Since the noise density function f has been assumed to be symmetric about the origin, and since the samples from each channel are independent and should be equally weighted, it is reasonable to assume that the regions are symmetric about the coordinate axes and the main diagonals. The set of weights (+1, −1, or 0) assigned to regions in the second (fourth) quadrant are therefore complements of those assigned to the corresponding regions in the first (third) quadrant. (See Fig. 1.)

The test statistic for the general three-level CC detector can now be written as

$$T_{CC} = \sum_{i=1}^{n} h(X_1^i, X_2^i) . \tag{10}$$

This can also be expressed as $T_{CC} = n_P - n_N$, where n_P, n_N are the numbers of observation pairs falling in the regions P, N, respectively. As shown in Refs. 6 and 12 we can

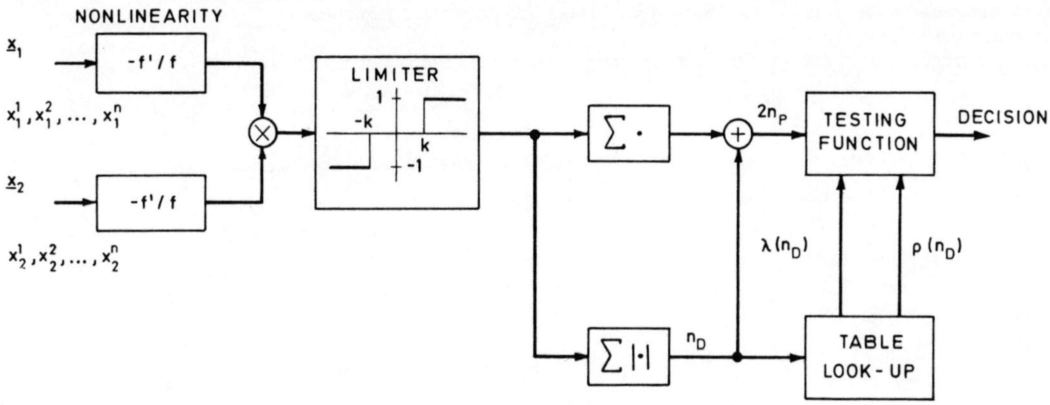

FIG. 2. Structure of the optimum three-level CC detector.

construct a nonparametric detector based on this test statistic by using the idea of conditional testing. Note that the distribution of T_{CC} under H_0 does depend on the noise density function f. However, *conditioned* on the number $n_D = n_P + n_N$, we have that n_P is binomial with parameters $p = \frac{1}{2}$ and n_D under H_0. This follows because if it is known that an observation did not fall in the region O, we can conclude that it has to be either in P or in N, with equal probability $p = \frac{1}{2}$ under H_0 because of symmetry. Thus the *conditional* distribution of $T_{CC} = 2n_P - n_D$, conditioned on n_D, is independent of f under H_0. Under the alternative in general $p > \frac{1}{2}$. Therefore appropriate threshold and randomization probabilities can be set as functions $\lambda(n_D)$ and $\rho(n_D)$, respectively, of n_D, to give a desired constant false-alarm probability. The variable n_D has a finite number of possible values, and such a system can be implemented with a table look-up arrangement to set the test parameters as soon as the data has been observed and n_D determined. For a detailed discussion of the theory of conditional testing the reader is referred to Refs. 9 and 13. It should be noted that the efficacy results we discuss in this paper remain valid whether or not a conditional test is used to obtain a constant false-alarm probability.

The efficacy $\mathcal{E}_{CC}$ for the test statistic T_{CC} is obtained by finding the mean and the variance of the statistic; details of the derivation are given in Appendix A. The result for the efficacy $\mathcal{E}_{CC}$ of a three-level CC is

$$\mathcal{E}_{CC} = \frac{[\int\int_P l(x)l(y)f(x)f(y)dxdy - \int\int_N l(x)l(y)f(x)f(y)dxdy]^2}{\int\int_{P\cup N} f(x)f(y)dxdy}, \tag{11}$$

where l is given by Eq. (3). This efficacy depends on how the regions P, N, and O are defined. We can also find the region P^* (thus N^* and O^*) which maximize the efficacy given by Eq. (11). The following theorem specifies the shape of the region for any assumed density f.

Theorem: The regions, P^*, N^*, and O^* which maximize the efficacy given by Eq. (11) are of the form

$$\begin{aligned} P^* &= \{(x_1^i, x_2^i):\ l(x_1^i)l(x_2^i) > k\}, \\ N^* &= \{(x_1^i, x_2^i):\ l(x_1^i)l(x_2^i) < -k\}, \\ O^* &= \{(x_1^i, x_2^i):\ -k \le l(x_1^i)l(x_2^i) \le k\}, \end{aligned} \tag{12}$$

where k is some non-negative constant.

This theorem specifies the form of the optimum partitioning. The optimum value of k can then be obtained to maximize the efficacy of Eq. (11). The proof is given in Appendix B.

Figure 2 shows the structure of the optimum three-level CC detector. It is obtained by the addition of a dead-zone limiter between the multiplier and the summer of the optimum (analog) correlation detector mentioned earlier. It is interesting to observe that the optimum partitioning of the two-dimensional data space follows the test structure of the optimum analog correlation detector.

The above result permits the evaluation of the best performance obtainable from a three-level CC detector. In Sec. III we will examine a special class of three-level CC detectors which includes the PCC and the DZC.

III. A SPECIAL CLASS OF THREE-LEVEL CC DETECTORS

Consider the form of partitioning and weighting of the data space shown in Fig. 3. Here we introduce a threshold function g to define the regions. Due to the

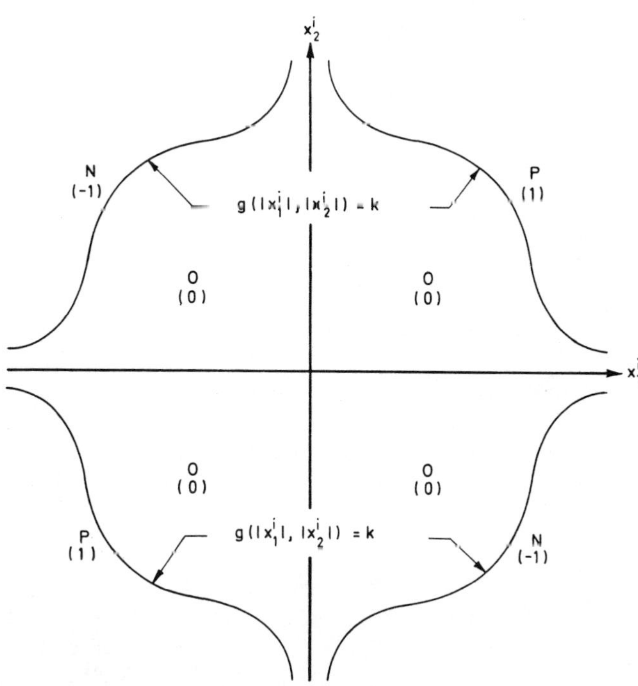

FIG. 3. Partitioning and weighting of input-data space for a special class of three-level CC detectors.

TABLE I. Efficacy of various three-level CC detectors. Numerical values are shown for Gaussian noise with zero mean and unit variance.

Description	Efficacy	Numerical results: Optimum partitioning parameter c^*	Numerical results: Optimum efficacy
PCC [Fig. 4(a)]			
$t(x)=0$	$16f^4(0)$	...	0.405
DZC [Fig. 4(b)]			
$t(x)=\begin{cases}\infty, & 0\le x<c\\ c, & c\le x\end{cases}$	$\dfrac{4f^4(c)}{[1-F(c)]^2}$	0.612	0.656
Square DZ CC [Fig. 4(c)]			
$t(x)=\begin{cases}c, & 0\le x<c\\ 0, & c\le x\end{cases}$	$\dfrac{4[2f(0)f(c)-f^2(c)]^2}{F(c)[1-F(c)]}$	1.08	0.545
Triangular DZ CC [Fig. 4(d)]			
$t(x)=\begin{cases}c-x, & 0\le x<c\\ 0, & c\le x\end{cases}$	$\dfrac{8[f(0)f(c)-\int_0^c f(c-x_1)f'(x_1)dx_1]^2}{F(c)-2\int_0^c F(c-x_1)f(x_1)dx_1}$	2.02	0.656
Quarter-circle DZ CC [Fig. 4(e)]			
$t(x)=\begin{cases}c-[c^2-(c-x)^2]^{1/2}, & 0\le x<c\\ 0, & c\le x\end{cases}$	$\dfrac{8[f(0)f(c)-\int_0^c f(t)f'(x_1)dx_1]^2}{F(c)-2\int_0^c F(t)f(x_1)dx_1}$	3.33	0.694
Optimum CC [Fig. 4(f)]			
$t(x)=c/x$	$\dfrac{8[\int_0^\infty f(t)f'(x_1)dx_1]^2}{1-2\int_0^\infty F(t)f(x_1)dx_1}$	0.820	0.710

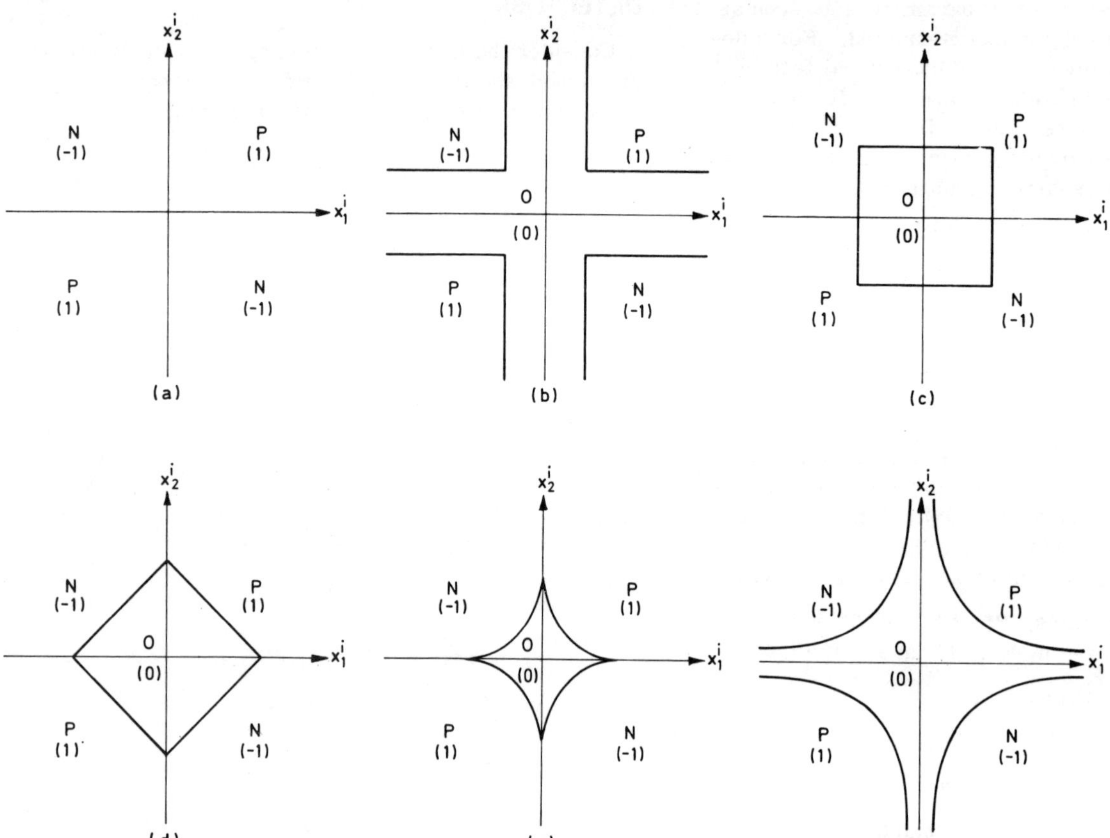

FIG. 4. Various types of partitioning and weighting of input data for a special class of three-level CC detectors. (a) PCC, (b) DZC, (c) square DZ CC, (d) triangular DZ CC, (e) quarter circle DZ CC, and (f) optimum CC (for Gaussian noise).

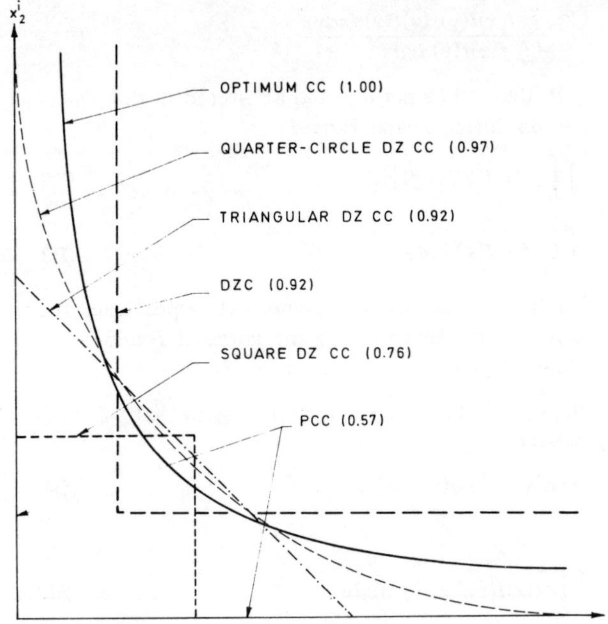

FIG. 5. Optimum partitioning of various three-level CC detectors for Gaussian noise with zero mean and unit variance. First quadrant only is shown.

assumed symmetry of the regions we will consider the function g in the first quadrant only. The boundary between the regions P and O is defined by an equation $g(x_1^i, x_2^i)=k$, where k is some constant; the region P is $\{(x_1^i, x_2^i): g(x_1^i, x_2^i)>k\}$. In particular we consider now only those three-level CC detectors where the equation $g(x, y)=k$ can be written as $y=t(x)$, a non-negative single-valued function of x defined for almost all x, with $g(x, y)>k$ being equivalent to $y>t(x)$. The definition of the regions P, N, and O now becomes

$$P=\{(x_1^i, x_2^i): |x_2^i|>t(|x_1^i|): \text{ and } \operatorname{sgn}(x_1^i)\operatorname{sgn}(x_2^i)=1\},$$
$$N=\{(x_1^i, x_2^i): |x_2^i|>t(|x_1^i|): \text{ and } \operatorname{sgn}(x_1^i)\operatorname{sgn}(x_2^i)=-1\}, \quad (13)$$
$$O=\{(x_1^i, x_2^i): |x_2^i|\le t(|x_1^i|)\}.$$

The test statistic T_t can be defined as in Eq. (10) with P, N, and O given by Eq. (13). We can see that we get the PCC with $t\equiv 0$ and the DZC detector with t defined as

$$t(x)=\begin{cases}\infty, & 0<x\le c\\ c, & c<x,\end{cases} \quad (14)$$

where c is some non-negative constant.

The efficacy $\mathcal{E}_t$ for the test statistic T_t is obtained from Eq. (11), with the definitions of P and N given by Eq. (13). In Appendix A it is shown that

$$\mathcal{E}_t=\frac{8[\int_0^\infty f(t)f'(x)\,dx]^2}{1-2\int_0^\infty F(t)f(x)dx}. \quad (15)$$

If we let $t(x)=0$ for all x, Eq. (15) becomes

$$\mathcal{E}_t=16f^4(0), \quad (16)$$

which is the well-known expression for the efficacy of the PCC. For t given by Eq. (14), we get from Eq. (15) that

$$\mathcal{E}_t=\frac{4f^4(c)}{[1-F(c)]^2}, \quad (17)$$

which is the efficacy of the DZC detector with threshold parameter c, and the result is consistent with the one obtained in Ref. 6.

The solution $t=t^*$ which maximizes the efficacy $\mathcal{E}_t$ satisfies the condition

$$[f'(t)/f(t)][f'(x)/f(x)]=c, \quad \text{for all } x, \quad (18)$$

where c is some constant which will be called the partitioning parameter. The proof of Eq. (18) is given in Appendix C. We can then substitute for t in Eq. (15) and obtain the optimum c. Since t was constrained to be single valued, Eq. (18) is valid only if f'/f is a monotonic nondecreasing (or nonincreasing) function, which would be the case if the density f were strongly unimodal (p. 15 of Ref. 14). Therefore, provided that f is strongly unimodal, the result agrees with the one obtained in Sec. II. The class of strongly unimodal densities is wide enough to contain many cases of practical interest. In general, however, the optimum CC for a particular form of partitioning may be of interest. If we constrain the detector structure so that t is a specific function of a partitioning parameter c, then for any density f we can find $c=c^*$ which maximizes the efficacy. In Sec. IV we consider several types of t which give simple-to-implement intuitive extensions of the PCC.

IV. NUMERICAL RESULTS

The various types of three-level CC detectors including the PCC and DZC can be compared explicitly on the basis of their efficacies. Table I shows the description of each detector we have considered and its efficacy. Figure 4 illustrates the partitioning employed by these various detectors. Numerical values of the optimum partitioning parameter c^* and the corresponding efficacies are shown in Table I. The efficacies shown are for Gaussian noise with zero mean and unit variance. The optimum partitionings of these detectors and their AREs relative to the optimum (for Gaussian noise) are shown in Fig. 5. In this figure we observe that the optimum partitioning for any given constrained detector structure tends to approximate the optimum unconstrained partitioning.

From the values obtained we see that the DZC detector which was considered in Ref. 6 has a performance close to that of the optimum three-level CC detector, although it maintains a simplicity of implementation comparable to that of the PCC. Note that the ARE of the DZC detector relative to the optimum is 0.92. We also note that the quarter-circle CC, which gives a close approximation of the optimum partitioning, shows nearly optimum performance. The ARE of the quarter-circle CC relative to the optimum is 0.97. The square CC is also easy to implement, but results in a poorer ARE of 0.76 with respect to the optimum. Thus it appears that the DZC detector[6] is the best compromise between performance and ease of implementation.

ACKNOWLEDGMENT

This research is supported by the Air Force Office of Scientific Research, Air Force Systems Command, USAF, under Grant AFOSR 77-3154.

APPENDIX A: DERIVATION OF EFFICACY EQUATIONS

Since the observations are assumed to be independent the mean and the variance of T_{CC} are the sums of the means and the variances of $h(X_1^i, X_2^i)$ for $i=1, 2, \ldots, n$. From Eq. (9), $h(X_1^i, X_2^i)$ takes on values 1, -1, and 0 with probabilities $p_P = \text{Prob}\{(X_1^i, X_2^i)\in P \mid H_0\}$, $p_N = \text{Prob}\{(X_1^i, X_2^i)\in N \mid H_0\}$, and $p_O = \text{Prob}\{(X_1^i, X_2^i)\in O \mid H_0\}$, respectively. It is obvious that the mean of T under H_0 is zero. Hence,

$$\text{Var}\{T \mid H_0\} = n[(1)^2 \text{Prob}\{(X_1^i, X_2^i)\in P \mid H_0\} + (-1)^2 \text{Prob}\{(X_1^i, X_2^i)\in N \mid H_0\}]$$

$$= n\left[\iint_{P\cup N} f(x)f(y)dxdy\right] \tag{A1}$$

and

$$E_\theta\{T \mid H_1\} = n \int_S dF_S(s) \left[\iint_P f(x-\theta s)f(y-\theta s)dxdy - \iint_N f(x-\theta s)f(y-\theta s)dxdy\right], \tag{A2}$$

where F_S is the distribution function of the signal S_t. Let f' and f'' be the first and second derivatives, respectively, of f. We assume that they exist almost everywhere. By applying L' Hôpital's rule in the intermediate stages, we get

$$\frac{d}{d\theta^2} E_\theta\{T \mid H_1\}\big|_{\theta=0} = n\left[\iint_P f'(x)f'(y)dxdy - \iint_N f'(x)f'(y)dxdy\right]. \tag{A3}$$

Thus, from Eq. (8), the efficacy $\mathcal{E}_{CC}$ is given by

$$\mathcal{E}_{CC} = \frac{[\iint_P f'(x)f'(y)dxdy - \iint_N f'(x)f'(y)dxdy]^2}{\iint_{P\cup N} f(x)f(y)dxdy}. \tag{A4}$$

For the special class of three-level CC detectors considered in Sec. III, the efficacy $\mathcal{E}_t$ can be obtained by applying to Eq. (A4) the definitions of P and N given in Eq. (13). This gives

$$\mathcal{E}_t = \frac{\left[2\int_0^\infty \int_t^\infty f'(x)f'(y)dxdy - 2\int_{-\infty}^0 \int_t^\infty f'(x)f'(y)dxdy\right]^2}{4\int_0^\infty \int_t^\infty f(x)f(y)\,dxdy}$$

$$= \frac{8[\int_0^\infty f(t)f'(x)\,dx]^2}{1 - 2\int_0^\infty F(t)f(x)dx}. \tag{A5}$$

APPENDIX B: PROOF OF THEOREM

Let P be the region which gives maximum efficacy, and let this be $\mathcal{E}'$. Due to the assumed symmetry of the regions it will be sufficient to consider P only. The efficacy $\mathcal{E}'$ of Eq. (11) may be written as

$$\mathcal{E}' = \frac{2[\iint_P l(x)l(y)f(x)f(y)dxdy]^2}{\iint_P f(x)f(y)dxdy}. \tag{B1}$$

Suppose P (thus N) is not defined as stated in the *Theorem*. Let us define A and B by

$$A = \iint_P l(x)l(y)f(x)f(y)dxdy, \tag{B2}$$

$$B = \iint_P f(x)f(y)dxdy. \tag{B3}$$

We will only consider the case where A is positive; for negative A we may interchange the roles of P and N, without loss of generality. Let $A/B = 2k$.

(i) Suppose there exists a region δ such that $\delta \subset P$ but $\delta \not\subset P_0$, where

$$P_0 = \{(x, y) : l(x)l(y) > k\}. \tag{B4}$$

Let

$$a = \iint_\delta l(x)l(y)f(x)f(y)dxdy \tag{B5}$$

and

$$b = \iint_\delta f(x)f(y)dxdy. \tag{B6}$$

Since $\delta \not\subset P_0$,

$$a/b \le k = A/2B. \tag{B7}$$

Let $\mathcal{E}_0'$ be the efficacy obtained when δ is deleted from P and assigned to O. Then the ratio of efficacies $\mathcal{E}'/\mathcal{E}_0'$ is given by

$$\frac{\mathcal{E}'}{\mathcal{E}_0'} = \frac{A^2}{B}\frac{B-b}{(A-a)^2} = [1-(b/B)]/[1-(a/A)]^2 < [1-(b/B)]/[1-(2a/A)] \tag{B8}$$

From (B7), $b/B \ge 2a/A$, and we know $b/B < 1$. Thus $\mathcal{E}'/\mathcal{E}_0' < 1$, which contradicts the hypothesis that $\mathcal{E}'$ is the maximum efficacy.

(ii) Suppose a region δ exists such that $\delta \not\subset P$ but $\delta \subset P_0$. Let $\mathcal{E}_0''$ be the efficacy obtained when δ is made part of P. Now (B7) becomes

$$a/b > k = A/2B. \tag{B9}$$

Then,

$$\frac{\mathcal{E}'}{\mathcal{E}_0''} = \frac{A^2}{B}\frac{B+b}{(A+a)^2} = [1+(b/B)]/[1+(a/A)]^2 < [1+(b/B)]/[1+(2a/A)]. \tag{B10}$$

From (B9), $2a/A > b/B > 0$, thus $\mathcal{E}'/\mathcal{E}_0'' < 1$.

APPENDIX C: PROOF OF EQUATION (18)

We assume that $0 < F(x) < 1$ for all finite x. Let

$$t(x) = t^*(x) + \gamma\epsilon(x), \tag{C1}$$

where $t^*(x)$ is the $t(x)$ which maximizes the detector efficacy, γ is an independent variable, and $\epsilon(x)$ is an arbitrary function of x defined for all x.

For an extremum of the efficacy $\mathcal{E}_t$, the following condition should be met:

$$\left.\frac{\partial \mathcal{E}_t}{\partial \gamma}\right|_{\gamma=0} = 0 \ . \quad \text{(C2)}$$

Thus we get

$$2\left[1 - 2\int_0^\infty F(t^*)f(x)dx\right] \int_0^\infty \epsilon(x) \frac{f'(t^*)}{f(t^*)} \frac{f'(x)}{f(x)} \times f(t^*)f(x)dx + \left[\int_0^\infty f(t^*)f'(x)dx\right] \times \int_0^\infty \epsilon(x)f(t^*)f(x)dx = 0 \ . \quad \text{(C3)}$$

Now we can see that $[1 - 2\int_0^\infty F(t^*)f(x)dx]$ is nonzero and finite valued. Hence, (C3) can be satisfied only if

$$\frac{f'(t^*)}{f(t^*)} \frac{f'(x)}{f(x)} = \frac{-\int_0^\infty f(t^*)f'(x)dx}{2[1 - 2\int_0^\infty F(t^*)f(x)dx]} = c \ , \quad \text{(C4)}$$

where c is some constant.

[1]S. S. Wolff, J. B. Thomas, and T. R. Williams, "The Polarity Coincidence Correlator: A Nonparametric Detection Device," IRE Trans. Inf. Theory IT-8, 5–9 (1962).

[2]S. S. Wolff and J. L. Gastwirth, "Robust Two-Input Correlators," J. Acoust. Soc. Am. 41, 1212–1219 (1967).

[3]J. B. Thomas and T. R. Williams, "On the Detection of Signals in Non-stationary Noise by Product Arrays," J. Acoust. Soc. Am. 31, 453–462 (1959).

[4]H. Ekre, "Polarity Coincidence Correlation Detection of a Weak Noise Source," IRE Trans. Inf. Theory IT-9, 18–23 (1963).

[5]J. J. Faran, Jr. and R. Hills, Jr., "Correlators for Signal Reception," Harvard U. Acoustic Res. Lab., Cambridge, MA, Tech. Memo 28 (1952).

[6]S. A. Kassam and J. B. Thomas, "Improved Nonparametric Coincidence Detectors," J. Franklin Inst. 303, 75–84 (1977).

[7]J. Capon, "Optimum Coincidence Procedures for Detecting Weak Signals in Noise," IRE Int. Conv. Rec., Pt. 4, 154–166 (1960).

[8]S. A. Kassam and J. B. Thomas, "Dead-zone limiter: An application of conditional tests in nonparametric detection," J. Acoust. Soc. Am. 60, 857–862 (1976).

[9]D. A. S. Fraser, *Nonparametric Methods in Statistics* (Wiley, New York, 1957).

[10]J. B. Thomas, "Nonparametric Detection," Proc. IEEE 58, 623–631 (1970).

[11]J. Capon, "On the Asymptotic Efficiency of Locally Optimum Detectors," IEEE Trans. Inf. Theory IT-7, 67–71 (1961).

[12]S. A. Kassam and J. B. Thomas, "Generalizations of the Sign Detector based on Conditional Tests," IEEE Trans. Commun. COM-24, 481–487 (1976).

[13]E. L. Lehmann, *Testing Statistical Hypotheses* (Wiley, New York, 1959).

[14]J. Hajek and Z. Sidak, *Theory of Rank Tests* (Academic, New York, 1967).

Reprinted with permission from *IEEE Trans. Inf. Theory* **IT-23**:708–721 (1977)

The Optimum Constant False Alarm Probability Detector for Relatively Coherent Multichannel Signals in Gaussian Noise of Unknown Power

PHILIP G. GRIEVE

I. INTRODUCTION

A. Problem Formulation

THIS PAPER is concerned with the detection of multichannel signals in Gaussian noise of unknown power. Specifically, the optimum constant false alarm probability (CFAP) detector for relatively coherent multichannel signals, corrupted by additive channel-correlated Gaussian noise of unknown power, is developed. The received signals are corrupted by the presence of additive wide-sense stationary Gaussian noise of unknown power. Additionally, each signal contains a common random phase angle (modeling lack of knowledge of the carrier's initial phase).

Typical situations that fit to this general model are diversity communication through a noisy medium and target detection by radar or sonar arrays in the presence of clutter or jamming. In these examples, a common set of noise sources cause a noise random process to be received at the receiving elements of an antenna array. Each element's noise output consists of a linear combination of time-delayed replicas of the common noise sources. The delays are proportional to the path lengths between noise sources and array elements. To illustrate this point, the reception of a single noise source by a two-element array as shown in Fig. 1 can be considered. Since only relative time delays are important, only the relative path length differences need be calculated. Assuming that the noise source is sufficiently distant that the paths of propagation are parallel, the noises received at the right and left elements are $n(t)$ and $n(t-a)$, respectively, where $a = v^{-1}d \sin \gamma$, and v is the velocity of propagation. This example is important because it indicates that at any fixed time instant, the set of noise voltages at the array outputs is a set of time samples obtained from a narrow-band Gaussian random process (in the example, there are two time samples, $n(t)$ and $n(t-a)$). This is typical of array outputs because normally the element outputs will be sums of delayed replicas of many narrow-band noise sources.

Manuscript received May 18, 1976; revised February 17, 1977.
The author is with the Hughes Aircraft Company, Culver City, CA 90230.

B. Mathematical Model

A mathematical description of the system model is given here. The output of each of M channels is assumed to be time-sampled to yield K real numbers. The entire set of system outputs thus may be written as

$$\{A_m(t_k)\cos(\omega_c t_k + \psi_m(t_k) + \theta) + V_m(t_k)\cos(\omega_c t_k + \phi_m(t_k))\},\quad k = 1,\cdots,K,\ m = 1,\cdots,M, \tag{1}$$

where in the mth channel;

- $A_m(t)$ known signal amplitude function,
- $\psi_m(t)$ known signal phase function,
- ω_c carrier frequency in rad/s,
- θ random phase angle, common to all channels,
- $V_m(t)$ amplitude of a narrow-band Gaussian random noise process,
- $\phi_m(t)$ phase of a narrow-band Gaussian random noise process.

In this case, the narrow-band waveforms may be resolved into their respective cosine and sine components, without loss of generality, in the usual manner for carrier systems. It is convenient to define M-dimensional vectors of these components as follows:

$$\mathbf{S}_{ck}(\theta) = [A_1(t_k)\cos(\psi_1(t_k)+\theta), \cdots, A_M(t_k)\cos(\psi_M(t_k)+\theta)]^t,$$

$$\mathbf{S}_{sk}(\theta) = [A_1(t_k)\sin(\psi_1(t_k)+\theta), \cdots, A_M(t_k)\sin(\psi_M(t_k)+\theta)]^t,$$

$$\mathbf{N}_{ck} = [V_1(t_k)\cos(\phi_1(t_k)), \cdots, V_M(t_k)\cos(\phi_M(t_k))]^t,$$

$$\mathbf{N}_{sk} = [V_1(t_k)\sin(\phi_1(t_k)), \cdots, V_M(t_k)\sin(\phi_M(t_k))]^t,$$

where the superscript t denotes transpose, and then t

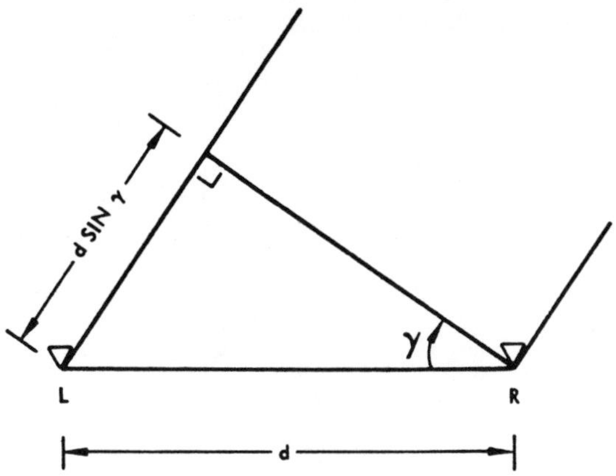

Fig. 1. Two-element array

define the $2M$-dimensional partitioned vectors

$$S_k(\theta) = [S_{ck}^t(\theta) | S_{sk}^t(\theta)]^t$$
$$N_k = [N_{ck}^t | N_{sk}^t]^t. \qquad (2)$$

Physically, $S_k(\theta)$ represents the signal waveform with a phase shift of θ radians.

The set of real numbers comprising the system output of (1) may now be written in terms of the $2M$-dimensional vectors as

$$X_k = S_k(\theta) + N_k, \qquad k = 1, \cdots, K.$$

The vector X_k is the set of sine and cosine array output time-samples obtained at the kth time instant.

The detection problem can now be written as a statistical hypothesis testing problem in which the null hypothesis

$$H: X_k = N_k, \qquad k = 1, \cdots, K$$

is tested versus the alternative

$$K: X_k = N_k + S_k(\theta), \qquad k = 1, \cdots, K.$$

It is assumed that the time-sampling intervals are sufficiently large (greater than the reciprocal of the noise bandwidth) so that the Gaussian noise vectors $\{N_k\}$ are uncorrelated and thus independent. It is further assumed that these vectors are zero mean and identically distributed with covariance matrix $E\{N_k^t N_k\} = \sigma^2 \Lambda$, where σ^2 is unknown and Λ is a known covariance matrix. The unknown parameter σ^2 is proportional to the noise power. As mentioned previously, the array noise outputs, sampled at a fixed time, are equivalent to K samples from a narrowband Gaussian random process. Hence, Λ can be assumed to have certain special properties that are discussed in Section III. The phase angle θ is assumed to be uniformly distributed over 0 to 2π, and independent of the noise distribution. The random phase angle models the lack of knowledge of the starting phase of the carrier. The construction of the optimum detector of H versus K will now be discussed.

C. Statistical Background

The detector to be developed is "optimum" in the sense that it yields the largest possible probability of detection for a given (fixed in advance) constant false alarm probability. If the noise power were known in advance, the method of obtaining the optimum detector would be a straightforward application of the well-known Neyman–Pearson likelihood ratio test theory, since the null hypothesis would contain a single probability distribution. When the noise power is unknown, however, the null hypothesis contains an uncountable number of possible distributions. A hypothesis of this type is said to be "composite." However, an optimum detector can still be found if candidate detectors are required to provide a constant false alarm probability. This constraint is often mandatory in practical systems.

The technique used to find the optimum CFAP detector consists of applying an existing theory from the statistical literature. This theory is most often called the theory of "most powerful similar tests" and has its roots in the early work of Neyman and Pearson. Applications of the theory to communication-theoretic problems is not new (cf. [3] and [17]), but apparently is not widely known. For this reason, the pertinent theory and methods are briefly described in Section II.

D. Summary of Results

In Section III, the technique of Section II is used to obtain the optimum detector structure. The detector structure is given in Fig. 2. The essential feature of the optimum detector is the comparison of a ratio statistic with a threshold. The numerator of the statistic is the sum of squares of the outputs of two matched filters: one matched to the signal and the other to the signal phase rotated by 90 degrees. The denominator is the sum of squares of the whitened observations. Thus the ratio tends to normalize out the unknown noise power.

The performance of the optimum detector is computed in Section IV. The detector's behavior is found to be governed by a probability distribution known as "Fisher's distribution C" in the statistical literature. The closed-form false alarm and detection probabilities obtained are used to generate typical performance curves of probability of detection versus signal-to-noise ratio for various false alarm probabilities and sample sizes (Figs. (3)–(6)). In the limit, as the number of data samples approaches infinity, both the structure and the performance of the optimum CFAP detector are found to converge to that of the detector which is optimum when the noise power is known. Fig. 7 illustrates the loss in performance (additional signal-to-noise ratio required) at a detection probability equal to 1/2 of the optimum CFAP detector for the unknown noise power case relative to the optimum detector for the known power case. The figure shows that for the range of false alarm probability (10^{-4} to 10^{-10}) considered, a data sample size of 40 or more will yield performance within 1 dB of the known noise power detector.

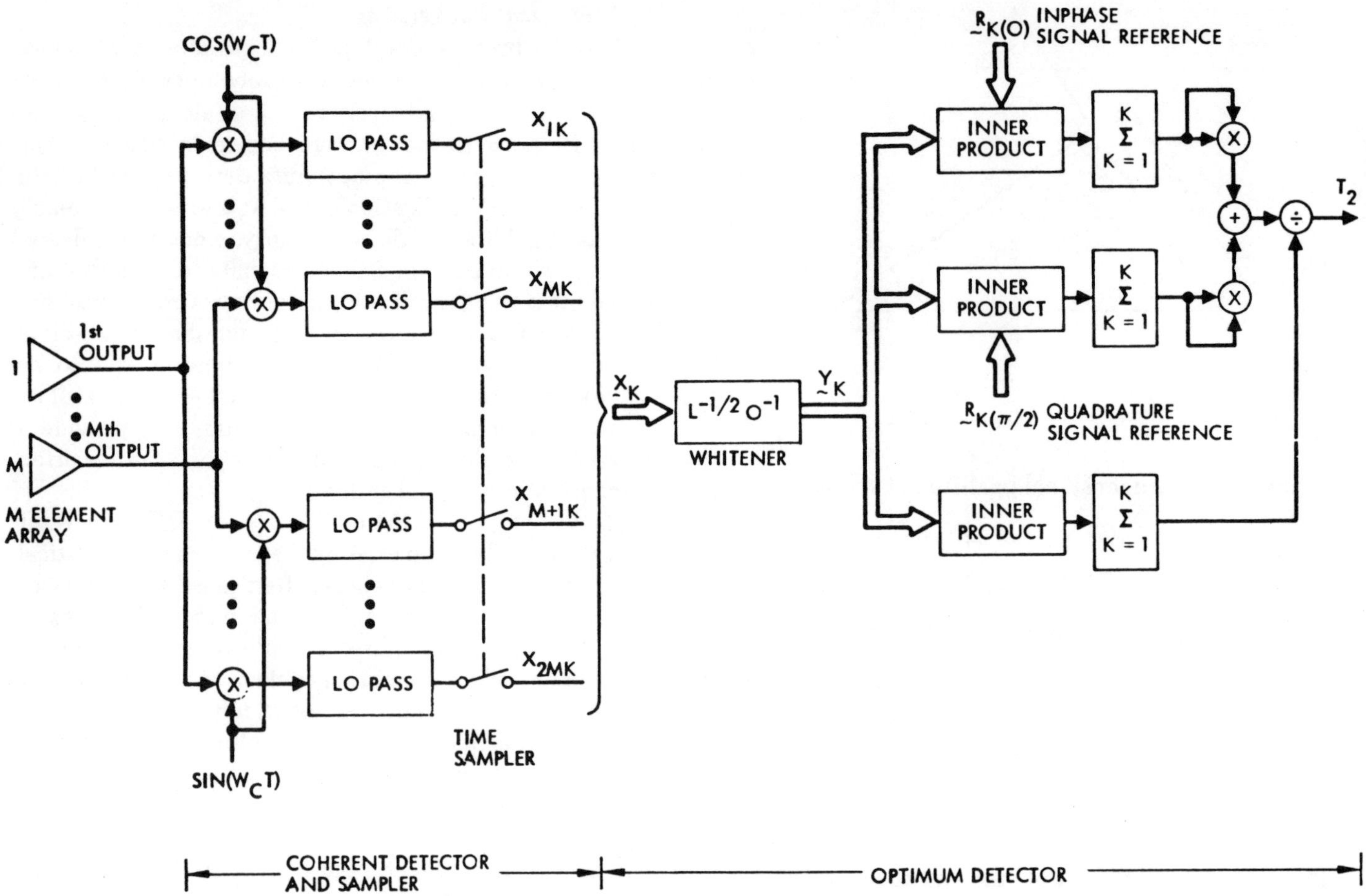

Fig. 2. Optimum CFAP detector.

In the final section, the techniques developed are applied to three illustrative examples. In the first, the optimum CFAP detector is found for a Gaussian noise-jammed signal received with a two-element array. The performance is computed and found to be a function of the angular location of the jammer. In the second example, the optimum CFAP detector is found for a single-channel signal in white Gaussian noise of unknown power. The detector compares the squared magnitude of a filter matched to the signal with a threshold proportional to the total energy in the data samples. The final example extends the single channel problem to the case when the noise is colored. In this case, the structure is found to be the same as when the noise was white, except that now a filter matched to the whitened signal is used and the energy threshold is taken to be proportional to the energy in the whitened observations.

II. Statistical Methods

A. Detectors for Composite Null Hypotheses

Two approaches generally have been taken in dealing with the problem of finding a CFAP detector when the null hypothesis distribution has unknown quantities, and a straightforward application of the usual Neyman and Pearson theory is not possible. The first approach is used when the null hypothesis distribution depends on an unknown parameter. A statistic, which is an accurate estimator of the unknown parameter, is constructed and the estimate used in place of the parameter in the likelihood ratio test. If the maximum-likelihood estimate is used, the result is called a generalized likelihood ratio test [14, p. 92].

A second approach is used in situations where the lack of knowledge of the null hypothesis distribution is not easily parameterized. For example, the null hypothesis may state only that the probability density function is continuous and with even symmetry. This type of hypothesis is termed "nonparametric" because parameterization in the usual sense is not possible. Detectors developed for this case are often analogs of appropriate existing hypothesis tests in the statistical literature. In general, these detectors extensively reduce the data, via nonlinear transformations, to statistics that are insensitive to the unknown quantities of the null hypothesis. The most common data reductions are to signs, to magnitudes, to orders, and to ranks. Detectors based on these quantities are called "nonparametric detectors" after the type of null hypothesis for which they were developed [2]–[8]. An extensive survey of nonparametric detectors is given in Carlyle [9, p. 293].

A systematic and unified approach to CFAP detector construction can be taken if it is assumed that the most desirable or "optimum" CFAP detector is the one with the

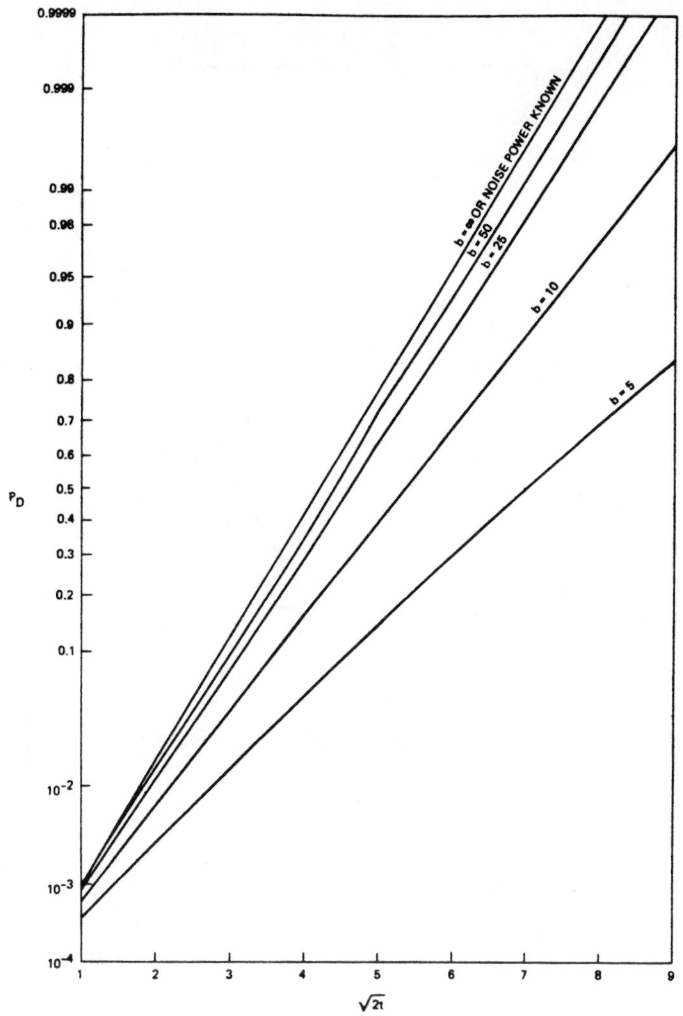

Fig. 3. $P_{FA} = 10^{-4}$; detection probabilities for optimum CFAP detectors.

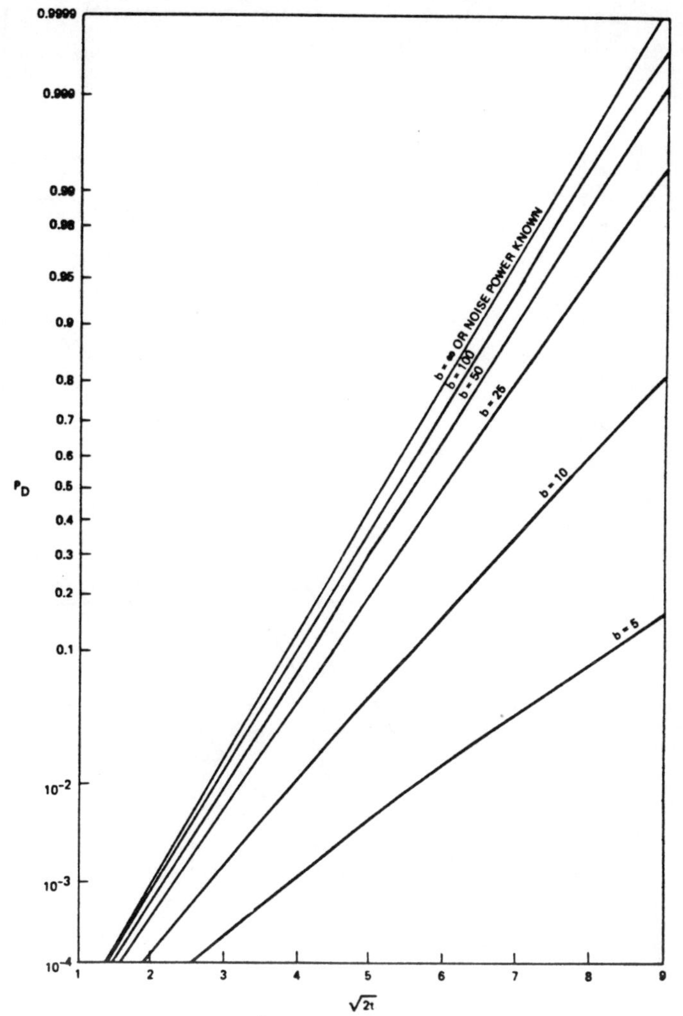

Fig. 4. $P_{FA} = 10^{-6}$; detection probabilities for optimum CFAP detectors.

largest possible detection probability for a given (constant) false alarm probability. Neither of the two previous approaches will necessarily yield the optimum detector. This third approach is identical to that taken in the optimization of detectors when the null hypothesis class of distributions is "simple" (contains a single distribution). The next section contains a discussion of the statistical methods which can be used to obtain the optimum CFAP detector.

Finally it should be noted that for some detection systems, the requirement can be relaxed for a constant false alarm probability; that is, it could be required only that the false alarm probability be bounded above by a known value, say α. Statistically speaking, a detector of this type is said to be of "level α." Since this constraint is weaker than that of CFAP, the class of possible detectors is larger (and includes CFAP detectors, of course). Statistical theory exists for obtaining most powerful "level α" detectors. This theory is essentially an extension of the Neyman–Pearson theory of composite hypotheses. Unfortunately, in the process of extension, the result becomes much less explicit, depending on an undetermined probability distribution (called a "least favorable distribution"). In specific problems, the determination of this distribution involves considerable difficulty and is the subject of ongoing statistical investigation. Since these "level α" tests are outside the scope of this work, they will not be considered further. The interested reader is referred to Lehmann and Stein [17] and Lehmann [12, p. 90].

B. Optimum CFAP Detector Theory

This section is concerned with the statistical theory that shows how to construct the optimum CFAP detector for a specified problem. First, the equivalence between CFAP detectors and similar hypothesis tests of the statistical literature is established. Next, it is shown how similar test theory can be used to characterize the class of CFAP detectors for the problem. Finally, the method by which similar test theory enables the selection of that detector in the CFAP class exhibiting greatest detection probability (i.e., the optimum one) is demonstrated.

The detection problems of interest here have a composite (rather than simple) null hypothesis class of distributions. Hence, the class contains a multiplicity of distributions, any of which could possibly describe the state

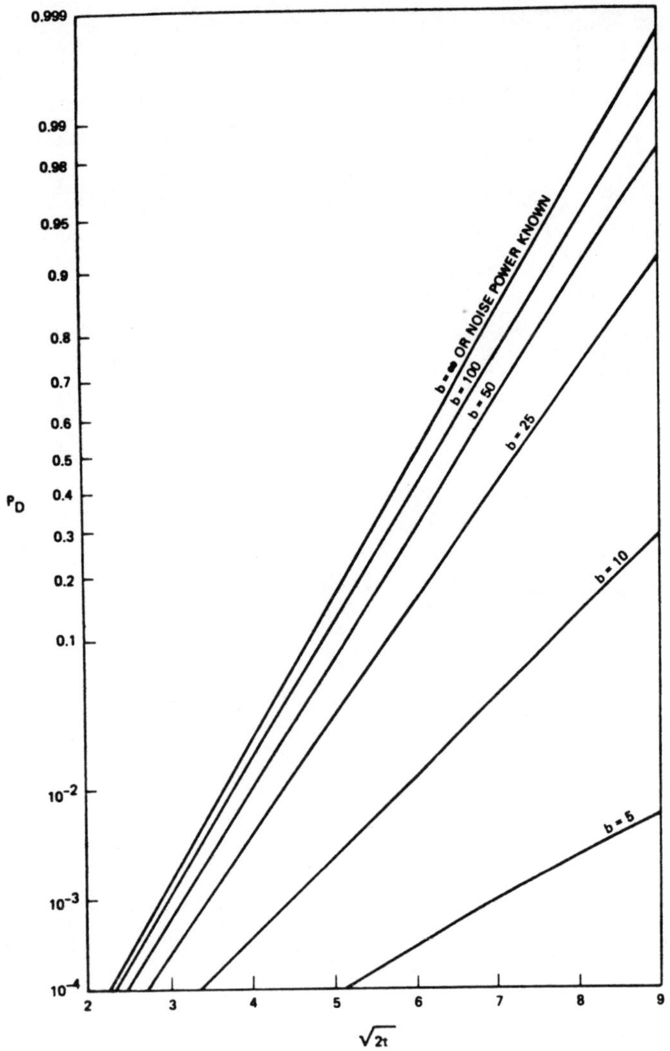

Fig. 5. $P_{FA} = 10^{-8}$; detection probabilities for optimum CFAP detectors.

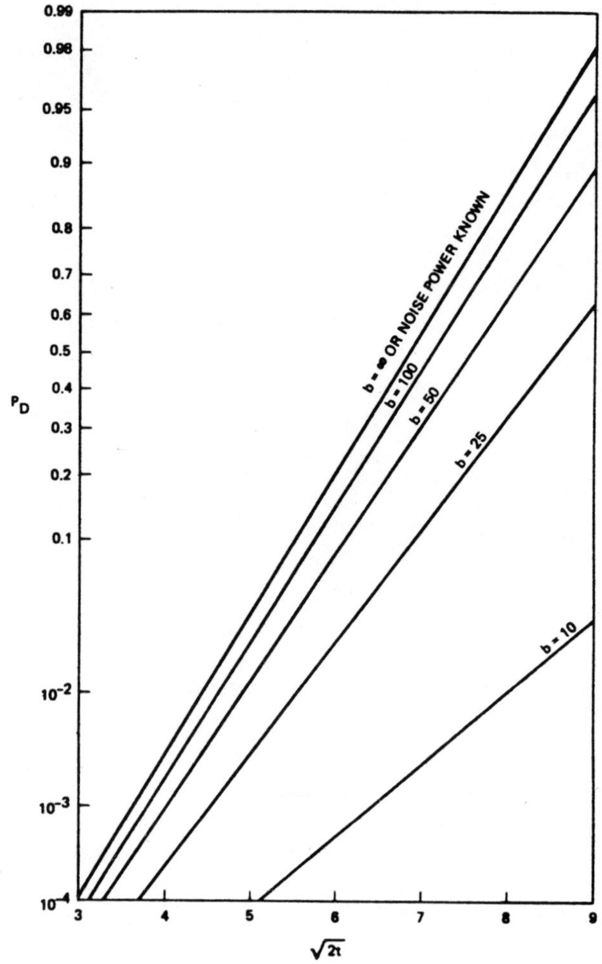

Fig. 6. $P_{FA} = 10^{-10}$; detection probabilities for optimum CFAP detecto

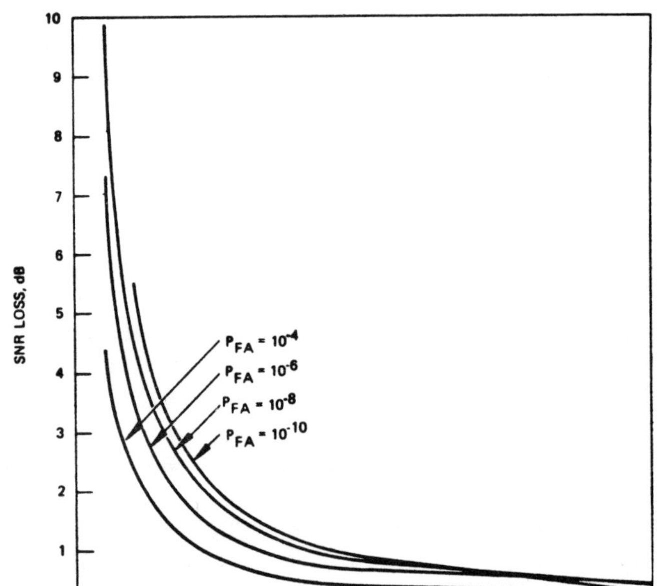

Fig. 7. Performance loss of optimum detector.

of affairs when the signal is absent. For these problems, it is convenient (and conventional) to adopt the following notation.

- The set of all probability distributions belonging to the composite null hypothesis class, denoted H, is indexed by the parameter θ, and the set of all θ is denoted by Ω.
- A detector (statistical hypothesis testing procedure) is characterized by its "test function"

  $$\phi(x) = \text{Prob}\{\text{reject null hypothesis} \mid X = x\},$$

  where X is a random variable and x a particular outcome of X. If $\phi(x)$ is unspecified except that $0 \leq \phi(x) \leq 1$, the testing procedure is said to be "randomized."

With this notation, the false alarm probability of a detector characterized by test function $\phi(x)$ is

$$P_{FA}(\theta) = \int \phi(x)\, dP_\theta(x) = E_\theta\{\phi(X)\}$$

which, in general, depends on which $P_\theta \in H$ is the true state of affairs. A CFAP detector exhibits a false alarm probability that is independent of which P_θ is true; that is, a "level α" CFAP detector satisfies

$$P_{FA} = E_\theta\{\phi(X)\} = \int \phi(x)\, dP_\theta(x) = \alpha, \qquad \text{all } \theta \in \Omega.$$

In the statistical literature, detectors are called hypothesis tests, and a "similar test" [12, p. 130] is identical to a CFAP detector as defined above (previously noted in [3]); that is, both similar tests and CFAP detectors have a false alarm probability that is the same constant value for every member distribution of the null hypothesis class.[1] The usefulness in noting the equivalence between similar tests and CFAP detectors is that a large body of statistical theory concerning similar tests exists. The concept of similarity was first introduced by Neyman and Pearson [10] in their work with composite null hypotheses. By requiring that only similar tests of a particular composite hypothesis be considered, they found that in some cases the composite problem could be reduced to an equivalent simple hypothesis problem. For them, however, the restriction to similar tests was an artificial requirement whereas, in the detection problems being considered here, it is not.

The eventual purpose of this discussion is to explain how the optimum (greatest detection probability) CFAP detector or similar test can be found. The method first requires the characterization of the totality of similar tests for the specific problem (then the best one of them can be selected). Lehmann and Scheffe [11] first showed that this could be done under certain conditions. Their work has been summarized in [12, p. 130], and will now be described after several definitions are made. In these definitions, the parameter set ω is arbitrary but will usually be interpreted as being equal to Ω, the parameter set of the null hypothesis class.

Definitions: A statistic $T(X)$ is said to be *sufficient* for the family $\mathcal{P}^X = \{P_\theta(x) | \theta \in \omega\}$ (if $\omega = \Omega$, then $\mathcal{P}^X = H$) if the conditional distribution of X, given $T(X) = t$, is independent of θ. An equivalent statement is that a factorization exists of the probability density function $p_\theta(x) = g_\theta(T(x))h(x)$, where the first factor may depend on θ, but depends on x only through $T(x)$, while the second factor is independent of θ.

Definition: A statement is said to *hold almost everywhere* (a.e.) $\mathcal{P}$ if it holds except on a set N with $P_\theta\{N\} = 0$, all $\theta \in \omega$.

Definition: A test function $\phi(x)$ is said to have *Neyman structure* with respect to the sufficient statistic $T(x)$ for the family $\mathcal{P}^X = \{P_\theta(x) | \theta \in \omega\}$, if

$$E\{\phi(X)/T(X) = t\} = \alpha \text{ almost every } \mathcal{P}^T,$$

where $\mathcal{P}^T$ denotes the induced family of distributions of $T(X)$ as θ ranges over ω. A Neyman structure test function is one whose conditional average is a constant function of the given value t of $T(x)$. Since the unconditional expectation of the Neyman structure test may be written

$$E_\theta\{\phi(X)\} = E_\theta\{E\{\phi(X)/T(X) = t\}\} = \alpha, \qquad \text{all } \theta \in \omega,$$

then it is clear that the set of all Neyman structure tests is a subset of the set of all similar tests. The Neyman structure test is characterized by its conditional false alarm probability being α, irrespective of the given value of $T(X)$. Since the distribution of X, given $T(X)$, is independent of θ, then the conditional distribution is simple and the conditional test can be optimized, for every value of $T(X)$ held fixed, in the usual manner. The resulting test is the Neyman structure test with greatest detection probability. However, it is not always true that every similar test has Neyman structure—the following definition allows the case when this is true to be characterized.

Definition: A family of probability distributions $\mathcal{P}$ is said to be *complete* if, for any measurable function $f(x)$,

$$E_\theta\{f(X)\} = \int f(x)\, dP_\theta(x) = 0, \qquad \text{all } P_\theta \in \mathcal{P}$$

implies

$$f(x) = 0 \text{ almost every } \mathcal{P}.$$

A weaker condition on the family, bounded completeness, requires that the above hold only for bounded functions; that is, for those that satisfy

$$|f(x)| \le M, \qquad \text{all } x$$

where M is any finite real number. Note that completeness implies bounded completeness, but not vice versa. The following important theorem shows the conditions under which every similar test has Neyman structure.

Theorem 1: A necessary and sufficient condition for all similar tests of $\mathcal{P}^X$ to have Neyman structure with respect to $T(x)$ is that $\mathcal{P}^T$ be bounded complete.

Proof: cf. [12, p. 134].

This theorem indicates that if the family of distributions of the sufficient statistic for the composite null hypothesis is bounded complete, then all similar tests have Neyman structure. If the sufficient statistic is not bounded complete, then similar tests will occur that do not have Neyman structure (see [11] for examples), and the optimization technique described above cannot succeed since it applies only to tests with Neyman structure. Reference [17] considers this situation in more detail; however, the case is beyond the scope of this report. Hence, in the following discussion, the property of bounded completeness is assumed—and the specific problem dealt with also exhibits this property.

By the assumption that the sufficient statistic has the property of bounded completeness, all possible similar test

[1] The nomenclature "similar" test stems from the following observation. If $\phi(x)$ describes a nonrandomized similar test, then this function takes on only the values zero or one. The region of the sample space where the null hypothesis is rejected (where one decides target present) may then be written

$$A = \{x | \phi(x) = 1\}.$$

The region A is then said to be "similar to the sample space" in that both regions have expected values that are independent of the actual value of $\theta \in \Omega$ (i.e., the expected value of the entire sample space is always one).

functions are characterized since they must have Neyman structure. Of these, it is now desired to find which similar test function is the "best" in the sense that it is most powerful (largest probability of detection) against a particular simple alternative. This includes the situation in which a composite alternative is actually present, and it is desired to see if a uniformly most powerful (UMP) similar test exists (that is, one test most powerful against every member of the alternative). The method used in this case is to assume a simple alternative formed from an arbitrary member of the alternative class, and then to observe whether the structure of the MP similar test depends on the chosen alternative. The solution of the optimum problem depends on the property of Neyman structure tests having constant false alarm probability on each locus of points of the sample space where $T(x) = t$, *i.e.*, on, $\{x \mid T(x) = t\}$. Since the distribution of X, given $T(X) = t$, is independent of θ under the null hypothesis, the composite null hypothesis becomes simple when conditioned on $T(X) = t$. Holding t fixed, the usual Neyman–Pearson result for simple hypotheses may then be applied. If this is then done for every value of t, the resulting likelihood ratio test is the optimum similar test of the composite null hypothesis versus the given simple alternative. The test structure is, in general, a function of the value $T(x) = t$ held fixed. These results are summarized as follows.

Theorem 2: Let $p_H(x \mid t)$ and $p_K(x \mid t)$ be the conditional probability densities of X, given $T(x) = t(T(X)$ is the bounded complete sufficient statistic for $H)$ under the null hypothesis and alternative, respectively. Then with

$$L_t(x) = \frac{p_K(x \mid t)}{p_H(x \mid t)},$$

the most powerful similar test of H versus K with $P_{FA} = \alpha$ is given by

$$\phi(x) = \begin{cases} 1, & L_t(x) > C(t,\alpha) \\ a(t,\alpha), & L_t(x) = C(t,\alpha) \\ 0, & L_t(x) < C(t,\alpha), \end{cases}$$

where $C(t,\alpha)$ and $a(t,\alpha)$ are chosen to satisfy $E\{\phi(X)/T(X) = t\} = \alpha$.

Proof: cf. [13].

The optimum similar test structure is a likelihood ratio test just as is the usual optimum test for the simple null hypothesis case. However, the test is different in that it depends on the function $T(x)$ of the data as well as on the data x themselves. The previous theorem is inconvenient in that the likelihood ratio is formed from the conditional probability densities. These can sometimes be formidable to obtain from the given unconditional ones. The following theorem shows that, under certain assumptions (valid in our examples), the test structure may be deduced from the given unconditional probability densities of the original data without actually having to find the conditional densities. This theorem uses the observation that a sufficient statistic can often be viewed as a single coordinate in an N-dimensional coordinate transformation of the sample space.

Theorem 3: Let $T(x)$ be the sufficient statistic for H and assume that $T(x)$ is one-dimensional and can be considered as an entry of $y = d(x)$, an N-dimensional coordinate (one-to-one) transformation of the sample space. Then the most powerful similar test of H versus a simple alternative K with $P_{FA} = \alpha$ is given by

$$\phi(x) = \begin{cases} 1, & L_t'(x) > C'(t,\alpha) \\ a'(t,\alpha), & L_t'(x) = C'(t,\alpha) \\ 0, & L_t'(x) < C'(t,\alpha), \end{cases}$$

where

$$L_t'(x) = \left. \frac{p_K(x)}{h(x)} \right|_{T(x)=t},$$

$p_K(x)$ unconditional density of X under K,

$h(x)$ factor of $p_H(x) = g_\theta(T(x))h(x)$ independent of θ (see the previous definition of sufficient statistic).

Proof: See Appendix C.

Example: As an example of application of the theory of this section, consider

$$p_H(x) = \exp\left\{\frac{x_1^2 + x_2^2}{2\theta}\right\}(2\pi\theta)^{-1}, \qquad \theta \text{ unspecified},$$

$$p_K(x) = \exp\left\{\frac{(x_1 - m)^2 + (x_2 - m)^2}{2\sigma^2}\right\}(2\pi\sigma^2)^{-1},$$

$$\sigma, m > 0 \text{ unspecified},$$

$T(x) = x_1^2 + x_2^2$ (sufficient for θ under H, note: $T(x)$ is the squared radius under a polar coordinate transformation)

$$g_\theta(x) = P_H(x), \qquad h(x) = 1.$$

Then, Theorems 2 and 3 show that the MP similar test of H versus K' (a simple alternative formed from K by fixing $m = m'$, $\sigma = \sigma'$) compares

$$\exp\left\{-\frac{1}{2\sigma'^2}(t - 2m'(x_1 + x_2) + m'^2)\right\}(2\pi\sigma'^2)^{-1}\Big|_{x_1^2+x_2^2=t} \gtrless C(t,\alpha)$$

or, equivalently, the test compares

$$x_1 + x_2\big|_{x_1^2+x_2^2=t} \gtrless C_1(t,m',\sigma',\alpha).$$

The structure of the test becomes apparent when the coordinate transformation

$$\cos(u) = x_1/\sqrt{t}$$
$$\sin(u) = x_2/\sqrt{t}$$

is made. The test then becomes

$$\cos(u) + \sin(u) \gtrless C_2(t,m',\sigma',\alpha).$$

Since the test structure is not a function of σ' or $m' > 0$ and since C_2 is chosen under H and so cannot depend on m' or σ', the test is thus the uniformly most powerful similar test of H versus the composite alternative K.

III. Optimum Detector Construction

The detection problem developed in Section I is to find the optimum CFAP detector of

$$H: \boldsymbol{X}_k = \boldsymbol{N}_k, \qquad k = 1, \cdots, K$$

versus

$$K: \boldsymbol{X}_k = \boldsymbol{N}_k + \boldsymbol{S}_k(\theta), \qquad k = 1, \cdots, K,$$

where $\boldsymbol{N}_k$ and $\boldsymbol{S}_k(\theta)$ are defined by (2), $\{\boldsymbol{N}_k\}$ are independent, identically distributed zero mean $2M$-dimensional Gaussian vectors with common covariance matrix $E\{\boldsymbol{N}_k^t \boldsymbol{N}_k\} = \sigma^2\Lambda$, Λ is a known covariance matrix for time-samples from a narrow-band Gaussian random process, and σ^2 is an unknown parameter, proportional to the noise power.

In Section II, it was indicated that the first step of the construction is to obtain the sufficient statistic for H. The probability densities of H are written

$$p_H(\boldsymbol{X}_1, \cdots, \boldsymbol{X}_K) = \prod_{k=1}^{K} [(2\pi\sigma^2)^{2M} \det \Lambda]^{-1/2} \cdot \exp\{-(2\sigma^2)^{-1}\boldsymbol{X}_k^t\Lambda^{-1}\boldsymbol{X}_k\}, \qquad \sigma^2 > 0$$

and the sufficient statistic for H is

$$T = T(\boldsymbol{X}_1, \cdots, \boldsymbol{X}_k) = \sum_{k=1}^{K} \boldsymbol{X}_k^t\Lambda^{-1}\boldsymbol{X}_k. \tag{3}$$

When H is true, T can be shown to be distributed as chi-square with $2MK$ degrees of freedom (Section IV). The family of chi-square distributions for T, obtained as σ^2 ranges over all nonzero values, is well-known to be complete [12, p. 132]. Since the family is complete, it is also bounded complete. Thus, the MP similar test of H versus a simple alternative (obtained by letting $\sigma = \sigma_1$ in K) can be found via Theorem 2. Theorem 3 indicates that the MP test is

$$(2\pi)^{-1} \int_0^{2\pi} \prod_{k=1}^{K} \sigma_1^{-1} \exp\{-(2\sigma_1^2)^{-1}(\boldsymbol{X}_k - \boldsymbol{S}_k(\theta))^t\Lambda^{-1}(\boldsymbol{X}_k - \boldsymbol{S}_k(\theta))\}\, d\theta|_{T=t} \gtrless C(t,\alpha). \tag{4}$$

Expanding the exponent and neglecting positive constants yields the equivalent form for (4)

$$\int_0^{2\pi} \prod_{k=1}^{K} \exp\{-(2\sigma_1^2)^{-1}(\boldsymbol{X}_k^t\Lambda^{-1}\boldsymbol{X}_k - 2\boldsymbol{X}_k^t\Lambda^{-1}\boldsymbol{S}_k(\theta) + \boldsymbol{S}_k^t(\theta)\Lambda^{-1}\boldsymbol{S}_k(\theta))\}\, d\theta|_{T=t} \gtrless C_1(t,\alpha). \tag{5}$$

The leftmost term in the exponent of (5) is proportional to t and thus is a constant. The rightmost term may be simplified by using the identity

$$\boldsymbol{S}_k(\theta) = \boldsymbol{S}_k(0)\cos\theta + \boldsymbol{S}_k(\pi/2)\sin\theta \tag{6}$$

so that

$$\boldsymbol{S}_k^t(\theta)\Lambda^{-1}\boldsymbol{S}_k(\theta) = \boldsymbol{S}_k^t(0)\Lambda^{-1}\boldsymbol{S}_k(0)\cos^2\theta + \boldsymbol{S}_k^t(\pi/2)\Lambda^{-1}\boldsymbol{S}_k(\pi/2)\sin^2\theta + 2\boldsymbol{S}_k^t(0)\Lambda^{-1}\boldsymbol{S}_k(\pi/2)\cos\theta\sin\theta. \tag{7}$$

In Appendix A, it is shown that the following identities hold because Λ is the covariance matrix of a narrow-band random process

$$\boldsymbol{S}_k^t(0)\Lambda^{-1}\boldsymbol{S}_k(0) = \boldsymbol{S}_k^t(\pi/2)\Lambda^{-1}\boldsymbol{S}_k(\pi/2)$$
$$\boldsymbol{S}_k^t(0)\Lambda^{-1}\boldsymbol{S}_k(\pi/2) = 0; \tag{8}$$

hence (7) simplifies to $\boldsymbol{S}_k^t(0)\Lambda^{-1}\boldsymbol{S}_k(0)$ which is a constant (independent of θ) and proportional to the signal energy. From (6), the middle term may be expanded so that the test of (5) now becomes (neglecting the left and right terms of the exponent)

$$(2\pi)^{-1} \int_0^{2\pi} \exp\left\{\sigma_1^{-2} \sum_{k=1}^{K} \boldsymbol{X}_k\Lambda^{-1}(\boldsymbol{S}_k(0)\cos\theta + \boldsymbol{S}_k(\pi/2)\sin\theta)\right\} d\theta|_{T=t} \gtrless C_2(t,\alpha). \tag{9}$$

The integral of (9) is well-known [14, p. 339] to equal $I_0((\sigma_1^{-4}T_1)^{1/2})$ where $I_0(\cdot)$ is the modified Bessel function and

$$T_1 = \left(\sum_{k=1}^{K} \boldsymbol{X}_k^t\Lambda^{-1}\boldsymbol{S}_k(0)|_{T=t}\right)^2 + \left(\sum_{k=1}^{K} \boldsymbol{X}_k^t\Lambda^{-1}\boldsymbol{S}_k(\pi/2)|_{T=t}\right)^2. \tag{10}$$

The modified Bessel function is monotonic-increasing in its argument so that the test of (9) simplifies to

$$T_1 \gtrless C_3(t,\alpha). \tag{11}$$

To better interpret T_1, it is convenient to replace $\boldsymbol{X}_k$ and $\boldsymbol{S}_k(\theta)$ by their "whitened" equivalents

$$\boldsymbol{Y}_k = L^{-1/2}O\boldsymbol{X}_k$$
$$\boldsymbol{R}_k(\theta) = L^{-1/2}O\boldsymbol{S}_k(\theta),$$

where O is the orthogonal matrix ($O^t = O^{-1}$) which brings Λ to the diagonal form $L(O\Lambda O^{-1} = L)$, and $L^{-1/2}$ is the inverse of the square root of L ($L^{-1/2}L^{-1/2} = L$). With this substitution, T_1 of (10) becomes

$$T_1 = \left(\sum_{k=1}^{K} \boldsymbol{Y}_k^t\boldsymbol{R}_k(0)|_{T=t}\right)^2 + \left(\sum_{k=1}^{K} \boldsymbol{Y}_k^t\boldsymbol{R}_k(\pi/2)|_{T=t}\right)^2, \tag{12}$$

and T of (3) becomes

$$T = \sum_{k=1}^{K} \boldsymbol{Y}_k^t\boldsymbol{Y}_k.$$

The implication of T being held fixed in (12) becomes clear if $2MK$-dimensional direction cosines are defined by $D_{mk} = Y_{mk}T^{-1/2}$, where Y_{mk} is the mth entry of $\boldsymbol{Y}_k$. Then let $\boldsymbol{D}_k = [Y_{1k}T^{-1/2}, \cdots, Y_{2Mk}T^{-1/2}]^t$, and (12) becomes

$$T_1 = t\left[\left(\sum_{k=1}^{K} \boldsymbol{D}_k^t\boldsymbol{R}_k(0)\right)^2 + \left(\sum_{k=1}^{K} \boldsymbol{D}_k^t\boldsymbol{R}_k(\pi/2)\right)^2\right]. \tag{13}$$

Now inserting (13) into (11) and neglecting the positive constant t, we find that (11) becomes

$$T_2 = \left(\sum_{k=1}^{K} Y_k^t Y_k\right)^{-1}\left[\left(\sum_{k=1}^{K} Y_k^t R_k(0)\right)^2 + \left(\sum_{k=1}^{K} Y_k R_k(\pi/2)\right)^2\right] \gtrless C_\alpha. \quad (14)$$

Equation (14) describes the desired optimum detector. Note that the detector's structure does not depend on the value of $\sigma = \sigma_1$ held fixed in K, and hence the test is the UMP-CFAP test of H versus K, as desired. Further, because T_2 is a ratio, the statistic, and hence its distribution, do not depend on the parameter σ. Under H, T_2 is thus statistically independent of T, the boundedly complete sufficient statistic for σ [12, p. 162]. Thus, for this particular problem, the MP-similar test is not conditional on T.

The statistic T_2 has a straightforward physical interpretation. The inner products of the whitened data vectors $\{Y_k\}$ with the whitened signal reference vectors $\{R_k(0)\}$ and $\{R_k(\pi/2)\}$ are evaluated by the righthand term of (14). These inner products can be thought of as spatial matched filters. Each reference vector pair is perpendicular, i.e.,

$$R_k^t(0)R_k(\pi/2) = S_k^t(0)\Lambda^{-1}S_k(\pi/2) = 0, \quad (15)$$

by the results of Appendix A, hence the sum of squares of the inner products removes the carrier phase θ. (To see this, substitute $Y_k = S_k(\theta)$ in (14).) The leading term of T_2 serves to normalize the unknown noise power. The optimum detector is depicted in Fig. 2. The distribution of T_2 and the performance of the optimum detector are found in the next section.

IV. Performance of the Optimum Detector

A. Finite MK Case

The optimum detector is given by (14) in which T_2 is compared with a threshold. The performance of the detector (achieved probability of detection (P_D) for a specified constant probability of false alarm (P_{FA})) will now be found by computing the distribution of T_2 under H and K, and then evaluating the appropriate probabilities. It is assumed first that $MK < \infty$.

To facilitate the computation, define the following $2MK$-dimensional partitioned vectors:

$$Y = [Y_1 | \cdots | Y_M]^t$$
$$R(0) = [R_1(0) | \cdots | R_M(0)]^t \quad (16)$$
$$R(\pi/2) = [R_1(\pi/2) | \cdots | R_M(\pi/2)]^t.$$

Then

$$T = Y^t Y \quad (17)$$

$$T_2 = T^{-1}[(Y^t R(0))^2 + (Y^t R(\pi/2))^2]. \quad (18)$$

Equation (15) can be used to show that the vectors $R(0)$ and $R(\pi/2)$ are perpendicular, i.e.,

$$R^t(0)R(\pi/2) = \sum_{k=1}^{K} R_k^t(0)R_k(\pi/2) = 0.$$

It is thus possible to introduce a coordinate rotation in which $R(0)$ is coincident with the $2MK$-dimensional unit vector $E_1 = [1,0,\cdots,0]^t$ and $R(\pi/2)$ is coincident with $E_2 = [0,1,0,\cdots,0]^t$. Let P be the $2MK$ dimensional orthogonal matrix representing the rotation. Then, after the rotation, the vectors of (16) are

$$Z = PY$$
$$Q(0) = PR(0) = C_0 E_1$$
$$Q(\pi/2) = PR(\pi/2) = C_{\pi/2} E_2. \quad (19)$$

The constants C_0 and $C_{\pi/2}$ are determined by

$$C_0^2 = Q_{(0)}^t Q_{(0)} = R_{(0)}^t R_{(0)} = \sum_{k=1}^{K} S_k^t(0)\Lambda^{-1}S_k(0)$$

$$C_{\pi/2}^2 = Q^t(\pi/2)Q(\pi/2) = R^t(\pi/2)R(\pi/2) = \sum_{k=1}^{K} S_k^t(\pi/2)\Lambda^{-1}S_k(\pi/2), \quad (20)$$

and (8) shows that $C_0 = C_{\pi/2}$. The results of (20) have used the fact that inner products are invariant under coordinate rotations. This fact enables (17) and (18) to be written as

$$T = Z^t Z \quad (21)$$

$$T_2 = C_0^2 T^{-1}[Z_1^2 + Z_2^2]. \quad (22)$$

T_2 is the ratio of sums of squares, each of which will turn out to be chi-square and distributed independently of the other. A well known statistic—the F statistic—is composed of the ratio of independent chi-square variates, each divided by its respective degrees of freedom. To write T_2 in terms of this known quantity, let the statistic F be defined by

$$F = \left[(2\sigma^2(KM-1))^{-1}\sum_{i=3}^{2KM} Z_i^2\right]^{-1}[(2\sigma^2)^{-1}(Z_1^2 + Z_2^2)], \quad 0 \le F \le \infty. \quad (23)$$

Then T_2 can be written in terms of F as

$$T_2 = C_0^2 2F[2F + 2(MK-1)]^{-1}, \quad 0 \le T_2 \le C_0^2. \quad (24)$$

The distribution of F is now found as an intermediate step in obtaining that of T_2.

The distribution of F depends on that of Z which is, in turn, induced by Y, and that of Y in turn by $\{Y_k\}$. Consider the distribution of $Y_k = L^{-1/2}OX_k$. Since Y_k is a linear combination of Gaussian variates, it is a Gaussian M-vector with covariance matrix

$$\operatorname{cov}\{Y_k\} = L^{-1/2}OE\{N_k N_k^t\}O^{-1}L^{-1/2} = \sigma^2 I.$$

That is, the entries of Y_k are independent Gaussian random variables with common variance σ^2. When H is true, the variables are zero mean; while when K is true, given θ, the vector of means is given by

$$E_{K/\theta}\{\boldsymbol{Y}_k\} = \boldsymbol{R}_K(\theta).$$

The distribution of the $2MK$-vector $\boldsymbol{Y}$ is now immediately evident because the $\boldsymbol{Y}_k$ are independent. Thus the entries of $\boldsymbol{Y}$ are independent Gaussian random variables with common variance σ^2, mean zero under H, and with the partitioned mean vector for K, given θ, being

$$E_{K/\theta}\{\boldsymbol{Y}\} = [\boldsymbol{R}_1(\theta)| \cdots |\boldsymbol{R}_K(\theta)]^t. \quad (25)$$

Now $\boldsymbol{Z}$ is related to $\boldsymbol{Y}$ through the rotation P. Rotations do not change the covariance matrices of random vectors whose entries are uncorrelated and have the same variance. Thus the entries of $\boldsymbol{Z}$ remain independent Gaussian with variance σ^2. The rotation does change the mean vectors, however. Under H, the means remain zero, while under K, given θ, the mean vector is

$$E_{K/\theta}\{\boldsymbol{Z}\} = PE_{K/\theta}\{\boldsymbol{Y}\}. \quad (26)$$

Equation (6) enables (25), and hence (26), to be simplified

$$E_{K/\theta}\{\boldsymbol{Z}\} = P(\boldsymbol{R}(0)\cos\theta + \boldsymbol{R}(\pi/2 \sin\theta) = C_0(\boldsymbol{E}_1 \cos\theta + \boldsymbol{E}_2 \sin\theta). \quad (27)$$

The distribution of F is induced by that of $\boldsymbol{Z}$. When H holds, the numerator and denominator of F are each the sum of squares of zero mean, unit variance, independent Gaussian random variables, and each sum is divided by the number of variables summed. Hence the numerator and denominator are each chi-square variates (of 2 and $2(KM - 1)$ degrees of freedom) divided by their respective degrees of freedom. F, the ratio of these independent sums, is known to be distributed with the F-distribution with 2 and $2(KM - 1)$ degrees of freedom. When K holds, given θ, the denominator remains a chi-square variate divided by its degrees of freedom (since $E_{K/\theta}\{Z_i\} = 0$; $i > 2$). However, Z_1 and Z_2 are mean $C_0 \cos\theta$ and $C_0 \sin\theta$, respectively. Thus the numerator is a noncentral chi-square random variable with noncentrality parameter $\psi^2 = \sigma^{-2}C_0^2$. Further, the numerator is distributed independently of the random variable θ. Hence, unconditionally under K, F is the independent ratio of a noncentral chi-square variate and a central chi-square variate, each divided by its degrees of freedom. This ratio is said to have the noncentral F distribution with 2 and $2(KM - 1)$ degrees of freedom and noncentrality parameter $\psi^2 = \sigma^{-2}C_0^2$.

Finally, the distribution of T_2 of (14) can be found from that of F. Equation (24) indicates that T_2 is proportional to the statistic

$$E^2 = \nu_1 F[\nu_2 + \nu_1 F]^{-1}, \qquad 0 \le E^2 \le 1$$

where $\nu_1 = 2$ and $\nu_2 = 2(KM - 1)$. The statistic E^2 is distributed with Fisher's "distribution C" [15] which is discussed in Appendix B. The distribution of T_2 is a special case of the Fisher distribution. Let the cumulative distribution function of Fisher's distribution C with parameters $a = \nu_1/2$ and $b = \nu_2/2$ and noncentrality parameter $t = \psi^2/2$ be

$$\Pr\{E^2 \le x\} = \Phi_C(x|a,b,\psi^2/2).$$

Then the P_D and P_{FA} for the optimum CFAP detector of (14) are

$$P_{FA} = \Pr\{T_2 > C_\alpha|H\} = 1 - \Phi_C(C_\alpha C_0^2|1,KM - 1,0))$$

$$P_D = \Pr\{T_2 > C_\alpha|K\} = 1 - \Phi_C(C_\alpha C_0^2|1,KM - 1,\sigma^{-2}C_0^2/2). \quad (28)$$

Appendix B shows that the test threshold is given by

$$C_\alpha = C_0^{-2}[1 - P_{FA}^{(KM-1)^{-1}}],$$

so that P_D is an increasing function of the parameter $\sigma^{-2}C_0^2/2$. This parameter is the ratio of the total energy in the whitened signals to the noise power parameter σ^2. Figs. 3–6 show P_D versus $\sqrt{2t} = C_0/\sigma$ for $P_{FA} = 10^{-4}, 10^{-6}, 10^{-8}$, and 10^{-10}, respectively. In the figures, various values of the degrees-of-freedom parameter, $b = KM - 1$, are assumed. The value $b = \infty$ corresponds to the performance of the limiting case which is the same as that obtained when the noise power is known (see next section).

B. Infinite MK Case

As was pointed out by a referee, when $MK \to \infty$, the performance of the optimum detector of (14) should approach the performance of the detector which is optimum when the noise power is known. This is true and will now be demonstrated.

Let the statistic $T_3 = MKT_2$ be defined. This statistic is another form of the optimum detector. Equation (24) shows that

$$\lim_{MK\to\infty} T_3 = C_0^2(\infty) \lim_{MK\to\infty} F,$$

where $C_0^2(\infty) = \lim_{MK\to\infty} C_0^2$ is proportional to the finite signal energy. The denominator of F is a chi-square variate divided by its number of degrees of freedom, $2(KM - 1)$. Hence, as $MK \to \infty$, the denominator converges to one. Thus the limiting form of the optimum detector statistic is the statistic which is proportional to the numerator of F, namely $Z_1^2 + Z_2^2$.

The optimum detector for the case when the noise power is known is derived next. The problem statement is identical to that at the beginning of Section III except that σ^2 is now assumed to be known. The likelihood ratio can thus be formed in the conventional manner. The simplification of the likelihood ratio is nearly identical to the steps leading to (11). The result is a test in which a statistic related to T_1 of (10) is compared with a threshold. The statistic is obtained from (10) by eliminating the $T = t$ constraint—that is, by removing the $|_{T=t}$ notation from (10). Equations (12), (19), and (22) show that the statistic is proportional to $Z_1^2 + Z_2^2$. Thus the optimum known noise power detector and the limiting ($MK \to \infty$) case of the optimum unknown noise power detector are identical.

The performance of the detector when $MK \to \infty$ is thus identical to that of the known noise power detector. This detector's performance is given in [14, p. 344]; the results there are related to those here by the identities $N_0/2 = \sigma^2$, $E_r^{1/2} = C_0$, and $d^2 = C_0/\sigma^2 = 2t$. The results of [14] have been included as the curve "$b = \infty$ or noise power known" in Figs. 3–6.

The performance loss of the optimum detector relative to the optimum known noise detector is shown in Fig. 7. The curves of the figure were computed in the following manner. For a given P_{FA}, the signal-to-noise ratio, $2t$, which yielded $P_D = 0.5$ was noted for the values $b = 5, 10, 25, 50, 100$, and ∞ (the known noise case). Then the loss, in dB, over the $b = \infty$ case, was plotted versus the appropriate value of B. The figure indicates that performance within 1 dB of the known noise performance is reached in all cases considered if $b \geq 40$.

V. Examples

A. Detection in the Presence of a Jammer

Consider the array of Fig. 1 and suppose that $n(t)$ is a noise jammer located at angle γ while the desired signal is located dead ahead ($\gamma = 0$). For simplicity, let the signal be constant in amplitude and phase (DC). Further, let the narrow-band Gaussian noise process have a carrier symmetric spectral density function $S(f)$. Then it is well-known [16, p. 162] that the second-order properties of the cosine and sine components of $n(t)$ are governed by $S(f)$, that is (using the notation of (1)),

$$R_c(\tau) = E\{V(t)\cos(\phi(t))V(t+\tau)\cos(\phi(t+\tau))\}$$
$$= E\{V(t)\sin(\phi(t))V(t+\tau)\sin(\phi(t+\tau))\}$$
$$= 2\int_0^\infty S(f)\cos((f-f_c)\tau)\,df$$
$$R_c(0) = E\{n^2(t)\} = \sigma^2$$
$$R_{cs}(\tau) = E\{V(t)\cos(\phi(t))/V(t+\tau)\sin(\phi(t+\tau))\}$$
$$= 2\int_0^\infty S(f)\sin((f-f_c)\tau)\,df.$$

Because of the assumed carrier symmetry of $S(f)$, $R_{cs}(\tau) \equiv 0$, which indicates that the sine and cosine processes are independent.

The elements of the partitioned vectors of (2) can now be evaluated; they are

$$S_{ck}(\theta) = A\cos\theta[1,1]^t$$
$$S_{sk}(\theta) = A\sin\theta[1,1]^t$$
$$N_{sk} = [V(t_k)\sin(\phi(t_k)),\ V(t_k-a)\sin(\phi(t_k-a)-\omega_c a)]^t$$
$$N_{sk} = [V(t_k)\sin(\phi(t_k)),\ V(t_k-a)\sin(\phi(t_k-a)-\omega_c a)]^t,$$

and the elements of Λ are

$$\Lambda_c = \begin{bmatrix} 1 & \rho(a)\cos(\omega_c a) \\ \rho(a)\cos(\omega_c a) & 1 \end{bmatrix}$$
$$\Lambda_s = \begin{bmatrix} 0 & -\rho(a)\sin(\omega_c a) \\ \rho(a)\sin(\omega_c a) & 0 \end{bmatrix},$$

where $\rho(\cdot) = \sigma^{-2}R_c(\cdot)$. The inverse of Λ is the symmetric matrix

$$\Lambda^{-1} = (\det\Lambda)^{-1/2}\begin{bmatrix} 1 & -\rho(a)\cos(\omega_c a) & 0 & \rho(a)\sin(\omega_c a) \\ & 1 & -\rho(a)\sin(\omega_c a) & 0 \\ & \text{symmetric} & 1 & -\rho(a)\cos(\omega_c a) \\ & & & 1 \end{bmatrix}$$

and

$$\det(\Lambda) = (1-\rho^2(a))^2.$$

The implementation of the optimum detector of (14) and Fig. 2 is straightforward. The performance is governed by (28). The P_D is proportional to the parameter $\sigma^{-2}C_0^2$ of (20). The parameter is a measure of signal-to-noise ratio (SNR). For this example,

$$\sigma^{-2}C_0^2 = \sigma^{-2}2KA^2(1-\rho(a)\cos(\omega_c a)). \tag{29}$$

The SNR value of (29) undergoes a sinusoidal variation as the value of $\omega_c a$ varies. The maxima are obtained at the minima of $\rho(a)\cos(\omega_c a)$, and vice versa. This behavior can be given the following physical interpretation.

First the product $\omega_c a$ may be simplified by noting that the time delay is $a = v^{-1}d\sin(\gamma)$, where v is the velocity of propagation and d and γ are defined in Fig. 1. Also $\omega_c = 2\pi v\lambda^{-1}$, where λ is the wavelength of the carrier. These substitutions yield $\omega_c a = 2\pi\lambda^{-1}d\sin(\gamma)$. Next, it is well-known that the voltage gain pattern of an unweighted two-element array is proportional to $|\cos(\pi\lambda^{-1}d\sin(\gamma))|$. A quick sketch will show that, for $\rho(a) > 0$, the maxima of (29) coincide with the nulls of the array pattern—that is, when the jammer is nulled. When $\rho(a) < 0$, the opposite is true because the negative correlation changes the sense of nulls and peaks in the antenna pattern for the jammer. This latter case is less meaningful because usual physical noise sources have spectral shapes that yield $\rho(\tau) \geq 0$.

Finally, it is of interest that the problem model given in this example may be generalized to model target detection in the presence of clutter. The technique involves assuming a spatial continuum of noise sources rather than just a single one. The details are given in [17].

B. Single Channel Detection in White Noise

When $M = 1$, (14) gives the optimum CFAP detector of a single phase incoherent signal in the presence of white (independent from sample to sample) noise. Noting that

$$Y_k^t Y_k = X_k^t\Lambda^{-1}X_k$$
$$Y_k^t R_k(\theta) = X_k^t\Lambda^{-1}S_k(\theta)$$

and that, for this case, Λ becomes

$$\Lambda = \begin{bmatrix} \rho_c & \rho_s \\ -\rho_s & \rho_c \end{bmatrix},$$

where (see [16, p. 162])

$$\rho_c = \sigma^{-2}E\{N_{ck}^2\} = \sigma^{-2}E\{N_{sk}^2\} = 1$$
$$\rho_s = E\{N_{ck}N_{sk}\} = 0,$$

we see that Λ reduces to a two-dimensional identity matrix. The optimum statistic of (14) simplifies to

$$T_2 = \left(\sum_{K=1}^{K} \boldsymbol{X}_k^t \boldsymbol{X}_k\right)^{-1} \left[\left(\sum_{K=1}^{K} \boldsymbol{X}_k^t \boldsymbol{S}_k(0)\right)^2 + \left(\sum_{K=1}^{K} \boldsymbol{X}_k^t \boldsymbol{S}_k(\pi/2)\right)^2\right], \quad (30)$$

where, from (2) and (A-5),

$$\boldsymbol{X}_k^t \boldsymbol{X}_k = X_{ck}^2 + X_{sk}^2$$

$$\boldsymbol{X}_k^t \boldsymbol{S}_k(0) = X_{ck} S_{ck}(0) + X_{sk} S_{sk}(0)$$

$$\boldsymbol{X}_k^t \boldsymbol{S}_k(\pi/2) = -X_{ck} S_{sk}(0) + X_{sk} S_{ck}(0).$$

An interesting interpretation can be given to (30) if the signal and noise are interpreted as complex vectors—with the cosine component interpreted as the real component of a complex vector, and the sine component as the imaginary. The complex reference vector samples are then ($i = \sqrt{-1}$)

$$S_{ck}(0) + iS_{sk}(0), \qquad k = 1, \cdots, K,$$

and the output of a filter matched to the reference vector is

$$\sum_{K=1}^{K} (X_{ck} + iX_{sk})(S_{ck}(0) - iS_{sk}(0)).$$

The magnitude squared of this filter output is equal to the right term of (30) while the left term is the total energy in the data samples. Thus for this case, the optimum detector compares the magnitude squared of the usual matched filter with a threshold proportional to the total energy in the observation vector. Detectors of this structure are commonly used to obtain CFAP operation; however, the optimality of the structure is apparently not widely known. The performance of the detector is discussed at the end of the next example.

C. Single Channel Detection in Colored Noise

As another example, the detection of a single phase-incoherent signal in time-correlated noise of unknown power will be examined. Even though the general problem statement of Section I has assumed the noise to be uncorrelated, this example can be handled by setting $K = 1$ and interpreting m as the index of time-samples of the noise and signal (rather than as an index of array elements). With this assumption, the covariance matrix of the M Gaussian noise samples is $\sigma^2\Lambda$, and the results of Sections II and III apply for the case $K = 1$.

The form of the optimum CFAP detector is now found. First, as in the jammer example, let the noise spectral density function be carrier symmetric. Then the sine and cosine random processes are independent, and Λ becomes

$$\Lambda = \left[\begin{array}{c|c} \Lambda_c & 0 \\ \hline 0 & \Lambda_c \end{array}\right].$$

The optimum detector of (14) may now be written entirely in terms of Λ_c^{-1} and the reference vector $\boldsymbol{S}(0)$ of (2), through the use of identity (A-5). The result is

$$T_2 = (\boldsymbol{X}^t\Lambda^{-1}\boldsymbol{X})^{-1}[(\boldsymbol{X}_c^t\Lambda_c^{-1}\boldsymbol{S}_c(0) + \boldsymbol{X}_s^t\Lambda_c^{-1}\boldsymbol{S}_c(0))^2 + (-\boldsymbol{X}_c^t\Lambda_c^{-1}\boldsymbol{S}_s(0) + \boldsymbol{X}_s^t\Lambda_c^{-1}\boldsymbol{S}_c(0))^2]. \quad (31)$$

An interpretation similar to that of the last example can be given to (31) if the signals and noise are again interpreted as complex vectors. The complex reference vector then becomes the complex vector

$$\Lambda_c^{-1}\boldsymbol{S}_c(0) + i\Lambda_c^{-1}\boldsymbol{S}_s(0),$$

and, with the partitioning of (2), the complex data vector is $\boldsymbol{X}_c + i\boldsymbol{X}_s$, where $\boldsymbol{X}_c$ and $\boldsymbol{X}_s$ are the data cosine and sine component vectors, respectively. The output of a complex filter matched to the reference vector is given by

$$(\boldsymbol{X}_c + i\boldsymbol{X}_s)^t(\Lambda_c^{-1}\boldsymbol{S}_c(0) - i\Lambda_c^{-1}\boldsymbol{S}_s(0)).$$

The magnitude squared of the matched filter output is equal to the rightmost term of (31), while the term $\boldsymbol{X}^t\Lambda^{-1}\boldsymbol{X}$ is the total energy in the whitened data vector. For this example, the optimum detector of (14) can again be interpreted as comparing the magnitude squared of the output of a filter matched to the signal (whitened) with a threshold proportional to the total energy in the observation vector. The extension of the optimum structure commonly used to obtain a constant false alarm probability when the noise is white and of unknown power (see last example) to the colored noise case is a new result.

The performance of the detector can be obtained from (28). The evaluation of the parameter C_0^2 goes beyond the scope of this paper, however. A detailed calculation is given in [17] where, for a sinusoidal signal, the SNR is found to be proportional to the ratio of the signal energy to the noise spectral density, measured at the signal frequency.

ACKNOWLEDGMENT

The author wishes to thank the referees for their helpful comments that led to a refinement of the paper.

APPENDIX A
SOME PROPERTIES OF Λ^{-1} AND $\boldsymbol{S}_k(\theta)$

The covariance matrix of $\sigma^{-1}\boldsymbol{N}_k$ has special symmetry because $\boldsymbol{N}_k$ is composed of time-samples of a narrow-band random process. Equation (2) has defined the $2M$-dimensional vector in terms of the M-dimensional vectors $\boldsymbol{N}_{ck}$ and $\boldsymbol{N}_{sk}$. Now define $M \times M$ covariance matrices

$$\sigma^2\Lambda_c = E\{\boldsymbol{N}_{ck}\boldsymbol{N}_{ck}^t\} = E\{\boldsymbol{N}_{sk}\boldsymbol{N}_{sk}^t\}$$

$$\sigma^2\Lambda_s = E\{\boldsymbol{N}_{ck}\boldsymbol{N}_{sk}^t\},$$

then Λ may be written as

$$\Lambda = \left[\begin{array}{c|c} \Lambda_c & \Lambda_s \\ \hline -\Lambda_s & \Lambda_c \end{array}\right].$$

It is well known that Λ_c is symmetric ($\Lambda_c^t = \Lambda_c$), and that Λ_s is skew symmetric ($\Lambda_s^t = -\Lambda_s$) [16, p. 162].

It will now be shown that Λ^{-1} also has the special symmetry of Λ. The inverse of a symmetric matrix is well-known to be symmetric, so Λ^{-1} can be written as

$$\Lambda^{-1} = \left[\begin{array}{c|c} A & B \\ \hline B^t & D \end{array}\right]; \; A = A^t, D = D^t$$

and must satisfy

$$\Lambda\Lambda^{-1} = I,$$

which yields four equations in the three unknown M-dimensional matrices

$$\Lambda_c A + \Lambda_s B = I \tag{A-1}$$

$$-\Lambda_s A + \Lambda_c B = O \tag{A-2}$$

$$\Lambda_c B + \Lambda_s D = O \tag{A-3}$$

$$-\Lambda_s B + \Lambda_c D = I. \tag{A-4}$$

Equations (A-1) and (A-2) give

$$A = A^t = (\Lambda_c + \Lambda_s \Lambda_c^{-1} \Lambda_s)^{-1}$$

$$B = -B^t = -(\Lambda_s + \Lambda_c \Lambda_s^{-1} \Lambda_c)^{-1},$$

while (A-1) and (A-4) give $A = D$; thus, Λ^{-1} has the same properties as Λ.

Next, (2) shows that

$$S_k(\pi/2) = [-S_{sk}^t(0) | S_{ck}^t(0)]^t$$

$$S_k(0) = [S_{ck}^t(0) | S_{sk}^t(0)]^t, \tag{A-5}$$

and thus

$$\begin{aligned} S_k^t(0)\Lambda^{-1}S_k(0) &= S_{ck}^t(0) A S_{ck}(0) \\ &+ S_{sk}(0) A S_{sk}(0) + 2 S_{ck}^t(0) B S_{sk}(0) \\ &= S_k^t(\pi/2)\Lambda^{-1}S_k(\pi/2), \end{aligned} \tag{A-6}$$

and

$$S_k^t(0)\Lambda^{-1}S_k(\pi/2) = S_{sk}^t(0) B S_{sk}(0) + S_{ck}^t(0) B S_{ck}(0) = 0. \tag{A-7}$$

The result of (A-7) uses the fact that if B is skew symmetric, then, for any vector X,

$$X^t B X = (X^t B X)^t = -X^t B X = 0.$$

APPENDIX B
FISHER'S DISTRIBUTION C

The probability density function for Fisher's distribution C of [18] is given by

$$p_c(u) = e^{-t}[(a-1)!(b-1)!]^{-1}(a+b-1)! u^{a-1} \cdot (1-u)^{b-1} \sum_{k=0}^{\infty} [k!(a)_k]^{-1}(a+b)_k (tu)^k. \tag{B-1}$$

The coefficients within the infinite sum are those of the confluent hypergeometric series

$$(a)_k = a(a+1)(a+2)\cdots(a+k-1), \qquad (a)_0 = 1$$

and the sum is equal to the confluent hypergeometric function $M(a+b,b,tx)$.

The technique for integrating (B-1) to obtain the cumulative distribution function is outlined in [18], while [17] gives a more detailed account. The result is

$$\begin{aligned} \Phi_c(x|a,b,t) &= \int_0^x p_c(u)\,du \\ &= e^{-t(1-x)}(a+b-1)! \sum_{r=0}^{b-1} [r!(a+b-1-r)!]^{-1} \\ &\cdot x^{a+b-1-r}(1-x)^r M(-r, a+b-r, -tx), \end{aligned} \tag{B-2}$$

where, in this case, the infinite summation in the definition of $M(\cdot)$ is actually finite because $(-r)_j = 0$ for $j > r$. This yields

$$-r, a+b-r, -tx) = \sum_{j=0}^{r} [(a+b-r)_j j!]^{-1}(-r)_j(-tx)^j.$$

The P_{FA} and P_D of (26) are evaluated using (B-2) for the specialization $a = 1$. The test threshold T is found from the relation

$$\begin{aligned} P_{FA} &= 1 - \Phi_c(T|1,b,0) \\ &= 1 - \sum_{r=0}^{b-1} [r!(b-r)!]^{-1} b!(1-T)^r T^{b-r} = (1-T)^b \end{aligned}$$

so that $T = 1 - (P_{FA})^{b-1}$. Graphs of (B-1) may be found in [17].

APPENDIX C
PROOF OF THEOREM 3

Let the event $M(\Delta t)$ be defined by

$$M(\Delta t) = \{T(x) | t \le T(x) \le t + \Delta t\}.$$

Then [20, p. 137] shows that the conditional probability densities of X and $Y = d(x)$, given $M(\Delta t)$ (under H or K), are related by

$$p(x|M(\Delta t)) = f(y|M(\Delta t))|_{y=g(x)} J(x), \tag{C-1}$$

where $J(x)$ is the Jacobian of the one-to-one transformation. Now [20, p. 178] indicates that

$$f(y|M(\Delta t)) = \begin{cases} \dfrac{f(y)}{F_T(t+\Delta t) - F_T(t)}, & t \le T(x) \le t + \Delta t \\ 0, & \text{elsewhere}, \end{cases} \tag{C-2}$$

where $f(\cdot)$ is the unconditional density of Y, and $F_T(\cdot)$ is the marginal distribution function of $T(x)$. Substitution of (C-2) into (C-1) and use of the identity

$$p(x) = f(y)|_{y=g(x)} J(x)$$

yields

$$p(x)|M(\Delta t)) = \begin{cases} \dfrac{p(x)}{F_T(t+\Delta t) - F_T(t)}, & t \le T(x) \le t + \Delta t \\ 0, & \text{elsewhere}. \end{cases} \tag{C-3}$$

Equation (C-3) can now be used to obtain the conditional likelihood ratio

$$L_t(x) = \frac{p_K(x|t)}{p_H(x|t)} = \lim_{\Delta t \to 0} \frac{p_K(x|M(\Delta t))}{p_H(x|M(\Delta t))}$$

in terms of the unconditional ratio $L(x)$ if it is noted that

$$f_T(t) = \lim_{\Delta t \to 0} \frac{F_T(t + \Delta t) - F_T(t)}{\Delta t},$$

the result being

$$L_t(x) = \frac{L(x)}{L(t)}\bigg|_{T(x)=t}, \qquad \text{(C-4)}$$

where

$$L(t) = f_T(t/K)/f_T(t/H).$$

Equation (C-4) shows that the conditional likelihood ratio is proportional to the unconditional ratio evaluated under the constraint $T(x) = t$. The constant of proportionality $L(t)$ is a function only of the value t held fixed. The unconditional ratio, evaluated at $T(x) = t$, may be simplified because $T(x)$ is the sufficient statistic for H; that is,

$$p_H(x)|_{T(x)=t} = g_\theta(t)h(x)|_{T(x)=t}$$

yielding

$$L(x)|_{T(x)=t} = \frac{p_K(x)}{h(x)}\bigg|_{T(x)=t} g_\theta(t)^{-1}. \qquad \text{(C-5)}$$

Now the structure of the most powerful similar test of Theorem 2 can be written in terms of the unconditional densities. Substitution of (C-4) and (C-5) into the test structure shows that the comparison

$$L_t(x) = \frac{L(t)^{-1}}{g_\theta(t)} \frac{p_K(x)}{h(x)}\bigg|_{T(x)=t} \gtrless C(t,\alpha)$$

is equivalent to the comparison

$$L'_t(x) = \frac{p_K(x)}{h(x)}\bigg|_{T(x)=t} \gtrless C'(t,\alpha,\theta)$$

where

$$C'(t,\alpha,\theta) = g_\theta(t)C(t,\alpha)L(t).$$

Theorem 2 indicates that the constants $a'(t,\alpha)$ and $C'(t,\alpha,\theta)$ of the simplified test are chosen to satisfy $E\{\phi(X)/T(X) = t\} = \alpha$. Thus C' is actually independent of θ because the distribution of X given $T(X) = t$ is independent of θ. Hence the final (simplified) form of the most powerful similar test is

$$\phi(x) = \begin{cases} 1; & L'_t(x) > C'(t,\alpha) \\ a'(t,\alpha); & L'_t(x) = C'(t,\alpha) \\ 0; & L'_t(x) < C'(t,\alpha). \end{cases}$$

REFERENCES

[1] M. I. Skolnik, Radar Handbook. New York: McGraw-Hill, 1970.

[2] C. B. Bell, "Automatic distribution-free statistical signal detection," U. S. Navy Electronics Laboratory, San Diego, California, Research Report 1245, Oct. 1964.

[3] J. Dabkowski, Jr., "Analysis of constant false alarm probability radar system," Ph.D. thesis in Electrical Engineering, Illinois Institute of Technology, Chicago, IL, 1969.

[4] R. F. Daly and C. K. Rushforth, "Nonparametric detection of a signal of known form in additive noise," *IEEE Trans. Inform. Theory*, vol. IT-11, pp. 70–76, Jan. 1965.

[5] R. A. Groenevald, "A nonparametric rank correlation method for detecting signals in additive noise," *IEEE Trans. Inform. Theory*, Vol. IT-13, pp. 315–316, Apr. 1967.

[6] R. E. Schwartz, "Statistical invariance and optimum CFAR detection," General Electric Company Electronics Laboratory, Syracuse, NY, Technical Information Series, Report R68ELS-62, June 1962.

[7] P. Swerling, L. E. Brennan, and I. S. Reed, "Optimum CFAR detection tests for fluctuating and non-fluctuating signals," Technology Service Corporation, Santa Monica, CA 1967.

[8] N. M. Woinsky and L. Kurz, "A new approach to nonparametric detection using dependent samples and continuous processing," in *Proc. of 2nd Ann. Princeton Conf. Information Sciences and Systems*, Princeton University, Princeton, NJ, 1968.

[9] J. W. Carlyle, "Nonparametric methods in detection theory," in *Communication Theory*, (A. V. Balakrishnan, Ed.). New York: McGraw-Hill, 1968.

[10] J. Neyman and E. S. Pearson, "On the problem of the most efficient tests of statistical hypotheses," *Phil. Trans. Roy. Soc. London*, series A, vol. 231, pp. 289–337, 1933.

[11] E. L. Lehmann and Henry Scheffé, "Completeness, similar regions, and unbiased estimation—Parts I and II," *Sankhyā: The Indian Journal of Statistics*, vol. 10, pp. 305–340, 1950, and vol. 15, pp. 219–236, 1955.

[12] E. L. Lehmann, *Testing Statistical Hypotheses*. New York: Wiley, 1959.

[13] E. L. Lehmann and C. Stein, "On the theory of some nonparametric hypotheses," *Annals of Math. Stat.*, vol. 20, pp. 28–45, 1949.

[14] H. L. Van Trees, *Detection, Estimation and Modulation Theory*, vol. 1. New York: Wiley, 1968.

[15] P. B. Patnaik, "The non-central chi-square and F distributions and their applications," *Biometrika*, vol. 36, pp. 202–232, 1949.

[16] W. B. Davenport, Jr. and W. L. Root, *An Introduction to the Theory of Random Signals and Noise*. New York: McGraw-Hill, 1958.

[17] P. G. Grieve, "A unified approach to the construction of constant false alarm probability radar detectors," Ph.D. Thesis in Engineering, Univ. Calif., Los Angeles, 1975.

[18] J. Wishart, "A note on the distribution of the correlation ratio," *Biometrika*, vol. XXIV, pp. 441–456, 1932.

[19] E. L. Lehmann and C. Stein, "Most powerful tests of composite hypotheses," *Annals of Mathematical Statistics*, vol. 19, pp. 495–516, 1948.

[20] A. Papoulis, *Probability, Random Variables, and Stochastic Processes*. New York: McGraw-Hill, 1965.

21

Reprinted with permission from *IEEE Trans. Inf. Theory* **IT-13**:341–342 (1967)

THE KOLMOGOROV-SMIRNOV TESTS IN SIGNAL DETECTION

John B. Millard and Ludwik Kurz

The application of nonparametric methods to the detection of signals in unknown or long-term nonstationary noise has been receiving considerable attention recently.[1]–[5] Among the many distribution-free tests available for use in signal detection is a group of nonparametric statistics generally classified as the Kolmogorov–Smirnov, Cramér–von Mises tests. As statistics, these tests make use of various functional forms of the difference between two empirical distribution functions.

In this correspondence, an application of statistics of the Kolmogorov–Smirnov, Cramér–von Mises class to the problem of nonparametric detection of signals in noise will be considered. In particular, a statistic of this class, termed the mean difference statistic, is presented and is shown to be equivalent to the Mann–Whitney test statistic. The distribution of the statistic under the null hypothesis, assumption of signal not present, is derived using a two-dimensional extension of the theory of partitions. Statistics within the class are compared on the basis of power versus significance level where distributions under the alternative hypothesis, signal present, are obtained by Monte Carlo methods.

Consider the problem of determining whether two sets of samples of size n_1 and n_2, respectively, come from the same distribution. Let $\mathbf{X}_1 = x_{11}, x_{12}, \cdots, x_{1n_1}$ and $\mathbf{X}_2 = x_{21}, x_{22}, \cdots, x_{2n_2}$ be vectors representing the samples from populations 1 and 2.[1] Define the empirical distribution function for the samples $\mathbf{X}_1$ and $\mathbf{X}_2$ by $S_{1n_1}(x)$ and $S_{2n_2}(x)$ as

$$S_{1n_1}(x) = \frac{1}{n_1}\sum_{i=1}^{n_1} u(x - x_{1i})$$

$$S_{2n_2}(x) = \frac{1}{n_2}\sum_{i=1}^{n_2} u(x - x_{2i})$$

where $u(x)$ is the unit step function.

A measure of the strength of the hypothesis that the vector samples $\mathbf{X}_1$ and $\mathbf{X}_2$ are from the same underlying distribution[2] is some function of the difference between the empirical distribution functions $S_{1n_1}(x)$ and $S_{2n_2}(x)$. Call the statistic, which represents some distance function D, such that

$$D = G\{S_{1n_1}(x)' - S_{2n_2}(x)\}$$

where G represents some functional over the range of the variable x. The choice of the functional form G is determined by the particular application.

Several forms of the functional G have been analyzed for their distribution under the null hypothesis, in particular the point forms of the statistic:[3]

$$D_1 = \max_{-\infty<x<\infty} \{S_{2n_2}(x) - S_{1n_1}(x)\} \tag{1}$$

$$D_2 = \max_{-\infty<x<\infty} |S_{2n_2}(x) - S_{1n_1}(x)|. \tag{2}$$

Another form of the Kolmogorov–Smirnov, Cramér–von Mises Test is obtained through an integration over the range of the variable rather than by using a point determination. The distance function considered is the weighted mean-squared difference in empirical distribution functions,

$$D = K\int_{-\infty}^{\infty} \{S_{2n_2}(x) - S_{1n_1}(x)\}^2 \psi\{h(x)\}\, dh(x)$$

where

$$h(x) = \frac{n_1 S_{1n_1}(x) + n_2 S_{2n_2}(x)}{n_1 + n_2}.$$

The distribution of this statistic has been investigated for the case $\psi = 1$.[4],[5] The form of the weighting function, $\psi(t) = 1/t(1-t)$, has been suggested and analyzed for the goodness of fit problem[6],[7] but has not been analyzed in this context. For the purpose of this paper the statistic D when $\psi(t) = 1$ will be termed D_4 and when $\psi(t) = 1/t(1-t)$ will be termed D_5.

The problem of detecting a flat-topped pulse in additive noise through a linear channel is a decision theory problem involving the mean of the distribution in question. For this particular problem the nonparametric Kolmogorov–Smirnov test is chosen to give results consistent with this formulation. In particular the form chosen is

$$D = K\int_{-\infty}^{\infty} \{S_{2n}(x) - S_{1n}(x)\}\psi(x)\, dh(x);$$

$$h(x) = \tfrac{1}{2}\{S_{1n}(x) + S_{2n}(x)\}.$$

The choice of this form of test statistic within the Kolmogorov–Smirnov, Cramér–von Mises class is based on a consideration of the power of the test.

For the remainder of the paper, the unweighted mean difference test is considered, that is, $\psi(x) = 1$. Under this condition the statistic will be designated D_3[5] as follows:

$$D_3 = \int_{-\infty}^{\infty} \{S_{2n}(x) - S_{1n}(x)\}\, dh(x).$$

Theorem[6]

The distribution of the statistic D_3 under the null hypothesis, signal absent, is given by

$$\Pr\{D_3 = k/n\} = \frac{P^u_{2n,n}(k')}{\binom{2n}{n}}$$

Manuscript received June 3, 1966; revised November 10, 1966. This research was supported in part by the Doctoral Support Plan of Bell Telephone Laboratories, Inc. and in part by the National Science Foundation under Grant GK-1075

[1] In signal detection problems populations 1 and 2 will represent noise and signal plus noise sample spaces, respectively.

[2] This measure is independent of the actual form of the underlying distribution, however.

[3] The distribution of these statistics are available in the literature and will not be given here.[6]

[4] This is equivalent to the unweighted mean-square difference, the mean being taken over the sample points rather than the continuous range.

[5] This statistic is linearly related to the Mann–Whitney statistic,[7] the latter being a special form of the Kolmogorov–Smirnov test. In this paper the distribution of the statistic is derived for finite sample size using a generalized theory of partitions.

[6] For a proof of this theorem see Millard and Kurz.[8]

where $P^u_{n2,n}(k')$ is the partition of k' into n unequal parts, the maximum of which is less than or equal to $2n$, and the relation between k and k' is given by

$$k = 2k' - n(2n + 1).$$

The two-dimensional generating function used to generate $P^u_{2n,n}(k')$ is given by

$$g^u_m(x, y) = \prod_{i=1}^{m} (1 + x^i y) = 1 + \sum_{k=1}^{\frac{1}{2}m(m+1)} \sum_{j=1}^{m} P^u_{m,j}(k') x^{k'} y^j$$

where $m = 2n$: The desired result is obtained when j is set equal to n in the resulting array.

In order to evaluate the relative performance of the five forms of the Kolmogorov–Smirnov test, curves of power (β) versus significance level (α) for the statistics in question are shown in Figs. 1 and 2. The noise is assumed Gaussian.

The significance level for the statistics D_1 and D_2 were computed from (1) and (2), respectively. For the statistics D_3 the generating function was used to generate tables of the distribution function on the 7090 computer. Tables have been computed for sample sizes up to $n = 25$. The significance levels for this form of the Kolmogorov–Smirnov test were taken from these tables.

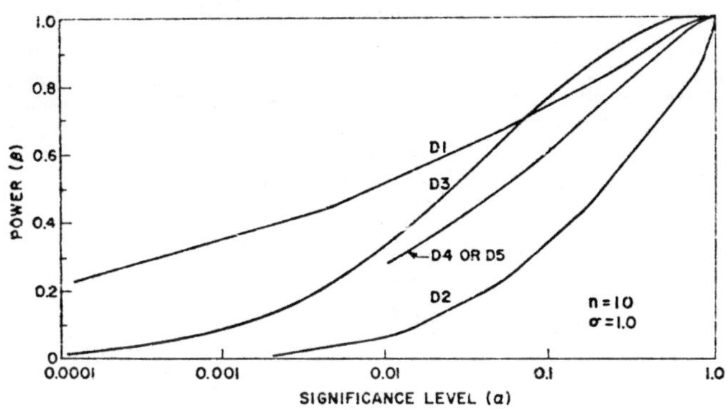

Fig. 1. Operating characteristics: Kolmogorov–Smirnov, Cramér–von Mises detectors; $\sigma = 1.0$.

The distributions of the statistics D_4 and D_5 under the null hypothesis and all statistics under the alternative hypothesis were found using a Monte Carlo method simulation of the system. The empirical distribution thus found, using one thousand computed values of the statistics, were used to compute the power of the various tests. These are the results plotted in Figs. 1 and 2.

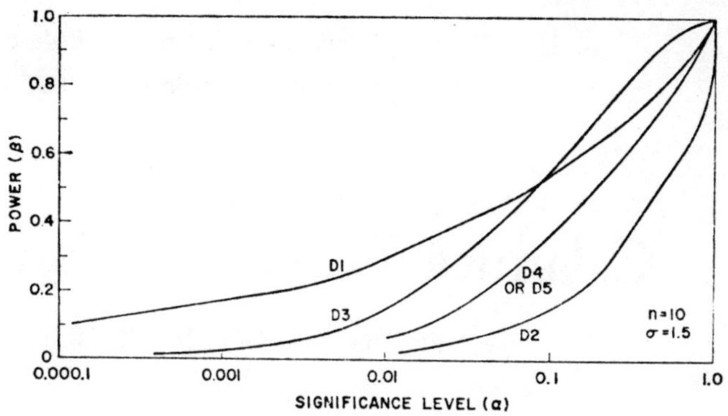

Fig. 2. Operating characteristics: Kolmogorov–Smirnov, Cramér–von Mises detectors; $\sigma = 1.5$.

The curves[7] of power versus significance level show that the mean difference statistic D_3 is a more powerful test, for the signal-to-noise ratios considered, than the mean square difference types, D_4 or D_5, as was originally conjectured. Moreover, in the range of significance levels for which the probabilities of Type I and Type II errors are approximately equal, the mean difference statistic is more powerful (Gaussian noise) than the best of the point forms of the Kolmogorov–Smirnov test, D_1, although for tests with a small significance level (at the noise levels for which the tests were evaluated) the statistic D_1 proves superior.

For the detection of a signal pulse in additive noise the mean difference statistic as a test function is the most powerful test among the tests considered in the Kolmogorov–Smirnov class in the most useful range of the statistic.

References

[1] J. Capon, "A nonparametric technique for the detection of a constant signal in additive noise," *IRE WESCON Conv. Rec.*, pt. 4, pp. 92–104, 1959.

[2] S. S. Wolff, J. B. Thomas, and T. R. Williams, "The polarity coincidence correlator: a nonparametric detection device," *IRE Trans. on Information Theory*,

[3] J. W. Carlyle and J. B. Thomas, "On nonparametric signal detectors," *IRE Trans. on Information Theory*, vol. IT-10, pp. 146–152, April 1964.

[4] R. F. Daly and C. K. Rushforth, "Nonparametric detection of a signal of known form in additive noise," *IEEE Trans. on Information Theory*, vol. IT-11, pp. 70–75, January 1965.

[5] J. B. Millard and L. Kurz, "Adaptive threshold detection of M-ary signals in statistically undefined noise," *1966 IEEE Conv. Rec.* (to be published).

[6] M. Fisz, *Probability Theory and Mathematical Statistics.* New York: McGraw-Hill. 1950.

[7] H. B. Mann and D. R. Whitney, "On a test of whether one of two random variables is stochastically larger than the other," *Ann. of Math. Stat.*, vol. 18, pp. 50–60, 1947.

[8] J. B. Millard and L. Kurz, "Nonparametric signal detection—an application of the kolmogorov-smirnov, Cramer-von Mises Tests," New York University Laboratory for Electroscience Research, New York, Tech. Rept. 400-127, January 1966.

[7] For additional curves see Millard and Kurz.[8]

Reprinted with permission from *IEEE Trans. Syst. Sci. Cybern.* **SSC-5**:150–155 (1969)

Calamity Detection Using Nonparametric Statistics

ERNEST G. HENRICHON, JR., MEMBER, IEEE, AND K. S. FU, MEMBER, IEEE

I. Introduction

Adaptive and learning control systems can be considered to have three features which distinguish them from conventional control systems [1]:

1) the ability to search "on-line" for an optimal action
2) the ability to estimate and store all pertinent information
3) the ability to recognize any change which may occur in the controlled plant-environment relation.

This paper is concerned with capability 3). In particular, the problem considered is that of detecting a sudden change or calamity in the normal operation of a control system for the case of a deterministic system (possibly unknown) with measurements corrupted by stationary noise with unknown statistics. This problem is not to be confused with the somewhat related problem of detecting when switching between known environments takes place in a stochastic control system.[1] The procedure to be proposed makes use of the statistics for nonparametric two-sample hypothesis testing. Two-sample tests have been used in signal detection investigations previously, e.g., [2]–[7], but little, if any, work has been done in detecting a fault in a dynamic system with unknown measurement noise statistics.

II. Two-Sample Tests

Let us consider two sets of independent samples $x_1,\cdots,x_n$ and $y_1,\cdots,y_m$ with continuous cumulative distribution functions (cdf) F and G, respectively. One of the celebrated problems in nonparametric statistics is to devise a test based on the two samples to test the hypothesis $H_0:F(x)=G(x)$ [8]. In the classical development of these tests, various alternative hypotheses have been considered:

$$H_1:F \neq G$$
$$H_2:F > G$$
$$H_3:F < G$$
$$H_4:F \neq G^k, \quad k > 0$$
$$H_5:F = f(G), \quad \text{where } f \text{ is a known function.}$$

Alternatives H_2 and H_3 are called single-sided alternatives, while H_4 and H_5 are known as Lehmann alternatives.

For purposes of this paper only, two-sided alternatives will be used, namely, H_1 or an equal tails test based on the combination of H_2 and H_3. Two of the better known tests which have been proposed for testing H_0 are briefly discussed in the following.

Kolmogorov–Smirnov (K–S) Test [4], [9], [10]

Let $F_n(x)$ and $G_m(y)$ be the empirical cdf's based on the two independent samples $x_1,\cdots,x_n$ and $y_1,\cdots,y_m$, where the empirical cdf $F_k(z)$ is defined as

$$F_k(z) = \frac{\text{number of } \{z_1,\cdots,z_k\} < z}{k}, \qquad -\infty < z < \infty. \tag{1}$$

The K–S test makes use of the statistic

$$D_{n,m} = \left(\sqrt{\frac{nm}{n+m}}\right) d \tag{2}$$

where

$$d = \sup_{-\infty<x<\infty} |F_n(x) - G_m(x)|.$$

The hypothesis H_0 is rejected when $D_{n,m}$ is too large. Smirnov [11] has shown that this test is consistent against H_1 as n,m approach infinity provided that the ratio n/m approaches a positive constant.

The distribution of $D_{n,m}$ under H_0 has been investigated by a number of authors [9] and critical values for various choices of n and m in the expression

$$P_{H_0}\left(D_{n,m} < \frac{k}{nm}\right) = \alpha$$

have been tabulated [12], [13].

It is important to note that not much work has been done for the distribution of $D_{n,m}$ under the alternative hypothesis H_1. Hence to compute the probability of acceptance of H_0 when the alternative is true, one has to know the form of the distributions in question. This has been done for certain particular distributions (e.g., for Gaussian cdf's with equal variances and unknown

Manuscript received September 3, 1968; revised December 27, 1968.
E. G. Henrichon, Jr., was with the School of Electrical Engineering, Purdue University, Lafayette, Ind. He is now with Information Research Associates, Inc., Waltham, Mass.
K. S. Fu is with the School of Electrical Engineering, Purdue University, Lafayette, Ind.
[1] This topic has been investigated by Newbold and Ho [18] for the case when the statistics describing each of the switched environments are known.

means) with the result that for these distributions, the power of the K–S test is not as good as the classical Neyman–Pearson test [14].

Wilcoxon-Mann-Whitney (WMW) Test [10], [15]

This test is generally used for testing H_0 versus H_2 or H_3 and hence is a one-sided test. If we test H_2 and H_3 together (in the form of an equal tails test), we have the WMW two-sided test. Note that this combination is not as general as H_1 but may be applicable in a great many instances, i.e., a shift in mean of the noise statistics.

Let $x_1,\cdots,x_n$ and $y_1,\cdots,y_m$ be two independent samples with continuous cdf's F and G, respectively. Let the WMW statistic be $T_{n,m}$, where

$$T_{n,m} = \sum_{i=1}^{n} \sum_{j=1}^{m} t(x_i,y_j) = \text{number of } y \leq x \qquad (3)$$

where

$$t(X,Y) = \begin{cases} 1, & \text{if } X \geq Y \\ 0, & \text{if } X < Y. \end{cases}$$

The mean and variance of $T_{n,m}$ are [15]

$$E(T_{n,m}) = nmp$$

and

$$\text{var}\,(T_{n,m}) = nm[(n-1)a_1^2 + (m-1)b_1^2 + p_1]$$

where

$$p = P(X \geq Y) = \int (1 - G(x))\, dF(x)$$

$$a_1^2 = a^2 - p^2$$

$$b_1^2 = b^2 - p^2$$

$$p_1 = p - p^2$$

$$a^2 = \int (1 - F(x))^2\, dG(x)$$

$$b^2 = \int G^2(x)\, dF(x).$$

Under H_0, $p = 1/2$ and

$$E(T) = \frac{nm}{2}$$

$$\text{var}\,(T) = \frac{nm(n+m+1)}{12}.$$

Mann and Whitney have shown [16] that

$$U = \frac{T - \frac{1}{2}nm + \frac{1}{2}}{\sqrt{\frac{1}{12}(nm)(n+m+1)}}$$

is approximately Gaussianly distributed with zero mean and unity variance even for $n = m = 8$. Again, as with the K–S statistic, not much work has been done for the distribution of $T_{n,m}$ under the alternative hypothesis.

III. Application to Calamity Detection

Let us consider a control system with known or unknown plant dynamics where the observed state vector is corrupted with measurement noise. We would like to be able to detect a malfunction or calamity of the system in as short a time as possible. To be more precise, consider as an illustrative example a continuous control system of the form

$$\begin{aligned} \dot{x}(t) &= Ax(t) + bu(t) \\ z(t) &= cx(t) + \eta(t) \end{aligned} \qquad (4)$$

where A is a 2×2 matrix, $b = (0,1)^T$, $c = (1,0)$, $x(t) = (x_1(t),x_2(t))^T$, and $\eta(t)$ is stationary white noise with unknown statistics. Let $Z = \{z(t_1),\cdots,z(t_n)\}$ be the set of sampled values corresponding to n samples of the process. On the basis of Z, we are to decide whether or not a calamity took place. Three cases will be considered.

Case 1

The plant dynamics are unknown but the system is in steady state with x_1 constant with time. Hence Z is a set of sampled values of the noise η with some unknown location parameter x_1 added to its cdf F.

Case 2

The plant dynamics and initial conditions are known. Thus the set $X = \{x_1(t_1),\cdots,x_1(t_n)\}$ can be calculated. Let $\Gamma = \{z(t_1) - x_1(t_1),\cdots,z(t_n) - x_1(t_n)\}$. Then Γ is a set of n samples of the unknown noise η. This problem is now a subclass of Case 1.

Case 3

The plant dynamics are unknown, but a collection of sets $Z_1,Z_2,\cdots,Z_k$ are available corresponding to normal operating conditions where

$$Z_j = \{z_j(1),\cdots,z_j(n_j)\} \qquad (5)$$

is a set containing n_j samples from the time interval $(t_{j-1},t_j]$.

Proposed Method for Cases 1 and 2

The approach to solving this problem obviously requires that a "standard" sample of the normal control system be obtained before calamity detection can begin. Let Z be the standard sample. (For Case 2, $Z = \Gamma$.) Assume that z_i comes from a distribution with continuous cdf F and that the z_i, $i = 1,\cdots,n$, are independent. During calamity detection, a test sample $Y = (y_1,\cdots,y_m)$ will be obtained in the same manner as Z. Assume that the cdf for the y_i is G. For each sample Y we will test the hypothesis H_0:$F = G$, based on one of the statistics described in the previous section.

One of the main shortcomings in the analysis of this procedure is that we can only specify the probability of rejection when H_0 is true (type 1 error). The power of the test depends on the type of distributions considered (parametric case) and, therefore, cannot be computed for a general nonparametric alternative. Hence we cannot specify the probability of a type 2 error (the probability that H_0 is accepted when the alternative hypothesis is true). Restating, we can only specify the probability that a normal operation is called a calamity. The probability that a calamity will not be detected cannot be specified unless a particular alternative parametric hypothesis is postulated. Note that this conclusion does not say that this approach is not useful; it only says that we do not know, in general, how useful it is.

Proposed Method for Case 3

Consider a control system which repeats its cycle at known instants, for instance, every T seconds. Let us partition this cycle time into J regions: $(0,t_1]$, $(t_1,t_2],\cdots$, $(t_{J-1},T]$. For each of the J regions R_i, $i=1,\cdots,J$, assume that we have a standard sample $Z_i=\{z_i(1),\cdots,z_i(n_i)\}$, where $z_i(k)$ is the sample value at time $t_{i-1}+(k/n_i)(t_i-t_{i-1})$, i.e., we take n_i samples equally spaced in time from each region R_i. If we treat this sample Z_i as coming from a population with a continuous stationary cdf F_i, we can proceed as in the proposed method for Cases 1 and 2 for testing each region R_i, $i=1,\cdots,J$. Without loss of generality, we shall assume that $n_i=n$, $i=1,\cdots,J$. Considering Z_i to be a set of n independent identically distributed random variables, any one observation drawn at random from this set must be considered as coming from a mixture distribution with probability density function (pdf) given by[2]

$$p_z(z) = \frac{1}{n} p_{i,1}(z) + \frac{1}{n} p_{i,2}(z) + \cdots + \frac{1}{n} p_{i,n}(z). \quad (6)$$

Now, letting $\tau_k=t_{i-1}+(k/n)(t_i-t_{i-1})$ and since $z(t)=x_1(t)+\eta(t)$, (6) can be rewritten as

$$p_z(z) = \frac{1}{n} p_\eta(z - x_1(\tau_1)) + \cdots + \frac{1}{n} p_\eta(z - x_1(\tau_n)).$$

Similarly, the cdf can be expressed as

$$F_z^{(n)}(z) = \frac{1}{n} F_\eta(z - x(\tau_1)) + \cdots + \frac{1}{n} F_\eta(z - x(\tau_n)). \quad (7)$$

Note that for $\Delta=t_i-t_{i-1}$ and for n approaching ∞, (7) becomes

$$\lim_{n\to\infty} F_z^{(n)}(z) = F_z(z) = \frac{1}{\Delta}\int_{t_{i-1}}^{t_i} F_\eta(z - x_1(t))\, dt. \quad (8)$$

This concept is illustrated in the following example.

[2] To avoid subscripted subscripts, the notation $p_{i,j}(z)$ has been introduced for $p_{z_i(j)}(z)$.

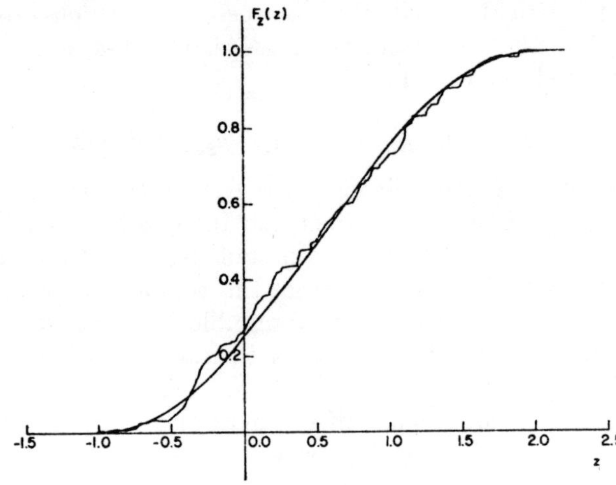

Fig. 1. Comparison of theoretical limiting cdf and empirical cdf based on 100 sample points.

Let

$$F_\eta(x) = \begin{cases} 0, & x \le -1 \\ \frac{1}{2}(x+1), & -1 < x < +1 \\ 1, & x \ge +1 \end{cases}$$

$$x_i(t) = t, \quad 0 < t \le 1.$$

Then from (8), we have

$$F_z(z) = \int_0^1 F_\eta(z-t)\, dt$$

or

$$F_z(z) = \begin{cases} 0, & z \le -1 \\ \frac{1}{4}(z+1)^2, & -1 < z < 0 \\ \frac{1}{4} + \frac{1}{2}z, & 0 \le z < 1 \\ z - \frac{1}{4}z^2, & 1 \le z < 2 \\ 1, & z \ge 2. \end{cases}$$

Fig. 1 shows a plot of this limiting distribution and a plot of an empirical cdf for $n=100$ which was generated by Monte Carlo methods on an IBM 7094 digital computer.

IV. Simulation Results

To indicate the feasibility of the method proposed in Section III, various simulations using Monte Carlo techniques were performed on an IBM 7094 digital computer. The experiments given in Table I consisted of distinguishing between two distributions involving n

TABLE I

EXPERIMENTAL RESULTS USING TWO-SAMPLE TESTS

Trial	Problem: Test $H_0: F = G$ versus $H_1: F \neq G$	Test	P(Decide H_1/H_0)	Sample Size	Experimental Results for P(Decide H_1/H_1) = 75%	Number of Iterations
1*	$N(0, 1) + Au_{-2}(t) + B$ $F: A = 1,\ B = 0$ $G: A = 1,\ B = 0(0.1)1$	K-S	0.05	$n = m = 40$	$B \approx 0.7$	100
2	same as trial 1	K-S	0.050	$n = m = 100$	$B \approx 0.5$	100
3†	$R(0, 2) + Au_{-2}(t) + B$ $F: A = 1,\ B = 0$ $G: A = 1,\ B = 0(0.1)1$	K-S	0.052	$n = m = 10$	$B \approx 1$	100 and 500
4	same as trial 3	K-S	0.054	$n = m = 40$	$B \approx 0.5$	100
5	same as trial 3	K-S	0.050	$n = m = 100$	$B \approx 0.3$	100
6	$R(0, 2) + Au_{-2}(t) + B$ $F: A = 1,\ B = 0$ $G: A = 1,\ B = 0(0.1)0.7$	WMW	≤ 0.05	$n = m = 10$	—	100
7	same as trial 6	WMW	≤ 0.05	$n = m = 40$	$B \approx 0.45$	100
8	same as trial 6	WMW	≤ 0.05	$n = m = 100$	$B \approx 0.25$	100
9	$N(0, 1) + Au_{-2}(t) + B$ $F: A = 0,\ B = 0$ $G: A = 0(0.1)1,\ B = 0$	WMW	≤ 0.05	$n = m = 40$	—	100
10	same as trial 9	WMW	≤ 0.05	$n = m = 100$	$A \approx 0.7$	100

* $N(a, b)$ represents a normal distribution with mean $= a$ and variance $= b$.
† $R(a, b)$ represents a rectangular distribution with mean $= a$ and interval length $= b$.

TABLE II

CALAMITY DETECTION RESULTS FOR A SIMPLE CONTROL SYSTEM

Calamity	Noise	Calamity Time (seconds)	Number of Samples (0.01 second apart)	Decision of Possible Intervals for Calamity (seconds)
a)	$N(0, 1)$	1.25	100	2–4
	$N(0, 0.25)$	1.25	100	2–5
	$N(0, 1)$	1.25	40	2.4–3.6, 4.0–4.4
	$N(0, 0.25)$	1.25	40	2.8–4.8
			5, 10	Results are poor
b)	$N(0, 1)$	2.25	100	4–5
	$N(0, 0.25)$	2.25	100	3–5
	$N(0, 1)$	2.25	40	1.2–1.6, 4.0–4.4
	$N(0, 0.125)$	2.25	40	0.8–1.2, 3.2–3.6, 4.4–4.8
c)	$N(0, 1)$	2.25	100	2–4
	$N(0, 0.25)$	2.25	100	2–5
	$N(0, 1)$	2.25	40	0.4–0.8, 2.0–2.4, 2.8–4.0
	$N(0, 0.25)$	2.25	40	0.0–0.4, 2.4–4.8
d)	$N(0, 1)$	2.25	100	2–5
	$N(0, 0.25)$	2.25	100	2–5
	$N(0, 1)$	2.25	40	2–4.8
	$N(0, 0.25)$	2.25	40	2.4–4.8
	$N(0, 0.25)$	2.25	10	0.8–0.9, 1.2–1.3, 1.2–4.3, 4.4–4.6, 4.8–5.0
	$N(0, 0.25)$	2.25	5	Results similar to 10 samples
e)	$N(0, 1)$	2.25	100	2–5
	$N(0, 0.25)$	2.25	100	2–5
	$N(0, 1)$	2.25	40	2–4.8
	$N(0, 0.25)$	2.25	40	2–4.8
	$N(0, 1)$	2.25	10	0.7–0.8, 2.2–4.9
	$N(0, 0.25)$	2.25	10	2.2–5.0
	$N(0, 1)$	2.25	5	0.56–0.80, 1.25–1.30, 2.25–4.25, 4.45–4.55, 4.65–4.85

equally spaced samples of a linear function of the form $x_1(t) = A \cdot u_{-2}(t) + B$, where $u_{-2}(t)$ is the unit ramp function and A and B are constants. These samples were then corrupted with the specified noise samples indicated in the second column of Table I. The fifth column indicates the minimum value of the critical parameter necessary to achieve 75-percent accuracy in deciding on the alternative hypothesis. The 75-percent level was chosen as a means of comparison only and has no special significance. As is to be expected, an increase in the value of the parameter in the fifth column yielded an increase in the detection level for all the experiments conducted here. For the tests considered, we might note that sample sizes should be greater than 10.

Table II gives the results of detecting a change in a dynamic control system using a WMW test. Samples of both the normal system and the faulty system were taken at every 0.01 second for 5 seconds. The tests were based on groups of n of these samples (i.e., detection time $\geq 0.01*n$ seconds). The control system under test is of the following form:

$$\begin{bmatrix} \dot{x}_1 \\ \dot{x}_2 \end{bmatrix} = \begin{bmatrix} 0 & 1 \\ -1 & -2\xi \end{bmatrix} \begin{bmatrix} x_1 \\ x_2 \end{bmatrix} + \begin{bmatrix} 0 \\ 1 \end{bmatrix} u$$

$$z = x_1 + \eta$$

where $u = 1$, $\xi = 1$, $x_1(0) = x_2(0) \equiv 0$, and is normally distributed$\sim N(0, \sigma^2)$.

The calamities investigated were the following:

a) ξ changed to 0.25 at time 1.25 seconds
b) ξ changed to 0.25 at time 2.25 seconds
c) x_1 and x_2 set to 0.0 at time 2.25 seconds
d) x_1 set at 2.5 at time 2.25 seconds
e) x_1 set at 5.0 at time 2.25 seconds.

A typical distribution is given in Fig. 2 for calamity a).

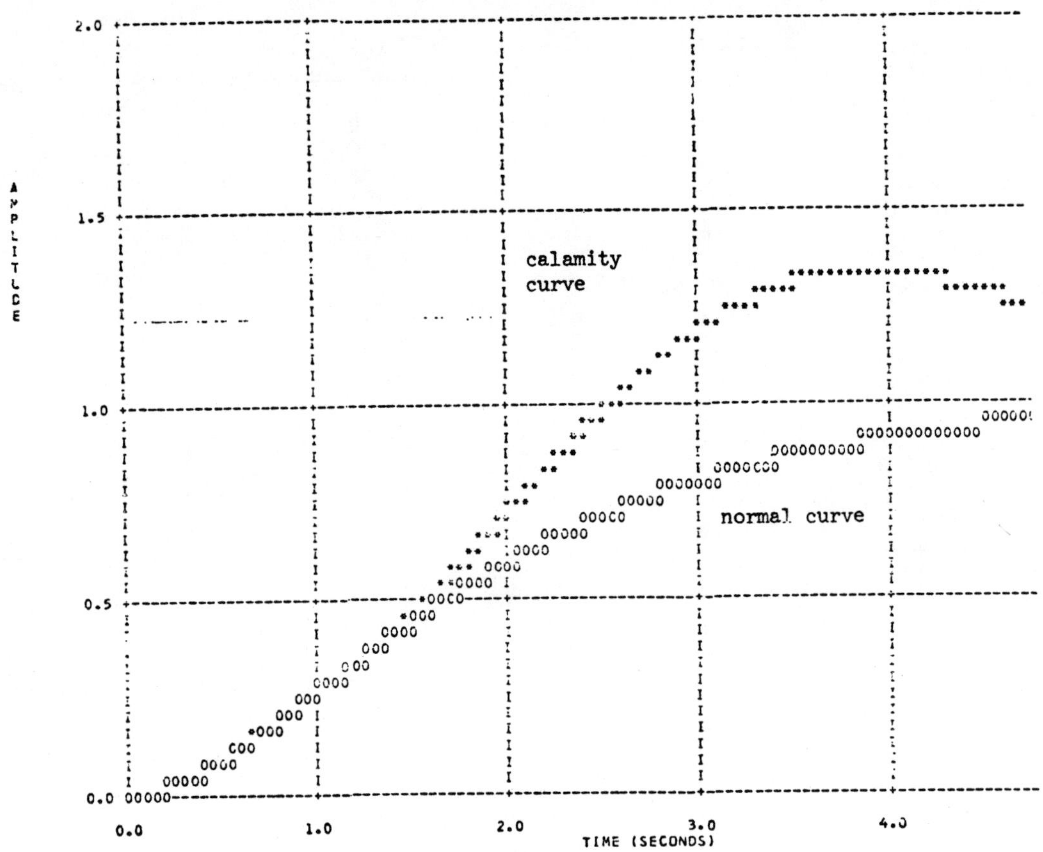

(a)

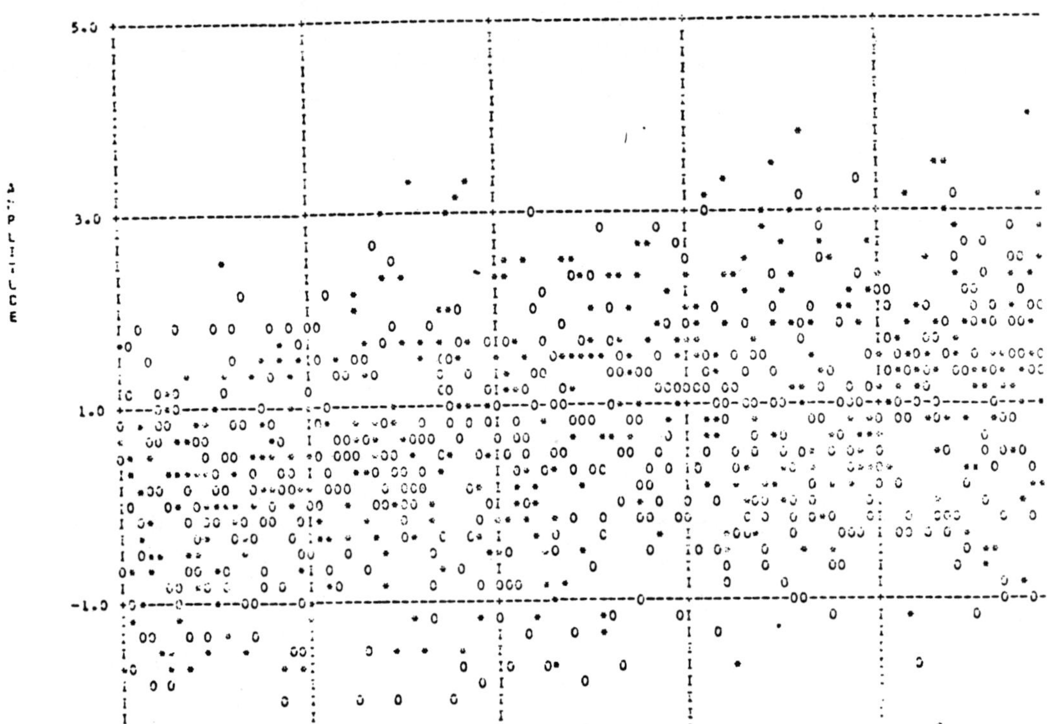

(b)

Fig. 2. (a) Normal and calamity curves without measurement noise. (b) Normal and calamity curves with measurement noise; $N(0,1)$.

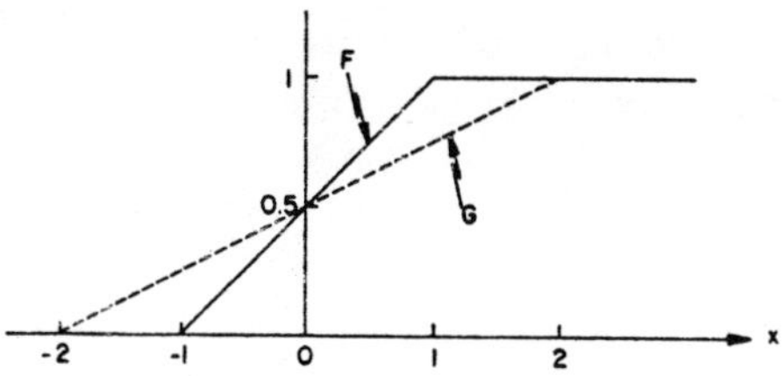

Fig. 3. Two distribution functions for which the WMW test is inadequate.

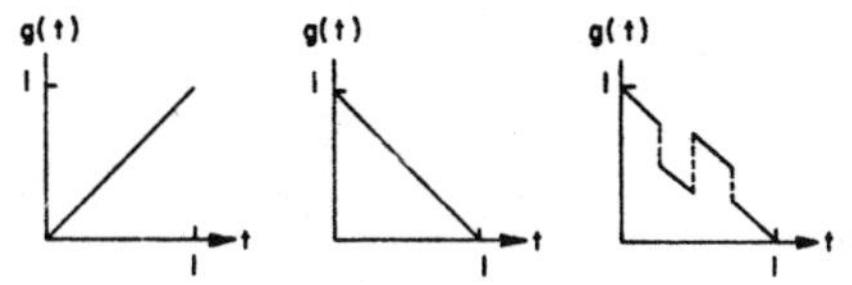

Fig. 4. Waveforms which cannot be distinguished by two-sample tests.

V. Conclusions

Only Gaussian and uniform noise sources were used in these simulations. The results of Tables I and II indicate that tests based on increased sample sizes have a much better rejection ratio. From Table I we note that the WMW test is more efficient than the K–S test for the simulations performed. This will not always be the case. To see this, consider the two distributions in Fig. 3. Using the WMW test with $H_0: F=G$ or, equivalently, $p=\int(1-G(x))dF(x)=1/2$ versus $H_1: p\neq 1/2$, we see that

$$p = \int (1 - G(x))\, dF(x)$$

$$= \int_{-1}^{+1} \frac{1}{2}\left(1 - \frac{1}{4}x - \frac{1}{2}\right) dx$$

$$= \frac{1}{2} - \int_{-1}^{+1} \frac{1}{4} x\, dx = \frac{1}{2}.$$

Thus this test cannot distinguish between F and G, whereas the K–S test can. We also note that the scheme presented for Case 3 in Section III could not distinguish between any of the output responses shown in Fig. 4, i.e., they all yield the same distribution function according to (7).

The concept of using a nonparametric statistic for analysis of a dynamic system with stationary measurement noise appears to be a useful technique. It is important to note that the procedure indicated in this paper is by no means limited to the second-order linear system used here only for purposes of illustration. The main point is that identical deterministic (as opposed to stochastic) systems operating under similar conditions will behave in similar fashions. If a change in the nominal operation of a deterministic system can be detected by observing the measured variables, then this change can also be detected when the measurements are corrupted by noise. (For vector-valued observations, nonparametric tests on each of the observed components could be performed independently.)

If the calamity state of operation were known or if a system were to operate in a switching environment, then the concept of a dual hypothesis test [17] might prove useful here. Basically, this is a sequential test for determining whether an unknown sample is the same as prototype sets of samples. The test continues as long as the comparison of the unknown sample and each of the stored prototypes yields similar results.

Acknowledgment

The authors wish to thank Prof. R. L. Kashyap for suggesting this problem to them.

References

[1] Z. J. Nikolic, "A study of learning systems operating in unknown stationary environments," Ph.D. dissertation, School of Electrical Engineering, Purdue University, Lafayette, Ind., Rept. TR-EE 67-1, January 1967.
[2] J. W. Carlyle and J. B. Thomas, "On nonparametric signal detectors," *IEEE Trans. Information Theory*, vol. IT-10, pp. 146–152, April 1964.
[3] J. B. Millard and L. Kurz, "Adaptive threshold detection of *M*-ary signals in statistically undefined noise," *IEEE Trans. Communication Technology*, vol. COM-14, pp. 601–610, October 1966.
[4] ——, "The Kolmogorov–Smirnov tests in signal detection," *IEEE Trans. Information Theory (Correspondence)*, vol. IT-13, pp. 341–342, April 1967.
[5] J. Capon, "A nonparametric technique for the detection of a constant signal in additive noise," *1959 IRE WESCON Conv. Rec.*, pt. 4, pp. 92–104.
[6] S. S. Wolff, J. B. Thomas, and T. P. Williams, "The polarity-coincidence correlator: a nonparametric detection device," *IRE Trans. Information Theory*, vol. IT-8, pp. 5–9, January 1962.
[7] R. F. Daly and C. K. Rushforth, "Nonparametric detection of a signal of known form in additive noise," *IEEE Trans. Information Theory*, vol. IT-11, pp. 70–76, January 1965.
[8] E. L. Lehmann, "The power of rank tests," *Ann. Math. Stat.*, vol. 24, pp. 23–44, 1953.
[9] D. A. Darling, "The Kolmogorov–Smirnov, Cramer–Von Mises tests," *Ann. Math. Stat.*, vol. 28, pp. 823–838, 1957.
[10] S. S. Wilks, *Mathematical Statistics*. New York: Wiley, 1963, pp. 452–468.
[11] N. V. Smirnov, "On the estimation of the discrepancy between empirical curves of distribution for two independent samples," *Bull. Math. Univ. Moscow*, vol. 2, no. 2, pp. 3–14, 1939.
[12] ——, "Table for estimating the goodness of fit of empirical distributions," *Ann. Math. Stat.*, vol. 19, pp. 279–281, 1948.
[13] F. J. Massey, "The distribution of the maximum deviation between two sample cumulative step functions," *Ann. Math. Stat.*, vol. 22, pp. 125–128, 1951.
[14] B. L. Van Der Waerden, "Order tests for the two-sample problem, pts. I and II," *Ned. Akad. Wetensch. Proc.*, ser. A56, pp. 303–316.
[15] H. Teicher, "Nonparametric methods in statistics," STAT 555 class notes, Purdue University, Lafayette, Ind., 1966.
[16] H. B. Mann and D. R. Whitney, "On a test of whether one or two random variables is stochastically larger than the other," *Ann. Math. Stat.*, vol. 18, pp. 50–60, 1947.
[17] H. D. Chadwick and L. Kurz, "Two sequential nonparametric detection procedures," Dept. of Electrical Engineering, New York University, N.Y., Rept. TR 400-155, March 1967.
[18] P. M. Newbold and Y. C. Ho, "Detection of changes in the characteristics of a Gauss–Markov process," *IEEE Trans. Aerospace and Electronic Systems*, vol. AES-4, pp. 707–718, September 1968.

23

Reprinted with permission from *IEEE Trans. Inf. Theory* **IT-16**:541–548 (1970)

A Distribution-Free Discrimination Procedure Based on Clustering

MARSHALL W. ANDERSON AND ROGER D. BENNING, MEMBER, IEEE

Abstract—A method is presented for constructing distribution-free tolerance regions for one distribution based on the clusters of the sample from a second distribution. The tolerance regions are constructed so as to minimize the probability of reserve judgment in a two-class discrimination procedure that allows the conditional error probabilities to be controlled within prescribed upper bounds. The method results in discrimination functions that are easily implemented on a computer. Although the method is not, in general, consistent with optimal procedures, it is appealing for high-dimensional problems with multimodal distributions.

I. Introduction

A DISTRIBUTION-free discrimination procedure is presented in an article by Quesenberry and Gessaman [1]. This procedure is appealing in that the conditional error probabilities can be controlled. To achieve this property, regions of partial decisions and reserve judgment are allowed. Quesenberry and Gessaman point out that the usefulness of their procedure depends upon the ability to control the probability measure of these regions. Since the regions of partial decisions and reserve judgment are formed from intersections of distribution-free tolerance regions, the usefulness of the procedure depends on the choice of ordering functions used to construct these tolerance regions.

For the two-class problem, the optimal procedure is defined to be the one that minimizes the probability of reserve judgment while controlling the conditional error probabilities within prescribed upper bounds. Quesenberry and Gessaman point out that for some families of distributions it is possible to select ordering functions that will make their distribution-free procedure consistent with the optimal procedure for the given family.

In practice, such a consistent distribution-free procedure often can not be constructed because the available information is not sufficient to suggest a parametric family that contains the distributions. In other cases, even if the parametric family can be suggested, optimal procedures will not be known. These limitations become more severe as the number of modes and dimensionality of the distributions increase.

The method presented in this paper avoids the above difficulties by achieving a suboptimum solution for the

Manuscript received July 10, 1969; revised February 10, 1970. This work was supported by USN Contract N00019-67-C-1216.
M. W. Anderson was with Bell Telephone Laboratories, Inc., Greensboro, N. C. He is now with North Carolina State University, Raleigh, N. C. 27607.
R. D. Benning is with Bell Telephone Laboratories, Inc., Greensboro, N. C. 27407

two-class discrimination problem, which requires only the assumption of continuity on the distributions. It bases the set of ordering functions used to form tolerance regions for the first distribution on the clusters of the sample drawn from the second distribution and vice versa. Although the resulting distribution-free procedure, in general, is not consistent with the optimal procedure, it should give satisfactory results for many problems. It has the added advantage of being easily implemented on a computer, whereas a consistent procedure would likely be difficult to use in practice.

Other distribution-free discrimination procedures have been suggested and analyzed by Fix and Hodges [2], [3], Stoller [4], Cover and Hart [5], Anderson [6], Kendall [7], Patrick [8], Patrick and Fisher [9], and Fu and Chien [10]. These are generally oriented to satisfying the Bayes' minimum-average-cost criterion. Kendall [7] also suggests allowing regions of reserve judgment. Fu and Chien [10] propose a sequential procedure based on Wald's sequential probability ratio test and a ranking procedure.

The distribution-free technique presented by Quesenberry and Gessaman [1] is summarized in Section II for discrimination between two classes before presenting the procedure based on clustering in Section III. The procedure is illustrated in Section IV by two examples. The first example considers nonoverlapping distributions for which perfect discrimination is possible. In the second example, the procedure is applied to some data taken from an article by Kendall [7]. The Appendix presents an extension of Fraser's [11] method of constructing distribution-free tolerance regions. This extension is used to show that the procedure based on clustering has the desired properties of Quesenberry and Gessaman's procedure.

II. General Model and Distribution-Free Procedure

General Model

An observation z is obtained on a random variable Z having either distribution P_1 or P_2, where P_1 and P_2 are distinct members of the class of continuous distributions on n-dimensional Euclidean space R^n. The problem is to decide whether z came from P_1 or P_2.

Let Δ denote the class of decisions with elements given by

δ_i: decide z came from P_i $\quad i = 1, 2$
δ_0: reserve judgment.

By reserve judgment it is meant that no decision is made as to whether z came from P_1 or P_2. A discriminant function d, which maps R^n to Δ, partitions the sample space into subsets given by

$$S_i = \{x: d(x) = \delta_i\} \qquad i = 0, 1, 2.$$

An error will be made if $d(z) = \delta_2$ and z came from P_1 or if $d(z) = \delta_1$ and z came from P_2. Then $q_1 = P_1(S_2)$ and $q_2 = P_2(S_1)$ are the conditional probabilities of an error given P_1 and P_2, respectively. The discrimination procedure d is said to be of size (α_1, α_2) if $q_j \leq \alpha_j$ for each $j = 1, 2$, where α_j is a fixed number in the open interval (0, 1). A procedure is exact size (α_1, α_2) if $q_j = \alpha_j$ for each $j = 1, 2$.

Distribution-Free Procedure

It is assumed that there is a sample $(x_{j1}, \cdots, x_{jn_j})$ available from each distribution P_j. Let a_j be the largest integer in the quantity $\alpha_j(n_j + 1)$. Using the theory of forming sample blocks summarized in the Appendix, construct a distribution-free tolerance region A_j containing a_j sample blocks on R^n for the distribution P_j. Let $\bar{A}_j = R^n - A_j$. The distribution-free discrimination procedure d^* is defined by

$$d^*(z) = \delta_i \qquad z \in S_i$$

where $S_1 = \bar{A}_1 A_2$, $S_2 = A_1 \bar{A}_2$, $S_0 = A_1 A_2 \cup \bar{A}_1 \bar{A}_2$. Since $S_2 \subseteq A_1$ and $S_1 \subseteq A_2$, the conditional error probabilities satisfy the inequalities

$$q_1 \leq P_1(A_1) \quad \text{and} \quad q_2 \leq P_2(A_2)$$

where $P_j(A_j)$ is a random variable having the beta distribution $Be[a_j, n_j - a_j + 1]$. For a particular α_j and n_j, Pearson's [12] tables or Murphy's [13] graphs can be consulted for the solution β_j to the probability statement

$$\Pr\{P_j(A_j) \leq \alpha_j\} = \beta_j.$$

Conversely, the tables can be used to determine the sample size n_j that is necessary to give an error size α_j with confidence β_j. An application of Chebyshev's inequality establishes that

$$P_j(A_j) \rightarrow_p \alpha_j \quad \text{as} \quad n_j \rightarrow \infty \qquad j = 1, 2.$$

That is, the distribution-free procedure is, with high confidence, size (α_1, α_2) for large sample sizes.

The above distribution-free procedure is valid for any set of ordering functions (satisfying mild conditions given in the Appendix) used to form the distribution-free tolerance regions. However, in an application one is interested in choosing the set that will minimize the probability of reserve judgment. To guide us towards this minimization, we observe a theorem in Quesenberry and Gessaman [1]. Under appropriate restrictions on the probability density functions f_1 and f_2 of the distributions P_1 and P_2, this theorem guarantees the existence of sets A_1^0 and A_2^0 such that the decision function d^0 defined by

$$d^0(z) = \delta_i \qquad z \in S_i^0,$$

where $S_1^0 = \bar{A}_1^0 A_2^0$, $S_2^0 = A_1^0 \bar{A}_2^0$ and $S_0^0 = A_1^0 A_2^0 \cup \bar{A}_1^0 \bar{A}_2^0$ minimizes the probability of reserve judgment. Although this theorem can only be used to determine ordering functions that will make the distribution-free procedure presented above consistent with d^0 when P_1 and P_2 are known to belong to specific types of distributions, the general content of the theorem serves as a useful guide. In essence, the theorem says that the set A_1^0 is chosen to include those values of x where the ratio $f_2(x)/f_1(x)$ is as large as possible, and A_2^0 is chosen to include those values of x where the ratio of $f_2(x)/f_1(x)$ is as small as possible, subject to the constraints $P_1(A_1^0) \leq \alpha_1$ and $P_2(A_2^0) \leq \alpha_2$. Using this guide, we present a method based on clustering analysis for determining the sequence of ordering functions such that the resulting distribution-free tolerance regions A_1 and A_2 tend to give minimum probability of reserve judgment.

III. Distribution-Free Procedure Based on Clustering

Our procedure first locates the clusters of the sample $(x_{21}, \cdots, x_{2n_2})$ from distribution P_2 (class 2). Ideally, a cluster-seeking technique [14], [15] locates those sets of sample points where the density is high compared to the density points in the surrounding regions. It is reasonable to expect the ratio $f_2(x)/f_1(x)$ to be relatively large in the class 2 cluster regions, so these regions are likely candidates for inclusion in the set A_1^0. A set of ordering functions for class 1 is chosen (see the last paragraph in this section) and one function associated, by determination of its parameters, with each of the "significant" clusters of class 2. One sample block (see the Appendix) is formed with each of the selected ordering functions. At each succeeding step, the choice of the ordering function is governed by an estimate of the ratio $f_2(x)/f_1(x)$ in the previously formed sample blocks. The process continues until the desired number a_1 of sample blocks for distribution P_1 have been formed. Set A_1 is the union of these sample blocks. The above steps are then repeated interchanging the roles of class 1 and class 2. This stepwise construction of the sets A_1 and A_2 is described more precisely by the following procedural statement.

Procedure 1

Sample sets $(x_{11}, \cdots, x_{1n_1})$ and $(x_{21}, \cdots, x_{2n_2})$, from distributions P_1 and P_2, respectively, and the functions $h_{11}(x; b_{11}), \cdots, h_{1m_2}(x; b_{1m_2})$ and $h_{21}(x; b_{21}), \cdots, h_{2m_1}(x; b_{2m_1})$, known except for the sets of parameters $\{b_{11}, \cdots, b_{1m_2}\}$ and $\{b_{21}, \cdots, b_{2m_1}\}$, define sets A_1 and A_2 as follows ($n_1 > m_1$, $n_2 > m_2$).

Step 1: Find the clusters $\{C_{21}, \cdots, C_{2m}\}$ for the class 2 sample set $(x_{21}, \cdots, x_{2n_2})$ using a cluster-seeking technique [14], [15]. The clusters are labeled so that if cluster C_{2i} is more compact (by an appropriate criterion) than cluster C_{2j}, then $i < j$. Let $m_2 = \min(m, a_1)$.

Step 2: Calculate the set of parameters $\{b_{11}, \cdots, b_{1m_2}\}$, where $b_{1i} = b_{1i}(x_{21}, \cdots, x_{2n_2})$. In practice, b_{1i} would usually depend only on a subset of the class 2 sample set; e.g., b_{1i} may depend on only those points in cluster C_{2i}. Note that b_{1i} may be a vector, and that b_{1i} and b_{1j} do not necessarily have the same dimensionality for $i \neq j$ (see the Appendix).

Step 3: Construct the sample blocks $B_{11}, \cdots, B_{1m_2}$ from the sample $(x_{11}, \cdots, x_{1n_1})$ and the ordering functions $h_{11}(x; b_{11}), \cdots, h_{1m_2}(x; b_{1m_2})$. This construction is described in the Appendix. If $m_2 = a_1$, then go to Step 5. If $m_2 < a_1$, then continue to Step 4.

Step 4: Sample blocks $B_{1(m_2+1)}, \cdots, B_{1a_1}$ are also formed from the functions $h_{11}, \cdots, h_{1m_2}$. The scheme described below tells which of these functions to use at each state.

a) Let ξ_i be the number of class 2 points in block B_{1i}. Form block $B_{1(m_2+1)}$ by using the ordering function used to construct block B_{1i_1}, where i_1 is defined by $\xi_{i_1} = \max \xi_i$, the maximum being taken over $i = 1, \cdots, m_2$.
b) To construct $B_{1(m_2+2)}$, use the ordering function used to form block B_{1i_2}, where i_2 is defined by $\xi_{i_2} = \max \xi_i$, the maximum being taken over $i = 1, \cdots, m_2 + 1$ and $i \neq i_1$.
c) To construct B_{1r}, $m_2 + 1 \leq r \leq a_1$, use the ordering function used to form B_{1i_r} where i_r is defined by $\xi_{i_r} = \max \xi_i$, the maximum being taken over $i = 1, \cdots, r - 1$ and $i \neq i_1, \cdots, i_{r-1-m_2}$.

Step 5: Set $A_1 = \bigcup_{i=1}^{a_1} B_{1i}$. After the set A_1 has been formed, it is not necessary to keep account of the individual sample blocks used to form A_1. Let k_i be the largest integer j for which h_{1i} was used to construct a sample block B_{1j} in Step 3 or 4. A sequence $\{\gamma_1, \cdots, \gamma_{a_1}\}$ is defined in the description of the sample block construction given in the Appendix. With this, the set A_1 can be described simply using only m_2 inequalities as

$$A_1 = \{x: h_{11}(x; b_{11}) < \gamma_{k_1} \quad \text{or} \quad \cdots h_{1m_2}(x; b_{1m_2}) < \gamma_{k_{m_2}}\}.$$

Step 6: Use the sample $(x_{21}, \cdots, x_{2n_2})$ from P_2 and the ordering functions $h_{21}(x; b_{21}), \cdots, h_{2m_1}(x; b_{2m_1})$ to construct A_2 by interchanging the roles of class one and class two in Steps 1–5.

In Steps 2 and 4 the ordering function used to construct sample block B_{1i}, $i = 1, \cdots, a_1$, is determined. In Step 2 a parameter in the functional form is evaluated from the cluster of class 2. In Step 4 the selection of the ordering function depends upon the previously constructed sample blocks and the class 2 sample. Similar statements hold for the ordering functions used to construct the sample blocks of class 2. In any case, the determination of the ordering functions fits the conditions of Theorem 1 in the Appendix. Thus, it follows from Theorem 1 that the coverage $P_j(A_j)$ has the beta distribution $Be[a_j, n_j - a_j + 1]$. As indicated in Section II, the sample size n_j and the number of sample blocks a_j can be adjusted to give an error size less than or equal to α_j with confidence β_j.

So far nothing has been said about the form of the ordering functions. A reasonable choice for many problems is just the Euclidean distance from the respective cluster center; e.g., $h_{1i}(x; b_{1i}) = (x - \bar{x}_{2i})^T(x - \bar{x}_{2i})$, where $\bar{x}_{2i} = b_{1i}$ is the mean of cluster C_{2i}. When the number of points in cluster C_{2i} is sufficient to make meaningful estimates of Γ_{2i}, the covariance matrix within cluster C_{2i}, a good choice for the ordering function might be

$$h_{1i}(x; b_{1i}) = (x - \bar{x}_{2i})^T \hat{\Gamma}_{2i}^{-1}(x - \bar{x}_{2i}),$$

which defines b_{1i} as the set of all elements in $\bar{x}_{2i}$ and $\hat{\Gamma}_{2i}^{-1}$. If there is some a priori knowledge concerning the distributions, it may be used to advantage in choosing the functional forms of the ordering functions. It is relatively easy to see how to use available information to improve the decision function for simple examples. However, as the dimensionality of the observation space increases or as the numbers of modes in the distributions increase, it becomes increasingly difficult to use the a priori knowledge to good advantage. For these complex problems, the hyperspherical and hyperelliptical ordering functions suggested above might be a good compromise between optimum performance with computational complexity and suboptimum performance with computation simplicity.

IV. Examples

Two examples are presented to illustrate our procedure and some of its properties. The first example illustrates that the procedure works well for discriminating between two nonoverlapping distributions. The second example applies the procedure to some data taken from Kendall [7] and compares the result to his.

Example 1

Let ρ_{11}, ρ_{12}, ρ_{13}, ρ_{21}, and ρ_{22} be positive real numbers and v_{11}, v_{12}, v_{13}, v_{21}, and v_{22} be fixed points in 2-dimensional Euclidean space R^2. Define sets V_{11}, V_{12}, V_{13}, V_{21}, and V_{22} as

$$V_{ij} = \{x \in R^2 : |x - v_{ij}| < \rho_{ij}\}.$$

Let continuous distributions P_1 and P_2 be defined on R^2 such that $P_1(V_{11} \cup V_{12} \cup V_{13}) = 1$ and $P_2(V_{21} \cup V_{22}) = 1$.

Consider using Procedure 1 to construct distribution-free tolerance regions A_1 and A_2 from samples drawn from P_1 and P_2. Hypothetical samples of size $n_1 = n_2 = 25$ from these distributions are illustrated in Fig. 1. Let $a_1 = a_2 = 4$, which allows the following confidence statements to be made (using Murphy's [13] graphs):

$$\Pr\{P_1(A_1) \leq 0.29\} = 0.95$$

$$\Pr\{P_2(A_2) \leq 0.29\} = 0.95.$$

Assume that a cluster-seeking program has identified a cluster C_{ij} in the region of V_{ij} and also calculated the parameter b_{ij}, which might be the cluster mean, for each i and j. The sets V_{ij} and parameters b_{ij} are shown in Fig. 1; the clusters can easily be visualized. Let the ordering functions be chosen as $h_{ij}(x; b_{ij}) = |x - b_{ij}|$, for each i and j. Here $\bar{\imath} = 2$ if $i = 1$ and vice versa.

Sample block B_{11} is given by $B_{11} = \{x: h_{11}(x; b_{11}) < \gamma_1\}$, where $\gamma_1 = \min_i h_{11}(x_{1i}; b_{11})$ and $(x_{11}, \cdots, x_{1(25)})$ is the sample from P_1. Similarly, B_{12} is constructed according to Step 3 of Procedure 1 using the ordering function h_{12}. Since there are 14 class 2 points in block B_{11} and only 10 in block B_{12}, h_{11} is chosen as described in Step 4 to construct block B_{13}. Function h_{12} is chosen to construct block B_{14} because there are 10 class 2 points in B_{12} and only one in B_{13}. The tolerance region A_1 is the union of B_{11}, B_{12}, B_{13}, and B_{14}, as stated in Step 5. Similarly, Steps 1–5 are followed to construct sample blocks B_{21}, B_{22}, B_{23}, and B_{24}, and set A_2 is the union of these. The resulting sample blocks are shown in Fig. 1. Finally, the decision regions are formed as $S_1 = \bar{A}_1 A_2$, $S_2 = A_1 \bar{A}_2$, and $S_0 = A_1 A_2 \cup \bar{A}_1 \bar{A}_2$ following the description in Section II. In view of the statements made there, it can be said that the resulting discrimination procedure is of size (0.29, 0.29) with 95 percent confidence.

It is interesting to pursue this example further. An inspection of the decision regions S_1 and S_2 in Fig. 1 shows that $S_1 \cap (V_{21} \cup V_{22}) = \phi$ and $S_2 \cap (V_{11} \cup V_{12} \cup V_{13}) = \phi$. Thus, it follows that $q_1 = P_1(S_2) = 0$ and $q_2 = P_2(S_1) = 0$; i.e., the true error probabilities are zero. One may inquire as to whether this was a chance happening or whether such good performance can be expected of Procedure 1 under special circumstances. To formalize further discussion, make the following additional assumptions.

1) The cluster-seeking program yields parameters b_{11}, b_{12}, b_{21}, b_{22}, and b_{23}, which converge in probability to the fixed points v_{21}, v_{22}, v_{11}, v_{12}, and v_{13}, respectively, as n_1 and n_2 get large.
2) The distance between v_{1i} and v_{2j} satisfies the inequality
$$|v_{1i} - v_{2j}| > \rho_{1i} + \rho_{2j}$$
for each $i = 1, 2, 3$ and $j = 1, 2$.
3) The ordering functions have the form
$$h_{ij}(x; b_{ij}) = |x - b_{ij}|.$$

Given fixed α_1 and α_2 and large sample sizes n_1 and n_2, the tolerance regions A_1 and A_2 are constructed using Procedure 1. From these, the decision regions S_1, S_2, and S_0 are formed as usual. No samples are shown, but a typical outcome of such a construction might have the appearance of Fig. 2. The assumptions made above ensure that for large sample sizes, the set A_1 will contain almost all of sets V_{21} and V_{22}. Similarly, set A_2 will contain almost all of sets V_{11}, V_{12}, and V_{13}. Thus, the performance for large sample sizes is given by

$$P_1(S_2) \xrightarrow[p]{} 0, \qquad P_2(S_1) \xrightarrow[p]{} 0$$

$$P_1(S_0) \xrightarrow[p]{} \alpha_1, \qquad P_2(S_0) \xrightarrow[p]{} \alpha_2.$$

If α_1 and α_2 are chosen to be small, the probability of reserve judgment is small. Procedure 1 results in a good choice of tolerance regions for this example.

If Procedure 1 is modified slightly by fixing a_1 and a_2 with $a_1 \geq 2$ and $a_2 \geq 3$, then both α_1 and α_2 approach zero as n_1 and n_2 approach infinity. Thus by slight modification, Procedure 1 can be made consistent with the optimal procedure, which gives perfect discrimination for this example.

This example can be generalized to show the tendency of Procedure 1 to work well on distributions that have their masses concentrated in distinct isolated clusters. This is a desirable property. We would be disappointed

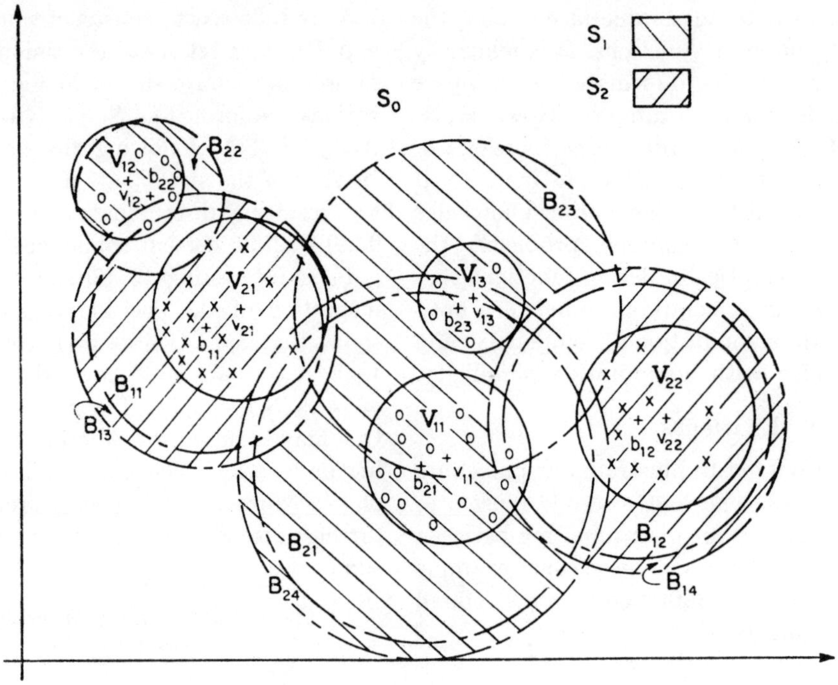

Fig. 1. Decision regions for Example 1 with sample sizes of 25.

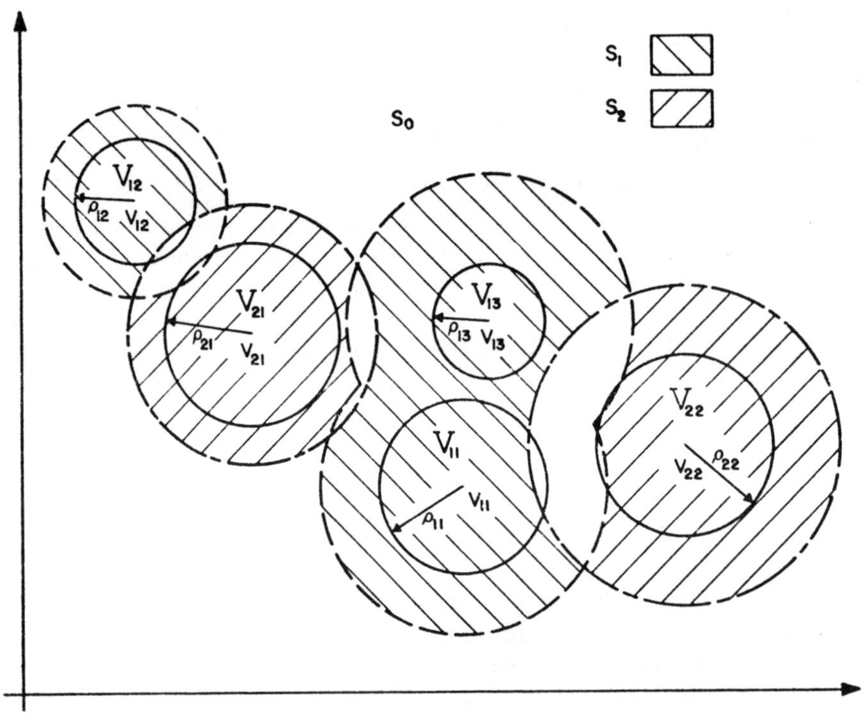

Fig. 2. Decision regions for Example 1 with large sample sizes.

in any procedure that did not perform well under such conditions since perfect or near-perfect discrimination is possible.

Example 2

Kendall [7] considers discrimination between two varieties of iris using petal length and petal width as features. In this example, Procedure 1 is applied to the same data. The results are compared to those obtained by Kendall using his order-statistic method, which is also a distribution-free procedure.

Prior to clustering, each coordinate is scaled by dividing through by the square root of the average of the within class sample variances in the coordinate direction. In this case petal length and petal width are divided by 0.51 and 0.24, respectively. The scaled data are plotted in Fig. 3.

Choosing $\alpha_1 = \alpha_2 = 0.05$, we find that $a_1 = a_2 = 2$ since $n_1 = n_2 = 50$ and a_i is the greatest integer in the quantity

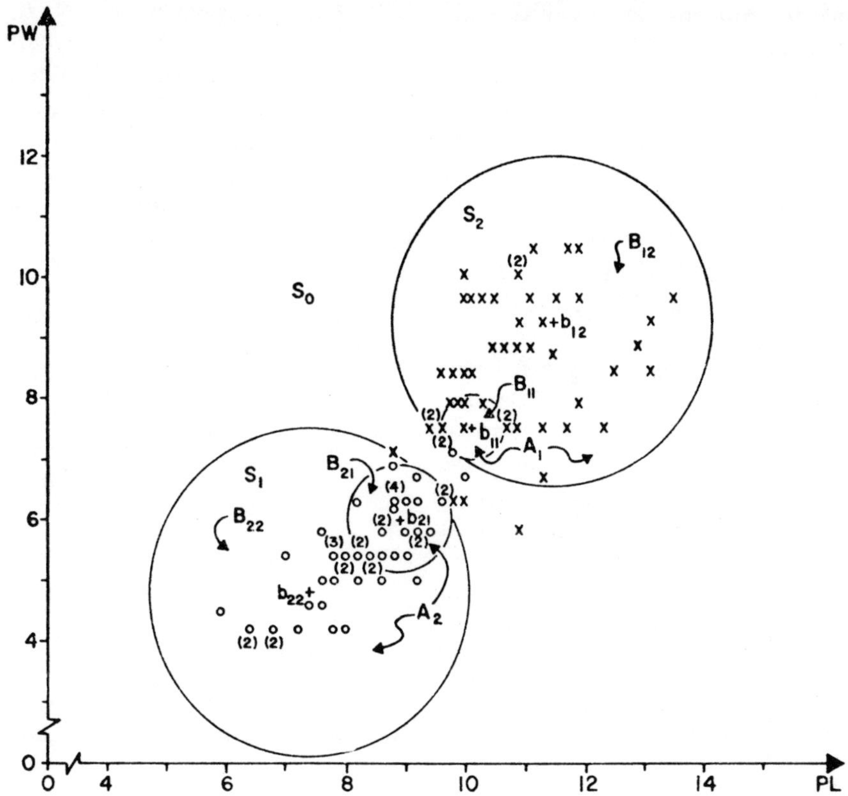

Fig. 3. Decision regions for Example 2 (the numbers in parentheses are the multiplicity of the data points). O— iris versicolor; X—iris virginica; +—parameters b_{ij}.

$\alpha_i(n_i + 1)$. Thus, we need, at most, two ordering functions and also can use, at most, two clusters for each class. The functional form of these ordering functions is chosen as

$$h_{ki}(x; b_{ki}) = |x - b_{ki}|,$$

with b_{ki} being the mean of the samples included in cluster $C_{\bar{k}i}$. Two clusters for each class were located using the diameter method of Johnson's [15] hierarchical clustering program. In this case, the two clusters were chosen so that they included all of the data from that class. The parameters b_{11}, b_{12}, b_{21}, and b_{22} are plotted in Fig. 3. Also shown in Fig. 3 are the tolerance regions A_1 and A_2 as constructed by Procedure 1. The decision regions S_1, S_2, and S_0 are formed from A_1 and A_2 as described in Section II. There are 2 cases of iris versicolor (class 1) and 6 cases of iris virginica (class 2) left undecided. One case of iris versicolor is misclassified. Letting q_1 and q_2 be the conditional error probabilities

$$q_1 = P_1(S_2) \quad \text{and} \quad q_2 = P_2(S_1),$$

the following confidence statements can be made:

$$\Pr\{q_1 \leq 0.09\} = 0.95$$

$$\Pr\{q_2 \leq 0.09\} = 0.95.$$

Kendall's [7] order-statistic method leaves 14 cases of versicolor and 8 cases of virginica undecided for this example. His method first discriminates on the basis of petal length (PL). Using the scaling of Fig. 3, the region where PL $\leq$ 8.6 is assigned to class 1 and the region where PL $\geq$ 10.1 to class 2. The region between these two values of petal length is further divided according to petal width (PW) as follows. Assign the region where PW $\leq$ 5.8 to class 1, the region where PW $\geq$ 7.9 to class 2, and in the remaining region make no decision. Since the region assigned to class 2 is contained in the union of two sample blocks of class 1 and the region assigned to class 1 is contained in the union of two sample blocks of class 2, Kendall's method also permits the above confidence statements to be made about the conditional error probabilities for this example.

In comparing our method to Kendall's, we do not conclude from this one example that our method is better in general. Both methods can be used to bound the error probabilitities by forming an appropriate number of sample blocks. The size of the reserve decision region for each depends on the form of the distributions. Our method compared favorably with his on this simple two-dimensional example because the tails of the distributions away from the clusters of the opposing class were insignificant. For other problems, this might not be so and Kendall's method would give smaller probability of reserve judgment. However, our method does have the advantage of being more easily applied to problems having multimodal distributions and/or more dimensions.

V. Discussion

Since Procedure 1 is based on clustering, it has some of the same advantages and limitations of cluster-seeking techniques. One of its advantages is that it is not greatly hampered by large dimensionality. Another is its capability of handling multimodal distributions. A limitation is the lack of a precise definition of what is and what is not a cluster.

Although Procedure 1, in general, is not consistent with the optimum procedure, it is appealing from the practical viewpoint in that it is easily implemented and should work reasonably well for samples of practical size. Since the procedure is structured to take advantage of any separation in multimodal distributions, it is most appealing when the distributions are multimodal.

The choice of ordering functions for Procedure 1 is not restricted to spherical or elliptical functions. If there is a priori knowledge about the distributions, it can be used to advantage in choosing the ordering functions. In some cases, this could result in Procedure 1 being consistent with the optimum procedure.

Appendix

The theory of coverages for sample blocks permits construction of distribution-free tolerance regions. An excellent introduction to the subject is given in a book by Wilks [16]. In the following, we summarize Fraser's [11] procedure for obtaining the sample blocks and then present an extension of his procedure. This extension shows that the sets A_1 and A_2 as constructed in Procedure 1 have the desired distribution-free properties.

To describe Fraser's technique, consider a random variable X whose distribution function F is a member of the class of continuous distributions on n-dimensional Euclidean space R^n. Let $h_1(x; b_1), \cdots, h_m(x; b_m)$ be a sequence of functions defined on R^n, $x \in R^n$, with vector valued parameters b_i such that the random variable $h_i(X; b_i)$, $i = 1, \cdots, m$ has a continuous distribution for each value of b_i. The functions $h_i(x; b_i)$ are referred to as ordering functions. The idea is to use a random sample $(x_1, \cdots, x_m)$ from F and the ordering functions to partition R^n in a particular way into $m + 1$ sets $B_1, \cdots, B_{m+1}$ such that if the random variable u_i is the probability measure from F contained in B_i, then the joint distribution function of $u_1, \cdots, u_m$ is independent of F. The set B_i is called a sample block and u_i is referred to as the coverage of the sample block B_i. Fraser [11] generates the sample blocks $B_1, \cdots, B_{m+1}$ in the following manner.

1) For the construction of B_1, let the parameter b_1 be given. Define $\gamma_1 = \min_i h_1(x_i; b_1)$, the minimum being taken over $i = 1, \cdots, m$. Then $B_1 = \{x \in R^n: h_1(x; b_1) < \gamma_1\}$. Let the integer i_1 be defined by $\gamma_1 = h_1(x_{i_1}; b_1)$.

2) For the construction of B_2, first determine[1] the parameter b_2 from γ_1; that is, $b_2 = b_2(\gamma_1)$. Now use $h_2(x; b_2(\gamma_1))$ and the points x_i, $i \neq i_1$, to partition $\tilde{B}_1$ into B_2 and $\tilde{B}_2$, as R^n was partitioned into B_1 and $\tilde{B}_1$ in 1); that is, $\tilde{B}_2 = \tilde{B}_1 - B_2$. Define $\gamma_2 = \min_i h_2(x_i; b_2(\gamma_1))$, the minimum being taken over $i = 1, \cdots, m$ and $i \neq i_1$. Then

$$B_2 = \{x \in \tilde{B}_1: h_2(x; b_2(\gamma_1)) < \gamma_2\}.$$

Let the integer i_2 be defined by $\gamma_2 = h_2(x_{i_2}; b_2(\gamma_1))$.

3) For the construction B_r, $1 < r \leq m$, first determine b_r from $\gamma_1, \cdots, \gamma_{r-1}$. Now use $h_r(x; b_r(\gamma_1, \cdots, \gamma_{r-1}))$ and the points x_i, $i \neq i_1, \cdots, i_{r-1}$, to partition $\tilde{B}_{r-1}$ into B_r and $\tilde{B}_r$. Define $\gamma_r = \min_i h_r(x_i; b_r(\gamma_1, \cdots, \gamma_{r-1})) = h_r(x_{i_r}; b_r(\gamma_1, \cdots, \gamma_{r-1}))$, the minimum being taken over $i = 1, \cdots, m$ and $i \neq i, \cdots, i_{r-1}$, which defines i_r. Then

$$B_r = \{x \in \tilde{B}_{r-1} : h_r(x; b_r(\gamma_1, \cdots, \gamma_{r-1})) < \gamma_r\}.$$

4) The block $B_{m+1} = \tilde{B}_m$.

Fraser [11] shows that the joint distribution of the coverages $u_1, \cdots, u_m$ is the m-variate Dirichlet distribution. It follows that the sum of any a of the coverages $u_1, \cdots, u_m$ has the beta distribution $Be(a, m - a + 1)$. Thus a distribution-free tolerance region whose coverage has the beta distribution $Be(a, m - a + 1)$, $a \leq m$ can be constructed from the sample blocks generated by the above procedure.

In Fraser's technique the parameter b_i is determined from the values $\gamma_i, \cdots, \gamma_{i-1}$ of the previously used ordering functions. We extend this by allowing the parameter b_i to also depend on a random variable Y that is independent of the random sample $(x_1, \cdots, x_m)$. To connect this discussion with Procedure 1 in Section III, associate the random variable X with one class and the random variable Y with the random sample from the other class. That is, when constructing blocks $B_{11}, \cdots, B_{1a}$, the realization y of Y is $y = (x_{21}, \cdots, x_{2n_2})$ and the random sample of X is $(x_1, \cdots, x_m) = (x_{11}, \cdots, x_{1n_1})$. Our extension of Fraser's method for constructing the sample blocks $B_1, \cdots, B_{m+1}$ is given in the following statement.

a) For the construction of B_1, first determine the value of b_1 from a realization y of Y. Then construct B_1 as described above in statement 1).

b) For the construction of B_r, $1 < r \leq m$, determine[2] the value of b_r from y and $\gamma_1, \cdots, \gamma_{r-1}$. Then construct B_r as described in statement 3).

c) Block B_{m+1} is again the complement of $\cup_{i=1}^{m} B_i$.

[1] Two illustrative examples of how the parameters b_r may depend upon the previous values $\gamma_1, \cdots, \gamma_{r-1}$ of the ordering functions are given in a paper by Fraser [17].

[2] In the procedure of Section III, the function $b_r(y, \gamma_1, \cdots, \gamma_{r-1})$ denotes an evaluation in Step 2 of a parameter from the clusters of the data y as well as a selection rule in Step 4 to determine which ordering function to use at stage r based on the cluster of the data y and the values $\gamma_1, \cdots, \gamma_{r-1}$ of the previously used ordering functions.

The following theorem shows that the joint distribution of the coverages $u_1, \cdots, u_m$ of these blocks is also the m-variate Dirichlet distribution.

Theorem 1

Let $u_1, \cdots, u_m$ be the coverages of the sample blocks $B_1, \cdots, B_m$ obtained by the procedure described in statements a)–c). Then the joint density $p(u_1, \cdots, u_m)$ is that of the m-variate Dirichlet distribution.

Proof: For a given value y_0 of the random variable Y, the procedure described in statements a)–c) reduces to that described in statements 1)–4). It follows from Fraser's [11] result that the conditional density function $p(u_1, \cdots, u_m \mid Y = y_0)$ is that of the m-variate Dirichlet distribution. Since this conditional density function is the same for each value of y_0, the desired result is established.

References

[1] C. P. Quesenberry and M. P. Gessaman, "Nonparametric discrimination using tolerance regions," *Ann. Math. Stat.*, vol. 39, no. 2, pp. 664–673, 1968.

[2] E. Fix and J. L. Hodges, Jr., "Discriminatory analysis, nonparametric discrimination: consistency properties," USAF School of Medicine, Project 21-49-004, Rept. 4, 1951.

[3] ——, "Discriminatory analysis, nonparametric discrimination: small sample performance," USAF School of Medicine, Project 21-49-004, Rept. 11, 1952.

[4] D. S. Stoller, "Univariate two-population distribution-free discrimination," *J. Amer. Stat. Assoc.*, vol. 49, pp. 770–777, 1954.

[5] T. M. Cover and P. E. Hart, "Nearest neighbor pattern classification," *IEEE Trans. Information Theory*, vol. IT-13, pp. 21–27, January 1967.

[6] T. W. Anderson, "Some nonparametric multivariate procedures based on statistically equivalent blocks," *Proc. Internatl. Symp. on Multivariate Analysis* (Dayton, Ohio), June 1965; also in *Multivariate Analysis*, P. R. Krishnaiah, Ed. New York: Academic Press, 1966, pp. 5–27.

[7] M. G. Kendall, "Discrimination and classification," *Proc. Internatl. Symp. on Multivariate Analysis* (Dayton, Ohio), June 1965); also in *Multivariate Analysis*, P. R. Krishnaiah, Ed. New York: Academic Press, 1966, pp. 165–185.

[8] E. A. Patrick, "Distribution free, minimum conditional risk learning algorithm," presented at the IEEE Internatl. Communications Conf. (Minneapolis, Minn.), June 1967.

[9] E. A. Patrick and F. P. Fisher, II, "A generalized k-nearest neighbor rule," *Inform. and Control* (to be published).

[10] K. S. Fu and Y. T. Chien, "Sequential recognition using a nonparametric ranking procedure," *IEEE Trans. Information Theory*, vol. IT-13, pp. 484–492, July 1967.

[11] D. A. S. Fraser, "Nonparametric tolerance regions," *Ann. Math. Stat.*, vol. 24, pp. 44–55, 1953.

[12] *Tables of the Incomplete Beta Function*, K. Pearson, Ed. New York: Cambridge University Press, 1934.

[13] R. B. Murphy, "Nonparametric tolerance limits," *Ann. Math. Stat.*, vol. 17, pp. 377–408, 1948.

[14] G. H. Ball, "Data analysis in the social sciences: what about the details?" *AFIPS Conf. Proc., Fall Joint Computer Conf.*, vol. 27, pp. 533–560, 1965.

[15] S. C. Johnson, "Hierarchical clustering schemes," *Psychometrika*, vol. 32, pp. 241–254, 1967.

[16] S. S. Wilks, *Mathematical Statistics*. New York: Wiley, 1962, pp. 234–244.

[17] D. A. S. Fraser, "Sequentially determined statistically equivalent blocks," *Ann. Math. Stat.*, vol. 22, pp. 372–381, 1951.

Reprinted with permission from *IEEE Trans. Comput.* **C-21**:1337–1347 (1972)

Distribution-Free Pattern Verification Using Statistically Equivalent Blocks

GUY W. BEAKLEY AND FRANZ B. TUTEUR

I. Introduction

PATTERN recognition typically involves a feature-extraction phase followed by a recognition or classification phase. Assuming the feature extraction to have been accomplished, the recognition problem can be formulated as follows. Suppose that x is an observation of one of k pattern classes, $i=1, 2, \cdots, k$, and suppose that if x belongs to class i, the cumulative probability distribution is $F_i(x)$. Suppose that n_i statistically independent observations known to be from class i are available for each $i=1, 2, \cdots, k$. These observations are referred to as training samples. Then a method is sought for classifying a new observation into one of the k classes.

In this paper the question to be answered is merely whether the new observation does or does not belong to a given class. Hence, $k=2$, and the decision is made by partitioning the space of all observations into two non-overlapping regions R_1 and R_2. Then x is classified as belonging to class i if it falls into region R_i.

Taking a Neyman–Pearson point of view, it would be desirable to minimize the probability of miss

$$P_M = \int_{R_1} dF_2(x) \tag{1}$$

subject to the condition that the probability of false alarm be no greater than a specified level:

$$P_{FA} = \int_{R_2} dF_1(x) \leq \alpha, \qquad 0 \leq \alpha \leq 1. \tag{2}$$

However, it is assumed in this paper that the probability distribution functions $F_i(x)$ are unknown. Then the following analog of the Neyman–Pearson criterion evolves naturally. Given n_i statistically independent observations from $F_i(x)$, $i=1, 2$, it is desirable to minimize

$$\int_{R_1} dF_2(x) \tag{3}$$

subject to either one of the conditions

$$P_r\left\{\int_{R_2} dF_1(x) \leq \beta\right\} = \gamma, \qquad 0 \leq \beta, \gamma \leq 1 \tag{4}$$

or

$$E\left[\int_{R_2} dF_1(x)\right] \leq \alpha, \qquad 0 \leq \alpha \leq 1. \tag{5}$$

Equations (4) and (5) can be established even though $F_1(x)$ is unknown. This is done through the theory of distribution-free tolerance regions [1]–[10].

Distribution-free tolerance regions have recently been used in classification problems. Anderson [19] proposed various multivariate statistical techniques based on the properties of statistically equivalent blocks. Patrick [20] and Patrick and Fisher [21] presented a general classification approach that is referred to as an empirical Bayes approach for distribution-free minimum conditional risk learning systems. Quesenberry and Gessaman [22] emphasized controlling the distribution of the conditional probabilities of error through the use of a region of reserve judgment. Anderson and Benning [23] presented an easily implemented, suboptimum, two-class procedure for controlling the distribution of the probabilities of error by using a region of reserve judgment.

This paper presents techniques in a two-class problem for controlling one error probability without requiring regions of reserve judgment. The techniques allow all F_2 training observations to be contained in R_2 in an attempt to minimize the miss probability. The approaches are not optimum from the Neyman–Pearson viewpoint, but they are easily implemented on the com-

Manuscript received December 14, 1971; revised June 26, 1972. This work was supported by General Dynamics/Electric Boat Company under Research Contract 8050-31-55001.
G. W. Beakley is with the David Sarnoff Research Center, RCA Corporation, Princeton, N. J. 08540.
F. B. Tuteur is with the Department of Engineering and Applied Science, Yale University, New Haven, Conn. 06520.

puter and have given good results in a speaker verification experiment.

II. The Ordering Procedure

Values for α, β, and γ in (4) and (5) can be determined from the number of statistically equivalent blocks formed using the F_1 observations and contained in R_2. The blocks should be constructed so that $\int_{R_2} dF_2(x)$ is maximized. It is assumed that the only information given about $F_1(x)$ or $F_2(x)$ is that they are continuous cumulative distribution functions. Therefore, given only a finite number of observations from each class, one can never be certain that $\int_{R_2} dF_2(x)$ is maximized. In the small sample case, a likely approach is to construct R_2 so that it contains as many F_2 observations as possible. All of the F_2 observations can be contained in R_2 if the F_1 observations $X_1^{(1)}, \cdots, X_{n_1}^{(1)}$ are ordered by functions that are in some sense centered about all n_2 of the F_2 observations $X_1^{(2)}, \cdots, X_{n_2}^{(2)}$. Then R_2 is made up of the first m blocks established by the ordering.

To accomplish this ordering, consider the continuous functions $d_{kj}(x, X_k^{(2)})$ of the arbitrary observation vector x, where $k=1, \cdots, n_2$, $j=1, \cdots, n_1$. These functions are basically "distance" functions that satisfy the following conditions:

$$d_{kj}(x, X_k^{(2)}) \geq 0$$
$$d_{kj}(X_k^{(2)}, X_k^{(2)}) = 0. \qquad (6)$$

A simple example is

$$d_{kj}(x, X_k^{(2)}) = |x - X_k^{(2)}|. \qquad (7)$$

The subscript k is used to label possibly differing functions that can be associated with each of the observations from the class F_2. The subscript j is used to label possibly differing functions used to form successive blocks. The need for such functions is illustrated in the next section.

The n_1+1 blocks $B_1, B_2, \cdots, B_{n_1+1}$ can now be defined as follows. First, let d_1 be the smallest distance between points from the two different classes, i.e.,

$$d_1 = \min_{1\leq k\leq n_2} \min_{1\leq i\leq n_1} d_{k1}(X_i^{(1)}, X_k^{(2)}). \qquad (8)$$

Define the regions L_{k1}, $k=1, \cdots, n_2$ by

$$L_{k1} = \{x: d_{k1}(x, X_k^{(2)}) \leq d_1\}. \qquad (9)$$

For two-dimensional vectors x and a metric as given by (7), the regions L_{k1} are seen to be circles centered at the points $X_k^{(2)}$ with radius d_1. The definition of d_1 is such that at least one of these circles contains a point from class 1 on its circumference. The probability that there are more than one such points is assumed negligible. This point is labeled $X_{(1)}^{(1)}$ and is said to be ordered. It is clear that for n-dimensional vectors and for the metric of (7), the regions L_{k1} are n-dimensional hyperspheres.

The first block B_1 is now defined as the union of all the regions L_{k1}:

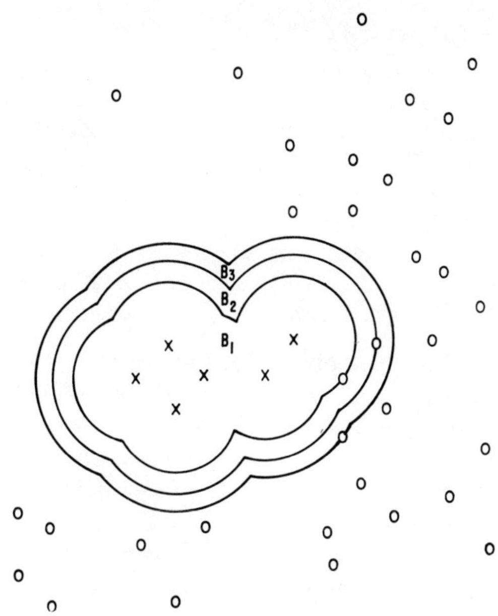

Fig. 1. All hyperspheres expand (AHE). O's belong to class 1. X's belong to class 2.

$$B_1 = \bigcup_{k=1}^{n_2} L_{k1}. \qquad (10)$$

For example, in the two-dimensional case pictured in Fig. 1, where the X's represent the F_2 observations and the O's represent the F_1 observations, block B_1 is the region inside the smallest diameter circles.

To obtain the second block, the distance d_2 is defined by

$$d_2 = \min_{1\leq k\leq n_2} \min_{\substack{1\leq i\leq n_1 \\ i\neq(1)}} d_{k2}(X_i^{(1)}, X_k^{(2)}). \qquad (11)$$

The regions L_{k2} are then defined as before by

$$L_{k2} = \{x: d_{k2}(x, X_k^{(2)}) \leq d_2\}, \qquad k = 1, \cdots, n_2. \quad (12)$$

If $d_{k2}(\cdot)$ and $d_{k1}(\cdot)$ are both of the form of (7), then for two-dimensional vectors, the regions L_{k2} are circles extending to the next closest point of class 1, which is labeled $X_{(2)}^{(1)}$, and which is the second class 1 point to be ordered.

The second block B_2 is now given by

$$B_2 = \left(\bigcup_{k=1}^{n_2} L_{k2}\right) \cap \overline{B}_1 \qquad (13)$$

where $\overline{B}_1$ is the set of points not contained in B_1. In Fig. 1, B_2 would consist of the union of all the annular areas between the circles of radius d_1 and d_2 less the intersection with the first block.

This procedure is continued, and therefore the rth block is given by

$$B_r = \left(\bigcup_{k=1}^{n_2} L_{kr}\right) \bigcap_{i=1}^{r-1} \overline{B}_i \qquad (14)$$

where

$$L_{kr} = \{x: d_{kr}(x, X_k^{(2)}) \leq d_r\}, \qquad k = 1, \cdots, n_2$$

and where

$$d_r = \min_{1\leq k\leq n_2} \min_{\substack{1\leq i\leq n_1 \\ i\neq(1)\cdots(r-1)}} d_{kr}(X_i^{(1)}, X_k^{(2)}).$$

The (n_1+1)th block is

$$B_{n_1+1} = \chi - \bigcup_{i=1}^{n_1} B_i = \chi \cap \left(\bigcap_{i=1}^{n_1} \overline{B}_i \right) \tag{15}$$

where χ is the entire observation space.

It is convenient to think of the blocks as being generated by hyperspheres [or other hypersurfaces, depending on the form of $d_{kr}(x, X_k^{(2)})$] *expanding* from the centers $X_k^{(2)}$, $k=1, \cdots, n_2$. The expansion continues until the first observation of class 1 is reached; this observation is thereby ordered [i.e., given rank order (1)] and the resulting volume is the first block. Subsequent blocks are generated by further expansions in search for other F_1 observations.

The region R_2 is the union of the first m blocks formed by this ordering:

$$R_2 = \bigcup_{i=1}^{m} B_i. \tag{16}$$

The value for m is obtained from the constraints on the design of the classifier. For example, suppose a classifier is to be designed for 95 percent confidence that the false alarm probability is less than 0.05. Then β and γ in (4) are equal to 0.05 and 0.95, respectively. One of the variables, n_1, the number of F_1 training observations, or m, the number of blocks used to construct R_2, can be selected. Then the value of the other variable can be found from graphs by Murphy [11] or from tables by Somerville [12]. For example, we find from Murphy for $\beta=0.05$, $\gamma=0.95$, and $n_1=210$ that $m=6$ blocks can be used to construct R_2. These numbers give an expected false alarm probability of 0.0284 with a standard deviation of 0.0114.

The general theory of distribution-free tolerance regions [1]–[12] does not cover the case where the regions depend on randomness outside of the F_1 observations.[1] A proof that the theory holds for additional randomness that is independent of the F_1 observations can be found in [23, Appendix]. Further demonstrations for the particular procedures of this paper can be found in [18, ch. 5].

III. Discussion of Practical Ordering Functions

Some simple ordering procedures based on the idea of expanding functions from the F_2 observations are discussed in this section. The relative merits of these procedures when applied to a problem with a limited sample size are investigated. For simplicity, let the ordering functions be defined as follows:

[1] The authors are indebted to a reviewer for pointing this out.

$$d_{kj}(x, X_k^{(2)}) = a_{kj} \left| x - X_k^{(2)} \right|$$
$$a_{kj} \geq 1, \qquad k = 1, \cdots, n_2; j = 1, \cdots, n_1. \tag{17}$$

Then the F_1 observations are ordered by hyperspheres that expand from the F_2 observations.

Distribution-free tolerance regions can be formed by any of the following three ordering procedures.

A. All Hyperspheres Expand (AHE)

Hyperspheres expand at the same rate from all F_2 observations until n_1+1 blocks have been formed (a_{jk} is equal to a constant for all j and k). The first m of these blocks make up region R_2. This procedure is illustrated in Fig. 1 for a two-dimensional vector x, $m=3$ blocks, $n_2=6$, $n_1=33$ with the observations $X_1^{(1)}, \cdots, X_{n_1}^{(1)}$ represented by O's and the observations $X_1^{(2)}, \cdots, X_{n_2}^{(2)}$ represented by X's.

Although this method guarantees that all of the F_2 training observations are contained in R_2, it is still possible that the total R_2 volume is too small to produce a reasonably low miss rate when used with new samples. This situation can be a direct result of having few F_2 training observations with at least m F_1 training observations being relatively close to the F_2 observations. In this case, a procedure, which assumes that for F_1 observations may be found in the vicinity of the F_1 observations that have previously been ordered, allows the regions centered by the F_2 observations to expand into regions that are connected and that have a larger volume than the regions produced by the AHE ordering procedure. The following two procedures use the information from previously ordered observations to allow region R_2 to expand faster in directions away from the clustered F_1 observations than toward them.

B. Ordered Hyperspheres Slowed

With this procedure, the hyperspheres that order the F_1 observations are not allowed to expand as rapidly in subsequent orderings as the other hyperspheres. The functions for ordering the first F_1 observation are the same as those for the AHE procedure. That is,

$$d_{k1}(x, X_k^{(2)}) = a_{k1} \left| x - X_k^{(2)} \right|, \quad k = 1, \cdots, n_2. \tag{18}$$

The functions for ordering the second F_1 observation are given by

$$d_{k2}(x, X_k^{(2)}) = a_{k2} \left| x - X_k^{(2)} \right|, \quad k = 1, \cdots, n_2 \tag{19}$$

where

$$\begin{aligned} a_{k2} &> a_{k1}, && k = (1) \\ a_{k2} &= a_{k1}, && \text{otherwise.} \end{aligned} \tag{20}$$

Note that $k=(1)$ is any k that satisfies

$$d_1 = a_{k1} \left| X_{(1)}^{(1)} - X_k^{(2)} \right|. \tag{21}$$

The increase of $a_{(1)1}$ to $a_{(1)2}$ may be viewed as a decrease in the rate at which the ordering hypersphere expands from the F_2 observation $X_{(1)}^{(2)}$ in search of the next F_1 observation.

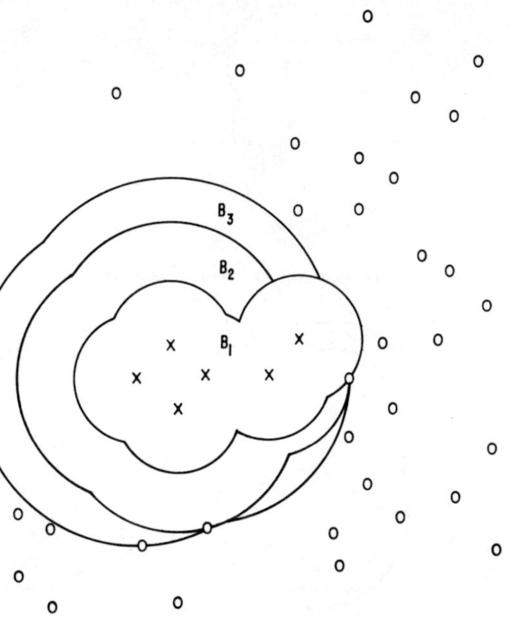

Fig. 2. Ordered hyperspheres constant (OHC). O's belong to class 1. X's belong to class 2.

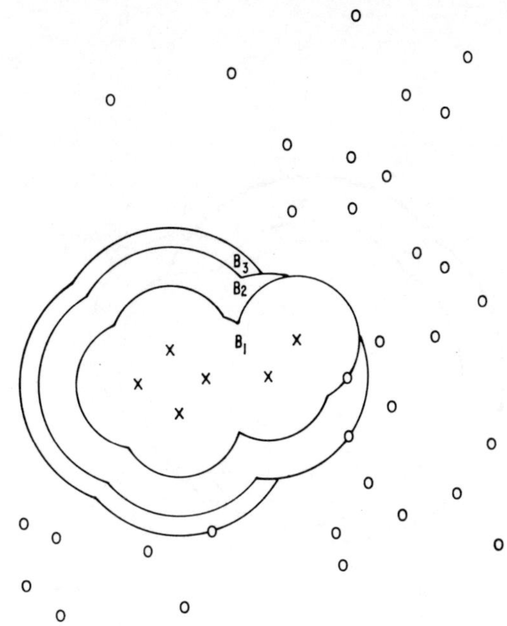

Fig. 3. Conditioned hyperspheres stop (CHS). O's belong to class 1. X's belong to class 2.

Similarly, the functions for ordering the rth F_1 observation are given by

$$d_{kr}(\mathbf{x}, \boldsymbol{X}_k^{(2)}) = a_{kr} \left| \mathbf{x} - \boldsymbol{X}_k^{(2)} \right| \tag{22}$$

where

$$a_{kr} > a_{k,r-1}, \qquad k = (r-1)$$
$$a_{kr} = a_{k,r-1}, \qquad \text{otherwise.}$$

$k=(j)$ is any k that satisfies

$$d_j = a_{kj} \left| \boldsymbol{X}_{(j)}^{(1)} - \boldsymbol{X}_k^{(2)} \right| . \tag{23}$$

The adjustment of the multiplicative constant a_{kr}, $k=(r-1)$ is quite arbitrary. In the speaker recognition experiment, which is discussed in Section V, the increase of $a_{(r-1),r-1}$ to $a_{(r-1),r}$ was made very large so that the differences in the three ordering procedures would become evident. Suppose a_{kr} is determined by

$$a_{kr} = (N a_{k,r-1})^N, \qquad k = (r-1)$$
$$a_{kr} = a_{k,r-1}, \qquad \text{otherwise} \tag{24}$$

where N is a large number. This essentially causes the hyperspheres that order the F_1 observations to stop expanding in relation to the hyperspheres that have not ordered an F_1 observation. The ordering procedure for this case will be called ordered hyperspheres constant (OHC).

Fig. 2 illustrates the OHC procedure for the same sample set as used in Fig. 1.

C. Conditioned Hyperspheres Slowed

With this procedure, the expansion of hyperspheres that intersect the F_1 observations is slowed even if these observations have been previously ordered by other hyperspheres. The ordering functions for $j \geq 2$ can be different from those of the previous two procedures because here an ordering hypersphere is slowed if it comes into contact with $\boldsymbol{X}_{(1)}^{(1)}$. Let $\boldsymbol{X}_{[1]}^{(1)}$ be the F_1 observation for which

$$d_2 = \min_{1 \leq k \leq n_2} \min_{1 \leq i \leq n_1} a_{k2} \left| \boldsymbol{X}_i^{(1)} - \boldsymbol{X}_k^{(2)} \right| \tag{25}$$

where

$$a_{k2} > a_{k1}, \qquad k = (1)$$
$$a_{k2} = a_{k1}, \qquad \text{otherwise}$$

and $k=(1)$ is any k that satisfies (21). During the second ordering, the expanding hyperspheres have intersected an F_1 observation $\boldsymbol{X}_{[1]}^{(1)}$. But this observation might well be the F_1 observation $\boldsymbol{X}_{(1)}^{(1)}$ that has already been ordered. Hence another F_1 observation has to be found before the second block is completed.

The search is continued until an F_1 observation that completes the block is found or until all hyperspheres have been slowed. In the latter case, the hyperspheres are allowed to pass the ordered F_1 observations so that the block can be completed. The extension of this procedure for the formation of m blocks is straightforward [18]. Fig. 3 illustrates the procedure for the same sample set as used in Figs. 1 and 2, for N equal to a very large number, and for

$$a_{kr} = (N a_{k,r-1})^N, \qquad k = [1], [2], \cdots, [j]$$
$$a_{kr} = a_{k,r-1}, \qquad \text{otherwise.} \tag{26}$$

This ordering is called conditioned hyperspheres stop (CHS).

The resulting decision regions for the three procedures are shown together in Fig. 4. In this particular example, it is seen that the AHE procedure produces a region R_2

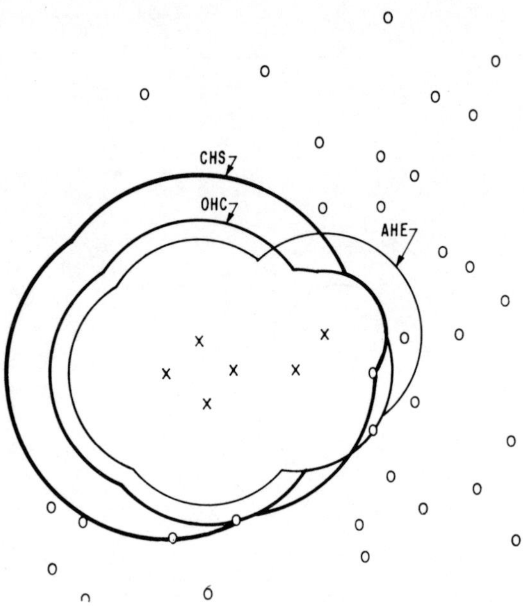

Fig. 4. Composite of the AHE, OHC, and CHS ordering procedures.

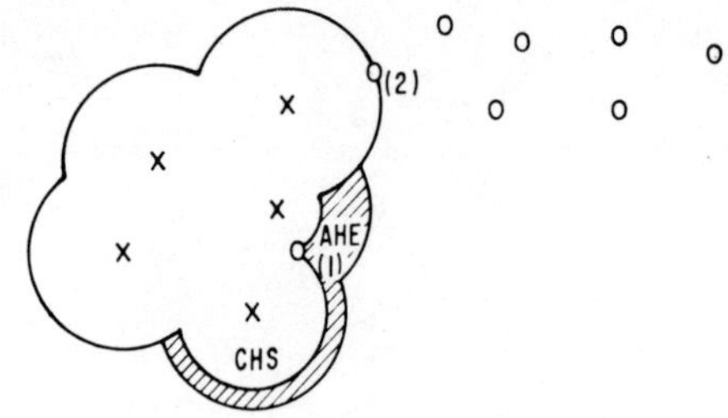

Fig. 5. Comparison in favor of the AHE procedure.

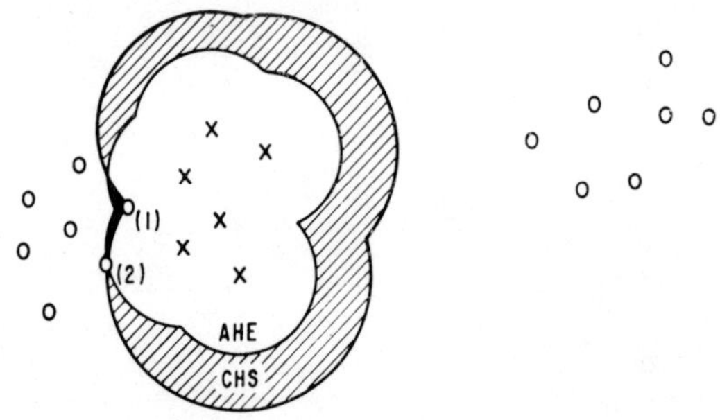

Fig. 6. Comparison in favor of the CHS procedure.

that has expanded into the O's, whereas the CHS procedure produces a region R_2 that has been stopped by the O's and has expanded in a direction away from the ordered O's.

A comparison of the three ordering procedures requires iteration of all possible sample sets. Nevertheless, some general observations can be made.

1) The AHE procedure is probably preferable to the OHC procedure, which is probably preferable to the CHS procedure, if spurious F_1 observations are involved. This is because hyperspheres in the AHE procedure, and in the OHC procedure to a lesser extent, continue to expand past the peripheral F_1 observations. For example, consider Fig. 5 where two blocks make up R_2. The resulting decision regions for the AHE and the CHS procedures are shown in the figure. The hyperspheres that expand from X_1 and X_2 stop when they intersect $O_{(1)}$ in the CHS procedure. They, of course, do not stop in the AHE procedure. The area of R_2 for the AHE procedure is equal to the area of R_2 for the CHS procedure plus the crosshatched area. Hence the miss probability in this case is less for the AHE procedure than for the CHS procedure.

2) The CHS procedure is probably preferable to the OHC procedure, which is probably preferable to the AHE procedure if the F_1 observations are tightly clustered in two or more clusters and the clusters are different distances from the F_2 observations. This is because the hyperspheres in the CHS procedure, and in the OHC procedure to a lesser extent, expand more in directions away from the ordered F_1 observations than do the hyperspheres of the AHE procedure. For example, consider Fig. 6 where two blocks are formed with the AHE and the CHS ordering procedures. The area of the CHS procedure is equal to the area of the AHE procedure plus the crosshatched region minus the shaded area. Since the crosshatched area is much larger than the shaded area, one may feel that the miss probability in this case is less for the CHS procedure than for the AHE procedure.

The AHE and CHS procedures have been compared using two univariate normal distributions and sample sizes 24, 49, 99, 250, 500, and 1000 [18]. The CHS procedure showed a clear superiority over the AHE procedure for all sample sizes.

Note that these three procedures are very easily programmed on a digital computer. One simply calculates the distance between every F_1 and F_2 training observation. A search is made for the smallest distance. Then the appropriate row or column is multiplied by a constant and a new search is made. This procedure is continued until m blocks are formed.

Note that when a new observation is to be classified, the following information must be stored in the computer for the various procedures.

1) AHE—all F_2 observations and the value of the mth order statistic.

2) OHC—all F_2 observations and the value of the m order statistics along with the indices of the F_2 observations corresponding to the m order statistics.

3) CHS—all F_2 observations and the values of the distances along with the corresponding indices of the F_2 observations found in the ordering process. This procedure requires at least as much storage as the OHC procedure.

D. Large Sample Properties

It is clear that the DFTR procedures approach a Neyman–Pearson test in the limit if the class probability densities are disjoint. However, for the more interesting case of overlapping densities, it appears that this is not the situation. For example, it can be shown easily that the AHE procedure does not approach a Neyman–Pearson test. Consider the univariate normal case. When the AHE procedure is used, the lengths of the intervals surrounding all F_2 observations are equal. As $n \to \infty$, the number of F_2 observations in a small interval of length ϵ around the maximum of $f_2(x)$ is greater than the number of F_2 observations in a small interval of length ϵ around any other point of $f_2(x)$. Hence if the intervals surrounding the F_2 observations coalesce, they would most likely coalesce in the region where $f_2(x)$ is a maximum. Using the likelihood ratio procedure, the smallest accept region is in the vicinity of the maximum of $f_2(x)/f_1(x)$. The maximum of $f_2(x)$ and the maximum of $f_2(x)/f_1(x)$ do not necessarily occur at the same point. Hence, in general, the AHE procedure does not approach the likelihood ratio procedure as n approaches infinity.

The results for the other DFTR procedures are not immediately apparent. An example was simulated on the computer using two univariate normal distributions and sample sizes of 24, 49, 99, 250, 500, and 1000 [18]. The simulation indicated that the other procedures did not approach the likelihood ratio test as the sample size became larger, even though they did perform better than the AHE procedure.

IV. Measure of the Miss Probability

A recognition system has been proposed that classifies with a given expected false alarm probability (or with a given confidence that the false alarm probability is less than a given quantity). It also correctly classifies all F_2 training observations. However, in most cases it does not minimize the miss probability.

If this classifier is to be used in a practical problem, a measure of the expected miss error is needed. Then if the expected miss error is much larger than desired, one can redesign the classifier by using more F_1 observations, by using more F_2 observations, or by using a different measurement space.

A rough estimate of the miss probability can be obtained by ordering the F_2 observations using procedures similar to those discussed in Section III. Such ordering results in blocks with respect to $F_2(\mathbf{x})$, not all of which are necessarily contained in the region R_2. However, the number of $F_2(\mathbf{x})$ blocks that *are* contained in region R_2 can be counted, and statements such as

$$E\left\{\int_{\bar{R}_2} dF_2(\mathbf{x})\right\} \leq \omega \tag{27}$$

or

$$P_r\left\{\int_{\bar{R}_2} dF_2(\mathbf{x}) \leq \phi\right\} \geq \nu \tag{28}$$

can be made. The quantities ω, ϕ, and ν are determined from n_2, the number of F_2 observations used to design the classifier, and from b, the number of $F_2(\mathbf{x})$ blocks inside R_2. This procedure may not be distribution free, as will be discussed later in this section.

Because of the procedure used to order the F_1 observations, all F_2 training observations must lie in R_2. Hence, all blocks formed by ordering the F_2 observations must have subsets that are contained in R_2. Therefore, good estimates of the cumulative distribution $F_2(\mathbf{x})$ in R_2 are made only if the ordering procedure is such to allow the blocks that are formed to be contained in R_2. The ordering procedure, however, should not use information about the location of the F_2 observations before they have been ordered, or information about the location of the boundary of R_2.

Consider the following procedure for ordering the F_2 observations. The procedure consists essentially of first choosing an arbitrary ordering function and locating an F_2 observation. A search is then made for another F_2 observation with a hypersphere that expands from the first F_2 observation. When the second F_2 observation is found, hyperspheres expand from both F_2 observations in search of a third F_2 observation. This process is continued until n_2+1 blocks are formed. The number of complete blocks contained in R_2 is counted, thereby giving numbers for ω, ϕ, and ν in (27) and (28).

The form of the first ordering function $h_1(\mathbf{x})$ is arbitrary. For simplicity, let $h_1(\mathbf{x})$ be given by

$$h_1(\mathbf{x}) = |\mathbf{x} - \boldsymbol{A}| \tag{29}$$

where $\boldsymbol{A}$ is a vector constant. Let $\boldsymbol{Y}_{(1)}^{(2)}$ be the F_2 observation that satisfies

$$h_1 = \min_{1 \leq i \leq n_2} |\boldsymbol{X}_i^{(2)} - \boldsymbol{A}|. \tag{30}$$

Here $\boldsymbol{Y}_{(1)}^{(2)}$ is used rather than $\boldsymbol{X}_{(1)}^{(2)}$ to avoid confusion with the latter symbol that was used when the F_1 observations were ordered. Then block C_1 is formed, where

$$C_1 = \{\mathbf{x}: h_1(\mathbf{x}) \leq h_1\}. \tag{31}$$

Block C_1 may or may not be contained in R_2, depending on the choice of the vector $\boldsymbol{A}$. For a distribution-free estimate of the miss probability, $\boldsymbol{A}$ should not be chosen using information about the location of the F_2 observations.

To form the second block, a search is made for an F_2 observation with a hypersphere that expands from $\boldsymbol{Y}_{(1)}^{(2)}$. Let the second ordering function be given by

$$h_{21}(\mathbf{x}, \boldsymbol{Y}_{(1)}^{(2)}) = |\mathbf{x} - \boldsymbol{Y}_{(1)}^{(2)}|. \tag{32}$$

Let $\boldsymbol{Y}_{(2)}^{(2)}$ be such that

$$h_2 = \min_{\substack{1 \leq i \leq n_2 \\ \boldsymbol{X}_i^{(2)} \neq \boldsymbol{Y}_{(1)}^{(2)}}} h_{21}(\boldsymbol{X}_i^{(2)}, \boldsymbol{Y}_{(1)}^{(2)}). \tag{33}$$

Then the second block is given by

$$C_2 = \{\mathbf{x}: h_{21}(\mathbf{x}, \boldsymbol{Y}_{(1)}^{(2)}) \leq h_2, h_1(\mathbf{x}) > h_1\}. \tag{34}$$

Let the functions for ordering the third F_2 observation be given by

$$h_{3j}(\mathbf{x}, \boldsymbol{Y}_{(j)}^{(2)}) = \left| \mathbf{x} - \boldsymbol{Y}_{(j)}^{(2)} \right|, \qquad j = 1, 2. \tag{35}$$

Let

$$h_3 = \min_{\substack{1 \le i \le n_2 \\ \boldsymbol{X}_i^{(2)} \ne \boldsymbol{Y}_{(1)}^{(2)}, \boldsymbol{Y}_{(2)}^{(2)}}} \min_{j=1,2} h_{3j}(\boldsymbol{X}_i^{(2)}, \boldsymbol{Y}_{(j)}^{(2)}) \tag{36}$$

and

$$K_{3j} = \{\mathbf{x}: h_{3j}(\mathbf{x}, \boldsymbol{Y}_{(j)}^{(2)}) \le h_3\}, \qquad j = 1, 2. \tag{37}$$

The third block is given by

$$C_3 = \left(\bigcup_{j=1}^{2} K_{3j} \right) \cap \left(\bigcap_{i=1}^{2} \bar{C}_i \right). \tag{38}$$

The functions for ordering the rth F_2 observation are given by

$$h_{rj}(\mathbf{x}, \boldsymbol{Y}_{(j)}^{(2)}) = \left| \mathbf{x} - \boldsymbol{Y}_{(j)}^{(2)} \right|, \qquad j = 1, \cdots, r-1. \tag{39}$$

Let

$$h_r = \min_{\substack{1 \le i \le n_2 \\ \boldsymbol{X}_i^{(2)} \ne \boldsymbol{Y}_{(1)}^{(2)}, \cdots, \boldsymbol{Y}_{(r-1)}^{(2)}}} \min_{1 \le j \le r-1} h_{rj}(\boldsymbol{X}_i^{(2)}, \boldsymbol{Y}_{(j)}^{(2)}) \tag{40}$$

and

$$K_{rj} = \{\mathbf{x}: h_{rj}(\mathbf{x}, \boldsymbol{Y}_{(j)}^{(2)}) \le h_r\}, \qquad j = 1, \cdots, r-1. \tag{41}$$

Then the rth block is given by

$$C_r = \left(\bigcup_{j=1}^{r-1} K_{rj}\right) \cap \left(\bigcap_{i=1}^{r-1} \bar{C}_i\right). \tag{42}$$

If the AHE procedure has been used to design the classifier, the regions L_{kr} (14) have the same radius for a particular k and for all $r = 1, \cdots, n_2$. Suppose m blocks were used to form R_2. If $h_2 \le d_m$, the entire block C_2 is contained in R_2. For the general case, it is easily seen that if $h_i \le d_m$, the entire block C_i, $i = 2, \cdots, n_2$ is contained in R_2. It is also seen that if $h_i > d_m$, the block C_i may or may not be contained in R_2. A conservative statement about the number of blocks contained in R_2 can be made by counting the number of times $h_i \le d_m$ for $i = 2, \cdots, n_2$.

Let the total number of blocks contained in R_2 be b. Then

$$E\left\{ \int_{\bar{R}_2} dF_2(\mathbf{x}) \right\} \le \frac{n_2 - b + 1}{n_2 + 1} \tag{43}$$

and the variance σ^2 is given by

$$\sigma^2 \left\{ \int_{\bar{R}_2} dF_2(\mathbf{x}) \right\} \le \begin{cases} \dfrac{b(n_2 + 1 - b)}{(n_2 + 1)^2 (n_2 + 2)}, & b \ge \dfrac{n_2 + 1}{2} \\[2ex] \dfrac{(b+1)(n_2 - b)}{(n_2 + 1)^2 (n_2 + 2)}, & b < \dfrac{n_2 + 1}{2}. \end{cases} \tag{44}$$

The values of ϕ and ν in (28) can be found by consulting Murphy's graphs [11].

To be certain that the F_2 blocks are distribution free, the ordering functions and region R_2 should be determined before the F_2 observations are taken. This, of course, cannot be done here because R_2 depends upon the F_2 observations. Hence this ordering procedure may produce an estimate of the probability of miss that is not distribution free. This procedure is used here only for obtaining a rough estimate of the probability of a miss, so that the classifier can be redesigned if the estimate is much poorer than the desired miss probability. We expect the estimate to perform this function adequately because it measures how well the hyperspheres making up R_2 are connected. For example, any time an F_2 block is counted as contained in R_2, we know that one of the hyperspheres making up R_2 contains the center of another of the hyperspheres making up R_2.

If the OHC or the CHS procedure has been used to design the classifier, all of the hyperspheres that make up R_2 do not have the same radius. Therefore, the number of F_2 blocks contained in R_2 varies with the F_2 observation with which the ordering starts. Furthermore, with classifiers of the OHC and CHS designs, the ith block is not necessarily contained in R_2 when $h_i \le d_m$. We reason here that our foremost concern is that we measure whether the hyperspheres are connected. Using this reasoning, we should use the radii of the hyperspheres that make up R_2 in ordering the F_2 observations. For example, the rth ordering function might be

$$h_{rj}(\mathbf{x}, \boldsymbol{Y}_{(j)}^{(2)}) = \frac{d_m}{d_j} \left| \mathbf{x} - \boldsymbol{Y}_{(j)}^{(2)} \right| \tag{45}$$

where d_m is the radius of the largest hypersphere making up R_2, and d_j is the radius of the hypersphere making up R_2 and centered by $\boldsymbol{Y}_{(j)}^{(2)}$. This equation simply alters the rate of expansion, so that the boundary of R_2 is simultaneously reached by all hyperspheres. This method gave better results in the speaker verification experiment than the method that did not use information about the radii of the hyperspheres making up R_2.

V. Automatic Speaker Verification

An automatic speaker verification experiment was used to test the hypersphere DFTR classification schemes. A speaker verification system tests the purported identity of a speaker from a sample of the speaker's voice. For example, suppose an automatic speaker verification system is to be used for allowing a person entrance through a company gate. The test subject might be required to push a button beside the name of the person he purports to be. The test subject then says a required phrase, for example, "My name is speaker X." A computer then identifies him as the *main speaker* (the speaker whose identity the test subject has assumed) or an *impostor*.

It is assumed that much of the information for the recognition of speakers is contained in the transition between phonemes as well as in the phonemes themselves. For this reason, a simple word that contained a diphthong was analyzed by calculating many short-term spectra over the length of the word. These spectra were used to form the measurement space in which the decision regions were constructed.

Recordings of 225 utterances by each of 3 speakers and 25 utterances by each of 26 impostors were furnished by Luck [13]. These utterances consisted of the sentence, "My code is (and then the speaker's initials)," digitized into 10-bit accuracy at 8000 samples per second. Only the word "my" was used in this analysis.

A short-term spectrum was calculated from 256 samples of the speech waveform by a fast Fourier transform [14]. Hence, each spectrum consisted of 128 unique frequency components. The questions: 1) How many spectra should we use? and 2) How coarse should each spectrum be? were investigated by forming three different 256-dimensional measurement spaces. The first measurement space was made up of 4 spectra, with each spectrum having 64 frequency components. The second space consisted of 8 spectra, with each spectrum having 32 components. The third space consisted of 16 spectra, with each spectrum having 16 components. The number of frequency components per spectrum was reduced from 128 to 64, 32, and 16 by simple averaging. The short-term spectra were spaced uniformly across the word "my" in all cases.

Let class 1 be the class of the impostors and let class 2 be the class of the main speaker. If a test observation falls in region R_2, it is classified as the main speaker. Otherwise it is classified as an impostor. The number of F_1 training observations and the number of F_1 blocks were chosen as 208 and 7, respectively. These numbers give the following parameters, where P_{FA} is the false alarm probability, $\Pr(\cdot)$ is the probability of $(\cdot)$, $E(\cdot)$ is the expected value of $(\cdot)$, and $\sigma(\cdot)$ is the standard deviation of $(\cdot)$:

$$\Pr(P_{FA} \leq 0.055) = 0.95$$
$$E(P_{FA}) = 0.0335$$
$$\sigma(P_{FA}) = 0.0124.$$

When using the DFTR approach on the speaker verification problem, it should be realized that assumptions are made about the *a priori* probability of occurrence of each impostor. If these *a priori* probabilities change, the probability distribution governing the impostor class changes. Hence, results will not be as expected. In the following experiment, both the training and test samples used equal *a priori* probabilities of occurrence of each impostor. Of course, when testing one speaker versus another speaker, this problem does not arise.

Table I shows the results that were obtained when the three DFTR procedures were trained and tested in the three different 256-dimensional spaces. The test false alarm rate, the test miss rate, and the total error rate are listed for each procedure. For example, for the AHE procedure in measurement space 1, 15 out of 439 impostor test observations were classified as the main speaker, and 7 out of 185 main speaker test observations were classified as an impostor. This gives a total error rate of 22/624 = 0.0352. The average error rate obtained for all three ordering procedures in measurement space 1 was 0.0347. This compared with an average error rate of 0.0251 for measurement space 2, and an average error rate of 0.0336 for measurement space 3.

TABLE I

Test Results for Three Different 256-Dimensional Measurement Spaces

Training Data

40 Main Speaker Training Samples 208 Impostor Training Samples

$\Pr(P_{FA} \leq 0.07) = 0.99$ $\Pr(P_{FA} \leq 0.055) = 0.95$
$E(P_{FA}) = 0.0335$
$\sigma(P_{FA}) = 0.0124$

Test Results

185 Main Speaker Test Samples 439 Impostor Test Samples

Procedure	False Alarm	Miss	Total Error
Measurement Space 1: 4 spectra, 64 components/spectrum			
1) AHE	15/439 = 0.0342	7/185 = 0.0379	0.0352
2) OHC	15/439 = 0.0342	7/185 = 0.0379	0.0352
3) CHS	14/439 = 0.0319	7/185 = 0.0379	0.0337
Measurement Space 2: 8 spectra, 32 components/spectrum			
1) AHE	12/439 = 0.0274	7/185 = 0.0379	0.0305
2) OHC	10/439 = 0.0228	6/185 = 0.0324	0.0256
3) CHS	9/439 = 0.0205	3/185 = 0.0162	0.0192
Measurement Space 3: 16 spectra, 16 components/spectrum			
1) AHE	7/439 = 0.0159	26/185 = 0.1405	0.0529
2) OHC	4/439 = 0.0091	14/185 = 0.0756	0.0288
3) CHS	8/439 = 0.0181	4/185 = 0.0216	0.0192

In measurement space 1, very little difference is noted in the total error rates for the three ordering procedures. In spaces 2 and 3, the CHS and the OHC procedures easily outperform the AHE procedure. This might have been expected, since the CHS and the OHC procedures use the information that becomes available during the ordering to select subsequent ordering functions. Hence the region R_2 for these procedures is somewhat shaped by both the F_1 and F_2 observations.

The error rate for the AHE ordering procedure averaged over the three spaces is 0.0395. This compares with 0.0299 for the OHC procedure and 0.0240 for the CHS procedure.

Note that the AHE procedure in measurement space 3 yielded a relatively poor miss rate. This was because the training observations from Impostor 6 were near those of the main speaker. In the training phase, the ordering hyperspheres for the AHE procedure continued to expand into the training observations of Impostor 6. Hence, the hyperspheres were stopped be-

TABLE II
TEST RESULTS FOR THE TWO MAIN SPEAKERS IN A 48-DIMENSIONAL SPACE

Decision Procedure	RF = Main Speaker			JD = Main Speaker		
	False Alarm	Miss	Total Error	False Alarm	Miss	Total Error
Pr $(P_{FA} \leq 0.055) = 0.95$ $E(P_{FA}) = 0.0335$						
1) AHE	0.0319	0.2972	0.1105	0.0297	0.3162	0.0900
2) OHC	0.0297	0.1837	0.0754	0.0297	0.2991	0.0863
3) CHS	0.0342	0.1243	0.0609	0.0297	0.2991	0.0863
Pr $(P_{FA} \leq 0.085) = 0.95$ $E(P_{FA}) = 0.0574$						
1) AHE	0.0550	0.2000	0.0977	0.0388	0.1710	0.0665
2) OHC	0.0455	0.0650	0.0512	0.0410	0.1625	0.0665
3) CHS	0.0387	0.0595	0.0448	0.0434	0.1368	0.0629
NN Rule	0.0205	0.0216	0.0209	0.0369	0.1111	0.0522

fore they became large enough to yield a good miss rate. For the OHC and CHS procedures, the hyperspheres that were expanding into the training observations of Impostor 6 were stopped, allowing the remaining hyperspheres to become much larger than the hyperspheres for the AHE procedure. Here is a case where the OHC and the CHS procedures clearly perform better than the AHE procedure.

A. Reduced Measurement Space

In order to compare the DTFR methods discussed thus far with others requiring much more computing time, and in order to use a second main speaker, the size of the measurement space was reduced.

The space that gave the best results in the 3-space experiment consisted of 8 spectra, with 32 components per spectrum. Since 8 spectra approximately covered the word "my" for most speakers, approximate coverage of the word was deemed important. Six spectra with 8 frequency components per spectrum were used to form the reduced space.

Table II shows the results for this space when two different main speakers were used. The first row shows the DFTR results with 7 blocks used to form R_2. The next row shows the results with 12 blocks used to form R_2. The results for the different DFTR procedures varied less for speaker *JD* than for speaker *RF*. The error rate averaged over all three DFTR procedures was approximately equal for each speaker. This error rate for the 7-block DFTR problem was 0.0823 for *RF* and 0.0875 for *JD*, and for the 12-block DFTR problem it was 0.0646 for *RF* and 0.0653 for *JD*.

The nearest neighbor (NN) rule [15]–[17] was then used to classify the observations. The NN rule performed better than the best DFTR procedure for both speakers. This might have been expected, since the relative shape of the decision regions for the NN rule is determined by the location of all main speaker and impostor training observations, whereas the relative shape of the decision regions for the DFTR procedures is determined by only part of the information about the impostor observations.[2] Note further that the DFTR procedures were designed for a given false alarm rate and not for total error (false alarm and miss). Hence a comparison on the basis of total error is somewhat more favorable to the nearest neighbor rule.

Several considerations favor the DFTR rule over the NN rule.

1) The NN rule took 5 times longer to test an observation than the DFTR rule for 40 main speaker training observations and 208 impostor training observations in a 48-dimensional space. Thus if the rules are compared on an equal computing time basis, the DFTR rules are, in fact, superior in performance for this particular case.

2) The CHS–DFTR procedure in measurement space 2 and the NN rule in the reduced space can be compared. The DFTR procedures in measurement space 2 took 50 percent less time to test than the NN procedure in the reduced space. Therefore, the CHS–DFTR procedure in measurement space 2 is superior in both error rate and computing time to the NN rule in the reduced space.

3) The DFTR rules allow the machine designer to know how well the machine is expected to perform, without assuming the form of the class probability distributions.

4) The DFTR rules require storage of the main speaker training observations and approximately m impostor training observations (more for the CHS procedure, less for the AHE procedure), whereas the NN rule requires storage of the main speaker training observations and all of the impostor training observations.

A theoretical comparison between the probability

[2] The size of the decision regions is determined by the location of the impostor observations, and this in turn affects the shape of the regions. However, the ordering procedures cannot use the information about the location of an impostor observation to influence the rate of expansion of each hypersphere until after the impostor observation is ordered. Hence, the information about the unordered observations cannot be used in the DFTR procedures as they are in the NN rule.

of error for the AHE–DFTR procedure and the probability of error for the NN rule has been made [18]. The following results for the miss probabilities were obtained without specifying the class probability distributions. These results hold as long as the same metric is used for both procedures, regardless of the dimensionality of the space.

When n_1 independent observations are available from class 1 and $n_2=1$ observation is available from class 2, the NN rule and the one-block DFTR procedures yield identical miss probabilities. Therefore, the miss probability for the m-block AHE–DFTR procedure, $m>1$, is less than or equal to the miss probability for the NN procedure.

For $n_2>1$, the miss probabilities for the NN and AHE–DFT procedures were compared on an intuitive basis. The result was that the miss probability for the NN rule is less than or equal to the miss probability for the one-block AHE–DFTR procedure. Comparison of the miss probabilities for $m>1$ block could not be made.

The false alarm probabilities cannot be compared without assuming the form of the class probability distributions. It can be shown that either rule may yield a lower false alarm rate than the other rule, depending on the underlying distributions.

B. Comparison of Test Error Rates with Those Predicted

The false alarm rate obtained from the test sample was compared to the 95 percent upper tolerance limit on the false alarm probability obtained from the DFTR approaches. All test false alarm rates fell below the 95 percent upper tolerance limit. For the 21 different cases studied, the average test false alarm rate was approximately equal to 80 percent of the average expected false alarm probability predicted from the DFTR approach. As expected, this percentage did not vary appreciably for the different ordering methods.

A comparison was made between the test miss rate and the measure of miss probability obtained from the tolerance regions approach. All test miss rates fell below the 95 percent upper tolerance limit. For the AHE procedure, the average miss rate was equal to 94 percent of the average expected miss rate. The corresponding percentages for the OHC and CHS procedures were 0.84 and 0.68, respectively.

VI. Conclusions

Some procedures have been presented for designing a pattern verification system with a given confidence that the false alarm probability is less than a desired quantity. These procedures maximize the number of main class training observations that are correctly classified. In addition, a method has been given for obtaining a measure of the miss probability. The procedures have been successfully applied to a speaker verification problem.

The advantages of the hypersphere DFTR classification procedure are the following. 1) It gives information about how well the classifier is expected to perform. This is done without knowledge of the form of the class probability distributions. 2) The procedure is able to form very complicated, unconnected decision regions. Hence it is useful when multimodal class probability distributions are involved. 3) The hypersphere DFTR procedure is very easily programmed on a digital computer. 4) The procedure is independent of the dimensionality of the measurement space. 5) It offers automatic reduction of the data that must be stored in the computer.

It should be noted, however, that this is a distribution-free procedure, and hence can be expected to be quite inefficient when compared to a procedure based upon *a priori* knowledge of the class probability distributions. The hypersphere DFTR procedure is most applicable to the situation in which: 1) the form of the probability distributions is unknown, and 2) information about the expected probability of error is desired without using a test sample.

Acknowledgment

The authors are very grateful to Prof. I. R. Savage for valuable discussions, and to Dr. J. E. Luck for furnishing the speaker data.

References

[1] S. S. Wilks, "Determination of sample sizes for setting tolerance limits," *Ann. Math. Stat.*, vol. 12, pp. 91–96, 1941.
[2] A. Wald, "An extension of Wilk's method for setting tolerance limits," *Ann. Math. Stat.*, vol. 14, pp. 45–55, 1943.
[3] H. Scheffé and J. W. Tukey, "Nonparametric estimation: I. Validation of order statistics," *Ann. Math. Stat.*, vol. 16, pp. 187–192, 1945.
[4] J. W. Tukey, "Nonparametric estimation: II. Statistically equivalent blocks and tolerance regions—The continuous case," *Ann. Math. Stat.*, vol. 18, pp. 529–539, 1947.
[5] ——, "Nonparametric estimation: III. Statistically equivalent blocks and multivariate tolerance regions—The discontinuous case," *Ann. Math. Stat.*, vol. 19, pp. 30–39, 1948.
[6] D. A. S. Fraser and R. Wormleighton, "Nonparametric estimation: IV," *Ann. Math. Stat.*, vol. 22, pp. 294–298, 1951.
[7] D. A. S. Fraser, "Sequentially determined statistically equivalent blocks," *Ann. Math. Stat.*, vol. 22, pp. 372–381, 1951.
[8] ——, "Nonparametric tolerance regions," *Ann. Math. Stat.*, vol. 24, pp. 44–55, 1953.
[9] ——, *Nonparametric Methods in Statistics.* New York: Wiley, 1957.
[10] J. H. B. Kemperman, "Generalized tolerance limits," *Ann. Math. Stat.*, vol. 27, pp. 180–186, 1956.
[11] R. B. Murphy, "Nonparametric tolerance limits," *Ann. Math. Stat.*, vol. 19, pp. 581–588, 1948.
[12] P. N. Somerville, "Tables for obtaining nonparametric tolerance limits," *Ann. Math. Stat.*, vol. 29, pp. 599–601, 1958.
[13] J. E. Luck, "Automatic speaker verification using cepstral measurements," *J. Acoust. Soc. Amer.*, vol. 46, pp. 1026–1032, Oct. 1969.
[14] J. W. Cooley and J. W. Tukey, "An algorithm for the machine calculation of complex Fourier series," *Math. Comput.*, vol. 19, pp. 297–301, 1965.
[15] E. Fix and J. L. Hodges, Jr., "Discriminatory analysis: Nonparametric discrimination; consistency properties," USAF School of Aviation Medicine, Randolph Field, Tex., Project 21-49-004, Rep. 4, Contract A S41 (128)-31, 1951.
[16] ——, "Discriminatory analysis: Nonparametric discrimination; small sample performance," USAF School of Aviation Medicine, Randolph Field, Tex., Project 21-49-004, Rep. 11, 1952.
[17] T. M. Cover and P. E. Hart, "Nearest neighbor pattern classification," *IEEE Trans. Inform. Theory*, vol. IT-3, pp. 21–27, Jan. 1967.
[18] G. W. Beakley, "Application of distribution-free tolerance

regions to pattern recognition," Ph.D. dissertation, Yale Univ., New Haven, Conn., Apr. 1970.

[19] T. W. Anderson, "Some nonparametric multivariate procedures based on statistically equivalent blocks," in *Proc. Int. Symp. Multivariate Analysis*, P. R. Krishnaiah, Ed. New York: Academic, 1966.

[20] E. A. Patrick, "Distribution-free minimum conditional risk learning systems," presented at the 1967 Princeton Conf. Inform. Theory; also, Purdue Univ., Lafayette, Ind., Rep. TR-EE-66-18, 1966.

[21] E. A. Patrick and F. P. Fisher, II, "Introduction to the performance of distribution-free, minimum conditional risk learning system," Purdue Univ., Lafayette, Ind., Rep. TR-EE-67-12, 1967.

[22] C. P. Quesenberry and M. P. Gessaman, "Nonparametric discrimination using tolerance regions," *Ann. Math. Stat.*, vol. 39, no. 2, pp. 664–673, 1968.

[23] M. W. Anderson and R. D. Benning, "A distribution-free discrimination procedure based on clustering," *IEEE Trans. Inform. Theory*, vol. IT-16, pp. 541–548, Sept. 1970.

25

DISTRIBUTION-FREE PROCESSORS BASED ON THE THEORY OF RUNS

V. R. Srinivasan

Abstract

This correspondence describes distribution-free processors based on the theory of runs. No assumption about either the functional form of the distribution of noise or the nature of its distribution is made. Hence the processors described are expected to yield constant false alarm (CFA) irrespective of the nature of noise encountered in the system.

Introduction

Distribution-free methods are widely attempted in recent times for signal detection in communication theory and radar [1]-[5], [7]. In [3], a distribution-free Doppler processor is described based on the ranks of the possible signal sample (PSS) among the known-noise sample (KNS).

The purpose of this correspondence is to describe distribution-free processors based on the theory of runs. Detector A and detector B are described. The distribution-free method suggested in this correspondence, besides being extremely general in the sense that it requires no assumption about the functional form of the distribution of noise, is also extremely simple. No complex analysis of distribution theory is needed: the simple binomial, Rayleigh, and Gaussian distributions provide necessary tools for the analysis. In addition, the nonparametric method suggested here is sensitive to both differences in shape and differences in locations between two distributions [6]. A single-path detection scheme and a two-path detection system are suggested for both detectors A and B.

The problem considered here is the same as the one in [3]. Since no assumption about the functional form of the distribution of the noise process is made, the method described here is strictly distribution free.

Detector A

A. Single Path Detection System

Let us first consider a single-path detection system which is described in Fig. 1. It is assumed that the $I(t)$ channel is sampled at times $0, T, 2T, \cdots, (M-1)T$, yielding a possible signal sample (PSS) $X_0, X_1, \cdots, X_{(M-1)}$. It is also assumed that corresponding to each $jT[j = 0, 1, 2, \cdots, (M-1)]$, a set of $(N-1)\cdots$ reference noise observations $u_{j1}, u_{j2}, \cdots, u_{j(N-1)}$ called known-noise sample (KNS) is also collected. The method of obtaining the KNS is described in [3].

Suppose that we have an array of MI registers (Fig. 2) which contain the totality of all observations:

$$A = \begin{array}{ccccccc} 1 & 2 & \cdots & j & \cdots & N \\ U_{01} & U_{02} & \cdots & [x_0] & \cdots & U_{0(N-1)} \\ U_{11} & U_{12} & \cdots & [x_1] & \cdots & U_{1(N-1)} \\ \vdots & \vdots & \cdots & \vdots & \cdots & \vdots \\ U_{(M-1)1} & U_{(M-1)2} & \cdots & [x_M] & \cdots & U_{(M-1)(N-1)} \end{array} \qquad (1)$$

where U_{ij} is the jth sample of the ith register and x_j is the observation at the center of the jth I register representing PSS.

Our null hypothesis (H_0) is that the PSS $(X_0, X_1, \cdots, X_{(M-1)})$ and the KNS $[U_{ik}; i = 0, 1, \cdots, (N-1); k = 0, 1, \cdots, (N-1); k \neq j]$ are samples from the same populations while the alternative hypothesis (H_1) is that H_0 is not true. This problem is analyzed through the theory of runs.

If all samples in the matrix A are in ascending order, we will get a sequence of the following type:

$$U_{i1}\, U_{j2}\, X_0\, X_1\, U_{21}\, U_{13}\, X_4\, X_5 \cdots. \qquad (2)$$

A run is a sequence of letters of one kind bounded by letters of the other kind. If U_{ij} and x_i are IID, then the samples in (2) will be well mixed, resulting in too many runs. If H_1 is true, then the number of runs in (2) is expected to be small. Based on this logic, denoting the total number of runs in the combined sample set by d, we decide that a target is present if

$$d \leqslant d_0.$$

Otherwise, we decide that noise alone is present.

In order to determine the total number of runs in a practical situation, the following method which can be accomplished either by means of a computer or through hardware is suggested.

The Method of Determining the Number of Runs

Let $\gamma_0, \gamma_1, \cdots, \gamma_{(M-1)}$ denote the ranks of $x_0, x_1, \cdots, x_{(M-1)}$ in the combined sample of PSS and KNS. Fig. 2 describes the method of determining the ranks of x_j for the case $M = 2$ which can be easily extended to any general case. The ranks $\gamma_0, \gamma_1, \cdots, \gamma_{(M-1)}$ are then ordered according to their magnitudes. Let the resulting ordered set of ranks be denoted by $(d_1, d_2, \cdots, d_M)$. In this set each subset of consecutive ranks is grouped into a distinct subset. The number of such distinct subsets is determined and let us represent it by n. Then the total number of runs is given by

$$\begin{aligned} d &= (2n+1) \text{ if } d_1 \neq 1 \text{ and } d_M \neq MN \\ &= 2n \quad \text{if } d_1 = 1 \text{ and } d_M \neq MN \\ &= 2n \quad \text{if } d_1 \neq 1 \text{ and } d_M = MN \\ &= (2n-1) \text{ if } d_1 = 1 \text{ and } d_M = MN. \end{aligned}$$

Determination of Threshold

Denoting the probability distribution of runs by $h(d)$, it is found from [6] that

$$h(d) = 2 \binom{M-1}{k-1} \binom{MN-M-1}{k-1} \Big/ \binom{MN}{M}; \; k = d/2 \qquad (3)$$

where (3) holds good for even values of d. Similarly for odd values of d, the distribution of d is given by

$$h(d) = \left[\binom{M-1}{k} \binom{MN-M-1}{k-1} + \binom{M-1}{k-1} \binom{MN-M-1}{k} \right] \Big/ \binom{MN}{M}; \quad k = (d-1)/2. \qquad (4)$$

The threshold d_0 can be determined through the relation

$$\sum_{d=2}^{d_0} h(d) = \alpha \qquad (5)$$

where α is the probability of false alarm.

The computation in (5) becomes quite cumbersome if M and N are very large. However, for large samples, d converges to a Gaussian distribution [6] with mean $E(d)$ and variance σ_d^2.

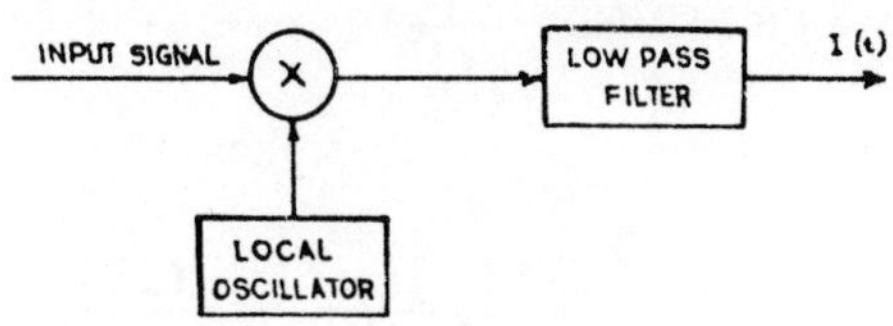

Fig. 1. Single-path detection system.

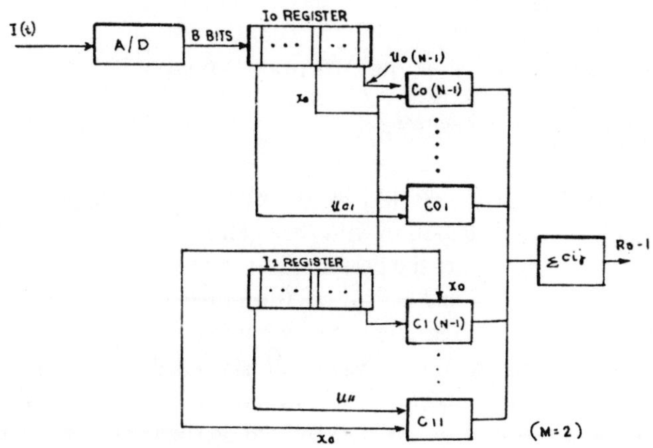

Fig. 2. Array of *M I* registers.

$$E(d) = 2n_1n_2/(n_1 + n_2) + 1 \tag{6}$$

$$\sigma_d^2 = 2n_1n_2(2n_1n_2 - n_1 - n_2)/(n_1 + n_2)^2(n_1 + n_2 - 1) \tag{7}$$

where $n_1 = M$ and $n_2 = M(N-1)$. It has been pointed out in [6] that this approximation holds good when $M > 10$ and $N \geqslant 2$.

A Two-Path Detection Scheme

A two-path detection scheme is described in Fig. 1. of [3]. In this type of system we assume that in addition to an array of M I registers, we are provided with an array of M Q registers where the signal is input after 90° phase shift. The M-array Q registers will contain the observations as

$$B = \begin{pmatrix} 1 & 2 & \cdots & j & \cdots & N \\ v_{01} & v_{02} & \cdots & [y_0] & \cdots & v_{0(N-1)} \\ v_{11} & v_{12} & \cdots & [y_1] & \cdots & v_{1(N-1)} \\ \vdots & \vdots & \cdots & \vdots & \cdots & \vdots \\ v_{(M-1)1} & v_{(M-1)2} & \cdots & [y_M] & \cdots & v_{(M-1)(N-1)} \end{pmatrix} \tag{8}$$

where v_{ij} is the jth sample of the ith Q register representing the KNS of the Q channel, and y_i is the observation at the center of the jth Q register representing PSS of the Q channel. Denoting the number of runs obtained in the I channel by d'_I and the number of runs obtained in the Q channels by d'_Q, we define

$$d_I = d'_I - E(d'_I)$$

$$d_Q = d'_Q - E(d'_Q).$$

It can be proved that the quantity

$$R \triangleq \sqrt{d_I^2 + d_Q^2}$$

can be approximated to a Rayleigh distribution with parameter σ_d^2 given by the relation in (7). Thus we decide that a signal is present, if

$$R \leqslant R_0$$

where R_0 is determined from the Rayleigh distribution for a given false alarm probability α.

Detector B

A Single-Path Detection Scheme

In this method, the number of runs in each set $\{x_j; u_{j1}, u_{j2}, \cdots, u_{j(N-1)}\}$ for $j = 0, 1, \cdots, (M-1)$ is determined. Let d_j denote the number of runs in the jth row of the matrix A represented by the relation of (1). From (3) and (4) it follows that

$$\text{Prob } \{d_j = 2\} = 2/N \tag{9}$$

$$\text{Prob } \{d_j = 3\} = 1-(2/N) \tag{10}$$

whenever H_0 is true.

Whenever a signal is present, the distribution of x_j is given by the density

$$f_1(x) = f_0(x - A \cos \overline{j\theta + \phi})$$

where $f_0(x)$ is the density of noise. Thus, the probability distribution of d_j whenever H_1 is true is given by

$$P_1\{d_j = 2\} = \int_{-\infty}^{\infty} [1-F_0(x)]^{N-1} f_0[x - A\cos(j\theta + \phi)]\, dx + \int_{-\infty}^{\infty} [F_0(x)]^{N-1} f_0[x - A\cos(j\theta+\phi)]\, dx$$

$$\neq (2/N) \quad \text{if } A\cos(j\theta + \phi) \neq 0 \tag{11}$$

$$P_1\{d_j = 3\} = 1 - P_1\{d_j = 2\}. \tag{12}$$

Let us now define a new random variable x_j which is defined as

$$x_j = 1 \text{ if } d_j = 2$$
$$x_j = 0 \text{ if } d_j = 3.$$

When noise alone is present, the number of 1s in M trials follows binomial distribution given by

$$P\left\{\sum_{j=0}^{(M-1)} x_j = r\right\} = \binom{m}{r} (2/N)^r (1-2/N)^{M-r}. \quad (13)$$

The total number of 1s in M sets of samples is counted and the following decision is made. If

$$\sum_{j=0}^{(M-1)} x_j \geqslant c_1 \text{ or } \sum_{j=0}^{(M-1)} x_j \leqslant c_2, \quad (14)$$

it is decided that a signal is present; otherwise we decide that noise alone is present. The thresholds c_1 and c_2 are determined by the relations

$$\sum_{r=c_1}^{M} \binom{M}{r} (2/N)^r (1-2/N)^{M-r} = \sum_{r=0}^{c_2} \binom{M}{r} (2/N)^r (1-2/N)^{M-r} = \alpha/2 \quad (15)$$

where α is the probability of false alarm. However, if M is large we can make use of the normal approximation to the binomial:

$$\binom{M}{r} (2/N)^r (1-(2/N))^{M-r} \simeq (1/\sqrt{2\pi}\sigma) \exp\{-[r - 2M/N]^2/2\sigma^2\} \quad (16)$$

where $\sigma^2 = 2M(N-2)/N^2$.

Note:

(i) In order that the distribution of the number of 1s is nondegenerate, we must have the condition $N \geqslant 3$.

(ii) If R_j denotes the rank of x_j, then $d_j = 2$ whenever $R_j = 1$ or $R_j = N$. If $R_j \neq 1$ and $R_j \neq N$, then $d_j = 3$. The method of determining R_j is explained in Fig. 7 of [3]. In fact, in the present method, all comparisons need not be made because once we get a combination of 0 and 1 as outputs from any two comparators the process of comparisons can be terminated. Thus, a considerable amount of time could be saved if a calculator for determining the number of runs is inserted in place of ranker in [3].

A Two-Path Detection Scheme

In addition to finding out the number of runs in each row of the A matrix (corresponding to $I(t)$ channel), we also determine the number of runs in each row of the B matrix (corresponding to the $Q(t)$ channel). Let e_j denote the number of runs in the jth row of the matrix B. When noise alone is present, the probability distribution of e_j is given by

$$\text{Prob}\{e_j = 2\} = 2/N \quad (17)$$

$$\text{Prob}\{e_j = 3\} = 1-2/N. \quad (18)$$

Let us define

$$U_j \triangleq d_j - [3-(2/N)] \text{ and } V_j \triangleq e_j - [3-(2/N)].$$

Let Z_j be a complex variable defined by

$$Z_j \triangleq u_j + iV_j$$

where $i = \sqrt{-1}$. Then the DFT of $\{Z_j\}$ is given by

$$W_h = \sum_{j=0}^{(M-1)} e^{-2\pi ijh/M} Z_j. \quad (19)$$

It may be proved that $|W_h| / \sqrt{M}$ asymptotically approaches the Rayleigh distribution with parameter $\sigma^2 = (2/N)[1-(2/N)]$. Since U_j and V_j are the deviations of the number of runs in the jth row of A and B matrices from their expected values (and they do *not* represent the central ranks [3]) a separate proof of convergence of $|W_h| / \sqrt{M}$ is needed. However, the proof is omitted because it is based on similar lines followed by Dillard et al. [3]. As in [3] based on the DFT coefficients, a decision may be made as to whether a target is present or not.

Acknowledgment

Thanks are due to Dr. R.P. Shenoy, Director, LRDE, Bangalore, for his guidance and kind permission to publish this paper.

References

[1] J.W. Carlyle, "Non-parametric methods in detection theory," *Communication Theory*, A.V. Balakrishnan, Ed. 1968.

[2] R.F. Daly and C.K. Rushforth, "Nonparametric detection of a known signal in additive noise," *IEEE Trans. Information Theory*, vol. IT-11, pp. 70-76, January 1965.

[3] G.M. Dillard and J.T. Rickard, " A distribution-free Doppler processor," *IEEE Trans. Aerospace and Electronic Systems*, vol. AES-10, pp. 479-86, July 1974.

[4] V.G. Hansen, "Detection performance of some nonparametric rank tests and an application to radar," *IEEE Trans. Information Theory*, vol. IT-16, pp. 309-18, May 1970.

[5] V.G. Hansen and V.A. Olsen, "Non-parametric radar extraction using a generalised sign test," *IEEE Trans. Aerospace and Electronic Systems*, vol. AES-7, pp. 942-50, September 1971.

[6] A.M. Mood and F.A. Graybill, *Introduction to the Theory of Statistics.* New York: McGraw-Hill, 1963.

[7] G.W. Zesli and T.S. Fong, "Performance of a two sample Mann-Whitney nonparametric detector in a radar application," *IEEE Trans. Aerospace and Electronic Systems*, vol. AES-7, pp. 951-9, September 1971.

Part V

DEPENDENT OBSERVATIONS IN NONPARAMETRIC DETECTION SCHEMES

Editors' Comments on Papers 26 Through 35

One of the major factors that have so far limited the applicability of nonparametric statistical tests is the assumption underlying most analyses of nonparametric techniques: the input data form a sequence of independent random variables. Although reasonable in many applications addressed by statisticians, the independence assumption cannot always be justified for signal-processing applications. The exact effects of dependent observations on the performance of nonparametric detectors designed under the independence assumption is generally difficult to specify, although it is usually simple to show loss of nonparametric performance or loss of the constant false-alarm probability characteristic. The formulation and analysis of efficient nonparametric detectors for operation on dependent observations continues to be an active area of research. The collection of papers in this last part represents the existing main ideas for the analysis and design of nonparametric detectors under conditions of dependent data.

The correspondence article by Wolff, Gastwirth, and Rubin (Paper 26) is one of the early contributions on this topic in the engineering literature. In this article, the authors consider the simple sign detector operating on a first-order autoregressive discrete-time process. They show that the variance of the sign test statistic depends on the nature of the univariate density function of the input process when the successive inputs are not independent. Conditions for asymptotic normality of the sign test statistic on autoregressive processes are also briefly considered. Such considerations are important to justify the use of the ratio of efficacies, for example, as a measure of relative performance (ARE).

In Paper 27, Gastwirth and Rubin have considered in further detail the effects of autoregressive dependence and of dependent observations from a class of Gaussian processes. Both the sign test and the Wilcoxon test are studied. The authors have reached some interesting conclusions about the effects of dependence on the sign test and the difficulty of circumventing these effects.

Paper 28 by Davisson, Feustel, and Modestino is concerned with dependence effects in rank tests, primarily the mixed tests defined earlier in Paper 16. The two principal characteristics of concern are the asymptotic distribution of the test statistics and the variation of false-alarm probability with different types of dependence (autocorrelation functions) at the inputs. This paper examines in particular the mixed Wilcoxon and the mixed Mann-Whitney and Kendall-tau (two-sample) tests. For sampling from Gaussian input processes with smoothness restrictions on the spectral density, the authors show that the mixed statistics are asymptotically normally distributed, which allows performances to be compared based on the ARE and also allows the variation in false-alarm probabilities to be studied for large sample sizes. One consequence of using a mixed statistic is that asymptotic normality can be proved as considered in this paper; for the fully ranked situations, asymptotic normality is harder to establish, although it is shown to be true in the case of the Mann-Whitney statistic.

A brief correspondence item by Cunningham (Paper 29) considers the effect of correlated clutter-noise samples in a radar detector. This particular detector compares an observation with N previous and following observations, and a decision is based on the rank of the observation from among these $(2N+1)$ values. The analysis is based on a Markov assumption for the data, and a particular Rician-correlated clutter distribution is used for specific numerical values. These show a considerable reduction in false-alarm probability over that obtained for independent samples. A further discussion of this detector has recently appeared in Figueiras-Vidal.[1]

Kanefsky and Thomas propose in Paper 30 a technique for obtaining essentially nonparametric operation of detectors based on dependent sampled data from continuous-time processes. They show that a given parametric test statistic can be modified by adding other functions of the observations (statistics) to it, so that the overall result is to yield a test statistic that is asymptotically normally distributed with known parameters for a large class of noise processes with a given spectral density. In essence two statistics need to be derived that compensate for the mean and variance of the original test statistic. Two simple illustrations of this technique are discussed in the paper, showing the usefulness of this approach.

Papers 31 and 32 describe and analyze techniques for maintaining asymptotically nonparametric performance characteristics (constant false-alarm probabilities), techniques that are based on the estimation of the variance of a test statistic that otherwise gives nonparametric performance only with independent data. Simakhin (Paper 31) computes the variance of several test statistics operating on samples from a T-dependent process (for which samples separated by a time interval larger than T are independent). By constructing consistent estimates of the variances that are distribution independent, the threshold of the detectors can be adjusted to maintain the false-alarm probability at a constant value in the asymptotic case and approximately constant for large sample sizes. This conclusion is based on the asymptotic normality of the test statistics considered (sign, Mann-Whitney and Kendall-tau statistics) for m-dependent processes. The asymptotic nonparametric characteristic of the detectors does not depend on a knowledge of the distributions and spectra of the inputs, although these do affect the performance improvements that can be obtained over the case of independent sampling. Asymptotic relative efficiencies are given for Gaussian distribution and Rayleigh distributions (incoherent detection) of the inputs and with double-exponential autocorrelation functions.

Sirvanci and Wolff (Paper 32) study the behavior of two-sample rank tests on general autoregressive input processes. Conditions for asymptotic normality of the test statistics are defined. Then consistent estimators for the asymptotic variance of the two-sample statistics can be used to normalize the test statistics, so that the resulting statistic is asymptotically normal with known mean and variance under the null hypothesis. The main factor governing the usefulness of this technique is the ease with which good estimates of the variance can be obtained from a finite number of dependent observations. The paper contains some interesting results of a simulation study of this procedure applied to the two-sample Wilcoxon or Mann-Whitney test.

Papers 33 and 34 present a different approach to the design of nonparametric detectors for dependent inputs. The correspondence article by Woinsky and Kurz (Paper 33) originates the basic idea of processing a sequence of groups of univariate samples or vector samples from the input by transformation of each group to a scalar variable. Although the components in each group may be dependent, if the multivariate groups or vectors are independent, then nonparametric performance can be obtained. This work considers the specific case of detection of a constant signal when a reference noise sample is available, using the Mann-Whitney test statistic. Asymptotic performance comparisons show that for the Gauss-Markov example, considerable improvement in performance can be obtained.

In Paper 34, Kassam and Thomas consider a general class of nonparametric detectors, which are modified versions of standard nonparametric detectors. The modified detectors operate at sampling rates higher than those that give rise to independent data. The data are then divided into groups with a sufficient number of observations skipped between groups, leading to a sequence of independent or almost-independent vectors of observations. After a multivariate-to-univariate transformation on each vector, the resulting independent univariate sequence can be processed by a standard nonparametric procedure. The paper also examines the possibilities of using different types of transformations. Asymptotic performance comparisons are made showing that considerable improvements in performance can be achieved using this technique, which allows the use of more observations from the input processes in the observation time interval. The nonparametric performance characteristic is obtained with no knowledge of the statistics of the input process (subject only to mild conditions), although performance improvements depend on the statistical description of the inputs.

More recently, in Paper 35, Al-Hussaini, Badran, and Turner have extended the techniques and analyses presented in Papers 33 and 34 to two-sample or two-input schemes with dependent data. Specifically, the generalized sign, the Mann-Whitney, and modified versions of the Savage and rank-squared statistics are considered, and asymptotic results showing improved performance characteristics are reported.

In addition to these and other papers, a recent book by Gibson and Melsa also contains a discussion of the general problem of dependent data.[2]

REFERENCES

1. A. R. Figueiras-Vidal. "Comments on 'False Alarm Rate for Rank Quantizer When Input Range Samples Are Not Independent.'" *IEEE Trans. Aerosp. Electron. Syst.* **AES-14**:703–705 (1978).
2. J. D. Gibson and J. L. Melsa. *Introduction to Nonparametric Detection with Applications*. Academic Press, 1975, pp. 181–190.

26

Reprinted with permission from *IEEE Trans. Inf. Theory* **IT-13**:311–313 (1967)

THE EFFECT OF AUTOREGRESSIVE DEPENDENCE ON A NONPARAMETRIC TEST

S. S. Wolff, J. L. Gastwirth, and H. Rubin

Abstract—**The sign test, which is nonparametric when used on independent data, is shown to lose its distribution-free property on data of the form $X_t = \varrho X_{t-1} + W_t$, $0 < |\varrho| < 1$, where $\{W_t\}$ is a sequence of independent and identically distributed random variables. Asymptotic normality of the sign-test statistic is proved in two cases. If the data $\{X_t\}$ are regarded as regularly spaced samples of a continuous-parameter lowpass process, then upon increasing the sampling rate indefinitely, the continuous-time sign test, an infinite clipper followed by an integrator, is also not distribution-free.**

I. Introduction

If the input to an infinite clipper consists of discrete samples X_t, $t = 1, \cdots, N$, statistically independent, from a population with distribution function $F(x)$, then

$$S'_N = \sum_1^N \operatorname{sgn} X_t, \qquad (1)$$

the sum of the clipper outputs, is a statistic that may be used to test the null hypothesis that the median of $F(x)$ is zero. The infinite clipper used this way performs the sign test, which is well known[1] to be distribution-free within the class of all absolutely continuous distributions $F(x)$. Thus the infinite clipper-plus-accumulator may be considered as a nonparametric detector for dc signals in noise.[2]

In this correspondence, we discuss the sign test on dependent data. Specifically, we assume the input to be a sample function of a simple first-order autoregression with independent errors or, equivalently, a linear process with exponential weighting function:

$$X_t = \rho X_{t-1} + W_t \qquad t = 1, \cdots, N \qquad (2)$$

Manuscript received December 16, 1965; revised June 14, 1966. This research was sponsored in part by the Office of Naval Research under Contracts Nonr-4010(13), Nonr-4010(09), and Nonr-2587(02).

or

$$X_t = \sum_{\nu=0}^{\infty} \rho^\nu W_{t-\nu} \qquad (3)$$

in which $0 \leq |\rho| < 1$ and $\{W_t\}$ is stationary white noise—a sequence of independent and identically distributed random variables—with zero mean and an even, bounded density function, except perhaps for nonzero probability at the origin. The strictly stationary sequence of observations $\{X_t\}$ might arise from the regular sampling of a continuous-parameter random process with spectral density proportional to $(\omega^2 + \alpha^2)^{-1}$.

For simplicity of analysis, we introduce the indicator random variables

$$Y_t = \begin{cases} 1 & X_t > 0 \\ 0 & X_t \leq 0 \end{cases}$$

so that an equivalent sign-test statistic is

$$S_N = \sum_1^N Y_t. \qquad (4)$$

In Section II we derive formulas for the mean and variance of S_N and illustrate them for the case that X_t is Gaussian and for the case that X_t has the double-exponential distribution; asymptotic results ($N \to \infty$) are also given. In Section III we prove that S_N is asymptotically normal. Section IV extends the material of Section II by passing to a continuous-time sign test (infinite clipper-plus-integrator) operating on continuous-parameter processes with spectral density $(\omega^2 + \alpha^2)^{-1}$ and Gaussian and double-exponential first-order probability distributions.

II. The Mean and Variance of S_N

Since the white-noise sequence $\{W_t\}$ has a symmetric distribution, it follows from (3) that the sequence of random variables $\{X_t\}$ also has a symmetric distribution; thus, $ES_N = E \sum_1^N Y_t = N/2$. We denote the event $Y_t = 1$ by A_t. Because the process is strictly stationary we may define $\varphi_k = P(A_t \cap A_{t+k}) - 1/4 = \varphi_{-k}$. It is easily shown that the variance of S_N is

$$\operatorname{var} S_N = \sum_{-N}^{N} (N - |k|)\varphi_k. \tag{5}$$

Example 1: The Double-Exponential Case

Here the random variables X_t of (2) have the marginal density function

$$\frac{dF(x)}{dx} = \tfrac{1}{2}e^{-|x|}. \tag{6}$$

For this to be true, it can be shown[3] by taking characteristic functions in (2) that the error-random variables W_t must be a mixture of the form

$$W_t = \begin{cases} 0 \text{ with probability } \rho^2 \\ e \text{ with probability } (1 - \rho^2) \end{cases} \tag{7}$$

where 0 denotes a random variable degenerate at the origin and e, independent of 0, has the double-exponential density of (6). We now show that $\varphi_k = \rho^k/4$. By iterating (2) we obtain

$$X_{t+k} = \rho^k X_t + \eta_{t+k}, \qquad k \geq 0, \tag{8}$$

where $\eta_{t+k} = \sum_0^{k-1} \rho^\nu W_{t+k-\nu}$ is independent of X_{t+k} and is also a mixture

$$\eta_{t+k} = \begin{cases} 0 \text{ with probability } \rho^{2k} \\ e \text{ with probability } (1 - \rho^{2k}). \end{cases} \tag{9}$$

Then

$$\begin{aligned} \varphi_k &= \Pr(X_t > 0,\, \rho^k X_t + \eta_{t+k} > 0) - 1/4, \qquad k \geq 0 \\ &= \int_0^\infty \tfrac{1}{2}e^{-x}\,dx\left[\rho^{2k} + (1 - \rho^{2k})\int_{-\rho^k x}^\infty \tfrac{1}{2}e^{-|y|}\,dy\right] - 1/4, \\ &= \rho^k/4. \end{aligned}$$

Thus, $\operatorname{var} S_N = \frac{1}{4}\sum_{-N}^{N} (N - |k|)\rho^k$, which after some manipulation sums to

$$\operatorname{var} S_N = \frac{1}{4}\left[N\frac{1+\rho}{1-\rho} - \frac{2\rho(1-\rho^N)}{(1-\rho)^2}\right]. \tag{10}$$

Asymptotically, of course,

$$\lim_{N\to\infty} \frac{\operatorname{var} S_N}{N/4} = \frac{1+\rho}{1-\rho}. \tag{11}$$

Example 2: The Gaussian Case $dF(x)/dx = (1/\sqrt{2\pi})e^{-x^2/2}$

We recall that if Y and Z are jointly Gaussian, zero mean with correlation coefficient ρ, then $\Pr(Y>0, Z>0) = \frac{1}{4} + (1/2\pi)\arcsin\rho$. From (8) $EX_tX_{t+k} = \rho^k$, $k \geq 0$, so that

$$\operatorname{var} S_N = \frac{1}{2\pi}\sum_{-N}^{N} (N - |k|)\arcsin\rho^k. \tag{12}$$

Using the power series expansion[4] of the arcsin function, we obtain

$$\operatorname{var} S_N = \frac{N}{4} + \frac{\rho}{\pi}\sum_{k=0}^\infty \frac{(1/2)_k(1/2)_k}{(3/2)_k} \cdot \frac{\rho^{2k}}{k!}\left[\frac{N}{1-\rho^{2k+1}} - \frac{1-\rho^{N(2k-1)}}{(1-\rho^{2k+1})^2}\right] \tag{13}$$

where $(a)_k = a(a+1)(a+2)\cdots(a+k-1)$.

Asymptotically, $(N \to \infty)$

$$\frac{\operatorname{var} S_N}{N/4} \to 1 + \frac{4\rho}{\pi}\sum_0^\infty \frac{(1/2)_k(1/2)_k}{(3/2)_k}\frac{\rho^{2k}}{k!}\frac{1}{1-\rho^{2k+1}}. \tag{14}$$

Moreover, the asymptotic variance is an upper bound for the variance at any finite N.

A brief comparison of the results of the preceding two examples is perhaps worthwhile. For independent data, the sign test is distribution-free for the class of absolutely continuous distributions. Indeed, when $\rho = 0$ both (10) and (13) show that var S_N reduces to $N/4$. In this case the Y_t are independent, so that S_N is a sum of N independent binomial random variables, and the result var $S_N = N/4$ is immediate. The effect of the autoregressive dependence ($|\rho| > 0$) is to destroy the nonparametric character of the test; specifically, the variance of S_N is no longer independent of the marginal distribution of X_t. Furthermore, we note that the asymptotic variance of S_N (normalized to $N/4$, its value for $\rho = 0$) when X_t is double-exponential is always greater than its value when X_t is Gaussian, if $\rho > 0$, or always less if $\rho < 0$. This phenomenon might have been expected from the outset: dependence among the observations means that past observations affect the value (and hence the sign) of the present observation. Since the double-exponential density has heavier tails than the normal and since large values of the observations have a greater effect on future observations than small values, the variance of S_N ought to be more affected by dependence when the process variables $\{X_t\}$ follow the double-exponential law than when they are normally distributed.

III. Asymptotic Normality of S_N

We now show that the random process $\{Y_t - \frac{1}{2}\}$, a clipped and centered version of the process $\{X_t\}$, satisfies conditions sufficient for the Blum-Rosenblatt Central Limit Theorem for strong mixing processes[5] to hold for the sequence of random variables S_N. We have the following theorem.

Theorem (Blum-Rosenblatt)

If conditions 1) to 3) are satisfied, there is a sequence of numbers $c_N \uparrow \infty$ such that $\sum_{t=1}^N c_N^{-1} Y_t$ is asymptotically normally distributed.

We state and verify the three conditions in turn.

Condition 1): Let A and B be events determined by conditions on $Y_{j_1}, \cdots, Y_{j_m}(j_1 < \cdots < j_m = 0)$ and $Y_r, Y_{r+k_2}, \cdots, Y_{r+k_n}(r > 0 < k_2 < \cdots < k_n)$, respectively. Then $|P(A \cap B) - P(A)P(B)| \leq f(r) \downarrow 0$ as $r \to \infty$. We put $B = B^* \cap C$ where B^* depends on $Y_{k_2}, \cdots, Y_{k_{r-1}}$, and C on Y_r alone. Since A, B^*, and C can be phrased in terms of the process variables $\{X_t\}$, we can write

$$P(A \cap B) - P(A)P(B) = \int_A dF(x_{j_1}, \cdots, x_0)[P(B \mid x_0) - P(B)] \tag{15}$$

where, on account of the Markov character of $\{X_t\}$, $P(B \mid x_{j_1}, \cdots, x_0)$ is replaced by $P(B \mid x_0)$. Then

$$P(B \mid x_0) - P(B) = P(B^* \mid C)[P(C \mid x_0) - P(C)]. \tag{16}$$

Since $P(B^* \mid C)$ is not zero, we need only show that $P(C \mid x_0) - P(C) \leq f(r) \downarrow 0$. Now C must be an event of the form $X > 0$ or $X \leq 0$; without loss of generality we assume the former. Using (8),

$$P(C \mid x_0) - P(C) = \int_{-\rho^r x_0}^\infty dH_r(\eta) - 1/2 \tag{17}$$

with $H_r(\eta)$ being the distribution function of η_r. Put $H_r(\eta) = H_{1,r}(\eta) + H_{2,r}(\eta)$, where $H_{1,r}(\eta)$ is absolutely continuous and $H_{2,r}(\eta)$ is a step at $\eta = 0$ of amplitude $dH_r(0)$; $\int_{-\infty}^\infty dH_{1,r}(\eta) = 1 -$

$dH_r(0)$. Then we can write (17) as

$$P(C \mid x_0) - P(C) = dH_r(0) + \int_{-\rho^r x_0}^{0} dH_{1,r}(\eta). \quad (18)$$

The first term in (18) is the probability that η_r is zero. Since $\eta_r = \sum_{\nu=0}^{r-1} \rho^\nu W_{r-\nu}$, Pr $\{\eta_r = 0\}$ = Pr {all W_ν, $\nu = 1, \cdots, r$ are zero} = $[\Pr\{W = 0\}]^r$, because the W_i are independent; $dH_r(0)$ tends to zero exponentially with r. The previous argument is valid provided that the W_i are not degenerate at the origin (in such a case, of course, the process variables would themselves be zero with probability 1). The boundedness of the density of W_i away from the origin implies $dH_{1,r}/d\eta\,|_{\eta=0} < \infty$. Hence the second term of (18) is bounded by $Kx_0\rho^r$, where $K < \infty$. Thus $[P(A \cap B) - P(A)P(B)]$ is bounded by a negative exponential function of r.

Condition 2): $E\,|\sum_1^N (Y_t - 1/2)|^2 = h(N) \uparrow \infty$ as $N \to \infty$. The results of section II show that this condition holds for the Gaussian and double-exponential processes.

Condition 3): $E\,|\sum_1^N (Y_t - 1/2)|^{2+\delta} = 0(h(N)^{1+\delta/2})$ for some $\delta > 0$. We choose $\delta = 2$; then the Schwarz inequality shows that

$$E\left|\sum_1^N (Y_t - 1/2)\right|^4 \geq E\left[\left|\sum_1^N (Y_t - 1/2)\right|^2\right]^2 = 0(N^2).$$

Thus the $\{Y_t\}$ processes under discussion satisfy the conditions for strong mixing. Hence by the Blum-Rosenblatt theorem, the sign-test statistic S_N (or S_N') is asymptotically normally distributed. Clearly, the sequence of constants c_N can be taken to be $(\text{var } S_N)^{1/2}$.

IV. Passage to Continuous Time

We now suppose that the data $\{X_k\}$ are regularly spaced samples from a segment of length T of a continuous-parameter random process with spectral density $(\omega^2 + \alpha^2)^{-1}$ and, therefore, normalized correlation function $e^{-\alpha|\tau|}$. Thus $X_k = X(kT/N)$ and

$$\rho = EX_kX_{k+1}/EX^2 = e^{-\alpha T/N}.$$

In analogy with the definitions in Section I, let

$$Y(t) = \begin{cases} 1 & X(t) > 0 \\ 0 & X(t) \leq 0 \end{cases}, \quad (19)$$

$$Y_k = Y\left(k\frac{T}{N}\right), \quad (20)$$

and

$$S_N^* = \frac{1}{N}\sum_1^N Y_k. \quad (21)$$

From (4), we note that var $S_N^* = 1/N^2$ var S_N. Now, keeping T fixed, we allow $N \to \infty$ so that $\rho \to 1$ and

$$S_N^* \to S^*(T) = \frac{1}{T}\int_0^T Y(t)\,dt. \quad (22)$$

When the marginal distribution of $X(t)$ is double-exponential (6), we may use (11) directly:

$$\text{var } S_N^* \to \frac{1}{4N}\frac{1+\rho}{1-\rho}$$

$$\to \frac{1}{4N}\frac{1+e^{-\alpha T/N}}{1-e^{-\alpha T/N}}$$

$$\to \frac{1}{2\alpha T} = \text{var } S^*(T). \quad (23)$$

In the Gaussian case, we proceed from (12) to get

$$\text{var } S_N^* = \frac{1}{2\pi N}\sum_{-N}^{N}\left(1 - \frac{|k|}{N}\right)\arcsin \rho^{|k|}$$

$$\to \frac{1}{\pi}\int_0^1 (1-z)\arcsin e^{-\alpha Tz}\,dz = \text{var } S^*(T). \quad (24)$$

This integral is difficult to evaluate analytically; however, as in the discrete-parameter case, T var $S^*(t)$ is bounded from above by its limiting value as $T \to \infty$. We have

$$\lim_{T\to\infty} T \text{ var } S^*(t) = \frac{1}{\pi\alpha}\,{}_3F_2(\tfrac{1}{2}, \tfrac{1}{2}, \tfrac{1}{2}; \tfrac{3}{2}, \tfrac{3}{2}; 1), \quad (26)$$

and by comparing the series expansion of the hypergeometric function ${}_3F_2$ with that of the arcsin, we see that var $S^*(T) < 1/2\alpha T$.

Since the results of the discrete-parameter analysis in Section II carry over to the continuous-parameter situation, the continuous-time sign test is not distribution-free within the isospectral class of processes having spectral density $(\omega^2 + \alpha^2)^{-1}$. We have not proved the normality of the continuous-time sign-test statistic, and direct application of the Blum-Rosenblatt theorem is not possible. We conjecture, however, that as $T \to \infty$, $T^{1/2}S^*(T)$ is asymptotically normal.

Finally, in order that our results for the continuous-parameter situation not be vacuous, we must be sure of the existence of the continuous-parameter processes we have analyzed. In the Gaussian case, the process is the well-known Gaussian (first-order) Markov process. The double-exponential process can be shown[6] to exist as a kind of filtered low-density shot noise.

V. Summary and Conclusions

We have shown that the sign test loses its distribution-free property when it is used on discrete- or continuous-parameter first-order autoregressions with independent errors. The asymptotic normality of the sign-test statistic has also been demonstrated in the discrete-parameter case.

When data are independent, the sign test is certainly one of the simplest nonparametric techniques available. The failure of even this simple procedure to remain nonparametric under dependence leads us to conjecture that the many nonparametric methods, well known in the statistical literature, that have been proposed for communication problems will not be nonparametric for dependent data.

Bibliography

[1] E. L. Lehman, *Testing Statistical Hypotheses*. New York: Wiley, 1959, pp. 147–149.

[2] M. Kanefsky and J. B. Thomas, "On polarity detection schemes with non-Gaussian inputs," *J. Franklin Inst.*, vol. 280, pp. 120–138, August 1965.

[3] J. L. Gastwirth and S. S. Wolff, "A characterization of the Laplace distribution," Dept. of Statistics, The Johns Hopkins University, Baltimore, Md., Tech. Rept. 28, 1965.

[4] *Handbook of Mathematical Functions* M. Abramowitz and I. A. Stegun, Eds. Washington, D. C.: NBS, publ. AMS-55, June 1964, p. 556.

[5] J. R. Blum and M. Rosenblatt, "A class of stationary processes and a central limit theorem," *Proc. Nat'l Acad. Sci.*, vol. 42, pp. 412–423, 1956.

[6] S. S. Wolff and J. L. Gastwirth, "On probability distributions for filtered white noise," *IEEE Trans. on Information Theory*, vol. IT-13, July 1967 (to be published).

Reprinted from *Am. Stat. Assoc. J.* **66**:816–820 (1971)

Effect of Dependence on the Level of Some One-Sample Tests

JOSEPH L. GASTWIRTH and HERMAN RUBIN*

This article studies the effect that serial correlation of the observations has on the distribution of the mean and two one-sample nonparametric tests, the sign test and the Wilcoxon test. It is shown that relatively slight dependence has a strong influence on the level of these tests.

1. INTRODUCTION

The usual one-sample t-test tests the null hypothesis that the mean, μ, of the population sampled is a specific value (which may be taken to be 0) against either a one-sided alternative ($\mu>0$) or a two-sided alternative ($\mu\neq 0$). For symmetric distributions, the sign test and Wilcoxon tests are nonparametric competitors of the t-test. Recently Pratt [6] studied the robustness of the level of several two-sample versions of nonparametric tests when the observations come from populations with different shapes. The purpose of the present article is to study the effect that dependence among the observations has on the asymptotic level of the one-sample tests.

Earlier the authors and Wolff [1] showed that the sign test is no longer distribution-free even when the observations are from two stationary processes with the same spectrum. In Section 4 we illustrate that the sign test can have the same distribution for processes with different spectra. This suggests that it will be difficult to obtain distribution-free tests for stationary processes. We prove, however, that for first order autoregressive processes with independent symmetric errors the variance of the sign test exceeds its value in the independent case when $\rho>0$. The reverse inequality holds when $\rho<0$.

The authors [3] studied the asymptotic behavior of various robust estimators on dependent processes. Using these methods we discuss the behavior of the level of the sign, Wilcoxon and t-test ($\overline{X}$) when the observations are from a completely regular [8] Gaussian process. The first order autoregressive process is studied in detail for both Gaussian and double-exponential marginal distributions. It turns out that on Gaussian data, with positive serial correlation, i.e., $\rho_k\geq 0$ for all k, the sign test is the least affected by dependence while $\overline{X}$ is most affected. On the first order autoregressive double-exponential process (FOADP) the changes in the levels of the sign test and $\overline{X}$ are identical while the level of the Wilcoxon test is more sensitive to dependence.

The relevant mathematical preliminaries are summarized in Section 2. It should be mentioned that Serfling [9] demonstrated that the two-sample Wilcoxon test is asymptotically normally distributed for dependent processes but did not study the effect of dependence on the Wilcoxon test in detail. There has been some interest in the engineering literature and we refer the reader to Modestino [5] for further references in this area.

2. PRELIMINARY MATERIAL

In this section we summarize the basic properties of the statistics and stochastic processes we shall study.

Let $\{X_i\}_{i=1}^n$ be a strictly stationary process [7] with continuous marginal distribution $F(x-\theta)$ where θ is a location parameter. The sign test statistic is defined in terms of the indicator random variables

$$Y_i = \begin{cases} 1 & X_i > 0, \\ 0 & X_i < 0 \end{cases} \tag{2.1}$$

and is

$$S_n = \textstyle\sum_{i=1}^n (Y_i - \frac{1}{2}). \tag{2.2}$$

The Wilcoxon statistic is based on the number of positive pairwise averages, where the pairs with distinct indices are counted twice, i.e., for each pair of observations X_i and X_j let

$$I_{i,j} = \begin{cases} 1, & (X_i + X_j) > 0 \\ 0, & (X_i + X_j) < 0, \end{cases} \tag{2.3}$$

then

$$\textstyle\sum_{\text{all pairs }(i,j)} I_{ij} = 2\sum_{i<j} I_{ij} + \sum_{i=1}^n I_{ii} \tag{2.4}$$

and the statistic W is expression (2.4) minus its expectation n^2. Of course, the sample mean is

$$\overline{X} = n^{-1}\textstyle\sum_{i=1}^n X_i. \tag{2.5}$$

---

* Joseph L. Gastwirth is visiting faculty advisor, Office of Statistical Policy, Office of Management and Budget, Executive Office of the President, Washington, D.C. 20503. Herman Rubin is professor, Department of Statistics, Purdue University, West Lafayette, Ind. 47907. The first author's research was supported by National Science Foundation Research Grant No. GP 12636 awarded to the Department of Statistics, The Johns Hopkins University. The second author's research was supported by the Office of Naval Research under contract NONR 2587 (02) awarded to the Department of Statistics, Michigan State University. The authors thank the editor and the referees for a variety of helpful suggestions.

When $\theta=0$ and $F(x)$ is symmetric all three statistics have mean 0, however, their variances depend on the structure of the stationary process $\{X_i\}$. For a strictly stationary process, the variance of the sign test is given by

$$n \operatorname{Var}(S_n) = \sum_{k=-n}^{n}\left(1-\frac{|k|}{n}\right)\varphi_k, \tag{2.6}$$

where

$$\varphi_k = P[X_0>0, X_k>0] - \tfrac{1}{4} = \operatorname{Cov}(Y_0, Y_k), \tag{2.7}$$

while the variance of $\overline{X}$ is

$$n \operatorname{Var}(\overline{X}) = \sum_{k=-n}^{n}\left(1-\frac{|k|}{n}\right)\rho_k, \tag{2.8}$$

where $\rho_k = \operatorname{Cov}(X_0, X_k)$. The asymptotic variance of the Wilcoxon test is given in [3] but the general formula is not simple. When $\{X_i\}$ is a completely regular Gaussian process with mean 0, variance 1 and correlation sequence $\rho_k = E(X_0X_k)$ satisfying $\sum|\rho_k|<\infty$, the asymptotic distribution of our three statistics is given by:

Theorem 2.1 [3]. On completely regular Gaussian processes such that $\sum|\rho_k|<\infty$, the three statistics are asymptotically normally distributed with mean 0 and asymptotic variances

$$nV(\overline{X}) \sim \sum_{-\infty}^{\infty} \rho_k = 1 + 2\sum_{1}^{\infty}\rho_k \tag{2.9a}$$

$$nV(S) \sim \frac{1}{2\pi}\sum_{-\infty}^{\infty} \arcsin \rho_k$$

$$= \frac{1}{4} + \frac{1}{\pi}\sum_{1}^{\infty}\arcsin \rho_k \tag{2.9b}$$

$$nV(W) \sim \frac{1}{2\pi}\sum_{-\infty}^{\infty}\arcsin(\tfrac{1}{2}\rho_k)$$

$$= \frac{1}{12} + \frac{1}{\pi}\sum_{1}^{\infty}\arcsin(\tfrac{1}{2}\rho_k). \tag{2.9c}$$

Another process of interest is the first order Markov process $X_{n+1} = \rho X_n + \epsilon_n$ with a double-exponential marginal distribution. Gastwirth and Wolff [2] showed that the errors ϵ_n are i.i.d. with the probability structure

$$\epsilon_n = \begin{cases} 0 & \text{with probability } \rho^2 \\ D_n & \text{with probability } 1-\rho^2, \end{cases}$$

where D_n is a sequence of i.i.d. double-exponential r.v.'s. The asymptotic behavior of the three statistics when the observations come from this process is given by:

Theorem 2.2. On first order autoregressive double-exponential processes (FOADP) the three statistics are asymptotically normally distributed with mean 0 and the following asymptotic variances:

$$nV(\overline{X}) \sim (1+\rho)/(1-\rho) \tag{2.10a}$$

$$nV(S) \sim \frac{1}{4}(1+\rho)/(1-\rho) \tag{2.10b}$$

$$nV(W) \sim \frac{1}{12} + \frac{3}{16}\left(\frac{\rho}{1-\rho} - \sum_{j=1}^{\infty}\frac{\rho^{3j}}{(2+\rho^j)^2}\right), \quad \rho>0 \tag{2.10c}$$

$$nV(W) \sim \frac{3}{32}\sum_{j=-\infty}^{\infty}\left(\rho^{|j|} - \frac{\rho^{3|j|}}{(2+|\rho|^{|j|})^2}\right) = \frac{1}{12} + \frac{3}{32}\left[\sum_{j=-\infty, \neq 0}^{\infty}\left(\rho^{|j|} - \frac{\rho^{3|j|}}{(2+|\rho|^{|j|})^2}\right)\right], \quad \rho<0. \tag{2.10d}$$

3. THE EFFECT OF DEPENDENCE ON THE LEVEL OF THE TEST

For any sequence of statistics T_n with means $E(T_n)$ and variances $V(T_n)$ such that $[T_n - E(T_n)]/V(T_n)^{\frac{1}{2}}$ is asymptotically normally distributed, the critical point K_α corresponding to a one-sided test of size α is determined by the relation

$$1 - \Phi(K_\alpha) = \alpha, \tag{3.1}$$

where $\Phi(x)$ is the c.d.f. of the standard normal distribution. When the observations are from a stationary process the variance of the statistic T_n differs from its value in the case of i.i.d. observations with the same marginal distribution. If we denote the asymptotic variance of T_n in the case of dependent data (stationary process) by D, its value in the case of i.i.d. observations by V and the ratio D/V by τ, then the critical region determined under the assumption of independent observations, i.e., reject H_0 if

$$\frac{T_n - E(T_n)}{V(T_n)^{\frac{1}{2}}} \geq K_\alpha, \tag{3.2}$$

has approximate probability

$$1 - \Phi(K_\alpha \tau^{-\frac{1}{2}}), \tag{3.3}$$

which is the asymptotic level of the test based on T_n when the observations are dependent. While one can always estimate the variance $V(T_n)$ of a statistic T_n, formulas (2.9) and (2.10) indicate that this procedure is tedious in the case of dependent observations. Thus one hopes to devise tests with the property that τ is close to one for a variety of stationary processes. Unfortunately, the levels of the common tests are sensitive to dependence. Throughout this section "the level" of a test will mean its asymptotic level.

For Gaussian processes with positive serial correlation our main result is:

Theorem 3.1. If $\{X_k\}$ is a completely regular Gaussian process whose autocorrelations satisfy $\sum\rho_k<\infty$, $\rho_k\geq 0$ for all k, then the level of the sign test changes less than the level of the Wilcoxon test which changes less than the level of the t test (based on $\overline{X}$). In all cases, the levels are greater than their value when the observations are independent.

The only process with some negative ρ_k we studied in

detail was the first order autoregression. In the Gaussian case the result is:

Theorem 3.2. On FOAGP's with negative ρ, the level of the sign test is less sensitive than the level of the Wilcoxon test which is less sensitive than the level of the test based on the mean, i.e.,

$$1 \geq \tau(S) \geq \tau(w) \geq \tau(\overline{X}). \tag{3.4}$$

Moreover, the ratios $\tau(S)/\tau(\overline{X})$ and $\tau(W)/\tau(\overline{X})$ are monotone increasing functions of $\eta = -\rho$.

Before presenting the proofs it is instructive to study the actual levels of the three tests considered. For nominal levels .05, .02 and .01 (set under the assumption of independence) when the observations are from a first order autoregressive Gaussian process ($\rho = \pm.1$, .2, .3 and .4). The results, given in the table, indicate that the levels of all the tests are *sensitive to dependence* and that the sign test is clearly the least sensitive of the three. As one expects, while the absolute change in level is larger at the larger nominal levels, the relative change is greater at the smaller significance levels.

EFFECT ON THE LEVEL OF THE TESTS ON FIRST ORDER AUTOREGRESSIVE GAUSSIAN PROCESSES

Nominal level	ρ: 0	.1	.2	.3	.4	-.1	-.2	-.3	-.4
					Sign test				
.05	.05	.0619	.0766	.0934	.1142	.0401	.0318	.0247	.0185
.02	.02	.0271	.0367	.0494	.0659	.0142	.0102	.0069	.0046
.01	.01	.0136	.0200	.0291	.0418	.006	.0039	.0024	.0014
					Wilcoxon test				
.05	.05	.0678	.0880	.1114	.1380	.0351	.0231	.0138	.0071
.02	.02	.0308	.0453	.0639	.0864	.0118	.0064	.0029	.0011
.01	.01	.0160	.0260	.0400	.0589	.0047	.0021	.0008	.0002
					Mean				
.05	.05	.0679	.0896	.1138	.1408	.0345	.0219	.0125	.0050
.02	.02	.0314	.0465	.0655	.0890	.0115	.0059	.0026	.0008
.01	.01	.0163	.0269	.0415	.0610	.0045	.0019	.0006	.0002

Proof of Theorem 3.1. From Theorem 2.1 it follows that

$$\tau(S) - 1 = (4/\pi) \textstyle\sum_{k=1}^{\infty} \arcsin \rho_k \tag{3.5a}$$

$$\tau(W) - 1 = (12/\pi) \textstyle\sum_{k=1}^{\infty} \arcsin (\rho_k/2) \tag{3.5b}$$

and

$$\tau(\overline{X}) - 1 = 2 \textstyle\sum_{1}^{\infty} \rho_k. \tag{3.5c}$$

The inequalities

$$(\tfrac{1}{3}) \arcsin x \leq \arcsin(x/2) \leq (\pi/6)x,\ 0 \leq x \leq 1 \tag{3.6}$$

imply that $\tau(S) \leq \tau(W) \leq \tau(\overline{X})$ and the result follows.

In order to prove Theorem 3.2 for FOAGP's with negative ρ recall that all three statistics are asymptotically normally distributed with asymptotic variances given by

$$nV(S) \sim (1/2\pi) \textstyle\sum_{k=-\infty}^{\infty} (-1)^k \arcsin(\eta^{|k|}), \tag{3.7a}$$

$$nV(W) \sim (1/2\pi) \textstyle\sum_{k=-\infty}^{\infty} (-1)^k \arcsin(\eta^{|k|}/2), \tag{3.7b}$$

and

$$nV(\overline{X}) \sim (1-\eta)/(1+\eta), \tag{3.7c}$$

where $\eta = |\rho|$. Since all variances in (3.7) are less than their corresponding values in the case of i.i.d. observations, the value of τ in formula (3.3) is less than one and the true level of the test is smaller than its value in the independent case.

Proof of Theorem 3.2. To derive the inequality on the far right of (3.4) note that

$$\begin{aligned} \tau(W) &= \frac{6}{\pi} \textstyle\sum_{k=-\infty}^{\infty} (-1)^k \sin^{-1}\left(\frac{\eta^{|k|}}{2}\right) \\ &= \frac{6}{\pi} \textstyle\sum_{j=1}^{\infty} \frac{c_j}{2^{2j-1}} \frac{1-\eta^{2j-1}}{1+\eta^{2j-1}}, \end{aligned} \tag{3.8}$$

after using the Taylor series $\arcsin x = \sum_{j=1}^{\infty} c_j x^{2j-1}$. Since $\tau(\overline{X}) = (1-\eta)/(1+\eta)$ formula (3.8) can be written as

$$\tau(W) = \frac{6}{\pi} \tau(\overline{X}) \textstyle\sum_{j=1}^{\infty} \frac{c_j}{2^{2j-1}} \Psi_j(\eta), \tag{3.9}$$

where

$$\Psi_j(\eta) = \frac{1-\eta^{2j-1}}{1+\eta^{2j-1}} \frac{1+\eta}{1-\eta}. \tag{3.10}$$

As the coefficients c_j are positive, the functions $\Psi_j(\eta)$ are monotone increasing functions of η. Since $\tau(W) = \tau(\overline{X}) = 1$ when $\eta = 0$, $\tau(W)/\tau(\overline{X})$ is a monotone increasing function of η and the inequality follows.

The relation $\tau(S) \geq \tau(W)$ is derived by expressing $\tau(S)$ as

$$\tau(S) = \frac{2}{\pi} \frac{1-\eta}{1+\eta} \textstyle\sum_{j=1}^{\infty} c_j \Psi_j(\eta). \tag{3.11}$$

As both expressions (3.11) and (3.9) are one when $\eta = 0$ and as $c_j > c_j 2^{-(2j-1)}$, (3.11) is always larger than (3.9). Again, $\tau(S)/\tau(\overline{X})$ is a monotone increasing function of η.

Remark. The approach of the level of the tests to zero as $\rho \to -1$ (or $\eta \to +1$) can also be obtained. From Theorem 6.4 of [3] it follows that the variance of the sign test goes to zero at a slower rate than the test based on $\overline{X}$ does so that the level of the sign test approaches zero at a slower rate than the level of a test based on $\overline{X}$. As η approaches one, the level of the test based on $\overline{X}$ is approximately

$$1 - \Phi(K_\alpha[(1+\eta)/(1-\eta)]^{\frac{1}{2}}) \tag{3.12}$$

while the level of the Wilcoxon test is approximately

$$1 - \Phi\left(K_\alpha \left[\frac{(1+\eta)}{(1-\eta)} \frac{\pi}{2.3^{\frac{1}{2}}}\right]^{\frac{1}{2}}\right). \tag{3.13}$$

As $\pi/2\sqrt{3}$ is less than one the level of the Wilcoxon test approaches zero at a slower rate than the level of the test based on $\overline{X}$. This behavior is consistent with Theorem 3.2.

In order to assess the effect that the marginal distribution has on the results, we studied the FOADP. From Theorem 2.2 it follows that the effect of dependence on the variance of the sign test statistic and on $\overline{X}$ is the same when the observations are from a double-exponential process. The analog of Theorems 3.1 and 3.2 is

Theorem 3.3. If $\{X_k\}$ is a FOADP with $\rho>0$, then the level of the Wilcoxon test changes *more* than the level of the sign test or the t-test. The levels of the sign and t-test ($\overline{X}$) change in the same manner. When $\rho<0$, the level of the sign test changes less than the level of the Wilcoxon test, i.e.,

$$1 \geq \tau(S) \geq \tau(W)$$

and the ratio $\tau(S)/\tau(W)$ is a monotone function of η.

Proof. From Theorem 2.2, when $\rho>0$ the relevant values of τ are

$$\tau(S) = \tau(\overline{X}) = \frac{1+\rho}{1-\rho} = 1 + \frac{2\rho}{1-\rho} \qquad (3.14)$$

and

$$\tau(W) = 1 + \frac{9}{4}\left(\frac{\rho}{1-\rho} - \sum_{j=1}^{\infty} \frac{\rho^{3j}}{(2+\rho^j)^2}\right). \qquad (3.15)$$

The first assertion is equivalent to $\tau(S) \leq \tau(W)$ and will follow once it is shown that

$$9 \sum_{j=1}^{\infty} \frac{\rho^{3j}}{(2+\rho^j)^2} \leq \frac{\rho}{1-\rho} \qquad \text{for } 0<\rho<1. \qquad (3.16)$$

$$\sum_{j=1}^{\infty} \frac{\rho^{3j}}{(2+\rho^j)^2} \leq \left(\frac{1}{4}\right) \sum_{j=1}^{\infty} \rho^{3j} = \frac{1}{4}\,\frac{\rho^3}{(1-\rho^3)}$$
$$= \frac{1}{4} \cdot \frac{\rho}{1-\rho} \cdot \frac{\rho^2}{1+\rho+\rho^2} \qquad (3.17)$$

and as the function $\rho^2[1+\rho+\rho^2]^{-1}$ increases on (0, 1) to its maximum value ($\frac{1}{3}$), the right side of (3.17) is

$$\leq (1/12)\rho[1-\rho]^{-1}, \qquad (3.18)$$

which implies (3.16).

4. FURTHER PROPERTIES OF THE SIGN TEST

In this section we generalize the qualitative nature of Theorem 3.3 to any first order autoregressive process with independent symmetric errors. Also we shall summarize some properties of the sign test statistic which suggest that it will be difficult to obtain "distribution-free" tests for stationary processes.

The main result of this section is

Theorem 4.1. If $\{X_k\}$ is a first order autoregressive process with independent symmetric errors, then the variance of the sign test statistic is greater than or equal to its independence value $n/4$ when $\rho>0$ and less than or equal to $n/4$ when $\rho<0$.

Proof. Using the representation (2.6) it follows that

$$n^{-1}V(S_n) = \tfrac{1}{4} + 2 \sum_{k=1}^{n} (1-k/n) \cdot [P(X_0>0, X_k>0) - \tfrac{1}{4}]. \qquad (4.1)$$

When $\rho>0$, $X_k = \rho^k X_0 + \epsilon^*$, where ϵ^* is independent of X_0. Since X_k and X_0 are positively regression dependent in Lehmann's sense [4] his Theorem 2 implies that

$$P[Y_0>0,\ Y_k>0] \geq \tfrac{1}{4} \qquad (4.2)$$

and the first assertion follows.

The proof of when $\rho<0$ uses Lemma 4.1 which is stated and proved at the end of the section. To prove the second assertion of Theorem 4.1, it must be shown that

$$\sum_{k=1}^{n} (1-k/n)[P(X_0>0, X_k>0) - \tfrac{1}{4}] \leq 0. \qquad (4.3)$$

If n is even, grouping the terms of the series in (4.3) in pairs yields

$$1/n \sum_{r=1}^{n} [P(X_0>0, X_{2r-1}>0) - \tfrac{1}{4}]$$
$$+ \sum_{r=1}^{n} (1-2r/n)[P(X_0>0, X_{2r-1}>0) \qquad (4.4)$$
$$+ P(X_0>0, X_{2r}>0) - \tfrac{1}{2}].$$

Since each triplet X_0, X_{2r-1} and X_{2r} satisfies the conditions of Lemma 4.1 each term in the second sum in (4.4) is ≤ 0. As X_0 and X_{2r-1} are negatively regression dependent each term in the first sum is ≤ 0 by Theorem 2 of [4]. When n is odd, the same argument is valid as one is adding a term of the form $P[X_0>0, X_{2m+1}>0] - \frac{1}{4}$, which is ≤ 0, to a sum of the form (4.4).

An example of an autoregressive process with $\rho>0$ where the variance of the sign test equals $n/4$ is provided by the uniform autoregressive with $\rho=\frac{1}{2}$. In this case

$$X_k = \tfrac{1}{2}X_{k-1} + \epsilon_k, \qquad (4.5)$$

where $\{\epsilon_k\}$ is a sequence of i.i.d. binomial r.v.'s assuming the values 0 and 1 with probability $\frac{1}{2}$. The sign test statistic is the number of $\{X_k\}$ which are greater than $\frac{1}{2}$ and this is precisely the number of $\{\epsilon_k\}$ which equal 1. Thus the null distribution of the sign test is that of the sum of n binomial r.v.'s which is its distribution in the case of independent observations.

Because of recent interest in second order stationary processes we would like to point out three properties of the sign test which indicate that it will be quite difficult to obtain "distribution-free" tests for stationary processes. These are:

> *Property 1:* The distribution of the sign test statistic can be different on two processes with the same spectrum (correlation sequence) and different marginal distributions.

> *Property 2:* The asymptotic distribution of the sign test statistic can be different for two processes with the *same spectrum* and the *same marginal* distribution.

Property 3: The sign test statistic can have the same distribution for processes with different spectra.

We conclude the article with the technical

Lemma 4.1. Consider three symmetric r.v.'s X^-, Y, Z with the structure

$$\begin{aligned} Y &= -\sigma X^- + \epsilon \qquad \eta, \sigma > 0 \\ X^- &= -\eta Z + \delta, \end{aligned} \tag{4.6}$$

where ϵ and δ are independent r.v.'s symmetric about the origin which are independent of X^- and Z, respectively. Then

$$P[Y > 0, X^- > 0] + P[Y > 0, Z > 0] \leq 1/2 \tag{4.7}$$

Proof. Letting $X^+ = -X^-$, the structure (4.6) becomes

$$Y = \sigma X^+ + \epsilon, \qquad X^+ = \eta Z - \delta$$

and $P[Y>0, X^->0] = \frac{1}{2} - P[Y>0, X^+>0]$ so assertion (4.7) is equivalent to

$$P[Y > 0, Z > 0] \leq P[Y > 0, X^+ > 0]. \tag{4.8}$$

Conditioning on X^+ and using the fact that Y and Z are related only via X^+ it follows that

$$\begin{aligned} P[Y > 0, Z > 0] &= E\{P[(Y > 0, Z > 0) \mid X^+]\} \\ &= E\{P(Y > 0 \mid X^+)P(Z > 0 \mid X^+)\}. \end{aligned} \tag{4.9}$$

Letting $f(x) = P[Y>0 \mid X^+ = x]$ and $g(x) = P[Z>0 \mid X^+ = x]$ assertion (4.8) can be expressed as

$$\int_{-\infty}^{\infty} f(x)g(x)dP(x) \leq \int_0^{\infty} f(x)dP(x) \tag{4.10}$$

where $P(x)$ is the c.d.f. of X^+. As the error r.v.'s are symmetric, $f(-x) = P[Y>0 \mid X^+ = -x] = 1 - P[Y<0 \mid X^+ = -x] = 1 - P[Y>0 \mid X^+ = x] = 1 - f(x)$ and $g(-x) = 1 - g(x)$ so that

$$\begin{aligned} &\int_{-\infty}^{\infty} f(x)g(x)dP(x) \\ &\quad = \int_0^{\infty} [f(x)g(x) + (1 - f(x))(1 - g(x))]dP(x). \end{aligned} \tag{4.11}$$

Substituting (4.11) in the left side of (4.10) shows that the assertion of the lemma is equivalent to

$$\begin{aligned} &\int_0^{\infty} [2f(x)g(x) + (1 - g(x)) - 2f(x)]dP(x) \\ &\quad = \int_0^{\infty} (1 - g(x))(1 - 2f(x))dP(x) \leq 0. \end{aligned} \tag{4.12}$$

Since $g(x) \leq 1$, $f(x) \geq \frac{1}{2}$ (as Y is positively regression dependent on X^+) and $dP(x)$ is a positive measure, the inequality on the right of (4.12) holds.

REFERENCES

[1] Gastwirth, J. L., Rubin, H. and Wolff, S. S., "The Sign Test on Autoregressive Data," *IEEE Transactions on Information Theory, IT-13*. 311–13, 1967.

[2] ——— and Wolff, S. S., "A Characterization of the Laplace Distribution," Technical Report No. 28, Statistics Department, The Johns Hopkins University, 1966.

[3] ——— and Rubin, H., "The Behavior of Robust Estimators on Dependent Data," Technical Report, Purdue University, 1970 (to be published in *Annals of Mathematical Statistics*).

[4] Lehmann, E. L., "Some Concepts of Dependence," *Annals of Mathematical Statistics* 37 (1966), 1137–53.

[5] Modestino, J. W., "Nonparametric and Adaptive Detection on Dependent Data," Technical Report 27, Department of Electrical Engineering, Princeton University, 1969.

[6] Pratt, J. W., "Robustness of Some Procedures for the Two-Sample Location Problem," *Journal of the American Statistical Association*, 59 (September 1964), 665–80.

[7] Rosenblatt, M., *Random Processes*, New York: Oxford University Press, 1962.

[8] Rozanov, Y. A., *Stationary Random Processes*, San Francisco: Holden-Day, 1967.

[9] Serfling, R. J., "The Wilcoxon Two-Sample Statistic on Strongly-Mixing Processes," *Annals of Mathematical Statistics* 39 (August 1968), 1202–09.

Reprinted with permission from *IEEE Trans. Inf. Theory* **IT-16**:32–41 (1970)

The Effects of Dependence on Nonparametric Detection

LEE D. DAVISSON, SENIOR MEMBER, IEEE, EDWARD A. FEUSTEL, MEMBER, IEEE, AND JAMES W. MODESTINO, MEMBER, IEEE

I. Introduction

Nonparametric tests have been proposed as the basis for practical detectors on sampled data when little information is available about the signal and/or noise distributions [1]–[8]. In theoretical studies, asymptotic relative efficiency (ARE) is the usual measure of optimality. Unfortunately, theoretical results for these tests have assumed statistical independence with the exception of [2], [3], and [10], which consider certain special cases. The purpose of this paper is to investigate the effects and ways of minimizing them of dependence on rank tests, in particular a class of nonparametric tests called "mixed" statistical tests [1]. This class of tests is a generalization of nonparametric statistics that provides a high level of efficiency with relatively little computation. A mixed statistic is formed in the following manner. The group of N observations from each of k channels (sometimes called "samples") is divided into p-equal groups of m observations (called the "group size") within each channel. A k-sample statistic called the intermediate statistic is used on mk observations to produce p values. The mixed statistic is formed by summing the p values. It is then compared with a threshold in the usual manner to decide between two hypotheses—the null and the alternative hypotheses. If the resultant value exceeds the threshold, the alternative is accepted; if it is less, the null hypothesis is accepted. The intermediate statistic is typically a rank statistic such as the Mann–Whitney (for $k = 2$ channels), which requires $O(m^2)$ operations. Thus the mixed test demands $O(m^2p = mN)$ operations compared with $O(N^2)$ for ranking all N values. This can be a significant reduction for large N. Of course, for $m = N$ (that is, $p = 1$), the usual nonparametric statistic results.

Manuscript received February 27, 1968; revised March 19, 1969. This work was partially supported under NASA grant NGR-31-001-068 and under USAF RADC Contract F 30602-68-C-0078.
L. D. Davisson is with Princeton University, Princeton, N. J.; on a leave of absence with the University of Southern California, Los Angeles, Calif. 90007.
E. A. Feustel is with Rice University, Houston, Tex. 77001.
J. W. Modestino was with Princeton University, Princeton, N. J. He is now with General Telephone & Electronics Laboratories, Waltham, Mass. 02154.

In this paper results are restricted to three well-known intermediate statistics, although generalization is possible. These are the one-sample Wilcoxon ($k = 1$ channel), the two-sample Mann–Whitney ($k = 2$), and the mixed Kendall $\tau(k = 2)$. The first two tests apply to detecting a sure signal in noise and the last to detecting a random signal in noise. As is usually done in nonparametric work, the tests are evaluated for nominally Gaussian data although some of the techniques described could be used on more general sources, e.g., autoregressive schemes. It is shown that for a large class of spectral densities, the mixed test statistics are asymtotically normal for fixed $m < \infty$ and that the fully ranked ($m = N$) Mann–Whitney statistic is asymptotically normal when the two channels are statistically independent of each other. For the fully ranked Wilcoxon and Kendall τ tests, asymptotic normality has not been established for the reasons discussed. The asymptotic normality of a test statistic justifies the use of ARE, given by the ratio of test efficacies, as a figure of merit.

The ARE with respect to an appropriate parametric test is found for various group sizes, sampling intervals, and methods of grouping. As the analysis is similar for the three statistics, the discussion is primarily given in terms of one of them, the Mann–Whitney. It is found that the ARE is larger for dependent than for independent data in some cases, and that the effects of dependence on ARE can be offset to some extent by appropriately grouping the sample values. If, on the other hand, a constant false-alarm rate is to be obtained, either the form of the dependence must be known, or some scheme such as adaptive thresholding must be devised.

II. Problems Arising from Dependence

It is obviously undesirable in many applications to sample so slowly that the sample values are independent. Hence there is a practical necessity of understanding the implications of dependence on false-alarm rate, power, and ARE. Dependent observations can in fact produce wide variations in false-alarm rate for a fixed threshold and varying amounts of dependence.

Derivations following the general style of Capon, Fraser, and Terry [6]–[8] indicate that rank and rank-like statistics can be designed to provide locally optimum results for a given distribution of the input data with or without dependence. It seems unlikely, however, that the distribution and, in particular, the form of the dependence would normally be known in situations where a nonparametric test would be useful. Thus, it is of interest to consider the effects of dependence on known statistics chosen under the assumption of independence, and to consider steps that can be taken to minimize the dependence effects within some general framework.

For the purpose of calculating asymptotic relative efficiency, a data model must be chosen. In the succeeding sections for ARE calculations, the data are taken to be stationary Gaussian with a nonzero continuous piecewise smooth spectral density for the sampled sequence. This implies absolute summability of the correlation function needed in certain of the limiting operations. Through the application of the Kolmogorov–Rozanov theorem [15], the sequence then has the strong mixing property, which is used to establish asymptotic normality. The restriction is sufficient rather than necessary, but is certainly satisfied in most practical situations.

III. Tests Based on the Mann–Whitney Statistic

This section discusses the results for tests based on the Mann–Whitney statistic. Detailed calculations are relegated to Appendix I. The applicable detection problem is the following. Let $y(t)$ and $x(t)$ be statistically independent "noise only" and "signal plus noise" channel outputs, respectively, for some range of values of t. It is desired to decide whether a signal is present in the latter channel; it is assumed to appear as a positive mean in $x(t) - y(\tau)$. Let

$$E[x(t) - y(\tau)] = \theta.$$

Then one is testing the hypothesis H_0: $\theta = 0$, versus the alternative H_1: $\theta > 0$.

If the data are sampled at intervals Δt, the optimum parametric test of mean shift on N sample values for an independent Gaussian process is the sample mean

$$s_\mu = \frac{1}{N} \sum_{i=1}^{N} [x(i\Delta t) - y(i\Delta t)] \gtrless C.$$

This paper is concerned only with asymptotic results and the sample mean is asymptotically efficient for the class of nonzero continuous spectral densities [12].

The Mann–Whitney test is a nonparametric test based on the algebraic ranks of the values in the signal channel; these can be most conveniently expressed as

$$S_{MW1}^N = \frac{1}{N^2} \sum_{i=1}^{N} \sum_{j=1}^{N} \operatorname{sgn} [x(i\Delta t) - y(j\Delta t)] \gtrless C \qquad (1a)$$

where C is a threshold determined a priori.

If the sampling interval Δt is such that the sampled values are statistically independent, then it is well known [4] that the ARE with respect to the sample mean is $3/\pi$ for Gaussian data. For dependent sample values, however, this does not hold. Let $\rho(n\Delta t)$ be the normalized data correlation function. The ARE with respect to the sample mean for sampled data (see Appendix I) is

$$E_{MW1} = \frac{1}{2} \sum_{-\infty}^{\infty} \rho(n\Delta t) \left\{ \sum_{-\infty}^{\infty} \sin^{-1} \left(\frac{\rho(n\Delta t)}{2} \right) \right\}^{-1}. \qquad (1b)$$

Note that for $\rho(n\Delta t) = \delta_{0n}$, the ARE becomes $3/\pi$. For an arbitrary covariance, the ARE may be larger or smaller than $3/\pi$, but never exceeds 1. This is illustrated by noting that

$$\sin^{-1}\left(\frac{x}{2}\right) \le \pi \frac{x}{6} \qquad 0 \le x \le 1$$

$$\ge \pi \frac{x}{6} \qquad -1 \le x \le 0.$$

Thus, if $\rho(n\Delta t) \ge 0$ for every n, $E_{MW1} \ge 3/\pi$. If $\rho(n\Delta t) \le 0$ for every n (except $n = 0$), $E_{MW1} \le 3/\pi$.

One of the major drawbacks of rank-type tests such as the Mann–Whitney is that the number of computations is $O(N^2)$. If the statistic is to be formed in real time as each sample is received, $2n - 1$ summands must be formed at the nth step. Thus for large values of N, the detector becomes impractical. A solution of this problem is based on a mixed statistic as described in Section I. The Mann–Whitney statistic is formed on each of the p groups, $mp = N$, where it is assumed that p is an integer for convenience. In [1], ARE's close to $3/\pi$ are found for m on the order of 5 to 10. The effectiveness of the method arises from the tendency of the Mann–Whitney intermediate statistics to be normal, on which the optimum operation is, of course, linear. If the subsets are taken sequentially, processing can proceed in real time with simple components.

The statistic for the preceding mixed test is

$$S_{MW2}^N = \frac{1}{m^2 p} \sum_{n=0}^{p-1} \left\{ \sum_{j=1}^{m} \sum_{k=1}^{m} \operatorname{sgn} [x(nm\Delta t + j\Delta t) - y(nm\Delta t + k\Delta t)] \right\}. \qquad (2a)$$

The ARE with respect to the sample mean for dependent samples is relatively complicated (see Appendix I),

$$E_{MW2}^m = \sum_{-\infty}^{\infty} \rho(n\Delta t) \cdot \left[\frac{1}{m} \sum_{n=-\infty}^{\infty} \sum_{j=-m}^{m} \sum_{k=-m}^{m} \left(1 - \frac{|j|}{m}\right)\left(1 - \frac{|k|}{m}\right) \cdot \sin^{-1} \tfrac{1}{2}(\rho[(nm + j)\Delta t] + \rho[(nm + k)\Delta t]) \right]^{-1}. \qquad (2b)$$

It can be verified that as $m \to \infty$, (1b) results. For moderate values of m, however, other terms enter in and must be evaluated.

While this sequential grouping is a natural technique for implementation and is as efficient as any for independent sample values, the discriminating ability of ranking in the presence of dependence is dimished as $\Delta t \rightarrow 0$ for fixed m. Thus an alternative would be to form the m groups from sample values p observations apart. This can be realized with a tapped delay line

$$S_{MW3}^{N} = \frac{1}{m^2 p} \sum_{n=1}^{p} \sum_{i=0}^{m-1} \sum_{j=0}^{m-1} \operatorname{sgn}\left[x(n\Delta t + ip\Delta t) - y(n\Delta t + jp\Delta t)\right]. \tag{3a}$$

The ARE (as $p \rightarrow \infty$, m fixed) is

$$E_{MW3}^{m} = \sum_{n=-\infty}^{\infty} \rho(n\Delta t) \cdot \left[\frac{1}{m} \sum_{n=-\infty}^{\infty} [\sin^{-1} \rho(n\Delta t) + 2(m-1) \sin^{-1} \tfrac{1}{2}\rho(n\Delta t)]\right]^{-1}. \tag{3b}$$

Again as $m \rightarrow \infty$, (1b) results. Note that $E_{MW3}^{m} \leq E_{MW1}$. Thus, full ranking of the data is better than either of the two mixed tests described. However, in (3b), (1b) is approached for values of m on the order of 10, as will be seen subsequently for a particular example.

Two other approaches to the problem can be suggested. The first would be to keep fixed the distance p between values in each group as well as the group size m. Then the total number N of observations would be divided into q groups of mp observations, $mpq = N$. This approach is useful for sequential-detection problems. Finally, one could keep p fixed and allow m to vary, $mp = N$. This approach involves the same difficulty as the ranking of all the data since the computational effort is proportional to m^2. The ARE can be calculated for these methods, but the calculations are omitted.

The above results are for sampled data, so that $N\Delta t = T$ is the total observation time. As $\Delta t \rightarrow 0$, $N \rightarrow \infty$ and tests for continuous data are obtained. Equations (1)–(3) now hold with integrals replacing sums as required in the obvious way for each test (i.e., in (1) $i\Delta t = t$; in (2), either p is fixed, $m\Delta t = \tau =$ constant, or $m\Delta t \rightarrow 0$ and $n\Delta t = t$; in (3) m is fixed and $p \rightarrow \infty$). The ARE's are obtained as $T \rightarrow \infty$. It will be of interest in the numerical results of Section V to present the ARE's of the various tests for sampled data with respect to the parametric test for continuous data. This provides a measure of the loss of efficiency in sampling. In this case, the first sum in (1b)–(3b) is replaced by

$$\frac{1}{\Delta t} \int_{-\infty}^{\infty} \rho(t)\, dt.$$

The Δt results from the proportionality $N\Delta t = T$.

IV. Other Statistics

The asymptotic relative efficiencies for the Wilcoxon and Kendall τ tests for the methods of grouping of (1)–(3) have been calculated and are relegated completely to the Appendixes for the sake of conciseness. In the case of Kendall's τ, the asymptotic relative efficiency is taken with respect to an energy detector, which may not be optimum for dependent data.

V. Numerical Results

A Gauss–Markov noise process was chosen as "typical" to illustrate numerical values,

$$\rho(t) = e^{-|t|}. \tag{4a}$$

In the case of the Kendall τ statistic where the signal correlation is a parameter for the second method of grouping, the same correlation is taken in one case and in the second case

$$\rho_s(t) = (1 + |t|)e^{-|t|}. \tag{4b}$$

In the first case, as is well known, singular detection (zero error probability) is possible on a finite segment of continuous data. Figs. 1–4 demonstrate the results of these computations for each of the statistics and methods of grouping on sampled data with respect to a parametric test for continuous data. The ARE's are given as a function of the sampling interval so that one can see the effects of sampling. A value of $m = 10$ was chosen for the mixed statistics. It is seen that the ARE's of the tests based on full ranking and of the tests based on the grouping of (3) increase monotonically as the sampling interval is decreased. This is the result of the considerations discussed following (1b). On the other hand, for the sequential grouping of (2), as $m\Delta t \rightarrow 0$, the intermediate statistic performs no better for any finite value of m than another. Thus, the Mann–Whitney becomes the polarity coincidence correlator, the Wilcoxon becomes the sign test, and the Kendall τ becomes a derivative polarity correlator. In Figs. 1–3 a maximum efficiency is attained at a nonzero value of Δt, which represents a tradeoff between the effectiveness of the ranking procedures and the number of samples used per unit time. In the singular detection problem of Fig. 4, however, the grouping of (2) results in an unbounded efficiency *with respect to the energy detector* as $\Delta t \rightarrow 0$. This is because the derivative polarity correlator results in singular detection. Hence, $m = 2$ might as well have been used.

Thus it can be seen that tests such as that of (3) can be designed to minimize the effects of dependence under some conditions, approaching both the parametric test designed for independent data and the fully ranked test in ARE without the much greater computational complexity of the latter. In certain cases, as in Fig. 4, it may be that a nonrank statistic is a more efficient form than the rank statistic. No complete answer is possible. Apparently estimates of the dependence must be used in detector design and applied in some adaptive fashion.

VI. Effect of Dependence on False-Alarm Rate

One of the characteristics motivating the use of nonparametric tests is the invariance of false-alarm rate α over a wide class of probability distributions. It is of interest to investigate the effect of dependence on this characteristic. S^N is asymptotically normal (see Section VII) so that

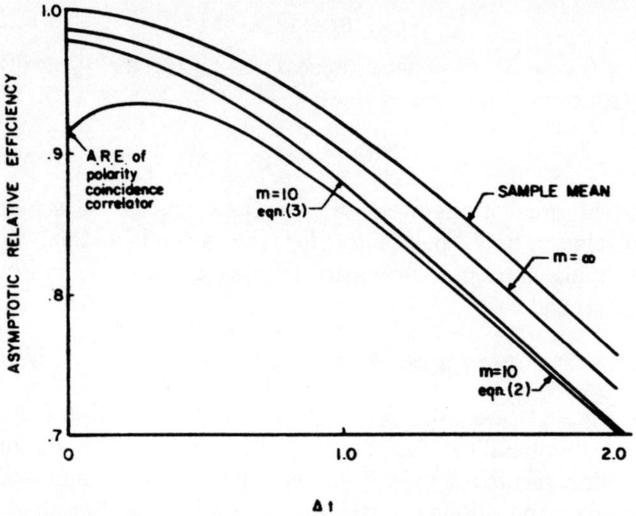

Fig. 1. The asymptotic relative efficiency with respect to a parametric test on continuous data for various Mann–Whitney detectors, $\rho_n(t) = e^{-|t|}$, Δt = sampling interval.

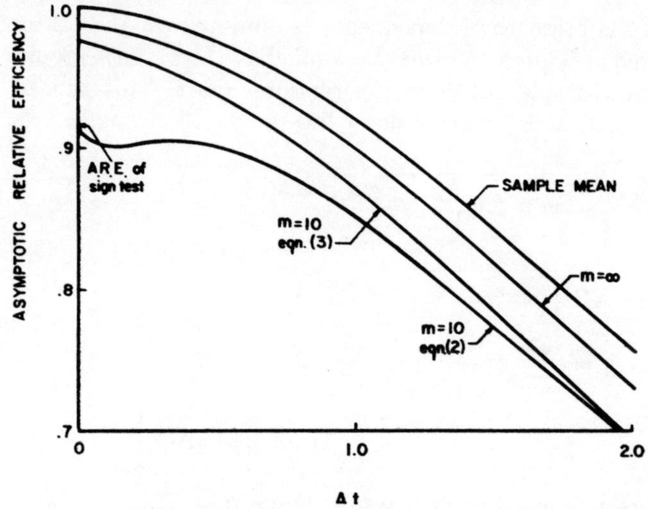

Fig. 2. The asymptotic relative efficiency with respect to a parametric test on continuous data for various Wilcoxon detectors, $\rho_n(t) = e^{-|t|}$, Δt = sampling interval.

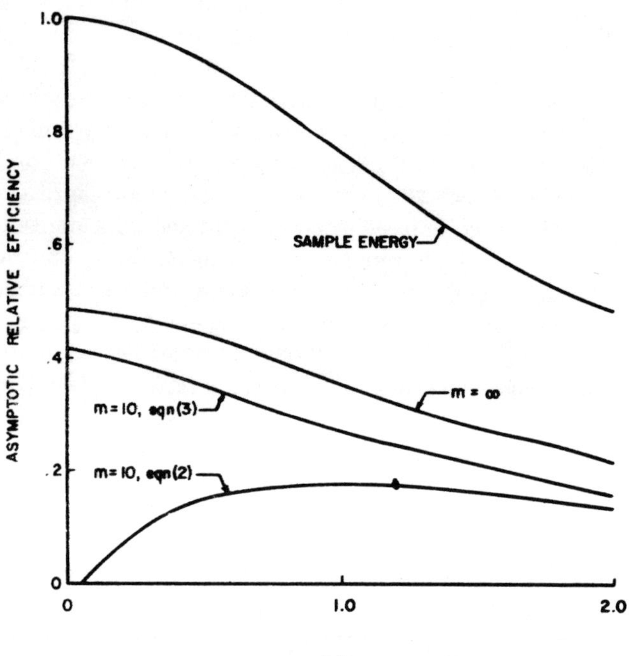

Fig. 3. The asymptotic relative efficiency with respect to a parametric test on continuous data for various Kendall τ detectors, $\rho_n(t) = e^{-|t|}$, $\rho_s(t) = (1 + |t|)e^{-|t|}$, Δt = sampling interval.

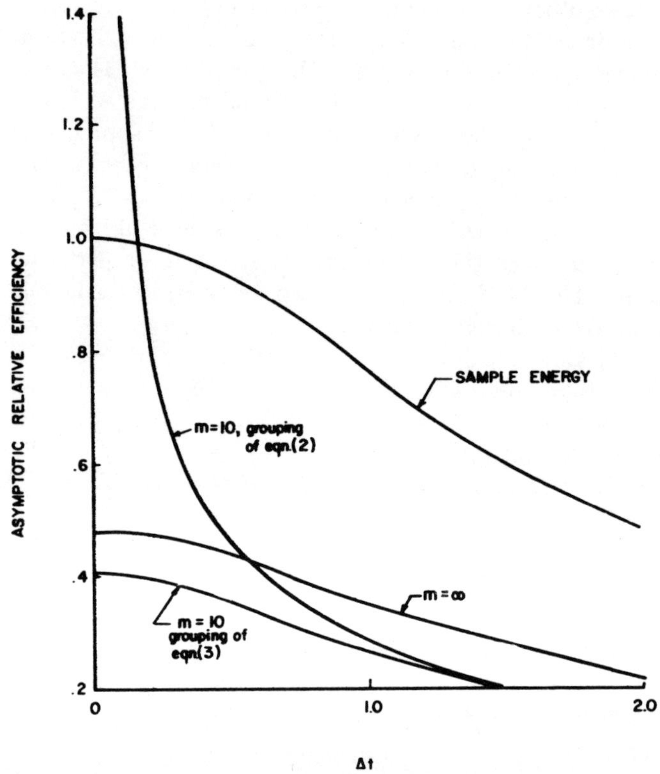

Fig. 4. The asymptotic relative efficiency with respect to a parametric test on continuous data for various Kendall τ detectors, $\rho_n(t) = \rho_s(t) = e^{-|t|}$, Δt = sampling interval.

as $N \to \infty$ for a fixed

$$1 - \alpha = \Phi\left[\frac{C_N}{\text{var}^{1/2}(S^N)}\right]$$

where $\Phi(\cdot)$ is the cumulative Gaussian distribution. Hence dependence affects the false-alarm rate through the variance of the test statistic. The amount can be measured by taking the ratio of this quantity to the variance with no dependence,

$$\frac{\text{var}(S^N)}{\text{var}(S^N)\,|_{\rho(n\Delta t) = \delta_{0n}}}.$$

From Figs. 1–4 and the Appendixes, it is seen that for any of the statistics under consideration, the ratio is approximately the same, namely

$$\frac{\text{var}\,[S_{i\mu}^N]}{\text{var}\,[S_{i\mu}^N]\,|_{\rho(n\Delta t)=\delta_{0n}}} \approx \left(\sum_{n=-\infty}^{\infty} \rho(n\Delta t)\right)^{-1} \qquad i = 1, 2$$

for the Mann–Whitney and Wilcoxon statistics, and

$$\left(\sum_{k=-\infty}^{\infty} \rho_n^2(k\Delta t)\right)^{-1}$$

for the Kendall τ statistics.

Thus as $\Delta t \rightarrow 0$, dependence results in a vastly different false-alarm rate from that designed for under the assumption of independence. For the numerical example of Section V the variance ratios become

$$\frac{1 - e^{-\Delta t}}{1 + e^{-\Delta t}}$$

for the Mann–Whitney and Wilcoxon tests.

Therefore, the tests cannot be said to be nonparametric in the sense of constant α under dependence if the dependence is of an unknown form. If the false-alarm rate is to be approximately correct for rapid sampling, some adaptive scheme as in [2], [3] is required, involving perhaps threshold learning if the decisions are recurrent.

VII. Asymptotic Distribution of Mixed Test Statistics

Implicit in the foregoing computations and subsequent conclusions is the assumption that the distribution of the test statistic is asymptotically normal. In this section this is established. Again for definiteness, the Mann–Whitney statistic is considered with the understanding that similar results hold for the Wilcoxon and Kendall τ statistics.

The theorem of Rosenblatt [14] is used to establish asymptotic normality of the mixed statistics for $m < \infty$. Let the process $\{x_t, t \in T\}$ be strong mixing. Clearly if $\{x_t\}$ is strong mixing, then so is any process $\{z_t\}$ derived from it if events defined in terms of z_ν, $\nu \leq t$ depend on the x_μ for values of μ in a finite union of intervals, and if events defined in terms of z_ν, $\nu \geq t' + \tau$, depend on the x_μ for μ in another finite union of intervals such that the minimum separation between the two sets of intervals becomes infinite with τ. In this paper the sampled sequences are associated with $\{x_t\}$, and the sums in the statistics are ordered so that a sequence $\{z_t\}$ has this separation property.

Define

$$H(r, s) = \sum_{t'=r}^{s} z'_t$$

$$\sigma^2(r, s) = \text{var}\,(H(r, s)).$$

Theorem (Rosenblatt [14])

Let $\{z_t, t \in T\}$ be a zero mean stochastic process. The statistic $\eta(r, s) = H(r, s)/\sigma(r, s)$ is asymptotically normal if $\{z_t\}$ is strong mixing and

$$\lim_{r-s\rightarrow\infty} \frac{E[H(r, s)]^4}{\sigma^4(r, s)} < \infty.$$

Let $x_{n+1} = \rho x_n + w_n$ where $\{w_n\}$ is an independent ergodic sequence. The strong mixing properties of this process follow trivially from the easily established fact [16] that ergodic Markov processes are strong mixing. Fortunately, for Gaussian processes it is possible to obtain more general criteria for strong mixing. In [15] it is shown that if $\{x_t, t \in T\}$ is a stationary Gaussian process possessing nonzero continuous piecewise smooth spectral density, then $\{x_t, t \in T\}$ is strong mixing.

Consider the mixed Mann–Whitney test statistic given by (2a).

$$S_{MW2}^N = \frac{1}{m^2 p} \sum_{n=0}^{p-1} z_n \tag{5}$$

with

$$z_n = \sum_{j=1}^{m} \sum_{k=1}^{m} \text{sgn}\,[x(nm\Delta t + j\Delta t) - y(nm\Delta t + k\Delta t)].$$

It is clear that $\{z_n\}$ is strong mixing.

Consider the grouping of (3a) with test statistic

$$S_{MW3}^N = \frac{1}{m^2 p} \sum_{n=0}^{p-1} z_n,$$

$$z_n = \sum_{j=1}^{m} \sum_{k=1}^{m} \text{sgn}\,[x(n\Delta t + jp\Delta t) - y(n\Delta t + kpt)].$$

The sequence $\{z_n\}$ is *not* strong mixing because z_0 and z_{p-1} use adjacent values of the $\{x\}$ and $\{y\}$ processes. However, if the statistic

$$T_{MW3}^N = \frac{1}{m^2 p} \sum_{n=0}^{p-(p)^{1/2}} z_n$$

is considered, rather than S_{MW3}^N, the strong mixing property of the statistic is preserved; and since

$$\frac{\lim_{N\rightarrow\infty} E[(S_{MW3}^N - T_{MW3}^N)^2]}{\text{var}^{1/2}\,(S_{MW3}^N)} \rightarrow 0,$$

S_{MW3}^N has the same asymptotic distribution as T_{MW3}^N.

Having established the strong mixing properties of the resultant test statistics operating on Gaussian data with the restricted spectral densities, it remains only to demonstrate the moment conditions. These are shown to be satisfied in Appendix IV, establishing the asymptotic normality of the mixed test employing the Mann–Whitney intermediate statistic for fixed finite $m < N$.

For the fully ranked case $m = N$, the test statistic cannot be put in the form of a single sum of random variables appropriate for the application of the standard central limit theorems. This is true even when the data are independent identically distributed. To circumvent this difficulty several ingenious limit theorems have been developed. Most notable in this regard are Hoeffding's limit theorems for U statistics [17], and the work of Chernoff and Savage [18]. These limit theorems succeed by approximating a given statistic T_N in mean square by another statistic T'_N that has the form of a single sum of random variables. If $\lim_{N\rightarrow\infty} E\{[T_N - T'_N]^2\} = 0$, then it follows that the two statistics have identical asymptotic distributions.

Obviously a similar approach can be employed in the dependent sample case. The Mann–Whitney statistic is a linear rank-order statistic that can be written in the form

$$T_N = \sum_{i=1}^{N} E_{Ni} z_{Ni} \tag{6}$$

where the E_{Ni} are given numbers in the interval (0, 1) and $z_{Ni} = 1$ if the ith smallest sample was from the signal channel and zero otherwise. For the Mann–Whitney statistic, $E_{Ni} = (i/N)$.

The asymptotic distribution of statistics of the form (6) operating on Gaussian data with restricted spectral density is established by an extension of the Chernoff–Savage theorem to be found in Appendix V. Unfortunately, it has not been possible to apply this result to the single-channel Wilcoxon detector or the Kendall τ detector. If fictitious X and Y channels are identified with the positive and negative samples, and the observations ranked with respect to magnitude, then the Wilcoxon statistic can be represented in the form (6) [19]. It would appear then that the extended Chernoff–Savage theorem is applicable to the single-channel Wilcoxon detector. The fallacy with this conclusion is the fact that while each of two sums called B_{1N} and B_{2N} in Appendix V are shown to be asymptotically normal, it has not been possible to show that they are *jointly* asymptotically normal, so that one cannot conclude that their *sum* $B_{1N} + B_{2N}$ is asymptotically normal. Consequently, it is not yet possible to establish the asymptotic normality of the fully ranked Wilcoxon detector. The situation with respect to the Kendall τ test presents another problem as it is not a linear rank-order statistic and hence cannot be written in the form (6). Nevertheless, when the sample values are independent, the Kendall τ statistic can be shown to converge in mean square to the Spearman Rank-Correlation test [18], which is a linear rank-order statistic. It has not been possible to show this when the sample values are dependent.

Appendix I

ARE's for the Mann–Whitney Tests

If the Gaussian normalized correlation function $\rho(t)$ is summable for a given Δt

$$\sum_{n=-\infty}^{\infty} |\rho(n\Delta t)| < \infty,$$

then the ARE of one test with respect to another can be expressed as the ratio of *efficacies* of the tests, where the efficacy of statistic S^N for the statistics in this paper is

$$\mathcal{E} = \lim_{N\to\infty} \frac{\left\{\dfrac{\partial^i}{\partial\theta^i} E[S^N]\right\}^2 \Big|_{\theta=0}}{N \operatorname{var}(S^N)\Big|_{\theta=0}} \tag{7}$$

where N is the number of samples in each channel, θ is the signal amplitude, and j is the smallest value such that the derivative is nonzero.

For the sample mean, the efficacy is readliy shown to be

$$\mathcal{E}_{2\mu} = \left(2\sigma^2 \sum_{k=-\infty}^{\infty} \rho(k\Delta t)\right)^{-1}$$

For the Mann–Whitney statistic of (1), as $\theta \to 0$,

$$\lim_{\theta\to 0} \frac{\partial}{\partial\theta} E[S_{MW1}^N] = \lim_{\theta\to 0} \frac{\partial}{\partial\theta} \{\Pr[x(i\Delta t) - y(j\Delta t) \geq 0 \mid \theta] - \Pr[x(i\Delta t) - y(j\Delta t) < 0 \mid \theta]\} = \sigma^{-1}\pi^{-1/2},$$

$$N \operatorname{var}[S_{MW1}^N]\,|_{\theta=0} = \frac{1}{N^3} \sum_{i=1}^{N}\sum_{j=1}^{N}\sum_{k=1}^{N}\sum_{l=1}^{N} E\{\operatorname{sgn}[x(i\Delta t) - y(j\Delta t)] \cdot \operatorname{sgn}[x(k\Delta t) - y(l\Delta t)]\}.$$

Now by the inverse sine rule, if x and y are Gaussian with mean zero and correlation coefficient ρ,

$$E[\operatorname{sgn}(x)\operatorname{sgn}(y)] = 2(\sin^{-1}\rho)/\pi.$$

Thus

$$N \operatorname{var}[S_{MW1}^N]\,|_{\theta=0} = \frac{2}{\pi N} \sum_{n=-N}^{N} \sum_{k=-N}^{N} \left(1 - \frac{|k|}{N}\right) \sin^{-1} \tfrac{1}{2}(\rho(n\Delta t) + \rho(k\Delta t)). \tag{8}$$

This sum can be evaluated asymptotically by noting that for any $\epsilon > 0$, there is an integer M independent of N such that

$$\left|\frac{2}{\pi N} \sum_{n=-N}^{N} \sum_{|k|\leq M} \left(1 - \frac{|n|}{N}\right)\left(1 - \frac{|k|}{N}\right) \cdot [\sin^{-1}\tfrac{1}{2}(\rho(n\Delta t) + \rho(k\Delta t)) - \sin^{-1}\tfrac{1}{2}\rho(n\Delta t)]\right| < \frac{\epsilon}{2}.$$

And for sufficiently large N

$$\left|\frac{2}{\pi N} \sum_{|n|<M} \sum_{|k|<M} \left(1 - \frac{|n|}{N}\right)\left(1 - \frac{|k|}{N}\right) \cdot \sin^{-1}\tfrac{1}{2}(\rho(n\Delta t) + \rho(k\Delta t))\right| < \frac{\epsilon}{2}.$$

Thus in the limit

$$\lim_{N\to\infty} N \operatorname{var}[S_{MW1}^N]\Big|_{\theta=0} = \frac{2}{\pi} \sum_{n=-\infty}^{\infty} 2\sin^{-1}\tfrac{1}{2}\rho(n\Delta t).$$

Thus the efficacy is

$$\mathcal{E}_{MW1} = \left(2\sigma^2 \sum_{n=-\infty}^{\infty} 2\sin^{-1}\frac{\rho(n\Delta t)}{2}\right)^{-1}$$

The asymptotic relative efficiency of the Mann–Whitney with respect to the sample-mean test is

$$\frac{\mathcal{E}_{MW1}}{\mathcal{E}_{2\mu}} = \frac{\sum_{n=-\infty}^{\infty} \rho(n\Delta t)}{\sum_{n=-\infty}^{\infty} 2\sin^{-1}\frac{1}{2}\rho(n\Delta t)}. \tag{9}$$

The ARE of (2b) follows from (8) with N replaced by m and the additional outer summation [over n in (2b)] to account for group-to-group correlation. A similar analysis results in (3b) for statistic S_{MW3}^N.

Appendix II

The Wilcoxon Statistic

The Wilcoxon is the analog of the Mann–Whitney for the one-channel case. Given the observation $x(t)$ with mean θ, it is desired to test H_0: $\theta = 0$ versus H_1: $\theta > 0$. It can be verified that the efficacy of the one-sample mean test agrees with that of Appendix I for the two-sample mean test as defined by (7) except for the factor of 2 in the denominator

$$\mathcal{E}_{1\mu} = \left(\sigma^2 \sum_{n=-\infty}^{\infty} \rho(n\Delta t)\right)^{-1}.$$

The Wilcoxon statistic is based on the sum of the ranks in magnitude of the positive sample values. It is most conveniently represented in the form

$$S_{W1}^N = \frac{1}{N^2} \sum_{i=1}^{N} \sum_{j=1}^{N} (1 + \delta_{ij}) \operatorname{sgn}[x(i\Delta t) + x(j\Delta t)] \gtrless C.$$

The efficacy is evaluated by proceeding as in Appendix I,

$$\begin{aligned}\lim_{\theta\to 0} \frac{\partial}{\partial\theta} E[S_{W1}^N] &= \frac{1}{N^2} \sum_{i=1}^{N} \sum_{j=1}^{N} \frac{1 + \delta_{ij}}{(1 + \rho[(i-j)\Delta t])^{1/2}} \frac{2}{\sigma\sqrt{\pi}} \\ &= \frac{1}{N} \sum_{n=-N}^{N} \left(1 - \frac{|n|}{N}\right) \frac{1 + \delta_{0n}}{(1 + \rho(n\Delta t))^{1/2}} \frac{2}{\sigma\sqrt{\pi}}.\end{aligned} \tag{10}$$

Here the correlation appears in the mean, as it did not for the Mann–Whitney, owing to the independence of the two channels there. However, using an argument similar to that in passing from (8) to (1b),

$$\lim_{N\to\infty} \lim_{\theta\to 0} \frac{\partial}{\partial\theta} E[S_{W1}^N] = 2\pi^{-1/2}\sigma^{-1}.$$

The variance also poses problems arising from dependence that were not encountered in Appendix I. A four-way sum has to be evaluated. Nonetheless, by an argument more involved algebraically, but similar to that in form employed on (8),

$$\lim_{N\to\infty} N \operatorname{var}(S_{W1}^N) = \frac{4}{\pi} \sum_{n=-\infty}^{\infty} 2 \sin^{-1} \tfrac{1}{2}\rho(n\Delta t). \tag{11}$$

Thus the efficacy is

$$\mathcal{E}_{W1} = \left(\sigma^2 \sum_{n=-\infty}^{\infty} 2 \sin^{-1} \tfrac{1}{2}\rho(n\Delta t)\right)^{-1}.$$

The ARE with respect to the sample mean is

$$\frac{\mathcal{E}_{W1}}{\mathcal{E}_{1\mu}} = \frac{\sum_{n=-\infty}^{\infty} \rho(n\Delta t)}{\sum_{n=-\infty}^{\infty} 2 \sin^{-1} \tfrac{1}{2}\rho(n\Delta t)}. \tag{12}$$

Thus the ARE is the same for the Wilcoxon as for the Mann–Whitney when the data is fully ranked. From this it can be inferred that channel-to-channel correlation does not affect the ARE of (1b). This is, in fact, true. It is *not* true, however, that the results for the Wilcoxon are the same as for the Mann–Whitney for mixed statistics as in (2) and (3), or that interchannel correlation leaves the mixed Mann–Whitney ARE unaffected. The efficacy of the mixed Wilcoxon with sequential grouping is a lengthy but easily found expression using (10) and the exact expression for the variance. By similar calculations, the efficacy for the third method of grouping is

$$\begin{aligned}\mathcal{E}_{W3} = [2^{-1/2} + m - 1]^2 &\cdot \left\{\sigma^2 \sum_{n=-\infty}^{\infty} [m \sin^{-1} \rho(\Delta t) + 2(m-1) \sin^{-1} 2^{-1/2}\rho(\Delta t)\right. \\ &\left. + 2(m-1)(m-2) \sin^{-1} \tfrac{1}{2}\rho(\Delta t)]\right\}^{-1}.\end{aligned}$$

Appendix III

The Kendall τ Statistic

The Kendall τ statistic is one that has application to the two-channel detection of an unknown signal in noise. Let $x(t)$ and $y(t)$ be the two-channel outputs

$$x(t) = \theta s(t) + n_x(t)$$

$$y(t) = \theta s(t) + n_y(t),$$

where $n_x(t)$ and $n_y(t)$ are independent noise processes and $s(t)$ is a common noise independent signal process. It is desired to test H_0: $\theta = 0$ versus H_1: $\theta \neq 0$. If the signal and noise processes are mean zero Gaussian, then the optimum parametric test for independent sample values is an energy detector

$$S_e^N = \frac{1}{N} \sum_{i=1}^{N} [x(i\Delta t) + y(i\Delta t)]^2 \gtrless C.$$

Let the normalized correlation function for the signal be $\rho_s(t)$ and the noise processes be $\rho_n(t)$, and let the noise variance be σ^2 and the signal variance be 1. Then the efficacy of the energy detector for dependent sample values is

$$\mathcal{E}_e = 8\left(\sigma^4 \sum_{k=-\infty}^{\infty} \rho_n^2(k\Delta t)\right)^{-1}.$$

The Kendall τ nonparametric test for correlation is based on the correlation between ranks in the two channels. The statistic is most conveniently expressed as

$$S_{\tau 1}^N = \frac{1}{N^2} \sum_{i\neq j} \operatorname{sgn}[x(i\Delta t) - x(j\Delta t)] \operatorname{sgn}[y(i\Delta t) - y(j\Delta t)].$$

By calculations similar to those in Appendixes I and II, the efficacy is found to be

$$\mathcal{E}_{\tau 1} = \left(\sigma^4 \sum_{k=-\infty}^{\infty} [\sin^{-1} \tfrac{1}{2}\rho_n(k\Delta t)]^2\right)^{-1}.$$

The asymptotic relative efficiency with respect to the energy detector then is

$$\frac{\mathcal{E}_{r1}}{\mathcal{E}_e} = \sum_{k=-\infty}^{\infty} \rho_n^2(k\Delta t)\left\{2 \sum_{k=-\infty}^{\infty} [2 \sin^{-1} \tfrac{1}{2}\rho_n(k\Delta t)]^2\right\}^{-1}. \quad (13)$$

It is interesting to note the similarity in form between this result and (1b) and (12). Note that the signal correlation is not a parameter.

The results for the second and third methods of groupings are obtained by computations similar to those in the preceding material. For the second method a lengthy expression results. For the third method,

$$\mathcal{E}_{r3} = \left\{\sigma^4 \frac{1}{m-1} \sum_{k=-\infty}^{\infty} [\tfrac{1}{2}\{\sin^{-1} \rho_n(k\Delta t)\}^2 + (m-2)\{\sin^{-1} \tfrac{1}{2}\rho_n(k\Delta t)\}^2]\right\}. \quad (14)$$

The signal correlation appears in the result for the sequential grouping but is removed in the third method of grouping.

Appendix IV

Rosenblatt Moment Conditions for Mann–Whitney Test

The Rosenblatt moment conditions will be demonstrated for the two-channel test employing the Mann-Whitney intermediate statistic and the grouping of (2a). Obvious analogous results hold for the other grouping and the single-channel Wilcoxon and Kendall τ tests. Consider the statistic S_{MW2}^N given by (2a). By Appendix I, var $(S_{MW2}^N) = O(N^{-1})$. Hence it is necessary to establish that $E[(S_{MW2}^N)^4] = O(N^{-2})$. In establishing this moment condition, use will be made of the following lemma.

Lemma

If $\{Y_i\}$ is a sequence of zero mean Gaussian random variables with normalized covariance ρ_{ij}, $\sum_{j=-\infty}^{\infty} |\rho_{ij}| < \infty$ for every i, then

$$\left|E \prod_{i=1}^{4} \text{sgn}\,(Y_{n_i})\right| \le K \sum_{i<j} |\rho_{n_i n_j}| \cdot [|\rho_{n_k n_l}| + \sum_{m,s=i,j} |\rho_{n_m n_k}|\, |\rho_{n_m n_l}|\, |\rho_{n_s n_l}|]_{k \ne l \ne i,j} \quad (15)$$

where K is a constant $0 < K < \infty$.

Proof: From Pawula [21]

$$E\left\{\prod_{i=1}^{4} \text{sgn}\,(Y_{n_i})\right\} = \frac{4}{\pi^2} \sum_{i<j} \rho_{n_i n_j} \int_0^1 \cdot \sin^{-1}\{\rho_{kl \cdot ij}(\alpha)\}(1 - \alpha^2 \rho_{n_i n_j}^2)^{-1/2}\, d\alpha \quad (16)$$

where $\rho_{kl \cdot ij}(\alpha)$ is the partial correlation coefficient defined as the correlation coefficient between the residuals of the random variables ξ_k and ξ_l when the best linear estimate in terms of ξ_i and ξ_j is removed from each, where the set $\{\xi_i\}$ has covariance

$$E[\xi_i \xi_j] = \alpha^{(1-\delta_{ij})} \rho_{n_i n_j}.$$

The magnitude of (16) is easily bounded as

$$\left|E \prod_{i=1}^{4} \text{sgn}\,(Y_{n_i})\right| \le \sum_{i<j} |\rho_{n_i n_j}| \max_{\alpha} |\rho_{kl \cdot ij}(\alpha)|. \quad (17)$$

By evaluating the minimum variance coefficients, calculating $\rho_{kl \cdot ij}(\alpha)$, and substituting in (17), an expression is obtained that can be bounded above (for $\sum_{j=-\infty}^{\infty} |\rho_{ij}| < \infty$) over all sets of $\{n_i; i = 1, \cdots, 4\}$ resulting in the inequality (15).

Now define the variables

$$Y_{i_k j_k} \triangleq x(n_k m\Delta t + i_k \Delta t) - y(n_k m\Delta t + j_k \Delta t)$$

for $k = 1, 2, 3, 4$, $i_k, j_k = 1, 2, \cdots, m$, and $n_k = 0, 1, 2, \cdots$. With

$$\rho_{kl} \triangleq \frac{E\{Y_{i_k j_k} Y_{i_l j_l}\}}{\sqrt{E\{Y_{i_k j_k}^2\}E\{Y_{i_l j_l}^2\}}} = \tfrac{1}{2}[\rho([n_k - n_l]\Delta t + [i_k - i_l]\Delta t) + \rho([n_k - n_l]\Delta t + [j_k - j_l]\Delta t)],$$

$$E_{H_0} |S_{MW2}^N|^4 = m^{-8}p^{-4} \sum_{n_1,n_2,n_3,n_4=0}^{p-1} \left[\sum_{i_1,i_2,i_3,i_4=1}^{m} \cdot \sum_{j_1,j_2,j_3,j_4=1}^{m} E_{H_0}\left\{\prod_{k=1}^{4} \text{sgn}\, Y_{i_k j_k}\right\}\right]. \quad (18)$$

From the preceding lemma the expectation can be bounded.

The summation on the right-hand side is over the six distinct combinations of four objects taken two at a time. By substituting this inequality into (18) and by suitable relabeling of indices for each of the six cases, it is found that

$$E_{H_0} |S_{MW2}^N|^4 = O(N^{-2}).$$

The moment conditions for the grouping represented by the statistic S_{MW3}^N followed by similar arguments, need not be repeated.

Appendix V

Asymptotic Normality of Fully Ranked Mann–Whitney Test

Let $\lambda_N = m/N$ and assume that for all N the inequalities $0 < \lambda_0 \le \lambda_N \le 1 - \lambda_0 < 1$ hold for some fixed $\lambda_0 \le \frac{1}{2}$. The quantity $F_m(x)$ is to be the empirical distribution function of the signal sample $\{X_i\}_{i=1}^m$, while $G_n(y)$ is the empirical distribution function of the reference sample $\{Y_i\}_{i=1}^n$. Define

$$H_N(x) \triangleq \lambda_N F_m(x) + (1 - \lambda_N)G_n(x),$$

the combined empirical distribution function while the quantity

$$H(x) \triangleq \lambda_N F(x) + (1 - \lambda_N) G(x)$$

where $F(x)$ and $G(x)$ are the actual signal and reference sample c.d.f.'s, is the combined population c.d.f. Instead of the linear rank-order statistics of the form (6), [18] considers statistics T_N of the form

$$T_N = \int_{-\infty}^{\infty} J_N[H_N(x)]\, dF_m(x), \tag{19}$$

which is easily seen to reduce to (6) by setting $E_{Ni} = J_N(i/N)$. Obviously, $J_N(\cdot)$ need only be defined on the lattice $1/N, 2/N, \cdots, N/N$, but it is convenient to extend its domain definition to $(0, 1]$ in such a way that $J_N(\cdot)$ is, in fact, continuous on this interval. The statistic T_N can be written as

$$T_N = A + B_{1N} + B_{2N} + \sum_{i=1}^{N} C_{Ni} \tag{20}$$

where the A, B, C terms represent constant first-order random and higher-order random terms, respectively. Furthermore,

$$\begin{aligned} B_{1N} &= \frac{(1 - \lambda_N)}{m} \sum_{i=1}^{m} [B(X_i) - E\{B(X_i)\}] \\ B_{2N} &= -\frac{(1 - \lambda_N)}{n} \sum_{i=1}^{n} [B^*(y_i) - E\{B(y_i)\}], \end{aligned} \tag{21}$$

where

$$\begin{aligned} B(x) &\triangleq \int_{x_0}^{x} J'[H(y)]\, dG(y) \\ B^*(x) &\triangleq \int_{x_0}^{x} J'[H(y)]\, dF(y) \end{aligned} \tag{22}$$

and where

$$J(\zeta) \triangleq \lim_{N\to\infty} J_N(\zeta) \qquad 0 < \zeta < 1,$$

is assumed to exist and is not constant. The quantity x_0 in (22) is determined arbitrarily, say by $H(x_0) = 1/2$. With some relatively mild regularity conditions added to the Chernoff–Savage theorem, it can be shown by an application of the Rosenblatt central limit theorem that both B_{1N} and B_{2N} are asymptotically normal, and since they are independent, so also is their sum. Furthermore, the higher order C_{Ni} terms can be shown to converge to zero in probability as $N \to \infty$, so that the asymptotic normality of T_N is established. In particular, the following is true.

Theorem

If $\{X_i\}_{i=1}^m$ and $\{Y_i\}_{i=i}^n$ are mutually independent sequences, each of which are obtained by homogeneous sampling of a stationary Gaussian process with absolutely integrable covariance function, and if

1) $J(H) = \lim_{N\to\infty} J_N(H)$ exists for $0 < H < 1$ and is not constant;
2)[1] $\int_{I_N} [J_N(H_N) - J(H_N)]\, dF_m = o_p(1/\sqrt{N})$ where $I_N \triangleq \{x : 0 < H_N(x) < 1\}$;
3) $J_N(1) = o(\sqrt{N})$;
4) $|J^{(i)}(H)| = |d^iJ/dH^i| \le K[H(1 - H)]^{-i-1/2+\delta}$ for $i = 0, 1, 2$ and for some $\delta > 0$;
5) $J_N(\cdot)$ is monotone increasing in its argument;
6) The limit function $J(\cdot)$ satisfies a uniform Lipshitz condition on $(0, 1)$, i.e., for each $x_0 \in (0, 1)$ there exists an $\delta(x_0) > 0$ and an M independent of x_0 such that

$$|J(x) - J(x_0)| < M\,|x - x_0|$$
$$\text{for} \quad |x - x_0| < \delta(x_0)$$

then for fixed F, G, and λ_N, the statistic T_N given by (19) is asymptotically nontrivially normally distributed as m, n, and hence $N \to \infty$.

The proof of this theorem is fairly lengthy and will not be given here. The conditions 1)–4) follow from the original Chernoff–Savage theorem while conditions 5)–6) have been added to allow application of the Rosenblatt central limit theorem and to establish the higher-order nature of the C_{Ni} terms when the data are no longer independent identically distributed. The conditons 1)–6) can all be shown to hold for the Mann–Whitney statistic.

References

[1] E. A. Feustel and L. D. Davisson, "The asymptotic relative efficiency of mixed statistical tests," *IEEE Trans. Information Theory*, vol. IT-13, pp. 247–255, April 1967.
[2] M. Kanefsky and J. B. Thomas, "On adaptive nonparametric detection systems using dependent samples," *IEEE Trans. Information Theory*, vol. IT-11, pp. 521–526, October 1965.
[3] M. Kanefsky, "An automatic decision threshold for polarity coincidence arrays," *1966 NEC Proc.*, pp. 321–326.
[4] J. W. Carlyle and J. B. Thomas, "On nonparametric signal detectors," *IEEE Trans. Information Theory*, vol. IT-10, pp. 146–152, April 1964.
[5] J. Capon, "Nonparametric technique for the detection of a constant signal in noise," *1959 IRE WESCON Conv. Rec.*, pt. 4, pp. 92–103.
[6] R. F. Daly and C. K. Rushforth, "Nonparametric detection of a signal of known form in additive noise," *IEEE Trans. Information Theory*, vol. IT-11, pp. 70–76, January 1965.
[7] K. S. Fu and Y. T. Chien, "Sequential recognition using a nonparametric ranking procedure," *IEEE Trans. Information Theory*, vol. IT-13, pp. 484–492, July 1967.
[8] M. N. Woinsky and L. Kurz, "An efficient nonparametric detector based on a three-sample classification model," *IEEE Trans. Information Theory*, vol. IT-14, pp. 744–751, September 1968.
[9] M. E. Terry, "Some rank-order tests which are most powerful against specific alternatives," *Ann. Math. Stat.*, vol. 23, pp. 346–366, 1952.
[10] D. A. S. Fraser, "Most powerful rank-type tests," *Ann. Math. Stat.*, vol. 28, pp. 1040–1043, 1957.
[11] J. Capon, "Asymptotic efficiency of certain locally most powerful rank tests," *Ann. Math. Stat.*, vol. 32, pp. 88–100, 1961.
[12] U. Grenander and M. Rosenblatt, *Statistical Analysis of Time Series.* New York: Wiley, 1957, p. 244.

[1] The notation $y = o_p(N)$ means $p - \lim_{N\to\infty} y/N = 0$.

[13] S. S. Wolff, J. L. Gastwirth, and H. Rubin, "The effect of autoregressive dependence on a nonparametric test," *IEEE Trans. Information Theory* (Correspondence), vol. IT-13, pp. 311–313, April 1967.

[14] M. Rosenblatt, "A central limit theorem and a strong mixing condition," *Proc. Natl. Acad. Sci.*, U.S.A., vol. 42, pp. 43–47, 1956.

[15] A. Kolmogorov and I. Rozanov, "On a strong mixing condition for stationary Gaussian processes," *Teor. Veroyatnost. i Primenen*, vol. 5, 1960.

[16] Yu. A. Rozanov, "An application of the central limit theorem," *Proc. 4th Berkeley Symp. on Mathematical Statistics and Probability.* Berkeley: University of California Press, 1961, pp. 445–454.

[17] W. Hoeffding, "A class of statistics with asymptotically normal distributions," *Ann. Math. Stat.*, vol. 19, pp. 293–325, 1948.

[18] H. Chernoff and I. R. Savage, "Asymptotic normality and efficiency of certain nonparametric test statistics," *Ann. Math. Stat.*, vol. 29, pp. 972–994, 1958.

[19] E. L. Lehmann, *Testing Statistical Hypotheses.* New York: Wiley, 1959, p. 257.

[20] M. G. Kendall and A. Stuart, *The Advanced Theory of Statistics.* New York: Hafner, 1961.

[21] R. F. Pawula, "A modified version of Price's theorem," *IEEE Trans. Information Theory*, vol. IT-13, pp. 285–288, April 1967.

Reprinted with permission from *IEEE Trans. Aerosp. Electr. Syst.* **AES-12**:65–67 (1976)

FALSE ALARM RATE FOR RANK QUANTIZER WHEN INPUT RANGE SAMPLES ARE NOT INDEPENDENT

E. P. Cunningham

Abstract

Under the assumption that the non-independent range samples fed into the rank quantizer form a first-order Markov sequence, the first threshold false alarm probability p_n is written in the form of a multiple integral, the multiplicity being equal to the number of samples or quantizer taps. The Rice model for narrow-band noise (clutter) is used to check the effect of correlation on p_n using the derived formula. For the limited case examined, correlation between samples decreases p_n as compared with the independent samples case.

Introduction

A rank quantizer forms an important part of many radar signal processing systems where it is used as a first threshold device, to convert the input analog video signal over one range sweep into digital data. In the analysis of the rank quantizer to date, the assumption has been made that the range samples are independent identically distributed random variables. Based on this assumption, it can be shown (Dillard and Antoniak [1]) that the false alarm rate p_n of the device does not depend on the particular probability distribution of the samples, i.e., it is distribution-free.

There is evidence that weather clutter is correlated in range as well as in azimuth. Taking range samples far enough apart to justify the assumption of independence might, in some cases, lead to loss of information. Hence it is desirable to relax the assumption of independent range samples.

In this paper, a method is developed whereby the false alarm rate may be obtained for a rank quantizer subjected to non-independent range samples.

Rank Order Quantizer

A schematic diagram of a rank order quantizer is shown in Fig. 1. The range video signal is sampled at intervals of T seconds. Each range sample (called the target tap) is compared with N samples immediately preceding and N samples immediately following it. If the target tap exceeds k of the $2N$ off-target taps, $k = 1, 2, \cdots, 2N$, i.e., has rank k, a hit (logical one) is counted. Otherwise a miss (logical zero) is registered. The false alarm rate is defined as the probability that a hit will be declared by the quantizer when there is no target actually present, i.e., the quantizer input consists of clutter alone.

Non-Independent Range Samples

With referenct to Fig. 1, consider the random process $X(t)$ and let $X_{-N}, \cdots, X_{-1}, X_0, X_1, \cdots, X_N$ be random variables denoting the $M = 2N + 1$ voltage taps, X_0 being the target

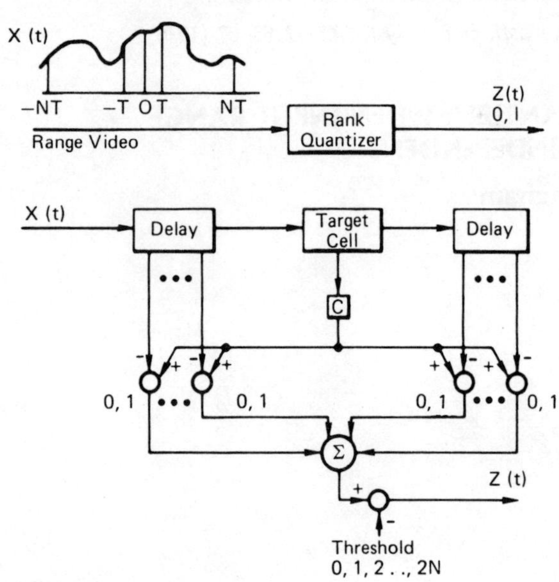

Fig. 1. Rank order quantizer.

tap. For brevity, write $f_{\mathbf{X}}(\mathrm{x})$ for the joint density of the M-dimensional random vector

$$\mathbf{X} = X_{-N}, \cdots, X_{-1}, X_0, X_1, \cdots, X_N . \tag{1}$$

Assume that the M random variable components of $\mathbf{X}$ are not independent. Consider the case where a hit is registered if the target tap voltage exceeds all other taps, i.e., $k = 2N$. Then

$$p_n = P\{X_0 > \text{the } 2N \text{ other taps}\}$$

$$= \int_{-\infty}^{\infty} dx_0 \int_{-\infty}^{x_0} dx_{-N} \cdots \int_{-\infty}^{x_0} dx_N f_{\mathbf{X}}(x_{-N}, \cdots x_{-1}, x_0, x_1 \ldots x_N). \tag{2}$$

Since the range samples cannot be negative, $f_{\mathbf{X}}(x) = 0$ if any one of the variables $X_i < 0$, $i = -N, \cdots, N$, so the lower limits $-\infty$ may be replaced by 0. In general, it is impractical to find the joint density $f_{\mathbf{X}}(x)$ for arbitrary values of M. Fortunately, many random processes in nature have the so-called Markov property, i.e.,

$$f_{\mathbf{X}}(\mathrm{x}) = f_N(x_N|x_{N-1}) f_{N-1}(x_{N-1}|x_{N-2}) \cdots f_1(x_1|x_0)$$

$$f_0(x_0|x_{-1}) \cdots \cdots f_{-N+1}(x_{-N+1}|x_{-N}) f_{-N}(x_{-N})$$

$$= f_{-N}(x_{-N}) \prod_{i=-N+1}^{N} f_i(x_i|x_{i-1}) . \tag{3}$$

In other words, knowledge of the present separates the past from the future. If the transitional density functions in (3) are all equal, i.e.,

$$f_i(x_i|x_{i-1}) = f_B(x_i|x_{i-1}),\ i = -N+1, \cdots, N \tag{4}$$

the random sequence $\mathbf{X}$ is said to be homogeneous. If, in addition, the "initial" probability density function $f_{-N}(x_{-N})$ does not depend upon where the sample X_{-N} is taken, i.e.,

$$f_{-N}(x_{-N}) = f_A(x_{-N}) \tag{5}$$

the sequence is said to be stationary.

We will assume that the range video samples entering the rank quantizer form a stationary Markov sequence. Then

$$f_{\mathbf{X}}(\mathrm{x}) = f_A(x_{-N}) \prod_{i=-N+1}^{N} f_B(x_i|x_{i-1}) . \tag{6}$$

Substituting for $f_{\mathbf{X}}(\mathrm{x})$ in (2) permits p_n to be computed.

If we are interested in the probability of the target tap exceeding k of the $2N$ other taps, $k = 0, 1, 2, \cdots, 2N$, we can proceed as follows.

The probability that X_0 exceeds $X_{-N}, \cdots, X_{-N+k-1}$ but is less than or equal to $X_{-N+k}, \cdots, X_{-N+k}, \cdots, X_N$ is

$$Q_{ki} = \int_0^{\infty} dx_0 \int_0^{x_0} dx_{-N} \cdot\cdot \int_0^{x_0} dx_{-N+k-1} \int_{x_0}^{\infty} dx_{-N+k}$$

$$\cdots \int_{x_0}^{\infty} dx_N f_{\mathbf{X}}(\mathrm{x}) \tag{7}$$

with $f_{\mathbf{X}}(\mathrm{x})$ given by (6). There are $L = \binom{M-1}{K}$ different orderings of the variables and hence L such terms which, when added together, give Q_k the probability that X_0 exceeds exactly k other taps. These terms will have different values in general because the integrand $f_{\mathbf{X}}(\mathrm{x})$ is unsymmetrical in the variables $x_{-N}, \cdots, x_N$. Hence they must be evaluated separately. Thus

$$Q_k = \sum_{i=1}^{L} Q_{ki} . \tag{8}$$

Then

$$p_n = P\{X_0 > \textit{at least } k \text{ other taps}\}$$

$$= \sum_{j=k}^{M-1} Q_j \tag{9}$$

the summand being given by (8) with k replaced by j.

Application to a Specific Correlated Clutter Distribution

The Rice model for narrow band Gaussian noise is given by (see [2])

$$f_A(\nu_1) = \begin{cases} \nu_1/\sigma_x^2 \exp[-\nu_1^2/2\sigma_x^2] & \nu_1 \geqslant 0 \\ 0 & \nu_1 < 0 \end{cases} \tag{10}$$

and

$$f_B(\nu_2|\nu_1) = \begin{cases} c\nu_2 \exp[-c/2(\nu_2{}^2 + \rho^2\nu_1{}^2)] \\ I_0(c\rho\nu_1\nu_2)\ \nu_1, \nu_2 \geqslant 0 \\ 0 \qquad \nu_1, \nu_2 < 0 \end{cases} \tag{11}$$

where

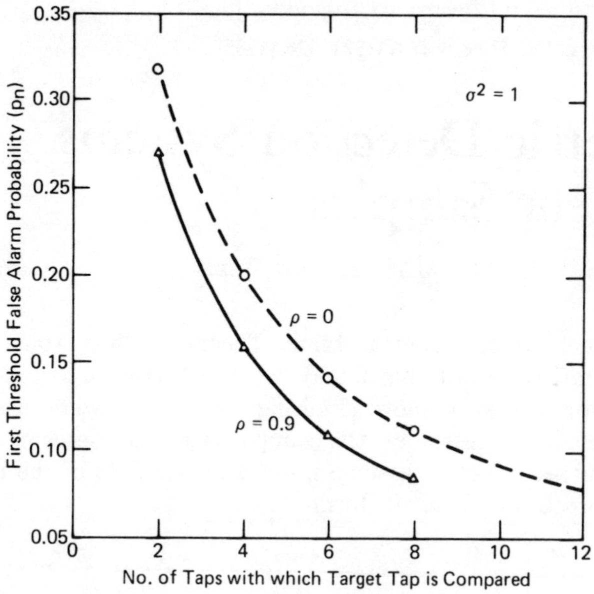

Fig. 2. Comparison of false alarm rates for correlated and uncorrelated samples.

$c = 1/[\sigma_x{}^2(1-\rho^2)]$ $\rho = R_c/\sigma_x{}^2$

R_c = covariance of the in-phase IF component

$\sigma_x{}^2$ = variance of the in-phase IF component

I_0 = zero order modified Bessel function of the first kind.

It will be noted that if the correlation coefficient ρ is zero, (11) reduces to the Rayleigh distribution (10) so that, *in this case,* zero correlation implies independence. This is not true in general, for an arbitrary distribution.

The above probability model was taken for the range clutter to check the effect of correlated samples on p_n in accordance with Equation (2) for various numbers of taps with $k = 2N$. A computer program was written using a multiple integration (FORTRAN) subroutine, which performs multiple integration in double precision using either 4, 8 or 16 point Gaussian quadrature. It was found that to go beyond 9 integrations was costly in computer time on the IBM 360 using this subroutine. Perhaps the reader with a more efficient multiple integration subroutine will be able to extend the results.

The results which were obtained up to that point are shown on Fig. 2 where p_n is plotted against $2N$, the number of taps with which the target tap is compared. It is apparent that with heavy correlation ($\rho = 0.9$) there is approximately a 20% to 30% decrease in p_n as compared with the uncorrelated ($\rho = 0$) case. A different clutter distribution model might indicate a different trend.

Summary and Conclusions

A method has been presented whereby the false alarm rate for a rank quantizer with correlated clutter input may be computed. The technique should be of value in comparison with experimentally derived false alarm rates.

Appendix

Independent Range Samples

For convenience, the method of computing the false alarm rate for the rank quantizer with independent range samples will be reviewed briefly here (see also [1]).

Consider the random process $X(t)$ and let $X_{-N}, \cdots, X_{-1}, X_0, X_1, \cdots, X_N$ be random variables denoting the $M = 2N + 1$ voltage taps, X_0 being the target tap. The random variables are assumed to be independent and identically distributed with probability density function $f_X(x)$ and cumulative probability distribution

$$F_X(x) = \int_{-\infty}^{x} f_X(x)\, dx .$$

If the order does not matter in which we pick the k random variables which are less than X_0, we have

$$P\{X_0 > \textit{exactly } k \text{ other taps}\} = \binom{M-1}{k} \int_{-\infty}^{\infty} f_X(x)[F_X(x)]^k [1-F_X(x)]^{M-k-1} dx \quad (12)$$

$$= \binom{M-1}{k} k!(M-k-1)!/M!$$

$$= 1/M. \quad (13)$$

The integral (12) is evaluated by noting that $f_X(x)\,dx = d\,F_X(x)$. Hence the probability of false alarm in a target-free environment is

$$P_n = P\,\{X_0 > \textit{at least } k \text{ other taps}\}$$

$$= M-k/M \quad (14)$$

$$= 1/M, \text{ if } k = M-1. \quad (15)$$

This implies that the quantizer is non-parametric or distribution-free, i.e., does not depend on the probability distribution of the samples. It will be noted, with reference to Fig. 1, that if the multiplicative constant C for the target tap were not unity it would be necessary to insert the particular distribution in (12) in order to obtain p_n thereby nullifying the non-parametric characteristics.

References

[1] G.M. Dillard and C.E. Antoniak, "A Practical Distribution-free Detection Procedure for Multiple-Range-Bin Radars," *IEEE Trans. Aerospace and Electronic Systems,* vol. AES-6, no. 5, September 1970.

[2] W.B. Davenport and W.L. Root, *An Introduction to the Theory of Random Signals and Noise.* New York: McGraw-Hill, 1958.

Reprinted with permission from *IEEE Trans. Inf. Theory* **IT-11**:521–526 (1965)

On Adaptive Nonparametric Detection Systems Using Dependent Samples

M. KANEFSKY, MEMBER, IEEE, AND J. B. THOMAS, FELLOW, IEEE

Introduction

Many detection systems belong to the class illustrated in Fig. 1. The system samples a (vector) input $\bar{x}(t)$ and calculates a test statistic $S(\bar{x})$ which is compared to a threshold W_0. If $S < W_0$, it is presumed that one situation H (hypothesis) exists with respect to the input, while if $S > W_0$, it is presumed that another situation K (alternative) exists. The *level* (*false alarm rate*) α of the detector is the probability of choosing K when H exists and is given by

$$\alpha = \text{Prob}_H \{S > W_0\}.$$

The *power* (*hit rate*) β of the detector is the probability of choosing K when K exists and is

$$\beta = \text{Prob}_K \{S > W_0\}.$$

If the level α can be fixed for some class of input processes, then the detector is said to be *nonparametric* (*or distribution-free*) with respect to that class.

Known nonparametric detection techniques are normally based on the assumption that independent samples of the processes involved are available [1], [2]. This assumption is frequently unrealistic. For narrow-band signals, for example, there is no sampling rate for which the independent sample assumption is valid, although a reasonable approximation can be obtained with sufficiently slow sampling. This procedure may be inefficient, given a finite decision time, and a nonparametric detector that could sample at a much faster rate would therefore be desirable.

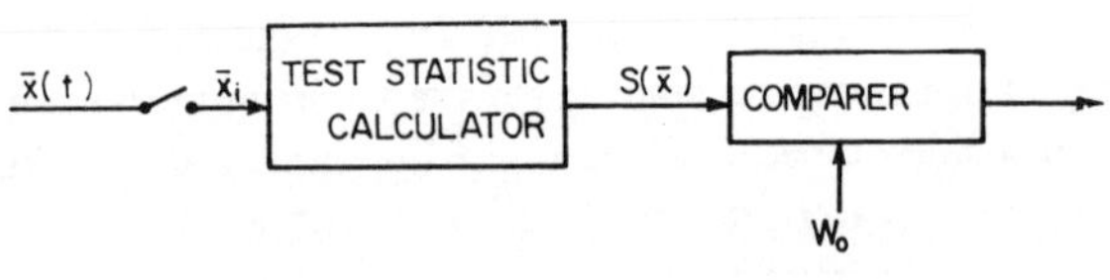

Fig. 1. Standard threshold detector.

With the present state of knowledge of nonparametric techniques, the alternative to the independent sample assumption seems to involve the use of some limit theorem, such as the Central Limit Theorem. Thus, some test statistic must be found whose distribution function approaches a known limiting form for a wide class of input processes as the sample size and decision time increase. Let us assume a test statistic $S(\bar{x})$ of the input processes $\bar{x}(t)$ of the form

$$S(\bar{x}) = \sum_{k=1}^{N} f\{\bar{x}(t + k\tau)\},$$

where τ is the sampling interval, N is the sample size, and $N\tau = T$ is the decision time. Further, assume that $y_k \triangleq f\{\bar{x}(t + k\tau)\}$ satisfies the following *strong mixing condition* [3], [4].

Let A, B be any events determined by conditions on the random variables y_k, $k \leq m$, and y_k, $k \geq n$, respectively, with $n > m$. The process $\{y_k\}$ is said to satisfy the strong mixing condition if

$$|P\{A \cap B\} - P\{A\}P\{B\}| \leq a(n - m)$$

for all such events A, B and some function $a(\tau)$ that decreases to zero as $\tau \to \infty$.[1] If in addition

$$0 < \lim_{T\to\infty} \frac{\text{Var}\{S(\bar{x})\}}{T} < \infty, \tag{1}$$

and

$$E\{|S(\bar{x}) - E\{S(\bar{x})\}|^4\} = O[\text{Var}^2\{S(\bar{x})\}], \tag{2}$$

then S is asymptotically normally distributed.

In order to set the threshold W_0, it is necessary to know that the test statistic is not only asymptotically normal, but also the values of its mean and variance. The problem, then, is to determine a function $f(\cdot)$ which is appropriate for a given detection problem and is such that the first two moments of the test statistic S can be

Manuscript received August 26, 1964; revised, June 15, 1965.
M. Kanefsky is with the School of Engineering, Yale University, New Haven, Conn. He was formerly with the Dept. of Electrical Engineering, Princeton University, Princeton, N. J.
J. B. Thomas is with the Dept. of Electrical Engineering, Princeton University, Princeton, N. J. He was supported during this research by the National Science Foundation through NSF Grant GP-1647.

[1] The strong mixing condition is satisfied by any ergodic Markov process, and by the stationary Gaussian process provided the correlation function approaches zero as its argument increases [4], [5]. Equating this condition on the correlation function to the strong mixing condition for the general case, is equivalent to the often-used engineering assumption that samples of a band-limited ergodic process that are sufficiently far apart can be considered independent. It has also been shown [6] that sums of certain nonlinear operations on normal random variables are asymptotically normal.

It may be of more practical importance to consider the "simple shot noise" process, which is a sum of independently occurring pulses of fixed shape [18]. If the pulses are of finite length, the process is obviously "strong mixing." For other cases some restrictions must be placed on the rate of decay of the pulses. This process has been used as a model of "impulse noise" [18] when the overlapping of pulses is small. Typical amplitude densities have been evaluated [18] and are decidedly non-Gaussian.

found (asymptotically at least) for a wide class of input processes. Throughout this paper it will be assumed that the input processes are stationary during the decision time (quasi-stationary). This assumption may not be necessary in general but it is necessary for the examples presented here.

It will now be demonstrated that adaptive techniques can be utilized to solve this problem and an algorithm will be given for finding an asymptotically nonparametric detector by beginning with some known parametric detector. Some examples are given and they are shown to be significantly more powerful than their standard nonparametric counterparts.

Theory

An m-input (threshold) detector D_s calculates a test statistic $S(\bar{x})$ of the input processes $\bar{x}(t)$ and compares this statistic with some threshold level W_0. Under the hypothesis H, the input vector is $\bar{x}(t) = \bar{n}(t)$ where $\bar{n}(t)$ is a set of m independent noise processes that are statistically identical. Under the alternative K, $\bar{x}(t) = \bar{n}(t) + \bar{s}(t)$ where $\bar{s}(t)$ is a signal vector whose components are statistically identical and independent with respect to the components of $\bar{n}(t)$. The signal is said to be present when $S(\bar{x}) > W_0$.

Let us assume an adaptive detector of the form shown in Fig. 2 where the test statistic (which is to be compared to the threshold W_0) has been written in the form

$$S(\bar{x}) = S_0(\bar{x}) - A(\bar{x}). \tag{3}$$

The level α of this detector is given by

$$\alpha = \text{Prob}_H \{S_0(\bar{x}) - A(\bar{x}) - W_0 > 0\}, \tag{4}$$

which becomes, asymptotically,

$$\alpha \rightarrow 1 - \Phi\left\{\frac{E_H\{A(\bar{x}) + W_0 - S_0(\bar{x})\}}{[\text{Var}_H \{S_0(\bar{x}) - A(\bar{x})\}]^{1/2}}\right\}, \tag{5}$$

where $\Phi\{\cdot\}$ is the unit normal cumulative distribution function. Equation (5) may be rearranged to give

$$E_H\{A(\bar{x})\} \rightarrow E_H\{S_0(\bar{x})\} - W_0 + \Phi^{-1}(1 - \alpha)[\text{Var}_H \{S_0(\bar{x}) - A(\bar{x})\}]^{1/2}. \tag{6}$$

The power β is given by

$$\beta = \text{Prob}_K \{S_0(\bar{x}) - A(\bar{x}) - W_0 > 0\}, \tag{7}$$

which becomes, asymptotically,

$$\beta \rightarrow 1 - \Phi\left\{\frac{E_K\{A(\bar{x}) + W_0 - S_0(\bar{x})\}}{[\text{Var}_K \{S_0(\bar{x}) - A(\bar{x})\}]^{1/2}}\right\}. \tag{8}$$

The advantage of representing the detector in this way is that for a given detection problem, one can start with a known detector D_{s_0} and then modify it, as in Fig. 2, by computing some statistic $A(\bar{x})$ that satisfies (6) for some class of input processes. The resulting detector is then asymptotically nonparametric with respect to that class. For this procedure to be meaningful it is necessary for the power of detection β to increase toward unity as either the sample size or the SNR increase. In other words,

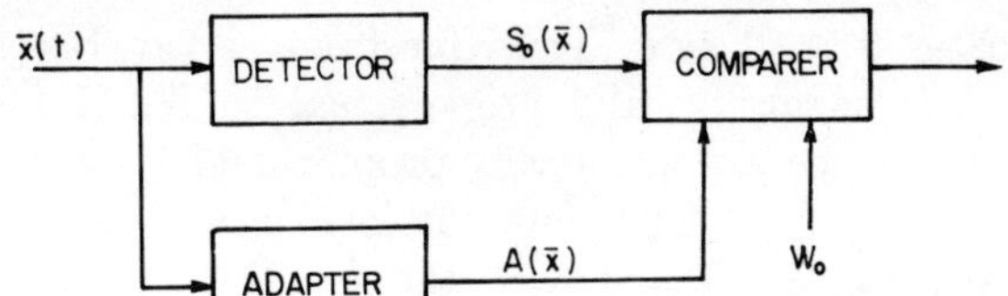

Fig. 2. Adaptive threshold detector.

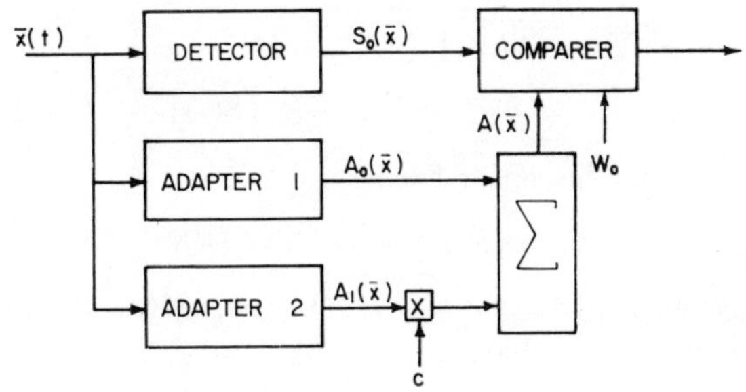

Fig. 3. Assumed adaptive structure.

we must have

$$\lim_{\substack{N\rightarrow\infty \\ \text{or} \\ \sigma_s/\sigma_n\rightarrow\infty}} \frac{E_K\{S_0(\bar{x}) - A(\bar{x})\}}{[\text{Var}_K \{S_0(\bar{x}) - A(\bar{x})\}]^{1/2}} = \infty \tag{9}$$

where σ_s^2/σ_n^2 is the signal-to-noise power ratio.

At this point let us assume that the adaptive part of the detector of Fig. 2 can be written as

$$A(\bar{x}) = A_0(\bar{x}) + cA_1(\bar{x}) \tag{10}$$

as in Fig. 3 where we require that

$$E_H\{A_0(\bar{x})\} = E_H\{S_0(\bar{x})\} - W_0. \tag{11}$$

This separates the adaptive structure into two components. The output of the first, proportional to the sample size N, must be asymptotically normal, and must satisfy (11). The second component will be shown to be asymptotically proportional to $N^{1/2}$. Rewriting (6) using (10) and (11), we have

$$\begin{aligned}\frac{c^2E_H^2\{A_1(\bar{x})\}}{[\Phi^{-1}(1 - \alpha)]^2} &\rightarrow \text{Var}_H \{S_0(\bar{x}) - A_0(\bar{x})\} + c^2 \text{Var}_H \{A_1(\bar{x})\} \\ &+ 2c[E_H\{A_0(\bar{x})A_1(\bar{x})\} - E_H\{A_0(\bar{x})\}E_H\{A_1(\bar{x})\}] \\ &- 2c[E_H\{S_0(\bar{x})A_1(\bar{x})\} - E_H\{S_0(\bar{x})\}E_H\{A_1(\bar{x})\}]\end{aligned} \tag{12}$$

If it is assumed that $[S_0(\bar{x}) - A_0(\bar{x})]$ is of the form $\sum_{i=1}^N g(\bar{x})$, it follows, assuming quasi-stationarity, that $\text{Var}_H\{S_0(\bar{x}) - A_0(\bar{x})\}$ is of the form

$$\text{Var}_H \{S_0(\bar{x}) - A_0(\bar{x})\} = Np^2(\bar{x})h^2(\bar{x}, N, \rho), \tag{13}$$

where $p^2(x) = \text{Var}_H\{g(\bar{x})\}$, $\rho(\cdot)$ is the normalized correlation function of the input processes, and

$$h^2(\bar{x}, N, \rho) = 1 + 2 \sum_{k=1}^{N-1} \left(\frac{N - k}{N}\right)R(\bar{x}, \rho, k\tau), \tag{14}$$

where $R(\bar{x}, \rho, k\tau)$ is the normalized correlation between $g[\bar{x}(t)]$ and $g[\bar{x}(t + k\tau)]$. With some loss of generality, we will assume that $R(\bar{x}, \rho, k\tau)$ decreases to zero as $\rho(k\tau)$

decreases toward zero. Thus, for fixed τ, $\mathrm{Var}_H\{S_0(\bar{x}) - A_0(\bar{x})\}$ is asymptotically proportional to N. It follows that c must be asymptotically proportional to $N^{-1/2}$ if it is to satisfy (12). Therefore the level α is fixed asymptotically if c, $A_0(\bar{x})$, and $A_1(\bar{x})$ satisfy

$$cE_H\{A_1(\bar{x})\} \to \Phi^{-1}(1-\alpha)[\mathrm{Var}_H\{S_0(\bar{x}) - A_0(\bar{x})\}]^{1/2}. \quad (15)$$

Alternatively, it follows from (6), (10), and (11) that

$$\Phi^{-1}(1-\alpha)[\mathrm{Var}_H\{S(\bar{x})\}]^{1/2} \to W_0 - E_H\{S(\bar{x})\} = cE_H\{A_1(\bar{x})\}.$$

From (2), we have, for fixed τ,

$$\mathrm{Var}_H\{S(\bar{x})\} \to O(T) = O(N)$$

and hence $cE_H\{A_1(\bar{x})\} \to O(N^{1/2})$. Also, on the assumption previously made that

$$S_0(\bar{x}) - A_0(\bar{x}) = S(\bar{x}) + cA_1(\bar{x}) = \sum_{k=1}^{N} g[\bar{x}(t + k\tau)],$$

then $A_1(\bar{x})$ must be of the form

$$A_1(\bar{x}) = \sum_{k=1}^{N} \mu_k.$$

Thus $E_H\{A_1(\bar{x})\} = O(N)$ and, hence, $c \to O(N^{-1/2})$.

If it happens that $A_0(\bar{x})$, in addition to satisfying (11), is such that $R(\cdot)$ is not a function of the input processes $\bar{x}(t)$, then clearly, choosing $A_1(\bar{x})$ such that $E_H\{A_1(\bar{x})\} = \lambda N p(\bar{x})$ and setting c equal to $\Phi^{-1}(1-\alpha)\, h(N, \rho)/\lambda N^{1/2}$ lead to asymptotic nonparametric detector. The constraint of (9) requires one final condition that $A_0(\bar{x})$ must satisfy, namely

$$E_K\{S_0(\bar{x})\} - E_K\{A_0(\bar{x})\} = M[N, \sigma_s^2/\sigma_n^2], \quad (16)$$

where $M[N, \sigma_s^2/\sigma_n^2]$ is a positive, monotonically increasing function in both variables, and $M(\cdot) = O\{N^{1/2+\delta}\}$, $\delta > 0$, for large N. This last equation would be satisfied, for example, if $E_K\{A_0(\bar{x})\} = E_H\{A_0(\bar{x})\}$.

To insure that the convergence of the false alarm rate is limited only by Central Limit Theorem considerations, it is sufficient that the multiplier c contain an additional term that is asymptotically proportional to N^{-1} and is a function of $\rho(t)$ only. Detectors that permit this solution are contained in the class that can be made asymptotically nonparametric via (15). Therefore a meaningful procedure is to attempt to find asymptotic nonparametric detectors and then check (12) to see if a correction term can be added. The asymptotic detectors with correction terms will be called *essentially* nonparametric.

Thus, the procedure for finding possible adaptive nonparametric detectors is as follows:

1) Given a detector D_{S_0}, find a test statistic $A_0(\bar{x})$ that satisfies

a) $E_H\{A_0(\bar{x})\} = E_H\{S_0(\bar{x})\} - W_0$,

b) $E_K\{S_0(\bar{x})\} - E_K\{A_0(\bar{x})\} = M[N, \sigma_s^2/\sigma_n^2]$,

c) $\mathrm{Var}_H\{S_0(\bar{x}) - A_0(\bar{x})\} = Np^2(\bar{x})h^2(\rho, N)$.

2) Find a test statistic $A_1(\bar{x})$ that has the property

$$E_H\{A_1(\bar{x})\} = \lambda N p(\bar{x}),$$

where λ is an arbitrary constant.

3) An asymptotic nonparametric detector is then obtained by multiplying $A_1(\bar{x})$ by the term c, where

$$c = \frac{\Phi^{-1}(1-\alpha)}{\lambda N^{1/2}} h(\rho, N).$$

4) Check (12) to see if a second-order correction term for the multiplier c can be found. We now illustrate this procedure by several examples.

Example 1

Consider the class of detectors D_{S_0} that have the property $E_H\{S_0(\bar{x})\} = 0$. A number of detectors can be found that possess this property if the class of noise processes are restricted to have either zero mean or zero median. If the test statistic $S_0(\bar{x})$ has the property

$$\mathrm{Var}_H\{S_0(\bar{x})\} = Np^2(\bar{x})h^2(\rho, N)$$

then an obvious solution for $A_0(\bar{x})$ is $A_0(\bar{x}) = 0$, $W_0 = 0$, and an asymptotic solution, at least, is probably obtainable. A specific example of such an m-channel detector is

$$S_0(\bar{x}) = \sum_{i=1}^{N} (x_{1i}x_{2i} \cdots x_{mi}), \quad (17)$$

where it is assumed that $E(\bar{x}) = 0$.

For $m = 1$, this detector corresponds to the Neyman-Pearson mean detector for finding a dc signal in additive Gaussian noise [7]. For $m > 1$, it corresponds to a correlation detector which is useful for detecting a random signal common to m channels that contain independent stationary noise sources. Correlation arrays have been considered for underwater sound detection [8]–[10].

For the test statistic of (17) it can be shown that

$$E_H\{S_0(\bar{x})\} = 0$$

and

$$\mathrm{Var}_H\{S_0(\bar{x})\} = N\sigma_n^{2m}\left[1 + 2\sum_{k=1}^{N}\left(\frac{N-k}{N}\right)\rho^m(k\tau)\right]. \quad (18)$$

It follows that a function $A_1(\bar{x})$ is required such that $E_H\{A_1(\bar{x})\} = \lambda N\sigma_n^m$. One possible solution for an asymptotic nonparametric m-input correlation detector is shown in Fig. 4 where

$$A_1(\bar{x}) = N\left[\frac{1}{N}\sum_{i=1}^{N}(x_{1i} - x_{2i})^2\right]^{m/2}. \quad (19)$$

Let us now check (12) to see if a correction term can be added. Solving this equation for c yields

$$c = \frac{\Phi^{-1}(1-\alpha)}{2^{m/2}N^{1/2}} h(\rho) + \frac{[\Phi^{-1}(1-\alpha)]^2}{2N} h^2(\rho)\,\delta_{m2} + O\{N^{-3/2}\}, \quad (20)$$

where

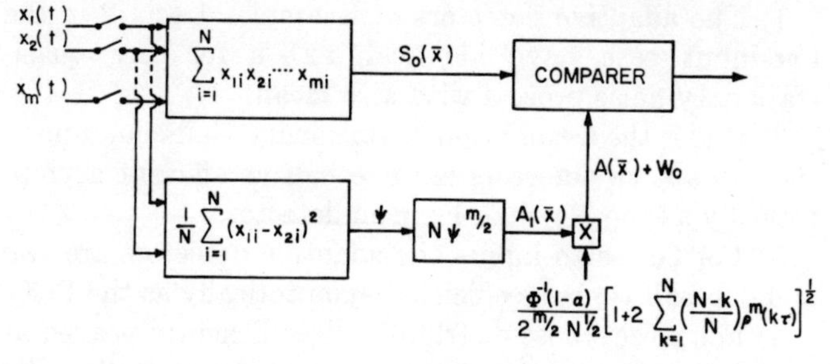

Fig. 4. A sampled m-input, correlation detector.

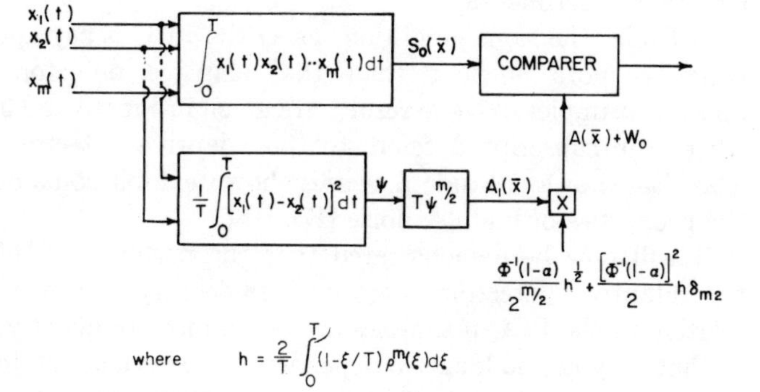

Fig. 5. A continuous, essentially nonparametric, m-input, correlation detector.

$$\delta_{m2} = \begin{cases} 1; & m = 2 \\ 0; & m \neq 2 \end{cases},$$

and where

$$h^2(\rho) = 1 + 2 \sum_{k=1}^{N-1} \left(\frac{N-k}{N}\right)\rho^m(k\tau). \tag{21}$$

The third term of (20) involves properties of the input statistics. It follows that modifying Fig. 4 to include the first two terms for c yields an essentially nonparametric detector with respect to the class of quasi-stationary input processes with zero means. Observe that c becomes independent of the spectral shape for large m.

As this detector can be sampled at any rate, let us consider the limiting case and replace the sums by integrals where

$$h^2(\rho)/N \to \delta \triangleq \frac{2}{T}\int_0^T \left(1 - \frac{\xi}{T}\right)\rho^m(\xi)\, d\xi. \tag{22}$$

Thus the final version of this detector, as shown in Fig. 5, is a continuous, essentially nonparametric, m-input, correlation detector.

Example 2

Consider the array detector, standard for underwater sound problems [11]–[13] for detecting a random signal common to m inputs each of which contain independent noise processes.

$$S_0(\bar{x}) = \sum_{i=1}^{N} (x_{1i} + x_{2i} + \cdots + x_{mi})^2. \tag{23}$$

For $m = 2$, this represents the Neyman-Pearson (optimum) detector when the processes are all Gaussian and the samples in each channel are independent [14]. Under the same conditions, and for $m > 2$, this detector approximates the maximum likelihood ratio for the small signal-to-noise case [13].

When $E(\bar{x}) = 0$, it can be shown that $E_H\{S_0(\bar{x})\} = mN\sigma_n^2$ and

$$\begin{aligned}\mathrm{Var}_H\{S_0(\bar{x})\} &= 2m(m-1)N\sigma_n^4 \\ &\cdot \left[1 + 2\sum_{k=1}^{N-1}\left(\frac{N-k}{N}\right)\rho^2(k\tau)\right] - mN\sigma_n^4 \\ &+ mN\mu_4\left[1 + 2\sum_{k=1}^{N-1}\left(\frac{N-k}{N}\right)\tilde{\rho}(k\tau)\right],\end{aligned} \tag{24}$$

where $\tilde{\rho}(t)$ is the normalized correlation function of $n^2(t)$ and μ_4 is the fourth moment of $n(t)$. If attention is restricted to the case where m is even, then a reasonable choice for the test statistic $A_0(\bar{x})$, for $W_0 = 0$, is

$$A_0(\bar{x}) = \sum_{i=1}^{N} (x_{1i} - x_{2i} + x_{3i} - \cdots - x_{mi})^2, \tag{25}$$

since $E_H\{A_0(\bar{x})\} = E_H\{S_0(\bar{x})\} = E_K\{A_0(\bar{x})\}$ and the first two requirements for $A_0(\bar{x})$ are satisfied. It can be shown that

$$\mathrm{Var}_H\{S_0(\bar{x}) - A_0(\bar{x})\} = 4m^2N\sigma_n^4\left[1 + 2\sum_{k=1}^{N-1}\left(\frac{N-k}{N}\right)\rho^2(k\tau)\right]. \tag{26}$$

It follows that $A_0(\bar{x})$ is a suitable choice and that a function $A_1(\bar{x})$ is required such that $E_H\{A_1(\bar{x})\} = \lambda 2m\sigma_n^2$. An obvious choice for $A_1(\bar{x})$ is $A_0(\bar{x})$ itself where $\lambda = \frac{1}{2}$. The resultant asymptotic nonparametric, m-input (m even), standard array detector is shown in Fig. 6. Observe that the output SNR

$$\{[E_K\{S_0(\bar{x}) - A_0(\bar{x})\}]^2/\mathrm{Var}_H\{S_0(\bar{x}) - A_0(\bar{x})\}\}$$

of this detector is proportional to m^2.

Substituting these test statistics into (12) and solving for multiplier c yield

$$c = \frac{2\Phi^{-1}(1-\alpha)}{N^{1/2}}h(\rho) + \frac{2[\Phi^{-1}(1-\alpha)]^2}{N}h^2(\rho) + O[N^{-3/2}], \tag{27}$$

where

$$h^2(\rho) = 1 + 2\sum_{k=1}^{N-1}\left(\frac{N-k}{N}\right)\rho^2(k\tau).$$

Modifying Fig. 6 to include the first two terms for the multiplier c yields an essentially nonparametric detector with respect to the class of quasi-stationary input processes with zero mean. By considering the limiting case of an infinite sampling rate, one obtains the continuous, essentially nonparametric, m(even) input, array detector shown in Fig. 7.

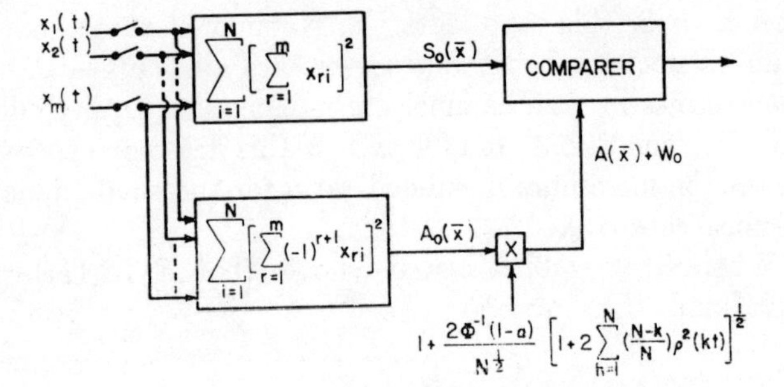

Fig. 6. A sampled m-input, standard array detector.

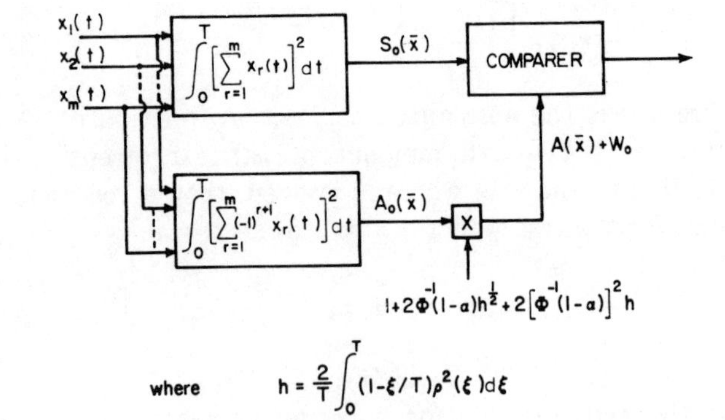

Fig. 7. A continuous, essentially nonparametric, m-input, standard array detector.

Comparison of Detectors for the Two-Input Case

The two-input adaptive detectors of Examples 1 and 2 can be compared with the Neyman-Pearson detector and the polarity coincidence correlator (PCC). The PCC is a two-input detector that is nonparametric with respect to inputs with known median, provided the samples are independent [15]. This detector calculates the statistic

$$S_{pcc} = \sum_{i=1}^{N} \text{Sgn}\,(x_{1i}x_{2i}), \tag{28}$$

where

$$\text{Sgn}\,(x) = \begin{cases} 1; & x > 0 \\ 0; & x < 0 \end{cases}.$$

The comparison technique will be the notion of *asymptotic relative efficiency* (a.r.e) [16]. If two detectors D_1 and D_2 require N_1 and N_2 samples, respectively, to attain the same power of detection for a given false alarm rate, then the a.r.e. of D_1 relative to D_2 is defined as

$$\bar{e}_{D_1/D_2} = \lim_{\substack{N_1, N_2 \to \infty \\ \sigma_s^2/\sigma_n^2 \to 0}} N_2/N_1. \tag{29}$$

$$\beta = \text{constant}$$

The following results have been obtained [17] under the assumption of independent samples.

1) The adaptive detectors of Examples 1 and 2 in the two-input case have identical a.r.e.'s for any quasi-stationary noise process with zero mean.

2) Under the assumption of stationary Gaussian inputs, these adaptive detectors are one half as efficient asymptotically as the Neyman-Pearson detector.

3) For Gaussian inputs the adaptive detectors are two and one-half times as efficient asymptotically as the PCC.

4) For noise whose distribution is sufficiently peaked at the origin, the adaptive detectors are asymptotically more efficient than the Neyman-Pearson detector whose threshold is set experimentally.

5) Under the same conditions as 4) the PCC is asymptotically more efficient than the adaptive detectors. Specific examples of these results are given in Figs. 8 and 9 where the subscript a refers to the adaptive detectors. The classes of amplitude densities shown exhibit some of the properties of impulse noise [17], [18].

It will now be demonstrated that the performance of the adaptive detectors may be significantly improved relative to the PCC if samples are taken more frequently, so that they are no longer independent. For both adaptive detectors, the power given by (8) approaches

$$\beta \to 1 - \Phi\left\{\frac{E_K\{S_1(\bar{x})\} - E_K\{S_0(\bar{x})\}}{[\text{Var}_H\,\{S_0(\bar{x}) - S_1(\bar{x})\}]^{1/2}}\right\}$$
$$= 1 - \Phi\left\{\frac{-k\sqrt{N}\,\sigma_s^2/\sigma_n^2}{h(\rho)}\right\}. \tag{30}$$

Since $h(\rho) \to 1$ as $\rho(t) \to 0$, the notion of an equivalent number of independent samples $N_{eq.}$ can be defined as

$$N_{eq} = N/h^2(\rho). \tag{31}$$

If N_m is defined as the maximum number of independent samples, then the a.r.e. calculated for independent samples can be increased by the multiplicative factor

$$I \triangleq \frac{N_{eq}}{N_m} \le \frac{T}{2N_m}\int_0^T (1 - \xi/T)\rho^2(\xi)\,d\xi \tag{32}$$

where the equality is approached for very fast sampling. For large T and very fast sampling, this ratio N_{eq}/N_m is closely approximated by

$$I \approx \frac{T}{2N_m}\int_0^\infty \rho^2(\xi)\,d\xi. \tag{33}$$

As an example, consider a noise spectrum determined by a single tuned circuit with large $Q(Q \ge 10)$ and center frequency $\omega_0/2\pi$. For this spectrum the normalized correlation function is given approximately by [19]

$$\rho(\xi) = e^{-\omega_0\xi/2Q}\cos\,(\omega_0\xi). \tag{34}$$

For large Q, (33) is closely approximated by $I \approx T\omega_0/N_mQ$. While independent samples can never be obtained with arbitrary finite sample intervals, let us assume that samples Q/f_0 apart are essentially independent $[|\rho(\tau)| \le$

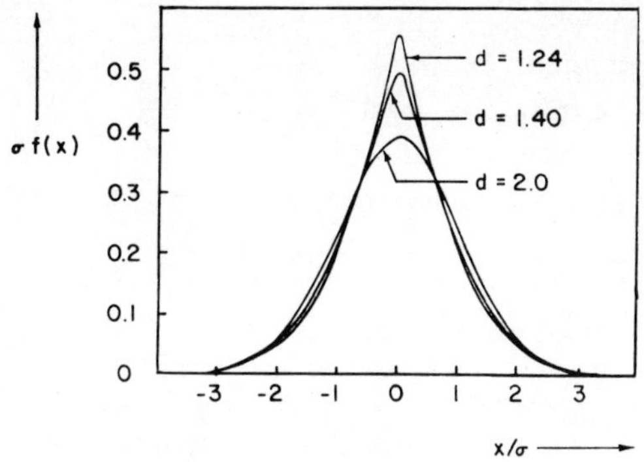

d	e_a/np	c_a/pcc
2.0	0.5	2.5
1.4	0.6	1.0
1.24	0.625	0.625
1.0	1.0	0.286

Fig. 8. Comparison of detectors where the noise amplitude density functions are of the form.

$$\frac{a}{\sigma} \exp. \left[-b \left| \frac{x}{\sigma'} \right|^{d} \right].$$

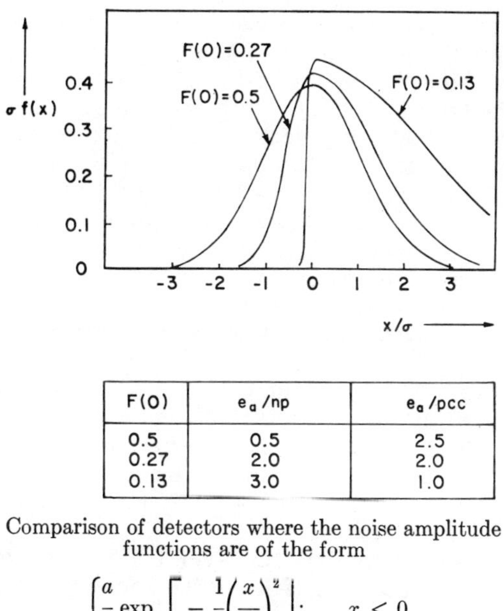

F(0)	e_a/np	e_a/pcc
0.5	0.5	2.5
0.27	2.0	2.0
0.13	3.0	1.0

Fig. 9. Comparison of detectors where the noise amplitude density functions are of the form

$$\begin{cases} \dfrac{a}{\sigma} \exp. \left[-\dfrac{1}{2} \left(\dfrac{x}{\sigma_-} \right)^2 \right]; & x < 0 \\ \dfrac{a}{\sigma} \exp. \left[-\dfrac{1}{2} \left(\dfrac{x}{\sigma_+} \right)^2 \right]; & x > 0. \end{cases}$$

0.043, $N_{eq} \geq 0.996N$]. Thus N_m is assumed to be Tf_0/Q and hence $I \approx 2\pi$. Therefore, rapid sampling increases the efficiency of the adaptive detectors by 2π relative to sampling at the bandwidth rate. It follows that the continuous adaptive detectors are of the order of 15 times as efficient asymptotically as the PCC for quasi-stationary Gaussian inputs that have single-tuned spectra.

Conclusions

An approach has been discussed for modifying a parametric detector to obtain an essentially nonparametric detector with respect to a wider class of inputs than that for which the detector was originally designed. A significant feature of such a device is that it does not depend on independent samples of the input processes. The major difficulty with this approach is that it does not lead to straightforward synthesis procedures. However, solutions do exist for a number of detectors and some examples have been given.

There are three disadvantages of the adaptive detectors relative to the standard nonparametric detector. These are as follows:

1) The adaptive detectors require large, but not unreasonable, sample sizes for the analysis to hold.

2) The quasi-stationary assumption is required, and

3) the detectors may be more difficult to implement. On the other hand, the adaptive detectors do not require the assumption of independent samples and, as a result, may significantly outperform their nonparametric counterparts.

References

[1] D. A. S. Frazer, *Nonparametric Methods in Statistics*, New York: Wiley, 1957.
[2] M. G. Kendall and A. Stuart, *The Advanced Theory of Statistics-Inference and Relationship*, vol. 2, New York, Hafner, 1961, pp. 465–535.
[3] M. Rosenblatt, "A central limit theorem and a strong mixing condition," *Proc. Nat'l. Acad. Sciences*, vol. 42, pp. 43–47, 1956.
[4] ——, "Independence and dependence," *1961 Proc. 4th Berkeley Symp.*, vol. 2, pp. 431–443.
[5] Y. A. Rozanov, "An Application of the central limit theorem," *ibid.*, pp. 445–454.
[6] T-C. Sun, "A central limit theorem for non-linear functions of a normal stationary process," *J. Math. Mech. (USA)*, vol. 12, pp. 945–977, November 1963.
[7] Kendall and Stuart, *op. cit.*, [2] p. 167.
[8] J. J. Faran and R. Hills, "The application of correlation techniques to acoustic receiving systems," Acoustics Research Lab., Harvard University, Cambridge, Mass., Tech. Memo. 28, November 1952.
[9] K. W. Goff, "The application of correlation techniques to some acoustic measurements," *J. Acoust. Soc. Am.*, vol. 27, pp. 236–246, March 1955.
[10] J. B. Thomas and T. R. Williams, "On the detection of signals in nonstationary noise by product arrays," *J. Acoust. Soc. Am.*, vol. 31, pp. 453–462, April 1959.
[11] P. Rudnick, "Small signal detection in the Dimus Array," *J. Acoust. Soc. Am.*, vol. 32, pp. 871–877, July 1960.
[12] V. C. Anderson, "Digital array phasing," *J. Acoust. Soc. Am.*, vol. 32, pp. 867–870, July 1960.
[13] F. Bryn, "Optimal signal processing by three-dimensional arrays operating on Gaussian signals and noise," *J. Acoust. Soc. Am.*, vol. 34, pp. 289–297, March 1962.
[14] J. Carlyle and J. B. Thomas, "On nonparametric signal detectors," *IEEE Trans. on Information Theory*, vol. IT-10, pp. 146–152, April 1964.
[15] S. S. Wolff, J. B. Thomas, and T. R. Williams, "The polarity-coincidence correlator: a nonparametric detection device," *IRE Trans. on Information Theory*, vol. IT-8, pp. 5–9, January 1962.
[16] E. L. Lehmann, *Testing Statistical Hypothesis*, New York: Wiley, 1959, p. 239.
[17] M. Kanefsky, "On sign tests and adaptive techniques for nonparametric detection," Ph.D. dissertation, Princeton University, Princeton, N. J., 1964.
[18] E. N. Gilbert and H. O. Pollak, "Amplitude distribution of shot noise," *Bell Sys. Tech. J.*, vol. 39, pp. 330–350, March 1960.
[19] D. Middleton, *An Introduction to Statistical Communication Theory*, New York: McGraw-Hill, 1960, p. 183.

31

Reprinted from *Telecommun. Radio Eng.* **27/28**:66–72 (1973)

Nonparametric Signal Detectors when the Signal Samples are Dependent

V. A. Simakhin

Adaptive detectors of stochastic and determinate signals buried in noise, such that the level of false alarms is asymptotically invariant for a wide class or noise distributions, are discussed for the case of dependent signal samples.

1. Introduction. The methods of nonparametric statistics [1] have been used for the detection of signals buried in noise the probability characteristics of which are not completely known. This has led to the development of so-called nonparametric detectors, in which the level of false alarms is invariant for a fairly wide class of noise distributions [2, 3]. When designing such detectors the samples are assumed independent; this assumption is not always justified, so that the practical application of such detectors is limited. For instance, the detector is often required to have fast operation, i.e., the time interval of the observations is restricted and the number of independent observations required to obtain a given efficiency may not be collected. At the same time, when we renounce the assumption of independent samples, the nonparametric detector loses the property whereby the level of false alarms is invariant to changes in the noise distribution [4, 5]. In this sense, the problem of nonparametric detector design for the case of dependent observations remains urgent. In [6, 7] the idea of detection threshold adptation (alignment) was used; while this method allows a nonparametric detector to be designed without assuming independent observations, the nonparametric circuits cannot then be synthesized directly (a solution is only obtained in simple particular cases). The approach below is also based on the idea of adaptation. However, as distinct from [6], we show that, if the noise belongs to the class of m-dependent processes (defined in the Appendix), the threshold adaptation unit can be implemented simply by evaluating the sampling correlation function of certain random variables. We compare such adaptive detectors with those operating on the assumption of independent samples, and show that, when the observation time is restricted, the adaptive detectors are more efficient.

2. Formulation of the problem. Let the channel output signal be y(t), representing either a mixture of the wanted signal with noise or else pure noise. It is required to detect, in the received sample y(t), the presence of the wanted signal, under the following assumptions:

a) the noise is a stationary m-dependent random process with unknown distribution function $F\{\vec{\xi}(t)\}$;

b) during the observation time T we can obtain m-dependent samples $y_1, \ldots, y_n$ of the random process at the channel output,

c) the distribution of the random process y(t) at the channel output differs, according to whether the wanted signal $\varphi(t)$ is present or absent, only in the signal-to-noise ratio parameter θ, accompanied additionally by a possible change in the distribution function $G_\theta\{\vec{\xi}(t)\}$;

d) the case of a weak wanted signals, i.e., $\theta \ll 1$, is considered.

3. Structure of the adaptive detector. We shall consider wanted signal detectors D, operating according to the scheme

$$D = \begin{cases} 1, & \text{if} \quad S > \lambda, \\ 0, & \text{if} \quad S \leqslant \lambda, \end{cases}$$

where λ is the threshold and S a statistic which is symmetric to a permutation of its arguments. Most of the statistics used in the design of nonparametric detectors [2], fall into this class. Investigation of them in the case of dependent observations [4, 5, 9] shows that they have an asymptotically normal distribution, with mean convergent to the mean of S in the case of independent samples (i.e., for nonparametric statistics the mean is independent of F), and variance which, when there is no signal, is a bounded function of the unknown noise distribution function F: $\sigma_0^2(S) = \psi(F)$. The probability of false alarm α in such a detector may then be written as

$$\alpha = P_{r_0}\{S > \lambda\} = 1 - \Phi\left\{\frac{\lambda - E_0(S)}{\sigma_0(S)}\right\}, \tag{1}$$

where $\Phi(x)$ is the standard normal distribution and E_0 denotes the mathematical expectation.

Given α, the threshold can be found from (1):

$$\lambda = \Phi^{-1}(1-\alpha)\sigma_0(S) + E_0(S). \tag{2}$$

It is clear from (2) that, if $F\{\vec{\xi}(t)\}$ is unknown, we cannot find λ from the given α, since $\sigma_0^2(S)$ depends directly on the unknown F. Hence, to design an asymptotically nonparametric detector, we replace $\sigma_0^2(S)$ by its consistent estimate $V_0^2(S)$, independent of F. The general theory of the construction of such estimates, based on U-statistics, can be found in [1], so that we shall only quote their final form below, for certain well-known detectors [2]. Then,

$$\Phi^{-1}(1-\alpha)V_0(S) + E_0(S) \to \lambda, \tag{3}$$

i.e., the threshold will adapt so that α is asymptotically constant for all types of noise belonging to the class of stationary m-dependent processes. Expression (3) defines the "adapter" structure, provided that the estimate $V_0^2(S)$ is known. If assumption d) is valid, it follows from [11] that, for single-channel detectors, and a direct extension of the Chernoff-Savage theorem [13] to the case of stationary m-dependent samples (i.e., for two-channel detectors), we have the convergence $\sigma_0^2(S) \to \sigma_0^2(S)$, where $\sigma_0^2(S)$ is the variance of S when a signal is present. This enables the threshold λ to adapt within the observation time regardless of whether pure noise or a signal-plus-noise mixture is present at the detector input.

The detector (whose circuit is given in Fig. 1) thus consists of a nonparametric detector for the case of independent observations, and an adaptation unit, which evaluates the threshold λ from (3). In Fig. 1, D is the discretizer, S the unit for evaluating the statistic, and TAU the threshold adaptation unit.

Below, in the context of given a priori information on the noise channel statistics and type of wanted signal, we consider the design of adaptive detectors based on well-known nonparametric detectors [2].

4. The median detector. With the assumptions a-d, let the only a priori information on the noisy channel statistic be the median M of the noise distribution function. With these assumptions, we can use for operation of the detector the statistic [10]: $S_1(M) = \frac{1}{n}\sum_{i=1}^{n} c(y_i - M)$, where $c(x) = 1$, if $x > 0$, and $c(x) = 0$ if $x \leqslant 0$.

Whether a signal is present or absent, S_1 satisfies all the requirements of the central limit theorem for stationary m-dependent sequences [8], i.e., it is asymptotically normal with the parameters

$$E_0(S_1) = \frac{1}{2};\quad E_\theta(S_1) = \frac{1}{2} + \theta\left[\frac{d}{d\theta}G_\theta(M)\big|_{\theta=0}\right]; \tag{4}$$

$$\sigma_0^2(S_1) = \frac{1}{n}\left[\xi_0 + 2\sum_{k=1}^{m}\xi_k - \frac{2m+1}{4}\right].$$

The expression for the variances $\sigma_0^2(S_1)$ is obtained in the Appendix. The consistent and F-independent estimate of $\sigma_0^2(S_1)$ is

$$V_0^2(S_1) = \frac{1}{n}\left[b_0 + 2\sum_{k=1}^{m} b_k - \frac{2m+1}{4}\right], \tag{5}$$

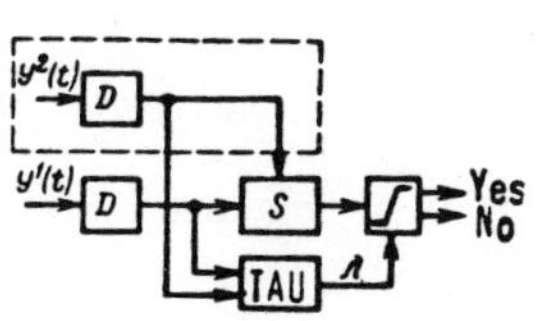

Fig. 1

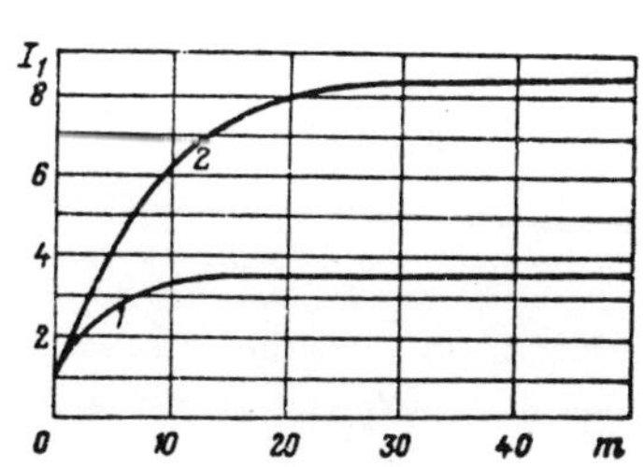

Fig. 2

where

$$b_k = \frac{1}{n-k}\sum_{i=1}^{n-k} c(y_i - M)c(y_{i+k} - M).$$

Substituting (5) and (4) in (3), we can obtain λ_1. The unit for evaluating the threshold λ_1 in fact makes up the adaptive part of the detector (the adapter). It follows from (3) and (5) that the adapter can be implemented in practice by using a digital correlation meter, operating with the binary random variables $c(y_i - M)$ and $c(y_{i+k} - M)$; these are easily obtained in practice by using threshold elements. The efficiency of the adaptive detector D_1 may be I_1 times that of the median detector D_1^* [10], which uses for its operation N independent observations, obtained during an interval T. We show in the Appendix that

$$I_1 = \frac{m+1}{1 + 8\sum_{k=1}^{m} \xi_k - 2m}.$$

When detecting a constant signal in additive Gaussian noise with correlation coefficient $\rho(\tau) = \rho(k\Delta t)$, we obtain

$$\xi_k = \frac{1}{2}\left\{1 - \frac{1}{\pi}\operatorname{arc\,tg}\left[\frac{\sqrt{1-\rho^2(k)}}{\rho(k)}\right]\right\}.$$

In the case of Rayleigh noise (incoherent detection),

$$\xi_k = \frac{1+\rho(k)}{2\sqrt{\rho(k)}}\, e^{-\frac{M^2}{2}}\left[1 - \operatorname{erf}\left(\sqrt{\frac{1-\rho(k)}{2(1+\rho(k))}}\, M\right)\right],$$

where erf(x) is the probability integral.

In Fig. 2 we show $I_1 = f(m)$ for $\rho(\tau) = e^{-\delta\tau}$; it has been assumed here that the observations $y(t)$ and $y(t \pm \tau_{ob})$ are independent, if $\rho(\tau_{ob}) = e^{-5}$, i.e., $\tau_{ob} = 5/\delta\, [\rho(k) = e^{-\delta\Delta t k} = e^{-\gamma k}$, where $\gamma = \delta\Delta t$ and $m = 5/\gamma$ for a discrete sample, e.g., with γ = 2, 1; 1/2; 1/5; 1/10, m = 1; 5; 10; 25;, 50 respectively]. When $m \to \infty$, we have $I_1 \approx 3.5$ for Gaussian noise (curve 1) and $I_1 \approx 8.5$ for Rayleigh noise (curve 2). The main increase in efficiency occurs when using 25 dependent observations (m = 25), falling within the interval τ_{ob}. Further increasing m does not significantly increase I_1.

Detector based on the sampling mean statistic. If, in addition to assumptions a-d, we assume that the mean value A of the noise distribution F is known a priori, the adaptive detector may be based on the sampling mean statistic

$$S_2 = \frac{1}{n}\sum_{i=1}^{n} y_i.$$

The fact that this statistic is asymptotically normal follows immediately from the central limit theorem for m-dependent stationary sequences [8]; here,

$$E_0(S_2) = A;\quad E_\theta(S_2) = A + \theta \int_{-\infty}^{\infty} y \frac{d}{dy}\left[\frac{d}{d\theta} G_\theta(y)\right]\Bigg|_{\theta=0}$$

and, see the Appendix, $\sigma_0^2(S_2) = \frac{\sigma_1^2}{n}\left[1 + 2\sum_{k=1}^{m}\rho(k)\right]$ where σ_1^2 and $\rho(k)$ are the noise variance and the correlation coefficient between y_i and y_{i+k} in when there is no signal. The consistent, F-independent estimate $V_0^2(S_2)$ is simply the sampling correlation function:

$$V_0^2(S_2) = \frac{1}{n}\left[\frac{1}{n}\sum_{i=1}^{n}(y_i - A)^2 + 2\sum_{k=1}^{m}\frac{1}{n-k}(y_i - A)(y_{i+k} - A)\right].$$

The adaptive nonparametric detector is designed according to the method of Sec. 3. It follows from the Appendix that the efficiency of using dependent observations, as compared with independent observations during the same time T, is given by

$$I_2 = \frac{m+1}{1 + 2\sum_{k=1}^{m}\rho(k)} \xrightarrow[m\to\infty]{} \frac{\tau_{ob}}{2\int_0^{\tau_H}\rho(\tau)\,d\tau}.$$

In Table 1 we give the values of I_2 (as $m \to \infty$) for the most typical normalized correlation functions $\rho(\tau)$, assuming that y_i and $y_i + \tau_{ob}$ are independent, if

$$|\rho(\tau_{ob})| \leqslant e^{-5} \approx 0{,}007.$$

Table

$\rho(\tau)$	I_2
$e^{-\alpha\tau}$	2,5
$(1+\alpha\tau)e^{-\alpha\tau}$	1,75
$e^{-\alpha\tau^2}$	1,26
$e^{-\frac{\omega_0\tau}{2Q}}\cos\omega_0\tau$	2,5
$e^{-\alpha\tau^2}\cos\omega_0\tau$	$1{,}26\, e^{\frac{\omega_0}{4\alpha}}$
$\delta(\tau)-\frac{1}{2\beta}e^{-\tau/\beta}$	$\beta\left(\ln\frac{1}{2\beta}+5\right)$; $I_{max}=24$

Detector with reference noise source. In addition to assumptions a-d, assume that we can obtain n_1 samples $x_1, \ldots, x_{n_1}$ of a reference noise source. In this case, when designing the detector, we can use the Mann-Witney statistic [2, 12]

$$S_3=\frac{1}{nn_1}\sum_{i=1}^{n}\sum_{j=1}^{n_1}c(y_i-x_j).$$

Extending the Chernoff-Savage theory to this case, we can show that S_3 is asymptotically normal, with the parameters

$$E_\theta(S_3)=\int_{-\infty}^{\infty}F(x-\theta)\,dF(x);\quad \sigma_0^2(S_3)=\frac{n+n_1}{nn_1}\left[\frac{1}{12}+2\sum_{k=1}^{m}\xi_k-\frac{m}{2}\right],$$

where

$$\xi_k=E_0\left\{c(y_i-x_j)\,c(y_{i+k}-x_l)|_{|j-l|>m}\right\}.$$

As above, to design the adaptive detector we replace $\sigma_0^2(S_3)$ by its consistent, F-independent estimate [9]

$$V_0(S_3)=\frac{1}{n^2n_1^2}\left\{\sum_{i=1}^{n}(R_i-i)^2+\sum_{i=1}^{n_1}(R_i^*-i)^2+2\sum_{k=1}^{m}\left[\frac{n}{n-k}\sum_{i=1}^{n-k}(R_i-i)\times\right.\right.$$
$$\left.\left.\times(R_{i+k}-i-k)+\frac{n_1}{n_1-k}\sum_{i=1}^{n_1-k}(R_i^*-i)(R_{i+k}^*-i-k)\right]-\frac{2m+1}{4}(n+n_1)\right\},$$

where R_i and R_j^* are the ranges of the samples y_i and x_j in the joint variational series of volume $n + n_1$. We show in the Appendix that the efficiency of the adaptive Mann-Witny detector D_3, compared to the detector D_3^* [12], is improved by a factor I_3, where

$$I_3=\frac{m+1}{1+24\sum_{k=1}^{m}\xi_k-\sigma_m};\quad \xi_k=\frac{1}{2}\left[1-\frac{1}{\pi}\operatorname{arc\,tg}\frac{\sqrt{1-\rho^2(k)}}{\rho(k)}\right];\quad \xi_k=\frac{1}{4-\rho^2(k)}$$

for Gasussian and Rayleigh noise respectively.

It is clear from the $I_3(m)$ curves of Fig. 3, plotted with $\rho(k) = e^{-\gamma k}$, that $I_3 \approx 2.5$ and $I_3 \approx 5.8$ for Gaussian noise (curve 1) and Rayleigh noise (curve 2) respectively as $m \to \infty$; the main increase in efficiency is observed with m = 25 coming within the τ_{ob} interval. The main difficulty in realizing the adaptive detector D_3 lies in finding the ranges of the random variables y_i and x_j, and evaluating the sampling correlation function of the ranges.

5. Correlation detector. Let the processes $y^{(1)}(t)$ and $y^{(2)}(t)$ be received at the channel outputs, and suppose that samples $y_1^{(1)}, \ldots, y_n^{(1)}$ and $y_1^{(2)}, \ldots, y_n^{(2)}$ can be obtained. We assume that, when there is no signal, the processes $y^{(1)}(t)$ and $y^{(2)}(t)$ are independent and that assumptions a-d hold for them. The wanted signal is a stationary m_1-dependent random process. The presence of a stochastic

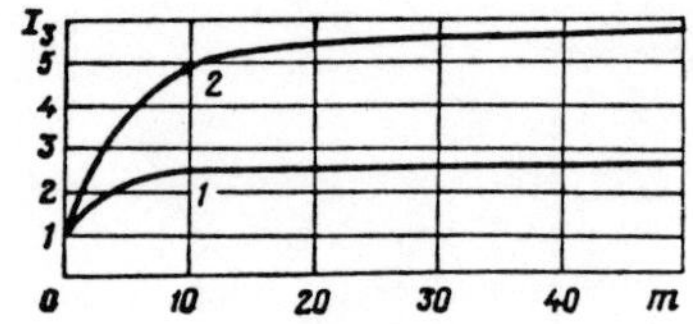

Fig. 3

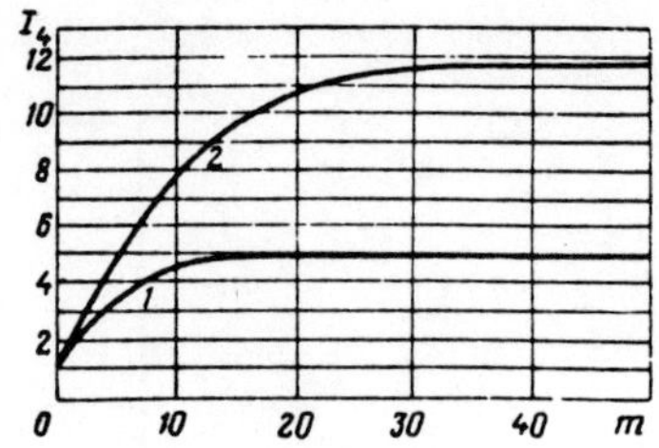

Fig. 4

wanted signal in both channels leads to a dependence appearing between them, so that we base the detector design on the statistic [14]

$$S_4 = \frac{2}{n(n-1)} \sum_{i=1}^{n-1} \sum_{j=i+1}^{n} \operatorname{Sign}(y_i^{(1)} - y_j^{(1)}) \operatorname{Sign}(y_i^{(2)} - y_j^{(2)}),$$

where Sign x is the sign function.

If the wanted signal is absent, S_4 is asymptotically normal [9] with parameters

$$E_0(S_4) = 0; \quad \sigma_0^2(S_4) = \frac{4}{n-m}\left[\frac{1}{9} - 2\sum_{k=1}^{m} \xi_k\right],$$

where

$$\xi_k = E_0\{\operatorname{Sign}(y_i^{(1)} - y_j^{(1)}) \operatorname{Sign}(y_{i+k}^{(1)} - y_l^{(1)}) \operatorname{Sign}(y_i^{(2)} - y_j^{(2)}) \times$$
$$\times \operatorname{Sign}(y_{i+k}^{(2)} - y_l^{(2)}) \mid |i+k-j| > m; \ |i+k-l| > m; \ |j-l| > m\}$$

Using the method of Sec. 3, to design the adaptive nonparametric detector D_4 we replace $\sigma_0^2(S_4)$ by its consistent and F-independent estimate. Using [9], we write this estimate as

$$V_0^2(S_4) = \frac{1}{n}\left\{\sum_{i=1}^{n} W_i^2 + 2\sum_{k=1}^{m} \frac{n}{n-k} \sum_{i=1}^{n-k} W_i W_{i+k}\right\},$$

where $W_i = \frac{1}{n-i-m} \sum_{j=i+m+1}^{n} \operatorname{Sign}(y_i^{(1)} - y_j^{(1)}) \operatorname{Sign}(y_i^{(2)} - y_j^{(2)})$. When the wanted signal is present, the mean value of S_4 tends to its mean for the case of independent observations [5]. The fact that S_4 is asymptotically normal can only be proved (by using the theory of U-statistics) for a particular case, when the mixture of wanted signal and noise is a stationary m_2-dependent random process. In other cases, the question of the distribution of the S_4 statistic still remains open. When the signal and noise are Gaussian random processes with different variances, it can be shown, see Appendix, that the increase I_4 of the D_4 detector efficiency over that of D_4^* (operating with N independent observations) [14] is

$$I_4 = \frac{m+1}{1 + 18\sum_{k=1}^{m} \xi_k},$$

where

$$\xi_k = \left[1 - \frac{2}{\pi} \operatorname{arc\,tg} \frac{\sqrt{1-\rho^2(k)}}{\rho(k)}\right]^2. \tag{6}$$

When the wanted signal and noise are Rayleigh random processes with different variances, we have

$$\xi_k = \left[\frac{\rho^2(k)}{4-\rho^2(k)}\right]^2. \tag{7}$$

It can be seen from (6) and (7) that I_4 is independent of the correlation properties of the wanted signal. The same result was obtained in [5]. It is clear from the $I_4(m)$ curves, plotted in Fig. 4 with $\rho(k) = e^{-\gamma k}$, that $I_4 \approx 5$ as $m \to \infty$ when the signal and noise are Gaussian random processes (curve 1), or $I_4 \approx 12$ when they are Rayleigh processes (curve 2).

Summary. Nonparametric signal detectors have been designed, which avoid the strict requirement for independent samples. In the case when T is restricted, which is of most practical interest, these detectors are more efficient than the nonparametric detectors operating under the assumption of independent samples. For the cases discussed, of Gaussian and Rayleigh noise with exponential correlation function, the efficiency of using dependent as against independent observations with $T < \infty$ fluctuates from 2.5 to 5 for Gaussian, and 5.5 to 12 for Rayleigh noise, in the various detection systems (see Figs. 2-4). The main increase in efficiency is achieved with $m = 15$ for Gaussian and $m = 25$ for Rayleigh noise. A further increase in m does not substantially improve the efficiency. Unfortunately, it is not yet possible to find the limits of variation of $I(m)$ for general types of dependence between the samples and the different F, but some idea of these limits may be gained from Table 1.

To implement the adaptive detector in practice, we have to add to the well-known detector a threshold adaptation unit, which evaluates the estimate of the variance of the statistic used in the design. Discrete correlometers may be used for this purpose. This method of adaptive detector design may also be applied to other well-known parameteric and nonparametric detectors, in the design of which a statistic symmetric to permutation of the initial observations is used.

APPENDIX

1. Let $x(t)$ be a random process, stationary in the narrow sense, satisfying the condition for strong intermixing, i.e., for any events A and B, connected respectively with the past $(-\infty, t)$ and the future $(t + \tau, \infty)$ of the process [8], $\sup\limits_{A,B} |P(AB)-P(A)P(B)|_{\tau\to\infty} \to 0$.

A stationary Gaussian process satisfies the condition for strong intermixing provided that its correlation function tends to zero as its argument increases. Replacement of this requirement on the correlation function by the condition for strong intermixing is equivalent in the general case to the assumption that independent observations can be obtained from a spectrally restricted ergodic process, provided that the interval τ_{ob} between them is sufficiently long. If $x_1, \ldots, x_n$ are random variables, obtained at the instants $t_1, \ldots, t_n$, with discretization period Δt, during a time interval T from the process $x(t)$, we shall assume henceforth that, during time T, we can obtain $N = T/\tau_{ob}$ independent observations. The sample volume here is $n = (m + I)N$, where $m = \tau_{ob}/\Delta t$ is the number of dependent observations falling within the interval τ_{ob}, i.e., we shall assume that the random quantities $(x_1, \ldots, x_i)$ are statistically independent of $(x_j, x_{j+1}, \ldots)$ when $|i-j|>|m$. Since the process is stationary in the narrow sense, the x_i have identical marginal distributions and the joint distribution of $x_i, \ldots, x_{i+j}$ is independent of i. A random process satisfying these conditions will be called a stationary m-dependent process [1, 8].

2. Since all the statistics S considered in Sec. 3 are asymptotically normal when a signal is present, the probabilities β_1 and β_2 of correct detection in the cases of independent and dependent observations respectively are

$$\beta_1 = 1 - \Phi\left\{\Phi^{-1}(1-\alpha) - \frac{\sqrt{N}\,[E_\theta(S) - E_0(S)]}{\gamma_0(S)}\right\},$$

$$\beta_2 - 1 - \Phi\left\{\Phi^{-1}(1-\alpha) - \frac{\sqrt{n}\,[E_\theta(S) - E_0(S)]}{\gamma_0(S)\,\psi(m)}\right\},$$

where $\gamma_0^2(S)$ is the variance of S in the case of independent observations and the function $\psi(m) \to 1$ as $m \to 0$ $(\Delta t \to \tau_{ob})$. Comparing β_1 and β_2, we find that the efficiency of the adaptive detector relative to the detector for independent samples is increased by a factor I, where

$$I = \frac{n}{\psi^2(m)\,N} = \frac{m+1}{\psi^2(m)}. \tag{8}$$

3. a) $$\sigma_0^2(S_1) = \frac{1}{n^2}\sum_{i,j}^{n} E_0(z_i z_j) - \frac{1}{4} = \frac{1}{n^2}\left\{\sum_{i=1}^{n} E_0(z_i^2) + \right.$$
$$\left. + 2\sum_{i=1}^{n-m}\sum_{j=i+1}^{i+m} E_0(z_i z_j) + \sum_{i=1}^{n-m-1}\sum_{j=i+m+1}^{n} E_0(z_i z_j) + \sum_{i=m+1}^{n}\sum_{j=1}^{i-m-1} E_0(z_i z_j)\right\} -$$
$$-\frac{1}{4} \xrightarrow[n\to\infty]{} \frac{1}{n}\left[\xi_0 + 2\sum_{k=1}^{m}\xi_k - \frac{2m+1}{4}\right]. \tag{9}$$

where $z_i = c(y_i - M)$; $\xi_k = E_0(z_i z_{i+k})$.

b) $$\sigma_0^2(S_2) = \frac{1}{n^2} E_0\left\{\sum_{i=1}^{n} z_i + 2\sum_{i=1}^{n-1}\sum_{j=i+1}^{n} z_i z_j\right\} = \frac{\sigma_1^2}{n}\left[1 \div \frac{2}{n}\sum_{i=1}^{n-1}\sum_{j=i+1}^{n}\rho(|i-j|)\right] = \frac{\sigma_1^2}{n}\left[1 + \frac{2}{n}\sum_{i=1}^{n-1}\sum_{k=1}^{n-i}\rho(k)\right] =$$
$$= \frac{\sigma_1^2}{n}\left[1 + \frac{2}{n}\sum_{k=1}^{m}(n-k)\,\rho(k)\right] \xrightarrow[n\to\infty]{} \frac{\sigma_1^2}{n}\left[1 + 2\sum_{k=1}^{m}\rho(k)\right],$$

where $z_i = y_i - A$; $\sigma_1^2 = E_0(z_i^2)$; $\rho(k) = E_0(z_i z_{i+k})$.

REFERENCES

1. D.A.S. Fraser, Nonparametric Methods in Statistics, Wiley, 1957.
2. J.W. Carlyle and J.B. Thomas, IEEE Trans., 1964, IT-10, No. 2.
3. B.R. Levin and A.F. Kushnir, Problemy Peredachi Informatsii, 1968, 4, No. 3.
4. S.S. Wolf, J.L. Gastwirth and H. Rubin, IEEE Trans., 1967, IT-13, No. 2.
5. I.D. Davisson et al., IEEE Trans., 1970, IT-16, No. 1.
6. M. Kanefsky and J.B. Thomas, IEEE Trans., 1965, IT-11.
7. M. Kanefsky, NEC Proc., 1966.
8. I.A. Ibragimov and Yu.B. Linnik, Nezavisimye i statsionarno.svyazannye velichiny (Independent and Stationarily Connected Variables), Fizmatgiz, 1965.
9. P.K. Sen, Calcutta Statist. Assoc. Bulletin, 1963, 12, No. 47.
10. D.G. Laniotis and J.K. Hancock, Zarubezhnaya Radioelektronika, 1965, No. 6.
11. A.F. Kushnir and A.I. Pinskiy, Teoriya Veroyatnostei i Yeye Primeneniya, 1971, 16, No. 2.
12. J. Capon, IRE Weson. Conv. Rev., 1959, Part 4, pp. 92-103.
13. H. Chernoff and I.R. Savage, Ann. Math. Statist., 1958, 29, No. 4.
14. R.A. Groeneveld, IEEE Trans., 1967, IT-13, No. 2.

Nonparametric Detection with Autoregressive Data

METE B. SIRVANCI AND STEPHEN S. WOLFF

I. Introduction

IT HAS BEEN shown [1], [2], [3] that rank tests, which are distribution-free with independent observations, lose this property when the data are dependent. In communications applications the assumption of independence cannot be easily justified if data are obtained by homogeneous sampling of a continuous parameter stochastic process.

If a detection statistic is asymptotically normal with zero mean, then its asymptotic false alarm rate can be written as $P_{FA} \cong 1 - \varphi(c/\sigma_0)$, where $\varphi(\cdot)$ is the standard normal cumulative distribution function (CDF), c is the detector threshold, and σ_0^2 is the asymptotic variance of the statistic under the null hypothesis. With independent data σ_0 does not depend on the distribution of the observations, but with dependence σ_0 is affected both by the distribution of data and by the underlying dependence scheme. Moreover, the statistic may no longer be asymptotically normal. Thus, dependence can cause considerable deviation from the preset false alarm probability.

In what follows we study the effects of autoregressive dependence on nonparametric detection. Following a brief presentation of autoregressive processes (Section II), we state in Section III a theorem for the asymptotic normality of a class of two-sample rank test statistics. In Section V we propose the method of "studentization" as a means of restoring (at least asymptotically) the distribution-free character of the rank tests.

For the case of weak signal detection with large sample sizes, an appropriate performance criterion is the asymptotic relative efficiency (ARE). In Section IV we illustrate the adverse effect of second-order autoregressive dependence on the ARE of the Mann–Whitney detector.

II. Preliminaries and the Data Model

In this paper we assume that the sampled raw data sequence satisfies an autoregressive model. The mth-(m finite)order stationary autoregressive process (SAP) $\{X_n\}$ is defined by the recurrence relation

$$X_n = \sum_{i=1}^{m} \lambda_i X_{n-i} + W_n, \qquad n = 0,\pm1,\pm2,\cdots \tag{1}$$

where $\{W_n\}$ is an independent identically distributed (i.i.d.) sequence with $\sigma_w^2 = \text{var}\,(W_n) < \infty$ and where the roots r_i of the polynomial $r^m - \lambda_1 r^{m-1} - \cdots - \lambda_m$ all lie in the unit circle, i.e., $|r_i| < 1$, for $i = 1,\cdots,m$. The moving average representation, which is obtained by successive substitutions in (1), namely,

$$X_n = \sum_{k=0}^{\infty} b_k W_{n-k}, \qquad \text{where } b_k = \sum_{i=1}^{m} a_i r_i^k \tag{2}$$

and where the a_i are constants, of autoregressive processes is more useful for our purposes.

Central limit theorems for dependent sequences require some kind of a mixing condition. One such condition is the ϕ-mixing condition. Let M_r^s be the σ-field generated by the random variables (r.v.'s) $X_r,\cdots,X_s, r \le s$. A sequence of r.v.'s $\{X_n\}$ is said to be ϕ-mixing if, for each k ($k = 0,\pm1,\pm2,\cdots$) and for each n ($n \ge 1$),

$$\sup_{A,B} |P(B|A) - P(B)| \le \phi(n),$$

where

$$\lim_{n\to\infty} \phi(n) = 0, \qquad A \in M_{-\infty}^k, \quad B \in M_{k+n}^{\infty}.$$

It has been shown that SAP's are not ϕ-mixing and that they are strong-mixing only for some special cases [4]. However, from (2) one observes that each member of an autoregressive sequence is a function of the ϕ-mixing $\{W_n\}$ sequence ($\{W_n\}$ is trivially ϕ-mixing, since it is an i.i.d. sequence).

Billingsley [5] has given the following formulation for stochastic processes that can be expressed as functions of ϕ-mixing sequences. Let $\{\xi_n\}$ be stationary and ϕ-mixing with mixing coefficient ϕ_n. Let f and f_l be measurable functions. Define

$$\eta_n = f(\xi_n,\xi_{n-1},\cdots), \qquad n = 1,2,\cdots,$$

$$\eta_{ln} = f_l(\xi_n,\xi_{n-1},\cdots,\xi_{n-l+1}), \qquad n = 1,2,\cdots,$$

$$v(l) = E(|\eta_0 - \eta_{l0}|^2), \qquad l = 1,2,\cdots,$$

$$S_n = \eta_1 + \eta_2 + \cdots + \eta_n, \qquad n = 1,2,\cdots.$$

Manuscript received August 26, 1975; revised March 24, 1976. This paper was presented at the 5th and 7th Princeton Conference on Information Science and Systems, Princeton, NJ, 1971 and 1973.

M. B. Sirvanci is with the Department of Operations Research and Statistics, Middle East Technical University, Ankara, Turkey.
S. S. Wolff is with the Ballistic Research Laboratories, Aberdeen Proving Grounds, Aberdeen, MD.

Then we state the following central limit theorem from [5].

Theorem 1: Suppose that $\{\xi_n\}$ is ϕ-mixing with $\Sigma_{n=1}^{\infty}\phi_n^{1/2} < \infty$ and that the sequence $\{\eta_n\}$ has mean zero and finite variance. Suppose further that $\Sigma_{l=1}^{\infty} v^{1/2}(l) < \infty$. Then the series

$$\sigma^2 = E(\eta_0^2) + 2\sum_{k=1}^{\infty} E(\eta_0\eta_k)$$

converges absolutely; if $\sigma^2 > 0$, then $(1/\sqrt{n})S_n$ converges in distribution to $N(0,\sigma^2)$.

For SAP's, we let

$$X_{ln} = \sum_{k=0}^{l-1} b_k W_{n-k} \qquad v_x(l) = E(|X_0 - X_{l0}|^2). \tag{3}$$

Lemma 1: For SAP's $\{X_n\}$, $\Sigma_{l=1}^{\infty} v_x^{1/2}(l) < \infty$.

Proof: In (2), let $a_p = \max_i |a_i|$ and $r_p = \max_i |r_i|$. Then $|b_k| \le m a_p r_p^k$. As a result

$$v_x(l) = \sigma_w^2 \sum_{k=l}^{\infty} b_k^2 \le K r_p^2$$

where K is a generic constant. Since $0 < r_p < 1$, the lemma follows.

Lemma 1 shows that the autoregressive sequence $\{X_n\}$ can be approximated by the sequence $\{X_{ln}\}$ in the sense above. Moreover, $\{W_n\}$ is a ϕ-mixing sequence with $\phi(n) = 0$, $n \ge 1$. Thus, the condition $\Sigma_1^{\infty}\phi_n^{1/2} < \infty$ is trivially satisfied and $(1/\sqrt{n})\ \Sigma_1^n X_i$ is asymptotically normal by Theorem 1.

III. Asymptotic Normality of Two-Sample Rank Tests Operating on Autoregressive Data

Let $\{X_i\}_{i=1}^m$ and $\{Y_i\}_{i=1}^n$ be observations from two random processes with CDF's $F(x)$ and $G(x)$, respectively. Assume that the two sequences are independent of each other. Let $N = n + m$, and let $Z_i = 1$ if the ith smallest observation in the combined sample is a member of $\{X_i\}_{i=1}^m$, otherwise let $Z_i = 0$. Then consider the statistic of the form

$$T_N = \frac{1}{m}\sum_{i=1}^{N} J_N[i/(N+1)]Z_i, \tag{4}$$

where $J_N[i/(N+1)]$ is the score corresponding to the ith smallest observation.

Let $F_m(x)$ and $G_n(x)$ be the empiric CDF's of the $\{X_i\}_{i=1}^m$ and $\{Y_i\}_{i=1}^n$ sequences, respectively. Define the combined CDF $H(x)$ and the combined empiric CDF $H_N(x)$ by $H(x) = \lambda_N F(x) + (1-\lambda_N)G(x)$ and $H_N(x) = \lambda_N F_m(x) + (1-\lambda_N)G_n(x)$, respectively where $\lambda_N = m/N$. Then (4) can be equivalently expressed as

$$T_N = \int_{-\infty}^{\infty} J_N[H_N(x)]\, dF_m(x). \tag{5}$$

We state now one of our main results, the asymptotic normality of the statistic T_N when the observation sequences $\{X_i\}$ and $\{Y_i\}$ are both autoregressively dependent.

Definition: Let $\mathcal{D}$ be the class of SAP's $\{X_n\}$ such that $E(X_n^2) < \infty$. Further assume that the distribution of X_{ln} defined in (3) is either absolutely continuous or is a mixture of absolutely continuous and discrete distributions where the discrete part has a finite number of atoms.

Theorem 2: If the sequences $\{X_i\}_{i=1}^m$ and $\{Y_i\}_{i=1}^n$ belong to the class $\mathcal{D}$ of autoregressive sequences and if

a) $J(H) = \lim_{N\to\infty} J_N(H)$ exists for $0 < H < 1$ and is not a constant;
b) $p\text{-lim}\ N^{1/2}\int_{0<H_N<1}[J_N(H_N) - J(H_N)]\, dF_m(x) = 0$;
c) $|J^{(i)}(H)| \le K[H(1-H)]^{-i/4-1/2+\delta}$, for $i = 0,1,2$ and for some $\delta > 0$, where $J^{(i)}(H)$ is the ith derivative of $J(H)$;
d) $J_N(1) = o(N^{1/2})$;
e) $|J[H(x)] - J[H(y)]| \le K|x - y|$, for each x and y;
f) $J(\cdot)$ is a nondecreasing function;

then, for fixed F, G, and λ_N,

$$\frac{T_N - \mu_N}{\sigma_N} \xrightarrow{D} N(0,1)$$

where

$$\mu_N = \int_{-\infty}^{\infty} J[H(x)]\, dF(x)$$

and

$$N\sigma_N^2 = 2(1-\lambda_N)\Bigg[\iint_{-\infty<x<y<\infty} \sigma_G^2(x,y) \cdot J'[H(x)]J'[H(y)]\, dF(x)dF(y) + \frac{1-\lambda_N}{\lambda_N}\iint_{-\infty<x<y<\infty} \sigma_F^2(x,y)J'[H(x)]J'[H(y)] \cdot dG(x)dG(y)\Bigg]$$

with

$$\sigma_F^2(x,y) = F(x)[1-F(y)] + 2\sum_{k=1}^{m-1}(1-k/m)[F_{X_i,X_{i+k}}(x,y) - F(x)F(y)]$$

and a similar expression for $\sigma_G^2(x,y)$, providing $\sigma_N^2 \ne 0$.

Proof: Chernoff and Savage [6] proved a similar theorem for independent observations under less strict assumptions for the score function $J(\cdot)$. They expressed T_N as

$$T_N = A + B_{1N} + B_{2N} + \sum_{i=1}^{6} C_{iN}$$

in [6], where it is shown that A is a constant, B_{1N} and B_{2N} are asymptotically zero mean normal, and the C_{iN} are asymptotically negligible. We follow their approach dealing only with problems arising from autoregressive dependence.

In [6] the proofs of the negligibility of the terms C_{4N}, C_{5N}, and C_{6N} do not depend on the assumption of the in-

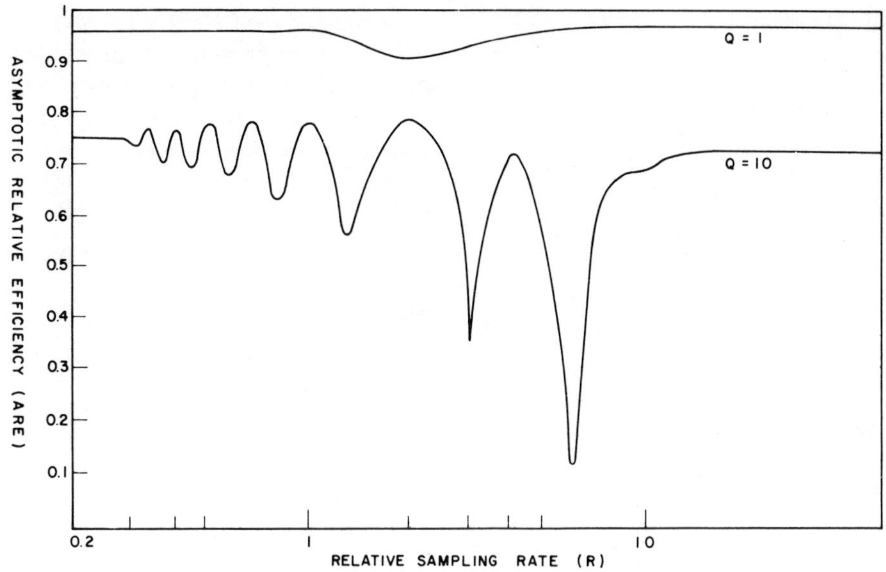

Fig. 1. ARE of Mann–Whitney detector for Q = 1 and Q = 10.

dependence of the observations. Our proofs for the negligibility of C_{1N}, C_{2N}, and C_{3N} in the case of autoregressive dependence are long for inclusion. In Appendix I, we sketch our approach briefly. In Appendix II we show that $B_{1N} + B_{2N}$ is asymptotically normal for autoregressive data.

Remark: From the expression for the variance σ_N^2, it is clear that the asymptotic variance depends on the distribution of the observations. Hence, the statistic T_N is no longer distribution-free. In Section V, we introduce the method of "studentization" to restore the distribution-free character of T_N for autoregressive data.

IV. The Effect of Dependence on ARE

As mentioned in Section I, ARE is a performance criterion for large sample sizes. The ARE of a certain test with respect to another can be expressed as a ratio of their efficacies if both test statistics are asymptotically normal in addition to satisfying certain other regularity conditions [7], [8]. To illustrate the adverse effect of dependence we present some numerical results for the ARE of the Mann–Whitney detector with respect to the sample-mean detector when the observed data are second-order autoregressive.

Suppose that the observation sequences $\{X_i\}$ and $\{Y_i\}$ are both obtained by a periodic sampling every τ s of the response of a single-stage *RLC* network driven by white noise. Let w_0 be the center frequency and Q be the bandwidth parameter of the network. Define $\beta = w_0/Q$, $w_1 = w_0(1 - 1/4Q^2)^{1/2}$, and the relative sampling rate $R = 2\pi/\tau w_0$. Then the observation sequences $\{X_i\}$ and $\{Y_i\}$ both satisfy a second-order autoregressive scheme with correlation function

$$\rho(k\tau) = \exp(-|k|\tau\beta/2)[\cos(k\tau w_1) + [1 - \exp(-\beta\tau)] \sin(|k|\tau w_1)/[1 + \exp(-\beta\tau)] \tan(w_1\tau)]. \quad (6)$$

If one assumes that the input to the network is Gaussian, then the observed output sequences $\{X_i\}$ and $\{Y_i\}$ are also Gaussian.

Consider the Mann–Whitney detector

$$T_{\text{MWN}} = \sum_{i=1}^{N} \sum_{j=1}^{N} \operatorname{sgn}(X_i - Y_j)/N^2.$$

With the above input data (second-order Gaussian autoregressive) the assumptions of Theorem 2 are satisfied for this detector. Hence we can express the ARE of the Mann–Whitney detector with respect to the sample-mean detector as a ratio of their efficacies as

$$\text{ARE} = \sum_{k=-\infty}^{\infty} \rho(k\tau)/2 \sum_{k=-\infty}^{\infty} \sin^{-1}[\rho(k\tau)/2],$$

where $\rho(k\tau)$ is the correlation function of the data sequence [2].

In Fig. 1, curves for ARE versus R (relative sampling rate) are presented for two types of networks with $Q = 1$ and $Q = 10$, respectively. We observe that a) the deterioration in performance is less for low Q networks than for high Q ones and b) performance is relatively better either for slow sampling or for fast sampling.

V. "Studentization" of Rank Tests for Autoregressive Data

In this section we first develop a consistent estimator S_N^2 of the asymptotic variance of the two-sample rank statistics T_N in (4). Then by forming the "student's ratio" $U_N = T_N/S_N$, we conclude that the "studentized" rank test statistic U_N is asymptotically normal with unit variance by an application of Slutsky's theorem [9].

From Theorem 2, it follows that, as $N \to \infty$,

$$\text{var}(T_N) \cong \text{var}(B_{1N}) + \text{var}(B_{2N}),$$

where, under the null hypothesis,

$$B_{1N} + B_{2N} = (1 - \lambda_N)\left[\sum_{i=1}^{m} [J[F(X_i)] - \bar{J}]/m + \sum_{i=1}^{n} [J[F(Y_i)] - \bar{J}]/n\right] \quad (7)$$

with $\bar{J} = \int_0^1 J(u)\,du$ and F is the common CDF of both of the samples $\{X_i\}_{i=1}^m$ and $\{Y_i\}_{i=1}^n$

Since B_{1N} and B_{2N} are independent and are similar in nature, we consider only B_{1N}. In location shift problems one usually has, $\bar{J} = 0$. Then we can write

$$\text{var}\,(B_{1N}) = [(1 - \lambda_N)^2/m] \sum_{k=-m}^{m} (1 - |k|/m)E(J[F(X_i)]J[F(X_{i+k})]). \quad (8)$$

As $m,N \to \infty$, we have

$$[m/(1 - \lambda_N)^2]\,\text{var}\,(B_{1N}) \to S_1 = \sum_{k=-\infty}^{\infty} E(J[F(X_i)]J[F(X_{i+k})]). \quad (9)$$

Now let

$$R(k) = E(J[F(X_i)]J[F(X_{i+k})]),$$

$$\hat{R}(k) = \sum_{i=1}^{m-|k|} J[F(X_i)]J[F(X_{i+k})]/(m - |k|), \quad (10)$$

$$\hat{R}_m(k) = \sum_{i=1}^{m-|k|} J[F_m(X_i)]J[F_m(X_{i+k})]/(m - |k|),$$

where $k = 0,\pm 1, \cdots, \pm(m - 1)$ and $F_m(\cdot)$ is the empiric CDF of the $\{X_i\}_{i=1}^m$ sequence.

We want to estimate $S_1 = \sum_{k=-\infty}^{\infty} R(k)$ by estimating each $R(k)$. Although $\hat{R}(k)$ is the natural estimator of $R(k)$, it contains the CDF F (presumably unknown). Instead we use $\hat{R}_m(k)$ to estimate $R(k)$ and $S_{1m} = \sum_{|k| \le C_m} \hat{R}_m(k)$ to estimate S_1.

Theorem 3: Assume that $|J'(u)| \le K < \infty$ a.e. and that $\{X_i\}$ belongs to the class $\mathcal{D}$ of autoregressive processes. Assume further that $E(X^4) < \infty$. Then

$$p - \lim S_{1m} = S_1, \qquad \text{for } C_m = Cm^{1/2-\delta}$$

where C is any positive constant and $0 < \delta < \frac{1}{2}$.

Proof: The proof of this theorem is lengthy. We indicate only the main steps in a sequence of lemmas.

Lemma 2: $E[F_m(x) - F(x)]^2 = 0(1/m)$ uniformly in x.

For i.i.d. sequences, $E(F_m - F)^2 = F(1 - F)/m$. For SAP's belonging to class $\mathcal{D}$, employing the approximation of X_n by X_{ln} of Section II one shows, in the proof Lemma 2, that $E(F_m - F)^2 \le K[F(1 - F)]^{1/2}/m$, K a positive constant.

Lemma 3: $E(J[F_m(X_i)]J[F_m(X_{i+k})] - J[F(X_i)] \cdot J[F(X_{i+k})])^2 = 0(1/m)$.

In the proof of this lemma, the result of Lemma 2 and the assumption that $|J'(u)| \le K$ a.e. are used.

Lemma 4: var $[\hat{R}(k)] = 0(1/m)$.

Lemma 4 is proven by approximating each $J[F(X_n)]$ by $J[F(X_{ln})]$, where X_{ln} is defined in (3).

Given the above lemmas, the proof of Theorem 3 is completed in Appendix II.

In a similar manner, we can construct a consistent estimator $S_{2n} = \sum_{|k| < C_n} \hat{R}_n(k)$ of

$$S_2 = \sum_{k=-\infty}^{\infty} E(J[F(Y_i)]J[F(Y_{i+k})]),$$

where

$$\hat{R}_n(k) = \sum_{i=1}^{n-|k|} J[F_n(Y_i)]J[F_n(Y_{i+k})]/(n - |k|).$$

Theorem 4: Under the null hypothesis that $F = G$,

$$U_N = [m/(S_{1m} + S_{2n})]^{1/2} T_N \xrightarrow{D} N(0,1).$$

Proof: Theorem 2 states the asymptotic normality of T_N, and Theorem 3 implies the consistency of $S_{1m} + S_{2n}$. Theorem 4 then follows by an application of Slutsky's theorem [9].

Theorem 4 indicates that certain two-sample rank detectors operating on autoregressive data are asymptotically distribution-free when "studentized." Since Theorem 3 does not completely specify the choice of C_m, for finite sample sizes we are faced with the problem of choosing an appropriate truncation constant C_m. We discuss this problem in the next section.

VI. Simulation Results for "Studentization"

In practice, with dependent data, the sample size is a critical issue for "studentization." If the empiric autocorrelation functions, $\hat{R}_m(k)$ and $\hat{R}_n(k)$, have the general character sketched in Fig. 2, the sample size is adequate. It is clear from Fig. 2 that the contributions to the sum $\sum_{k=-\infty}^{\infty} R(k)$ of terms with $|k| > k_1$ are small enough to be neglected for purposes of "studentization." The presence of negligible $\hat{R}(k)$ in the range $k_1 < |k| < k_2$ indicates that the sample size is adequate. The fluctuations in $\hat{R}(k)$, for $|k| > k_2$, are statistical and due to the usual effective decrease in sample size when estimating correlation for large lags. For the above case, one takes $C_m = k_1$.

When the autocorrelations $\hat{R}_m(k)$ and $\hat{R}_n(k)$ are considerably more irregular, so that a more-or-less well-defined k_1 beyond which the correlations are negligible is not obvious, we conclude that the sample size is not adequate and that effective "studentization" is unobtainable.

We now present some computer simulation results to illustrate the rate of convergence of the "studentized" versions of rank tests to $N(0,1)$. The two-sample Wilcoxon test was chosen for illustration because the score function of this test satisfies all the assumed conditions. The sizes of both samples were set to 100, i.e., $n = m = 100$ and $N = 200$. The data were generated with autoregressive dependence so that the observations in both samples satisfy either a) a first-order SAP, $X_n = rX_{n-1} - W_n$, where r is the correlation parameter and $|r| < 1$ or b) a second-order SAP with correlation function as given in (6). For first-order SAP's, we varied r, and for second-order SAP's, we varied

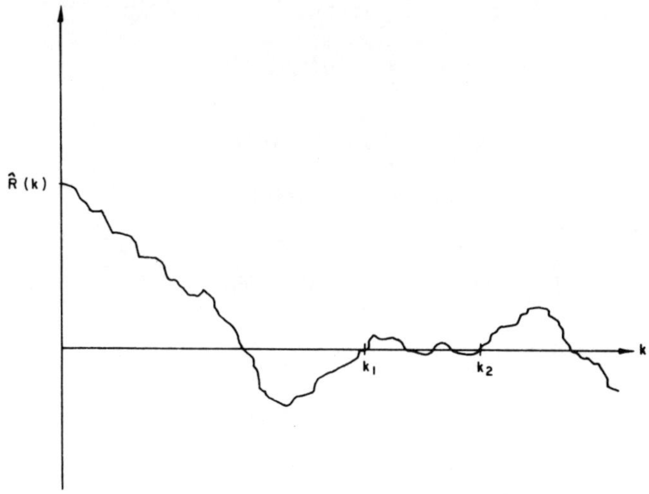

Fig. 2. Empiric autocorrelation function.

R and Q to change the correlation between the observations.

To test that the "studentized" version of the two-sample Wilcoxon test statistic U_N ($N = 200$) is distributed $N(0,1)$, we employed a χ^2-test statistic Q_{19} with 19 degrees of freedom (20 cells). We generated 200 independent U_N for forming each Q_{19}. The χ^2-test for the case previously mentioned is to reject the null hypothesis (U_N is distributed $N(0,1)$) if $Q_{19} > C$, where C is a constant.

The rate of convergence of the "studentized" statistic U_N depends on such factors as the CDF $F(\cdot)$ of the data, the correlation between the observations, and the choice of the truncation points C_m and C_n in the estimate of the variance of T_N. For $N = m + n = 200$ the most successful rule for choosing C_m was

$$C_m = \min_k \{k: \ \hat{R}_m(k) < 0\} - 1, \tag{11}$$

and analogously for C_n. Using the rule in (11), we computed the χ^2-test statistic Q_{19} for various choices of the sample CDF $F(\cdot)$ and for several values of the correlation parameters. We list the results in two tables: Table I for first-order SAP's and Table II for second-order SAP's. At an approximate significance level $\alpha = 0.05$, we rejected the null hypothesis for $Q_{19} > 30.1$.

Our results indicate that, unless the dependence is very weak, the sample size of $m = n = 100$ is not adequate for estimating the correlation variables $R(k)$. We observe from Tables I and II that Q_{19} increases with stronger dependence and as the data CDF deviates more from normality.

In Table III we present additional numerical results indicating that, even in small samples the "studentization" technique is effective in reducing the false alarm rate. Our results in this case are for the Wilcoxon test operating on second-order Gaussian autoregressions with sample sizes $n = m = 32$. Thresholds were set for the "nonstudentized" Wilcoxon test T and the "studentized" version U (assuming that the data are normally distributed) corresponding to a preset false alarm rate $\alpha = 0.05$. Then the

TABLE I

C.D.F. of the $\{W_n\}$ sequence	r Correlation Parameter	Q_{19} χ^2-statistic
N(0,1)	.5	19.6
N(0,1)	.7	24.8
N(0,1)	.75	26.2
N(0,1)	.8	32.2
N(0,1)	.9	61.4
Double Exponential Mean Zero Variance 1	.7	27.6
Cauchy Zero Median Scale Parameter 1	.7	37.2

TABLE II

C.D.F. of the $\{W_n\}$ Sequence	Correlation Parameters	Q_{19} χ^2-statistic
Double Exponential Mean Zero Variance 1	R=1 Q=1	15.4
Double Exponential Mean Zero Variance 1	R=10 Q=1	28.6
N(0,1)	R=10 Q=1	27.0
N(0,1)	R=1 Q=10	27.0

TABLE III

Correlation Parameters	$P_a(T)$	$P_a(U)$
R=1 Q=1	0.03	0.02
R=1 Q=5	0.23	0.08
R=1 Q=10	0.31	0.16

sample variances of T and U were obtained in 100 trials and were used in estimating an "achieved" false alarm probability P_a.

VII. Conclusion

These results indicate that, for large samples, "studentization" can be used to make the false alarm rate of a class of rank tests independent of the autoregressive dependence in the data. The sample size required for accurate "studentization" depends on the distribution of the data and the correlation between observations. For highly correlated data the sample size required may be so large that ranking procedures will be unattractive because of the computational burden. For small samples, if some unspecified amount of autoregressive dependence is suspected, the results indicate that "studentization" may be used as a routine procedure to guard against excessive false alarms.

In this paper, "studentization" has been proposed as a means of restoring the distribution-free character of rank tests operating on dependent data. Our results illustrate that "studentization" is effective in combating at least one adverse effect of dependence, i.e., the deviation from the preset false alarm probability. The adverse effect of dependence on performance level was shown in Section IV. The effect of "studentization" on the performances (specifically ARE's) of rank tests was not considered and is a topic which needs further investigation.

Appendix I
Negligibility of the C_{iN} Terms

In [6], those terms in which the proofs of asymptotic negligibility depend on the independence assumption are

$$C_{1N} = \lambda_N \int_{-\infty}^{\infty} (F_m - F)J'(H)\, d[F_m(x) - F(x)]$$

$$C_{2N} = (1 - \lambda_N) \int_{-\infty}^{\infty} (G_n - G)J'(H)\, d[F_m(x) - F(x)]$$

$$C_{3N} = \frac{1}{2} \int_{0<H_N(x)<1} (H_N - H)^2 J''[\gamma H_N + (1-\gamma)H]\, dF_m(x)$$

where $0 < \gamma < 1$. These terms can be expressed in terms of $[F_m(x) - F(x)]^r$ or $[G_n(x) - G(x)]^r$, where $r \geq 2$, or they can be analyzed as in [6]. In the former case, we make use of the result in Lemma 2, $E[F_m(x) - F(x)]^2 = 0(1/m)$, and show that $E[|C_{iN}|] \leq K/(N^{1/2} + \delta)$, where K is a constant, $i = 1,2,3$, and $\delta > 0$.

Appendix II
Proof of the Asymptotic Normality of the Sum $B_{1N} + B_{2N}$

From [6], we have

$$B_{1N} + B_{2N} = (1 - \lambda_N) \left[\sum_1^m (B(X_i) - E[B(X_i)])/m + \sum_1^n (B^*(Y_i) - E[B^*(Y_i)])/n \right]$$

where

$$B(x) = \int_{x_0}^{x} J'[H(y)]\, dG(y)$$

$$B^*(x) = \int_{x_0}^{x} J'[H(y)]\, dF(y)$$

with x_0 determined somewhat arbitrarily.

Consider

$$B_{1N} = (1 - \lambda_N) \sum_{i=1}^{m} (B(X_i) - E[B(X_i)])/m.$$

Let

$$Z(X_i) = B(X_i) - E[B(X_i)]$$

$$Z(X_{li}) = B(X_{li}) - E[B(X_{li})]$$

$$Z(\delta_{li}) = B(\delta_{li}) - E[B(\delta_{li})]$$

where

$$B(X_{li}) = \int_{x_0}^{X_{li}} J'[H(y)]\, dG(y)$$

$$B(\delta_{li}) = \int_{X_{li}}^{X_i} J'[H(y)\, dG(y)$$

Let

$$v_z(l) = E(|Z(X_i) - Z(X_{li})|^2).$$

Then

$$v_z(l) = E(|B(\delta_{li}) - E[B(\delta_{li})]|^2) \leq 4E[|B(\delta_{li})|^2].$$

From assumptions e) and f) of Theorem 2, we have

$$B(\delta_{li}) \leq K \left| \int_{X_{li}}^{X_i} J'(H)\, dH \right| = K|J[H(X_i)] - J[H(X_{li})]| \leq K|X_i - X_{li}| = K \left| \sum_{k=l}^{\infty} b_k W_{n-k} \right|.$$

Hence, by Lemma 1 we have

$$v_z^{1/2}(l) \leq K\, v_x^{1/2}(l)$$

where K is a generic constant.

We observe now that all the conditions of Theorem 1 are satisfied for the sequence of r.v.'s $\{Z(X_i)\}$ (i.e., the $Z(X_i)$ are functions of the ϕ-mixing process $\{W_n\}$ such that $E[Z(X_i)] = 0$, $E[Z^2(X_i)] < \infty$, and $\Sigma_{l=1}^{\infty} v_z^{1/2}(l) < \infty$). Thus, Theorem 1 is applicable to the sum $\Sigma_{i=1}^{m} Z(X_i)/m^{1/2}$. Therefore, we have

$$m^{1/2} B_{1N}/(1 - \lambda_N) \xrightarrow{D} N(0,\sigma^2)$$

with

$$\sigma^2 = E[Z(X)] + 2 \sum_{k=1}^{\infty} E[Z(X_0)Z(X_k)].$$

Consideration of the term B_{2N} is analogous. Since B_{1N} and B_{2N} are mutually independent, the asymptotic normality of $B_{1N} + B_{2N}$ follows immediately.

Appendix III
Proof of Theorem 3

Consider

$$|S_{1m} - S_1| \leq \left| \sum_{|k| \leq C_m} \hat{R}_m(k) - \sum_{|k| \leq C_m} \hat{R}(k) \right| + \left| \sum_{|k| \leq C_m}^{\infty} \hat{R}(k) - \sum_{-\infty}^{\infty} R(k) \right|. \quad (12)$$

Each term on the right in (12) will be shown to converge to zero in probability.

Consider the first term. By the Markov inequality

$$P\left(\left|\sum_{|k|\le C_m}[\hat{R}_m(k)-\hat{R}(k)]\right|\ge\epsilon\right)\le E\left[\left|\sum_{|k|\le C_m}[\hat{R}_m(k)-\hat{R}(k)]\right|\right]\Big/\epsilon\le\sum_{|k|\le C_m}E(|\hat{R}_m(k)-\hat{R}(k)|)/\epsilon. \quad (13)$$

Substituting for $\hat{R}(k)$ and $\hat{R}_m(k)$ from (10) and employing Lemma 3, we have

$$E(|\hat{R}_m(k)-R(k)|)\le K/m^{1/2}.$$

Thus, it follows that the last sum (13) is overbounded by $Km^{-1/2}(2C_m+1)$. Hence, the convergence of the first term on the right in (12) follows.

Next, consider the second term on the right in (12),

$$\left|\sum_{|k|\le C_m}\hat{R}(k)-\sum_{-\infty}^{\infty}R(k)\right|\le\left|\sum_{|k|\le C_m}[\hat{R}(k)-R(k)]\right|+\left|\sum_{|k|>C_m}R(k)\right|. \quad (14)$$

No probabilistic argument is necessary for the second term on the right in (14) because it is not random and its contribution vanishes as $m\to\infty$.

On the other hand, employing first the Markov inequality and then Schwartz's inequality, we have

$$P\left[\left|\sum_{|k|\le C_m}[\hat{R}(k)-R(k)]\right|\ge\epsilon\right]\le E\left(\sum_{|k|\le C_m}|\hat{R}(k)-R(k)|\right)\Big/\epsilon\le\sum_{|k|\le C_m}\mathrm{var}^{1/2}[\hat{R}(k)]/\epsilon. \quad (15)$$

However, Lemma 4 states that var $[\hat{R}(k)]=0(1/m)$. Thus, the last sum in (15) is overbounded by

$$\sum_{|k|\le C_m}K/m^{1/2}\le K/m^{\delta}.$$

Hence, the second term on the right in (12) also converges to zero in probability and

$$S_{1m}\xrightarrow{P}S_1.$$

ACKNOWLEDGMENT

The authors wish to express their sincere appreciation of the valuable comments made by the reviewers.

REFERENCES

[1] S. S. Wolff, J. L. Gastwirth, and H. Rubin, "The sign test on autoregressive data," *IEEE Trans. Inform. Theory,* vol. IT-13, pp. 311–313, April 1967.

[2] L. D. Davisson, E. A. Feustal, and J. W. Modestino, "The effects of dependence on nonparametric detection," *IEEE Trans. Inform. Theory,* vol. IT-16, pp. 32–41, Jan. 1970.

[3] M. Kanefsky and J. B. Thomas, "On polarity detection schemes with non-Gaussian input," *J. Franklin Inst.,* vol. 280, pp. 120–138, 1965.

[4] J. L. Gastwirth and H. Rubin, "The asymptotic distribution theory of the empiric CDF for mixing stochastic processes," *Ann. Statist.,* pp. 809–824, vol. 3, 1975.

[5] P. Billingsley, *Convergence of Probability Measures.* New York: Wiley, 1968, pp. 166–178.

[6] H. Chernoff and I. R. Savage, "Asymptotic normality and efficiency of certain nonparametric test statistics," *Ann. Math. Stat.,* vol. 29, pp. 972–994, 1958.

[7] E. J. G. Pitman, "Lecture Notes on Nonparametric Statistics," Columbia University, New York, NY, 1948.

[8] G. E. Noether, "On a theorem of pitman," *Ann. Math. Stat.,* vol. 26, pp. 64–68, 1955.

[9] M. Fisz, *Probability Theory and Mathematical Statistics.* New York: Wiley, 1963, pp. 236–238.

33

Reprinted with permission from *IEEE Trans. Inf. Theory* **IT-16:**355–358 (1970)

NONPARAMETRIC DETECTION USING DEPENDENT SAMPLES

Melvin N. Woinsky and Ludwik Kurz

***Abstract*—A new general approach to the formulation of a nonparametric detector using dependent samples is introduced and applied to a space-diversity system employing dc signaling. A comparison based on a form of asymptotic relative efficiency is made between the new detector and a Mann–Whitney detector. Under certain conditions the new procedure demonstrates an improvement in transmission efficiency.**

INTRODUCTION

A binary signal detector using a reference noise waveform can be formulated in terms of a multivariate nonparametric hypothesis test to allow for the use of dependent samples. Assume that the detector has available a sample waveform $N(t)$ from the stationary noise process and that a group of p equally spaced dependent samples over a time period T (decision-interval length) are obtained from this waveform to form the vector sample

$$(Y_1^{(1)}, Y_2^{(1)}, \cdots, Y_p^{(1)}) = \boldsymbol{Y}^{(1)}.$$

This process is repeated n times with sufficient delays between sampling groups so that finally n independent vector samples $\boldsymbol{Y}^{(1)}, \cdots, \boldsymbol{Y}^{(i)}, \cdots, \boldsymbol{Y}^{(n)}$ are obtained. The vector independence is not a severe limitation since the noise-vector samples are taken on a periodic reference interval that comprises a small part of the total transmission period. During normal transmission, p equally spaced dependent samples $(X_1, \cdots, X_p) = \boldsymbol{X}$ are obtained on the decision interval T. The following hypothesis-testing problem results:

$$H\text{: noise: } \boldsymbol{Y}^{(1)}, \cdots, \boldsymbol{Y}^{(n)}, \boldsymbol{X} \in F(x_1, \cdots, x_p)$$

$$K\text{: signal and noise: } \boldsymbol{Y}^{(1)}, \cdots, \boldsymbol{Y}^{(n)} \in F(x_1, \cdots, x_p)$$
$$\boldsymbol{X} \in G(x_1, \cdots, x_p),$$

where $F(x_1, \cdots, x_p)$ and $G(x_1, \cdots, x_p)$ are p-variate cumulative distribution functions. The detector is then equivalent to a 2-sample p-variate nonparametric hypothesis test. During the decision interval, m independent vector samples $\boldsymbol{X}^{(1)}, \cdots, \boldsymbol{X}^{(i)}, \cdots, \boldsymbol{X}^{(m)}$ can be obtained if an m-channel space- or time-diversity system is used. In a space-diversity system a reference noise sample $\boldsymbol{Y}^{(i)}$ could be obtained separately on each channel, thereby eliminating the spacing problem in obtaining the independent reference vector samples when $n = m$. For some situations, however, such as in an active sonar array, space diversity does not insure independent noise processes on each channel.

This general formulation has been introduced to stimulate further study of nonparametric methods with dependence by following a multivariate approach. Very little work has been done on the problem of dependence [1] in signal detection. Some multivariate nonparametric tests have been considered in the mathematical literature [2], [3]. However, the forms of these tests are such that their implementation for a signal-detection system would be quite complicated and, possibly, of limited use. In the present work a simple procedure that transforms the multivariate data to univariate data and then applies a univariate nonparametric test is applied to signal detection.

CONVERSION TO UNIVARIATE DATA

Let L be a fixed transformation from p-dimensional space to one-dimensional space. Then

$$X^{(i)} = L(\boldsymbol{X}^{(i)})$$

and

$$Y^{(i)} = L(\boldsymbol{Y}^{(i)})$$

are all univariate random variables. Since L is a fixed transformation, if

$$\boldsymbol{X}^{(i)} \text{ and } \boldsymbol{Y}^{(i)} \in F(x_1, \cdots, x_p),$$

then $X^{(i)}$ and $Y^{(i)}$ have the same univariate distribution, say $J_F(x)$. After the transformation, the multivariate problem reduces to a univariate 2-sample nonparametric problem with independent samples, and a suitable 2-sample detector can be used. The overall detector then consists of a transformation or predetector L that combines the dependent samples on the decision interval and a nonparametric detector that uses the combined samples. The overall operation is nonparametric with respect to the false-alarm rate α. The transformation L reduces the data, and in some cases this lowers the efficiency. The proper selection of L should minimize this effect. A "reasonable" choice for the predetector L is that device that is optimum when the noise is Gaussian and white, for then there is at

Manuscript received May 9, 1968; revised March 12, 1969. This work was supported in part by the Doctoral Support Plan of Bell Telephone Laboratories, Inc., and in part by the National Science Foundation, under Grant GK-1075. This paper is based on part of a thesis submitted to the Graduate Division of the School of Engineering and Science of New York University in partial fulfillment of the requirements for the degree of Doctor of Philosophy.

least no loss in efficiency due to predetection in this nominal noise environment.

Two questions arise concerning this technique. First, what is the loss in efficiency due to using the transformation L to reduce the data? Second, what is the improvement in efficiency due to this procedure as compared with a nonparametric detector that must employ a limited sampling rate to obtain nearly independent samples? These two questions are considered further in the following application of the procedure.

Comparisons Using Transmission Asymptotic Relative Efficiency

In order to make asymptotic comparisons, consider a diversity or multichannel system with dc signaling in stationary additive noise. The proposed system 1 is shown in Fig. 1(a) with the function L defined as the sample mean. Since the samples may be dependent, in the equivalent continuous system, the samples and summing device can be replaced by

$$L = T_1^{-1} \int_0^{T_1} x(t)\, dt.$$

The proposed system is compared with system 2 shown in Fig. 1(b). In this case the v samples $x_1^{(i)}, \cdots, x_v^{(i)}$ taken on each of the m channels during the decision interval T_2 are used *directly* by the same 2-sample detector as in system 1. The samples $x_j^{(i)}$ must in this case all be independent for system 2, in order to provide an invariant α. Therefore, there is some maximum value of v that will still result in approximately independent samples. There are n predecision groups of samples. Then in system 2 the 2-sample detector receives vn reference samples and vm decision samples. Because of the data reduction in system 1 the 2-sample detector in that case receives n reference samples and m decision samples.[1] In system 1, $J_F(x)$ is the distribution of the output of the summing device or integrator when only noise is present; and $J_G(x) = J_F(x - s)$, where $J_G(x)$ is the distribution when signal is present. For system 2, let $F(x)$ be the marginal distribution of any sample with noise present; then $F(x - s)$ is the marginal distribution of a sample if signal and noise are present. It will be assumed that both $F(x)$ and $J_F(x)$ are continuous.

In making the comparison, a form of asymptotic relative efficiency (ARE) similar to Pitman's ARE is used [4]. First, let Δ_1 be the sampling interval in system 1 if the summing device is used, and let Δ_2 be the sampling interval in system 2. Consider a sequence $\{s_k\}$ of signal strengths such that $\lim_{k\to\infty} s_k = 0$ and two increasing sequences of integers $\{n_k\}$, $\{m_k\}$ such that

$$0 < \lim_{k\to\infty} n_k/(n_k + m_k) < 1.$$

A sequence of systems previously described as systems 1 and 2 is then constructed for each k. In each system m_k represents the number of channels and n_k the number of predecision groups. For system 1, T_1 or Δ_1 and p remain fixed, and the threshold used by the 2-sample detector is adjusted so that for each k the false-alarm rate α remains the same. For system 1 we then obtain a sequence of powers $\{\beta^{(1)}(m_k, n_k)\}$. For system 2 we have the following two sequences of increasing integers $\{v_k n_k\}$ and $\{v_k m_k\}$. In system 2, Δ_2 remains fixed, and the threshold used is similarly adjusted so that for each k the value of α is the same as in system 1. In system 2 we obtain a sequence of powers $\{\beta^{(2)}(v_k m_k, v_k n_k)\}$. Construct the sequence $\{v_k\}$ in such

[1] The resulting test is related to the "mixed" statistical tests of Feustel and Davisson [5], although, having a different purpose, they do nonparametrics first, then averaging.

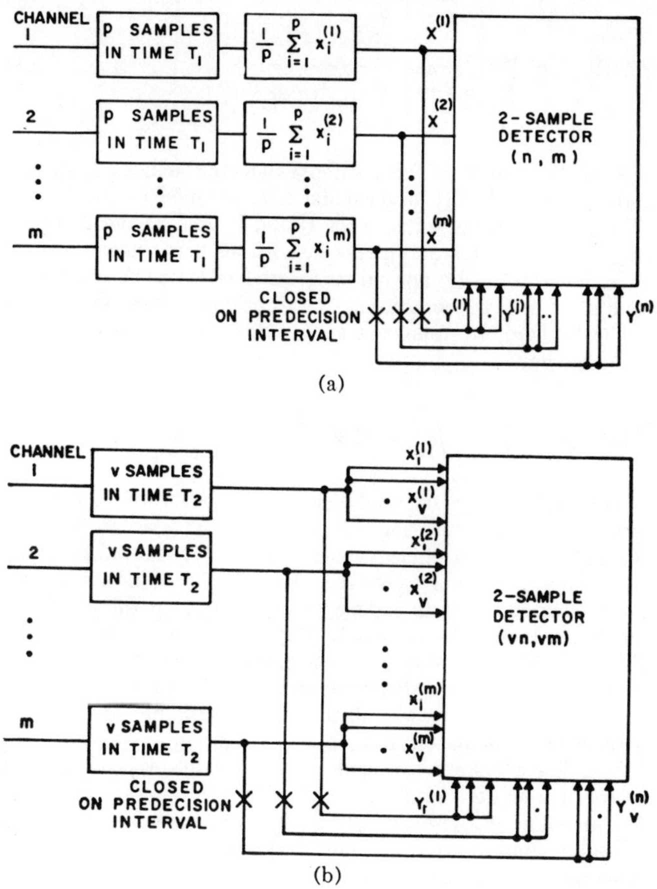

Fig. 1. Diversity system. (a) System 1. (b) System 2.

a way that

$$\lim_{k\to\infty} \beta^{(2)}(v_k m_k, v_k n_k) = \lim_{k\to\infty} \beta^{(1)}(m_k, n_k) \tag{1}$$

with the two limits existing and not equal to 0 or 1. Corresponding to the sequence $\{v_k\}$, construct the sequence of decision lengths in system 2 as

$$\{T_2(k)\} = \{\Delta_2(-1 + v_k)\}.$$

Define the transmission asymptotic relative efficiency (TARE) of system 1 with respect to system 2 as ($T_1 > 0$)

$$\text{TARE}_{1/2} = \lim_{k\to\infty} T_2(k)/T_1 = \Delta_2 \lim_{k\to\infty} (-1 + v_k)/T_1 \tag{2}$$

if the limit exists and is the same for all sequences $\{n_k\}$, $\{m_k\}$, $\{v_k n_k\}$, $\{v_k m_k\}$ satisfying (1). The $\text{TARE}_{1/2}$ is the relative transmission rate of the two systems for the same α and β and small-signal levels when the number of channels and predecision groups are large.

The Pitman ARE $[V/V^*]$ is defined in the literature [4] for comparing two different 2-sample detectors $V^*_{n^*,m^*}$ and $V_{n,m}$ with $n^*/m^* = n/m$. The $\text{TARE}_{1/2}$ can be written in terms of this efficiency by modifying the interpretation of V and V^*. In this case, V and V^* are the same 2-sample statistic, but for V the underlying distribution of the samples is $J_F(x)$, and for V^* the underlying distribution is $F(x)$. It then follows that

$$\text{TARE}_{1/2} = \frac{\Delta_2}{T_1} (\text{ARE}\, [V(J_F(x))/V(F(x))] - 1). \tag{3}$$

Subject to certain regularity conditions [4], if the asymptotic distribution of $V(J_F(x))$ and $V(F(x))$ is normal, then

$$\text{TARE}_{1/2} = \frac{\Delta_2}{T_1}\left\{\lim_{n,m\to\infty}\left[\frac{\left.\frac{d}{ds}E_K[V[J_F(x)]]\right|_{s=0}}{\left.\frac{d}{ds}E_K[V[F(x)]]\right|_{s=0}}\right]^2 - 1\right\}. \quad (4)$$

Since the input to the 2-sample detector in both system 1 and 2 is stochastically larger under K than under H, the Mann–Whitney statistic is suitable for V and is used as the 2-sample portion of the detector. In this case, it has been shown [4] that V is asymptotically normal and satisfies Pitman's regularity conditions for all continuous distributions. Therefore, (4) is applicable and, by following a procedure similar to that used in [4], we have

$$\text{TARE}_{1/2} = \frac{\Delta_2}{T_1}\left\{\left[\int_{-\infty}^{\infty} \mathcal{G}^2(x)\,dx \Big/ \int_{-\infty}^{\infty} f^2(x)\,dx\right]^2 - 1\right\} \quad (5)$$

where $f(x) = dF(x)/dx$ and $\mathcal{G}(x) = dJ_F(x)/dx$ are the corresponding density functions. Equation (5) is used to measure the improvement in efficiency due to the fact that rapid sampling is possible in system 1, while system 2 must limit its sample rate in order to obtain independent samples. In this case, the integrator is used in system 1, $\mathcal{G}(x)$ is the density function of $T_1^{-1}\int_0^{T_1} x^{(i)}(t)\,dt$, and Δ_2 is the minimum sampling interval in system 2 providing effectively independent samples.

Consider the case where all the samples obtained are actually independent no matter how fast the sampling is carried out. In making a comparison under these conditions assume that system 1 uses the summing device with a sampling interval Δ_1. In this case, $\mathcal{G}(x)$ is the density function of $p^{-1}\sum_{i=1}^{p} x_i^{(i)}$. Since all samples are independent, system 2 can sample at the same rate as system 1. Hence, let $\Delta_2 = \Delta_1$ and $T_1 = \Delta_1(p-1)$. In the Gaussian case it is readily shown that under these conditions $\text{TARE}_{1/2} = 1$ for all p, and there is no loss in efficiency. In general, one would expect a loss in efficiency in system 1 when the samples are actually independent. Consider the general case where the mean and variance of $f(x)$ exist and are given by μ and σ^2, respectively. From the stated conditions, $\mathcal{G}(x)$ exists with mean μ and variance σ^2/p. If we can assume that for large p, on the basis of the central limit theorem, $\int_{-\infty}^{\infty}\mathcal{G}^2(x)\,dx$ can be approximated by

$$\int_{-\infty}^{\infty} p[2\pi\sigma^2]^{-1}\exp[p(x-\mu)^2/\sigma^2]\,dx,$$

we get

$$\text{TARE}_{1/2} \doteq W^{-1}$$

where

$$W = 4\pi\sigma^2\left[\int_{-\infty}^{\infty} f^2(x)\,dx\right]^2.$$

The function W is proportional to the Pitman ARE of the Mann–Whitney test with respect to the t test and has been previously studied [4]. From [4] we have $4\pi(9/125) \le W < \infty$, so that $0 < \text{TARE}_{1/2} < 1.108$ approximately, for large p. The loss in efficiency due to averaging can then be very large for some distributions, while in other cases the efficiency can be improved slightly by averaging the independent samples before using them in the 2-sample Mann–Whitney detector. However, for independent samples and "near" Gaussian noise the loss would not be significant. The procedure then has the advantage of providing an invariant α with dependent samples, but its efficiency characteristics are partly determined by L.

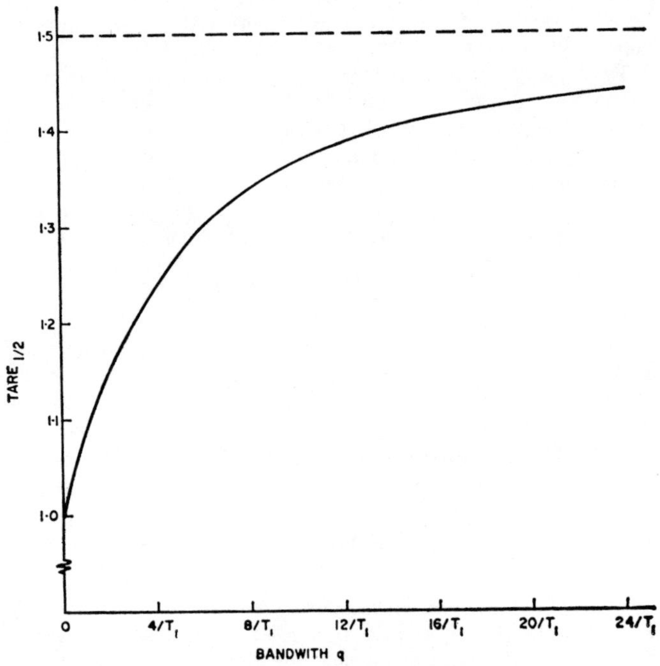

Fig. 2. Transmission asymptotic relative efficiency.

As an example to show the improvement in system 1 due to the removal of the sampling-rate restriction, consider the case where the noise $n(t)$ is a Gauss–Markov process. Then

$$E[n(t)n(t+\tau)] = \sigma_n^2 \exp(-q\,|\tau|),$$

where σ_n^2 = noise power, q = noise bandwidth. Assume also that $E[n(t)] = 0$ and that the noise is stationary.[2] Under these conditions $f(x)$ is $N(0, \sigma_n^2)$. Since an integrator is used in system 1, $\mathcal{G}(x)$ is the distribution of

$$L = T_1^{-1}\int_0^{T_1} x(t)\,dt$$

and is Gaussian. The efficiency can be obtained from (5). The correlation between the closest two samples is $\exp(-q\Delta_2)$, and if $\Delta_2 = 3/q$, the correlation is 0.05 and all samples are nearly independent. If $\Delta_2 < 3/q$, the correlation increases rapidly, and system 2 no longer provides a specified α. Therefore, letting[3] $\Delta_2 = 3/q$, we find that

$$\text{TARE}_{1/2} = \frac{1.5}{qT_1}\left[\frac{q^2T_1^2 - 2qT_1 + 2(1-e^{-qT_1})}{qT_1 - (1-e^{-qT_1})}\right]. \quad (6)$$

The curve of $\text{TARE}_{1/2}$ as a function of q is shown in Fig. 2. From this result it is evident that there is an improvement in efficiency, the magnitude of which depends upon the noise bandwidth.

Conclusions

The formulation of a nonparametric multivariate test has been proposed as a general method of dealing with dependence in signal-detection problems. A procedure based on a conversion to univariate data was applied to a dc space-diversity system. It was shown that an improvement in efficiency can be obtained

[2] The zero-mean assumption is not necessary, but it simplifies the calculations
[3] This is similar to an assumption used by Kanefsky [1] in making a comparison

when compared with a nonparametric detector with a sampling rate limited to obtain effectively independent samples.

REFERENCES

[1] M. Kanefsky and J. B. Thomas, "On adaptive nonparametric detection systems using dependent samples," *IEEE Trans. Information Theory*, vol. IT-11, pp. 521–526, October 1965.

[2] M. L. Puri and P. K. Sen, "On a class of multivariate multisample rank order tests," *Sankhya*, vol. 28A, pp. 353–376, 1966.

[3] P. K. Sen and M. L. Puri, "On the theory of rank order tests for location in the multivariate one sample problem," *Ann. Math. Stat.*, vol. 38, pp. 1216–1228, 1967.

[4] J. Capon, "A nonparametric technique for the detection of a constant signal in additive noise," *1959 IRE WESCON Rec.*, pt. 4, pp. 92–103.

[5] E. A. Feustel and L. D. Davisson, "The asymptotic relative efficiency of mixed statistical tests," *IEEE Trans. Information Theory*, vol. IT-13, pp. 247–255, April 1967.

A Class of Nonparametric Detectors for Dependent Input Data

SALEEM A. KASSAM, STUDENT MEMBER, IEEE, AND JOHN B. THOMAS, FELLOW, IEEE

I. Introduction

IT IS WELL KNOWN that standard nonparametric detectors lose their nonparametric performance characteristics when the input data is a sequence of dependent random variables [1], [2]. When the received information is in the form of a continuous-time random process, this consideration limits the sampling rate that can be used at the detector input and thus limits detection power that can be achieved in a given time. To improve performance through the use of dependent input samples without losing their nonparametric nature, modifications have to be made to the structures of standard nonparametric detectors. Different types of adaptive procedures for use in conjunction with these detectors have been proposed as a means for maintaining nonparametric operation under dependent inputs [3]–[7]. Although they are useful in improving performance in some cases, they may not always be effective and, in most cases, add significantly to the complexity of detector implementation.

In this paper we consider a class of detectors obtained by the addition of an easily implemented feature to the structures of standard nonparametric detectors. The detectors in this class are nonadaptive and retain nonparametric performance under conditions of dependent input data. They convert the input data samples into a sequence of multivariate n-component samples $\{Z_i\}$ obtained by grouping a fixed number n of consecutive input samples. A transformation, or reduction, to a single input data point is then effected, the resulting sequence of variables $\{Z^k\}$ being finally processed according to a standard nonparametric detection procedure. In this paper we shall be interested mainly in a simple summing operation on the n components of the multivariate samples. For a fixed transformation, we then have a one-to-one correspondence between the class of standard nonparametric detectors and the class of modified detectors we have defined.[1] Woinsky and Kurz [8] have originated this approach for dependent samples and have applied it to the specific situation where a reference noise channel is available.

As specific examples of detectors in the class defined here we will consider the sign detector and the Wilcoxon detector modified according to the preceding description. To facilitate comparison of the performances of these detectors with those of the corresponding standard detectors we will initially make some simplifying assumptions about the dependence structure of the input random processes and about the manner in which the multivariate samples Z_i are obtained. Eventually these restrictions will be removed, and the results derived will be shown to be applicable in cases where only a strong mixing condition is satisfied.

II. Preliminaries

The detection problem in this paper is the simple one of detecting a constant positive signal in additive noise with a single observation channel. Extensions to two- and multiple-channels, and to the detection of known and random signals (for example, the case of the polarity coincidence correlator (PCC)), can be approached in a similar way. We will assume the following about the noise process $N(t)$.

i) $N(t)$ is zero-mean second-order stationary mean-square continuous and has diagonally symmetric probability density functions (pdf's) of all finite orders, i.e.,

$$f_N(x_1,\cdots,x_k) = f_N(-x_1,\cdots,-x_k) \quad (1)$$

for all finite k, where $f_N(x_1,\cdots,x_k)$ is any k-variate pdf of $N(t)$.

The assumption of stationarity can be relaxed as we will see. Diagonal symmetry of the noise pdf's is a reasonable assumption in almost all cases of interest.

ii) There exists a finite maximum sampling rate $r_0 = 1/T$ such that for all rates $r \leq r_0$, the samples of the observed process can be assumed to be almost independent, that is, their mutual dependence can be considered negligible for the purpose of nonparametric detector performance analysis [4], [10]. The Gauss–Markov process is an example of such a process, where sampling at a rate equal to the

Manuscript received November 6, 1974; revised January 23, 1975. This work was supported in part by the National Science Foundation under Grant GK-24187 and in part by the U.S. Army Research Office under Contract DAHC04-69-0012.

The authors are with the Department of Electrical Engineering, Princeton University, Princeton, N.J. 08540.

[1] It should be noted that the detectors considered here are distinct from the "mixed" statistical tests of Feustel and Davisson [9]; these would in fact be characterized as belonging to the class of standard nonparametric detectors.

double-sided bandwidth gives rise to samples that can be assumed to be independent [4].

If the noise process is strong mixing, as in the case of the Gauss–Markov process, then the second assumption is satisfied. In the Appendix some mixing properties of processes are defined, and results derived from them are given. In order to obtain exact results in what follows we will assume initially that the noise process is in fact T-dependent. This will imply that the normalized autocorrelation function $\rho_N(\tau)$ of the noise process vanishes outside of a finite interval $(-T,T)$ and that samples separated by an interval greater than or equal to T are independent. A large class of processes satisfy this T-dependence condition; examples are memoryless nonlinear transformations of Gaussian processes with autocorrelation functions of finite support, finite moving average processes, and the simple shot noise process with a finite pulse width. We will also discuss later how these results are to be interpreted when the noise process is not T-dependent but satisfies some weaker mixing condition such as strong mixing. It will be shown that, in many such cases of interest, the results will remain valid as very good approximations to actual performances.

To compare the performances of two detectors we will use their asymptotic relative processing time (ARPT), obtained from their asymptotic relative efficiency (ARE). The ARE, which is a weak-signal large sample performance measure, can be computed readily in the applications here by the application of the Pitman–Noether theorem [11]. The ARPT of detector D_1 relative to detector D_2 is defined by [4]

$$\mathrm{ARPT}_{1,2} = \mathrm{ARE}_{1,2} \times [t_2/t_1] \tag{2}$$

where $\mathrm{ARE}_{1,2}$ is the ARE of detector D_1 with respect to D_2 and t_i is the time between samples of detector D_i. Thus ARPT is a comparison of the lengths of observation times required by two detectors to achieve the same performance as the observation times approach infinity.[2]

It will be assumed that the test statistics are asymptotically normally distributed. Under the assumption of T-dependence and from the detector structures to be formulated this will be true in almost all cases. When the conditions on the input process are relaxed, and the structures of the detectors are simplified, as we will consider later, this asymptotic normality does not follow immediately. However, in many cases of interest, it can be proved.

III. Detector Structure and Performance

From the assumptions made previously, it follows that for a continuous-time observed process a sampling rate r_0 results in a sequence of independent samples $X_1,X_2,\cdots,X_N$, where $NT = T_{\mathrm{obs}}$, the length of the observation interval. Let D_X denote a standard nonparametric detector operating on this sequence in some specific way. If the sampling rate is increased by a factor γ, where $\gamma = 1,2,\cdots$, we get a $(\gamma - 1)$-dependent sequence of samples $Y_1, Y_2,\cdots, Y_{N\gamma}$. Consider grouping n consecutive Y samples as one n-variate sample, with $(\gamma - 1)$ samples skipped between groups. This gives a sequence of independent n-variate samples $\boldsymbol{Z}_1,\boldsymbol{Z}_2,\cdots,\boldsymbol{Z}_{N\gamma/(n+\gamma-1)}$ (assuming $N\gamma/(n + \gamma - 1)$ is an integer), where

$$\boldsymbol{Z}_i = [Y_{(n+\gamma-1)(i-1)+1}, Y_{(n+\gamma-1)(i-1)+2},\cdots, Y_{(n+\gamma-1)(i-1)+n}]. \tag{3}$$

Each standard nonparametric detector D_X can be modified to operate on the input samples by treating them as a sequence of multivariate samples according to the preceding description. Let D_Z denote such a modified version of the detector D_X. We restrict D_Z to perform a reduction to univariate values of each multivariate sample, and then process the resulting sequence of independent univariate values in the same way as D_X processes its input. Thus we have a class of detectors $\{D_Z\}$ once an appropriate transformation is defined.

One of the simplest nontrivial transformations that can be used is defined by the summing operation where the n components of each multivariate sample are summed to obtain the reduction to univariate values. Although we will be mainly concerned with this transformation, the possibility of the use of more efficient transformations should be noted. We will briefly analyze the case of a nonlinear transformation with one modified detector to illustrate this possibility.

Let D_Y denote the adaptive threshold detector [4] based on D_X. A comparison of D_Z with respect to D_X and D_Y will now be made for the sign detector and the Wilcoxon detector. The comparison of D_Z with respect to D_Y will show how well D_Z performs relative to the adaptive detector, which has a more complex structure.

A. Modified Sign Detector

With the multivariate-to-univariate transformation being achieved through the summing operation we get a modified sign detector D_Z with test statistic

$$S_Z = [N\gamma/(n + \gamma - 1)]^{-1} \sum_{i=1}^{N\gamma/(n+\gamma-1)} \mathrm{sgn}\,(Z^i) \tag{4}$$

where

$$\mathrm{sgn}\,(x) = \begin{cases} 1, & x > 0 \\ 0, & x = 0 \\ -1, & x < 0 \end{cases}$$

and Z^i equals the sum of the n components of $\boldsymbol{Z}_i$.

Using the Pitman–Noether theorem, we find that for the sign detector

$$\mathrm{ARPT}_{Z,X} = n^2 \left[\frac{f_Z^2(0)}{f_N^2(0)}\right] \frac{\gamma}{n + \gamma - 1} \tag{5}$$

where $f_Z(z)$ is the common univariate pdf of the Z^i under the hypothesis of no signal present and $f_N(x)$ is the univariate noise pdf.

[2] In [8] the notion of transmission ARE is introduced where observation times remain finite and the number of channels approaches infinity.

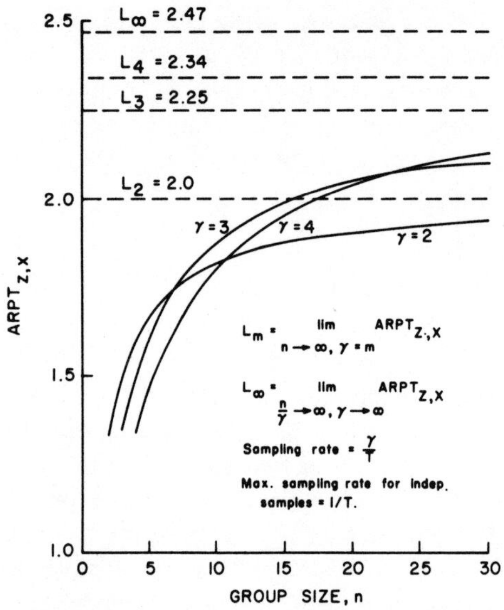

Fig. 1. Performance of modified sign and Wilcoxon detectors for Gaussian noise with normalized autocorrelation function

$$\rho_N(\tau) = \begin{cases} (1 - |\tau|/T)\cos(\pi\tau/T), & |\tau| \le T \\ 0, & |\tau| > T. \end{cases}$$

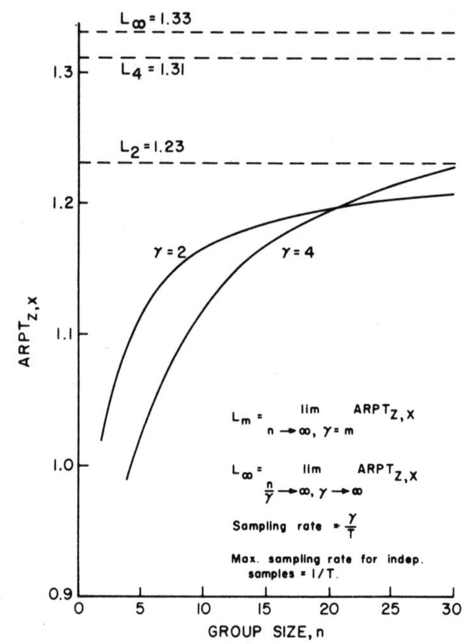

Fig. 2. Performance of modified sign and Wilcoxon detectors for Gaussian noise with normalized autocorrelation function

$$\rho_N(\tau) = \begin{cases} (1 - |\tau|/T)^2[1 + |\tau|/(2T)], & |\tau| \le T \\ 0, & |\tau| > T. \end{cases}$$

If the n-variate noise pdf is known, then $f_Z(z)$ can be found, and (5) can be evaluated. In the case of Gaussian noise we can evaluate the ARPT of (5) to find that

$$\mathrm{ARPT}_{Z,X} = \left(\frac{n^2}{n + 2 \sum_{i=1}^{\min(n,\gamma)-1} (n-i)\rho_N(iT/\gamma)} \right) \frac{\gamma}{n + \gamma - 1} \tag{6}$$

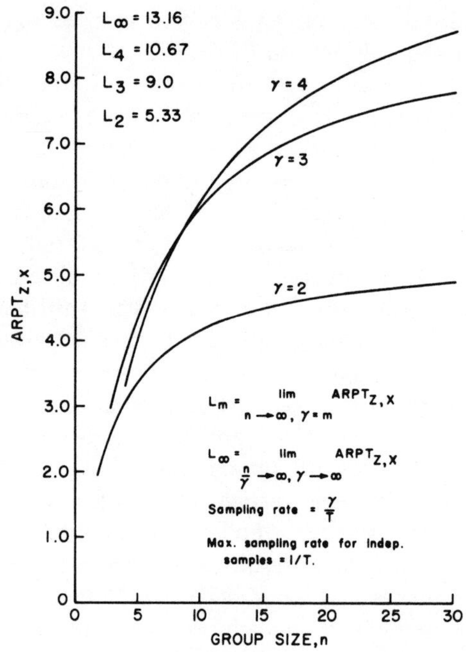

Fig. 3. Performance of modified sign and Wilcoxon detectors for Gaussian noise with normalized autocorrelation function

$$\rho_N(\tau) = \begin{cases} (1 - |\tau|/T)^2[1 + |\tau|/(2T)]\cos(2\pi\tau/T), & |\tau| \le T \\ 0, & |\tau| > T. \end{cases}$$

where $\rho_N(\tau)$ is the normalized autocorrelation function of the noise process. In the limit as $n \to \infty$ (increasing group size) for fixed γ we get

$$\lim_{n\to\infty} \mathrm{ARPT}_{Z,X} = \frac{\gamma}{1 + 2\sum_{i=1}^{\gamma-1} \rho_N(iT/\gamma)} \tag{7}$$

and, if we let $\gamma \to \infty$ also, we get

$$\lim_{\substack{\gamma\to\infty \\ n/\gamma\to\infty}} \mathrm{ARPT}_{Z,X} = \frac{T/2}{\int_0^T \rho_N(\tau)\, d\tau}. \tag{8}$$

It can be shown [12] that $\int_0^T \rho_N(\tau)\, d\tau \le T/2$, for arbitrary $\rho_N(\tau)$ of a T-dependent process, so that in the limiting case of high sampling rate and very large group size we have $\mathrm{ARPT}_{Z,X} \ge 1$. For finite values and for some common $\rho_N(\tau)$, the results are shown in Figs. 1–3. Also Tables I–III give corresponding numerical values of $\mathrm{ARPT}_{Y,X}$ for comparison. The equations for $\mathrm{ARPT}_{Y,X}$ can be derived by modifying those given in [4] to apply to the cases in this paper. Here D_Y is the adaptive threshold sign detector with perfect adaptation equivalent to an exact knowledge of the variance of the test statistic. The parameter γ has the same meaning for D_Y and D_Z, while there is no group-size parameter n for D_Y.

The results show that detector D_Z can be significantly better than D_X in the case of sign detection and compares favorably with D_Y, the adaptive threshold detector, which is not implemented as easily. In some cases, as is evident from Fig. 3 and Table III, D_Z performs even better than D_Y.

For a large class of non-Gaussian processes satisfying the earlier assumptions, the Z^i can be shown to be asymptotically

TABLE I
RELATIVE PERFORMANCE OF ADAPTIVE THRESHOLD SIGN DETECTOR D_Y FOR GAUSSIAN NOISE WITH

$$\rho_N(\tau) = \begin{cases} (1 - |\tau|/T)\cos(\pi\tau/T), & |\tau| \le T \\ 0, & |\tau| > T \end{cases}$$

γ	2	3	4	$\lim_{\gamma \to \infty}$
$ARPT_{Y,X}$	2.0	2.46	2.69	3.19

TABLE II
RELATIVE PERFORMANCE OF ADAPTIVE THRESHOLD SIGN DETECTOR D_Y FOR GAUSSIAN NOISE WITH

$$\rho_N(\tau) = \begin{cases} (1 - |\tau|/T)^2[1 + |\tau|/(2T)], & |\tau| \le T \\ 0, & |\tau| > T \end{cases}$$

γ	2	3	4	$\lim_{\gamma \to \infty}$
$ARPT_{Y,X}$	1.42	1.59	1.68	1.85

TABLE III
RELATIVE PERFORMANCE OF ADAPTIVE THRESHOLD SIGN DETECTOR D_Y FOR GAUSSIAN NOISE WITH

$$\rho_N(\tau) = \begin{cases} (1 - |\tau|/T)^2[1 + |\tau|/(2T)]\cos(2\pi\tau/T), & |\tau| \le T \\ 0, & |\tau| > T \end{cases}$$

γ	2	3	4	$\lim_{\gamma \to \infty}$
$ARPT_{Y,X}$	3.36	5.25	6.72	12.83

normally distributed. In this case we can obtain an approximation for the ARPT for large n as follows:

$$\underset{(\text{large } n)}{ARPT_{Z,X}} \approx \left(\frac{n^2}{n + 2\sum_{i=1}^{\gamma-1}(n-i)\rho_N(iT/\gamma)}\right) \cdot \left(\frac{1}{2\pi\sigma_N^2 f_N^2(0)}\right)\frac{\gamma}{n+\gamma-1} \tag{9}$$

where σ_N^2 is the noise variance. For the specific case where the noise process has a univariate uniform pdf we then obtain

$$ARPT_{Z,X} \approx (6/\pi) \times [ARPT_{Z,X} \text{ for the corresponding Gaussian case}].$$

On the other hand, for the double-exponential noise pdf, we have

$$ARPT_{Z,X} \approx (1/\pi) \times [ARPT_{Z,X} \text{ for the corresponding Gaussian case}].$$

The approximations can be shown to be very good for moderate values of n.

For the modified sign detector D_Z an odd predetection nonlinearity does not change the false alarm probability α of the detector as long as $f_Z(z)$ has no discrete component at $z = 0$. The nonlinearity may be time varying, and the noise may be nonstationary so long as the assumptions of diagonal symmetry of the pdf's and of the previously stated dependency structure of the process hold. For the case $n = 2$, if the nonlinearity is strictly monotonically increasing as well as odd, the power β of the detector is also preserved. Otherwise, odd nonlinearities can be used to increase the power of D_Z, as discussed in Section III–C.

The performance of the modified sign detector relative to the linear detector for dependent samples (and for moderate values of n and n/γ) is close to the performance of the simple sign detector relative to the linear detector for independent sampling. However, it may be more realistic to assume, even in the case of Gaussian noise, that the autocorrelation function of the noise process is not known. One possibility then would be to use the linear detector operating on independent samples (since the process is assumed to be T-dependent). However, the modified sign detector compares very favorably as an alternative detection procedure, achieving better detection performance for many types of autocorrelation functions. As will be seen in a later section, the use of a nonlinear transformation in D_Z can further improve the relative performance of D_Z.

B. Modified Wilcoxon Detector

This modified one-channel rank detector D_Z is based on the signed ranks of the independent transformed univariate values; the transformation is again assumed to be effected through a summing operation. For this detection scheme it can be shown easily that

$$ARPT_{Z,X} = n^2\left[\frac{\int_{-\infty}^{\infty} f_Z^2(z)\,dz}{\int_{-\infty}^{\infty} f_N^2(x)\,dx}\right]^2 \frac{\gamma}{n+\gamma-1}. \tag{10}$$

Here D_X is the Wilcoxon detector operating on the independent samples $X_1,\cdots,X_N$ of the observed waveform.

For the case of Gaussian noise, it can be shown that this expression is exactly the same as that obtained in Section III–A for the sign detector and Gaussian noise. This means that (6)–(8) continue to hold for the modified Wilcoxon detector D_Z. In addition, the adaptive threshold detector D_Y based on the Wilcoxon test statistic does not perform as well relatively as in the case of the sign detector, so that D_Z here is even more attractive.

For non-Gaussian processes and large n we may once again use the Gaussian approximation for the Z^i, and we get

$$\underset{(\text{large } n)}{ARPT_{Z,X}} \approx \left(\frac{n^2}{n + 2\sum_{i=1}^{\gamma-1}(n-i)\rho_N(iT/\gamma)}\right) \cdot \left(\frac{1}{4\pi\sigma_N^2[\int_{-\infty}^{\infty} f_N^2(x)\,dx]^2}\right)\frac{\gamma}{n+\gamma-1}. \tag{11}$$

For the case of a noise process with a uniform univariate pdf we have from the preceding that

$$ARPT_{Z,X} \approx (3/\pi) \times [ARPT_{Z,X} \text{ for the corresponding Gaussian case}]$$

and, for the double-exponential noise pdf, we have

$$\mathrm{ARPT}_{Z,X} \approx (2/\pi) \times [\mathrm{ARPT}_{Z,X} \text{ for the corresponding Gaussian case}].$$

Here again it can be shown that, where it can be applied, the approximation becomes very good for moderate values of n, especially if the correlation coefficient between adjacent samples is not too large.

One important advantage of the modified Wilcoxon detector D_Z is that it requires a smaller number of data values to be ranked. For example, suppose $N = 100$ and we are using $\gamma = 2$ and $n = 3$. Then $N\gamma/(n + \gamma - 1) = 50$. In this case we have to rank only 50 numbers instead of 100, a considerable saving.

Instead of basing the test statistic on the sums of the components of the multivariate samples Z_i other transformations could be used, for example, one that results in equal ranks being attached to hyperspheres instead of the hyperplanes we have used. The analysis of such a scheme is, however, more involved, although improved performance can be expected from such schemes as will be illustrated for the modified sign detector in the next section.

C. Nonlinear Transformations

To explore the possibility of obtaining further performance improvements through the use of nonlinear transformations in the modified detectors, let us consider the modified sign detector $D_{Z,h}$ where the transformation to Z^i is now achieved by summing a nonlinear function $h(\cdot)$ of each component of Z_i, that is, if $Z_i = (U_1{}^i, U_2{}^i, \cdots, U_n{}^i)$ then $Z^i = \sum_{j=1}^n h(U_j{}^i)$.

For the case where $h(\cdot) = \mathrm{sgn}(\cdot)$, an odd function that is not strictly monotonic, we can expect to change the power β when $n = 2$ (and $\gamma = 2$). For the resulting detector $D_{Z,\mathrm{sgn}}$ the test statistic S_Z as defined by (4) does not result in a nonparametric test. The reason is that Z^i can be zero with a nonzero probability dependent on the underlying probability distribution. However, a simple modification to this test statistic results in a test that is *conditionally nonparametric* [13]. Define the random variable C by

$$C = \sum_{i=1}^{2N/3} |\mathrm{sgn}\,(Z^i)| \tag{12}$$

(cf. (4) with $n = 2$, $\gamma = 2$). The statistic $\frac{1}{2}((2N/3)S_Z + C)$ is the number of the positive Z^i, and C is the total number of the nonzero Z^i, so that $\frac{1}{2}((2N/3)S_Z + C)$ conditioned on C has a binomial distribution with parameters p and C, where $p = \frac{1}{2}$ under the hypothesis of no signal present and $p > \frac{1}{2}$ under the alternative. Thus the statistic S_Z (or the normalized statistic $(2N/3)S_Z/\sqrt{C}$) can be used to form a conditional nonparametric test for all N, the threshold being a function of C as well.[3] It is interesting to note that the resulting detector is then an implementation of the bivariate sign test for location described in [13] where it is derived through the application of principles of invariance and sufficiency. For large N, the unconditional test becomes exactly nonparametric, and this fact can be exploited to obtain simpler tests when the sample size is large.

The efficacy $E_{Z,\mathrm{sgn}}$ for this modified sign detector $D_{Z,\mathrm{sgn}}$ when $n = 2$ can be shown to be

$$E_{Z,\mathrm{sgn}} = \frac{2f_N{}^2(0)}{F_N(0,0)} \tag{13}$$

where $F_N(\cdot,\cdot)$ is the bivariate probability distribution function of the noise process for a time separation of $T/2$. For D_Z, the modified sign detector using the linear transformation, the corresponding efficacy E_Z is

$$E_Z = 16\left[\int_{-\infty}^{\infty} f_N(-x,x)\,dx\right]^2 \tag{14}$$

where $f_N(\cdot,\cdot)$ is the bivariate density corresponding to $F_N(\cdot,\cdot)$. By the Pitman–Noether theorem, the ARE of $D_{Z,\mathrm{sgn}}$ with respect to D_Z can be obtained as a ratio of their efficacies. For the Gaussian case, when $n = 2$, this ratio $R = E_{Z,\mathrm{sgn}}/E_Z$ is given by

$$R = \frac{1 + \rho_N(T/2)}{1 + 2/\pi(\arcsin\,[\rho_N(T/2)])}. \tag{15}$$

Thus we have $R \geq 1$, for $0 \leq \rho_N(T/2) \leq 1$, and $R = 1$, for $\rho_N(T/2) = 0$ or 1, so that an improvement in performance is obtained from $D_{Z,\mathrm{sgn}}$.

To model a case where the univariate density function of the noise process is double exponential, we may assume that the bivariate samples $Z_i = (U_1{}^i, U_2{}^i)$ (in the case $n = 2$ and $\gamma = 2$) satisfy the autoregression scheme

$$U_2{}^i = \rho_N(T/2)U_1{}^i + W^i \tag{16}$$

where W^i is the independent residual term and $U_1{}^i$ and $U_2{}^i$ have double-exponential pdf's. We would get (16) if the noise samples form an autoregressive process with double-exponential marginal densities. The autoregressive process is not T-dependent; however, its use as a noise model is justified from considerations treated in the next section. We can then show that

$$R = \frac{[1 + \rho_N(T/2)][2 + \rho_N(T/2)]^2}{2[1 + \rho_N(T/2) + \rho_N{}^2(T/2)]^2} \tag{17}$$

so that R increases monotonically from one to two as $\rho_N(T/2)$ decreases from one to zero. Thus a significant improvement in performance can be obtained.

With $\gamma = 2$, using the Gaussian approximation for Z^i, we can show that, for Gaussian noise as $n \to \infty$, R decreases for all $0 \leq \rho_N(T/2) \leq \frac{1}{2}$ from its value for $n = 2$. In the limiting case ($n \to \infty$), we have $R \geq 2/\pi$, equality being achieved for $\rho_N(T/2) = 0$. For the autoregressive model with double-exponential marginal densities, R increases to two for all $0 \leq \rho_N(T/2) < 1$ as $n \to \infty$. These results are consistent with what one might expect from results on signal-to-noise ratios of infinitely clipped random variables.

For finite values of $n > 2$ similar conditional nonparametric tests can be constructed, which also become unconditionally nonparametric as $N \to \infty$. For odd n, of

[3] Of course, the threshold cannot be set until the value of C is known, i.e., until the end of the data sequence.

course, the test based directly on S_Z is nonparametric. Instead of the function sgn $(\cdot)$, a soft-limiter can be used to effect the nonlinear transformation to Z^i. In this case the loss in performance for larger n with Gaussian noise can be reduced at the expense of some reduction of performance gain in cases of noise with heavier-tailed density functions. An advantage in using the function sgn $(\cdot)$ is that the invariance properties of the detector are enhanced; α and β remain constant for any sign-preserving predetection nonlinearity.

This brief treatment on the use of a specific nonlinear transformation shows that, in general, the modified detectors may be further improved (but may be degraded under certain conditions) if simple nonlinear transformations are allowed in their structure. This leads naturally in some cases to the type of conditional test described here. Such tests are not hard to implement and, for large sample sizes, can be assumed to be nonparametric even in the unconditional form. Conditional tests can be avoided if the nonlinear function is appropriately chosen; for the sign detector, this means that the Z^i take on the value of zero with probability zero.

IV. Departures from Ideal Assumptions

A. Nonskipped Case

The skipping of $\gamma - 1$ samples between groups of length n was introduced in the construction of D_Z since, for simplicity of analysis, the multivariate samples had to be independent. When the noise is T-dependent, skipping achieves this desired independence. In most cases it would not be necessary to skip samples, however, because if large enough groups of samples are summed the correlations between adjacent group-sums become small enough to be negligible. In the Appendix it is shown that for a large class of processes such negligible correlations are sufficient to ensure near-independence of the transformed random variables $\{Z^i\}$. In many cases, n also does not have to be very large before this condition is obtained. The ARPT values we have derived would then represent conservative values for the relative performances because there would be no loss due to skipping.

B. Strong Mixing Noise Process

When the noise process is not T-dependent but strong mixing, there may not exist a sampling rate resulting in independent samples. However, there always exists a maximum sampling rate $r_0' = 1/T'$ for which the mutual dependence between samples can be considered to be negligible. We can then obtain the performances of the modified detectors D_z relative to the standard detectors D_x that use this sampling rate. The previous analysis, carried out under the assumption of T-dependence, still provides results that are very close approximations in the strong mixing case when T is replaced by T'.

As a specific example, consider the stationary Gauss–Markov process where $\rho_N(\tau) = \exp(-a|\tau|)$. It is well known that, for sampling at the rate a/π, the samples can be assumed to be independent for nonparametric detection applications [4], [10]. With a/π as the base sampling rate it can be shown that the correlation coefficient c between consecutive summed samples Z^i and Z^{i+1} with $\gamma - 1$ skipped samples between groups is

$$c = \frac{e^{-\pi}(1 - e^{-n\pi/\gamma})^2}{n(1 - e^{-2\pi/\gamma}) + 2e^{-\pi/\gamma}(e^{-n\pi/\gamma} - 1)}. \tag{18}$$

For $n = 2$ and $\gamma = 2$, we find that $c = 0.028$, which is negligible. If samples are not skipped, the value of c would be larger. Even so, for $n = 5$ and $\gamma = 2$, we have $c \approx 0.05$. This is again small enough to be negligible.

The Gauss–Markov process is perhaps the best-known example of a strong mixing process. However, as mentioned earlier, for any strong mixing process there exists a sampling rate r_0' for which the samples can be assumed to be independent and for which the previous analysis gives valid results. Some examples of autoregressive noise sequences that are strong mixing are given in [14] and, in the Appendix, the possibility of avoiding skipped samples for some strong mixing processes is indicated.

For the Gauss–Markov example we find that, on using (5) appropriately modified (that is, with no skipping loss) for $n = 5$ and $\gamma = 2$, the value for the $\text{ARPT}_{Z,X}$ is 1.45, a very significant improvement.

V. Conclusion

Simple modifications to standard nonparametric detectors lead to a class of detectors that can retain nonparametric performance under conditions of dependent input samples. Particular examples of detectors in this class have shown that their performances with continuous-time processes can be significantly better than those of corresponding standard nonparametric detectors. In many cases sampling rates that are between two and four times the rate for approximately independent samples are sufficient to produce this improvement. It has been shown further that even higher efficiencies can be obtained by detectors in which the multivariate-to-univariate reduction is achieved through a nonlinear operation on the components of the multivariate samples.

Although only one-channel detectors have been considered in this paper, similar performance improvements can be established for two channel and multichannel detectors in the general class defined here. This class is, therefore, a viable alternative to the standard nonparametric detectors with their restriction to independent sampling.

Appendix

Let $X(t)$, $t \in (-\infty,\infty)$, be a stationary random process on the probability space (Ω, M, P). Denote by M_r^s the σ-algebra generated by the process $X(t)$, for $r \le t \le s$.

Definition 1: The family of σ-algebras M_r^s and, by extension, the random process $X(t)$ is *T-dependent* if the σ-algebras $M_{-\infty}^t$ and $M_{t+\tau}^{\infty}$ are independent, for $\tau \ge T$, i.e.,

$$\sup_{A \in M_{-\infty}^t,\, B \in M_{t+\tau}^{\infty}} |P(A \cap B) - P(A)P(B)| = \alpha(\tau) \tag{A.1}$$

where $\alpha(\tau) = 0$, for $\tau \ge T$.

Definition 2: The family of σ-algebras M_r^s and, by extension, the random process $X(t)$ is *strong mixing* [15] if

$$\sup_{A \in M_{-\infty}^t, B \in M_{t+\tau}^\infty} |P(A \cap B) - P(A)P(B)| = \alpha(\tau) \quad (A.2)$$

where $\alpha(\tau) \to 0$, as $\tau \to \infty$.

Similar definitions exist for discrete-time processes.

Definition 3: Given the σ-algebras M_1 and M_2, the *maximal correlation coefficient* $m(M_1,M_2)$ between M_1 and M_2 is defined by [16]

$$m(M_1,M_2) = \sup_{X \in U_1, Y \in U_2} \frac{|E\{(X - E\{X\})(Y - E\{Y\})\}|}{E\{(X - E\{X\})^2\}^{1/2} E\{(Y - E\{Y\})^2\}^{1/2}} \quad (A.3)$$

where U_1 and U_2 are the collections of all random variables of finite second moments that are measureable with respect to the σ-algebras M_1 and M_2, respectively.

Correlation Properties of Strong Mixing Processes

Let $X(t)$ be stationary and strong mixing with $E\{X(t)\} = 0$.

Lemma 1: If $\exists \delta > 0$ such that $E\{|X(t)|^{2+\delta}\} \leq c < \infty$, then

$$|\rho(\tau)| \leq (4 + 6c)[\alpha(\tau)]^{\delta/(2+\delta)} \quad (A.4)$$

where $\rho(\tau)$ is the autocorrelation function of $X(t)$. For a proof see [17]. Stronger bounds can be obtained under more restrictive assumptions.

Lemma 1 implies that, for a strong mixing process, the autocorrelation function $\rho(\tau)$ falls off at a rate that is bounded from above by a function of $\alpha(\tau)$ provided that the process has moments of order higher than two.

Lemma 2:

$$\alpha(\tau) \leq m(M_{-\infty}^t, M_{t+\tau}^\infty). \quad (A.5)$$

This follows from Definitions 2 and 3.

When $X(t)$ is a Gaussian process, with normalized autocorrelation function $\rho_N(\tau)$, then

$$m(M_{-\infty}^t, M_{t+\tau}^\infty) = \sup_{s \geq \tau} |\rho_N(s)|. \quad (A.6)$$

Equation (A.6) follows from theorems proved in [16]. This equation also holds for the case where $X(t)$ is a second-order process obtained by passing a Gaussian process with normalized autocorrelation function $\rho_N(\tau)$ through a zero-memory nonlinearity. For other processes, the maximal correlation coefficient may be obtained from its bivariate density functions.

A sequence of random variables $\{Z^i\}$ that is obtained by summing groups of n consecutive samples of a strong mixing process is also strong mixing. Let $\{\alpha_n\}$ be the discrete-time equivalent of $\alpha(\tau)$ for $\{Z^i\}$. If Z^i is formed from $\boldsymbol{Z}_i$, where $\boldsymbol{Z}_i$ and $\boldsymbol{Z}_{i+1}$ are an interval T apart, then α_1 is no larger than $\alpha(T)$. Lemma 2 also enables bounds to be obtained for α_n when samples are not skipped between the $\boldsymbol{Z}_i$. When $X(t)$ is a Gaussian process, α_1 is bounded above by the absolute value of the correlation coefficient between Z^i and Z^{i+1}, assuming that the absolute value of the correlation coefficient between Z^i and Z^{i+k} is a nonincreasing function of k. Thus a small enough value of this coefficient indicates that the Z^i may be assumed to be independent. For the case where $X(t)$ is a second-order process obtained as the output of a zero-memory nonlinearity driven by a Gaussian process $X_G(t)$, the bound on α_1 is the same as for the case of the Gaussian process $X_G(t)$ itself.

References

[1] S. S. Wolff, J. L. Gastwirth, and H. Rubin, "The effect of autoregressive dependence on a nonparametric test," *IEEE Trans. Inform. Theory* (Corresp.), vol. IT-13, pp. 311–313, Apr. 1967.

[2] L. D. Davisson, E. A. Feustel, and J. W. Modestino, "The effects of dependence on nonparametric detection," *IEEE Trans. Inform Theory*, vol. IT-16, pp. 32–41, Jan. 1970.

[3] M. Kanefsky and J. B. Thomas, "On adaptive nonparametric detection systems using dependent samples," *IEEE Trans. Inform. Theory*, vol. IT-11, pp. 521–526, Oct. 1965.

[4] J. W. Modestino, "Nonparametric and adaptive detection of dependent data," Ph.D. dissertation, Dep. Elec. Eng., Princeton Univ., Princeton, N.J., 1969.

[5] V. A. Simakhin, "Nonparametric signal detectors when the signal samples are dependent," *Telecommun. Radio Eng.* (USSR), vol. 28, pp. 66–72, Jan. 1973.

[6] M. Kanefsky, "An automatic decision threshold for polarity coincidence arrays," in *Proc. Nat. Electronics Conf.*, vol. 22, pp. 2–6, 1966.

[7] S. S. Wolff and M. B. Sirvanci, "'Studentizing' rank tests for dependent data," in *Proc. 7th Princeton Conf. Information Sciences and Systems*, pp. 63–66, 1973.

[8] M. N. Woinsky and L. Kurz, "Nonparametric detection using dependent samples," *IEEE Trans. Inform. Theory* (Corresp.), vol. IT-16, pp. 355–358, May 1970.

[9] E. A. Feustel and L. D. Davisson, "The asymptotic relative efficiency of mixed statistical tests," *IEEE Trans. Inform. Theory*, vol. IT-13, pp. 247–255, Apr. 1967.

[10] M. Kanefsky and J. B. Thomas, "Polarity coincidence correlation using dependent samples," in *Proc. Nat. Electronics Conf.*, vol. 20, pp. 714–719, 1964.

[11] G. E. Noether, "On a theorem of Pitman," *Ann. Math. Statist.*, vol. 26, pp. 64–68, 1955.

[12] R. P. Boas and M. Kac, "Inequalities for Fourier transforms of positive functions," *Duke Math. J.*, vol. 12, pp. 189–206, Mar. 1945.

[13] S. K. Chatterjee, "A bivariate sign test for location," *Ann. Math. Statist.*, vol. 37, pp. 1771–1782, 1966.

[14] M. B. Sirvanci, "A study of autoregressive dependence in nonparametric detection," Ph.D. dissertation, Johns Hopkins Univ., Baltimore, Md., 1972.

[15] M. Rosenblatt, "A central limit theorem and a strong mixing condition," *Proc. Nat. Acad. Sci. U.S.*, vol. 42, pp. 43–47, 1956.

[16] A. N. Kolmogorov and Y. A. Rozanov, "On strong mixing conditions for stationary Gaussian processes," *Theory Prob. Appl.* (USSR), vol. 5, pp. 204–209, 1960.

[17] I. A. Ibragimov, "Some limit theorems for stationary processes," *Theory Prob. Appl.* (USSR), vol. 7, pp. 349–383, 1962.

35

Reprinted from *J. Franklin Inst.* **304**:265–278 (1977)

Two-Sample Non-parametric Detectors Using Dependent Samples

by EMAD AL-HUSSAINI and FARID BADRAN
Electronics and Communication Department,
Faculty of Engineering,
Cairo University, Cairo, Egypt

and LAURENCE TURNER
Department of Electrical Engineering
Imperial College of Science and Technology
South Kensington, London, SW7 2BT, England

ABSTRACT: *The performance of two-sample non-parametric detectors using dependent samples is considered. It is shown that in the situation in which square-law envelope detection is employed, the signals to be detected fluctuate in accordance with a chi-square distribution and the background interference is additive Gaussian noise; the two-sample Generalized Sign (GS), Mann-Whitney (MW), Modified Savage (MS) and Modified Rank Square (MRS) detectors have superior asymptotic processing times to those that are obtained when independent samples are used. Also, it is shown that in addition to the advantage of improved performance, detection using dependent samples can have implementation advantages as compared with detection based on the use of independent samples.*

I. Introduction

The use of non-parametric detection methods using independent samples has been described previously (**1**) and it has been shown that one of the main advantages and distinguishing features of non-parametric detection is that when using independent samples a constant false alarm probability (CFAR) is maintained for a wide variety of background noise distributions. However, more recently, it has been shown (**2**, **3**) that when dependent samples are used non-parametric methods lose their CFAR property. One method of maintaining a CFAR is to use an adaptive threshold (**4**, **5**), but the implementation of these adaptive techniques tends to be somewhat complicated. An alternative way of obtaining a CFAR has been suggested by Woinsky and Kurz (**6**), and by Kassam and Thomas (**7**). The suggestion of (**6**) and (**7**) involves the grouping of dependent samples and the use of independent groups. By this technique it is possible to obtain a CFAR performance with conventional non-parametric techniques, without having to use adaptive procedures.

In (**7**), Kassam and Thomas consider the case of the single-sample detection of constant positive signals imbedded in Gaussian noise, when coherent detection is employed. In the present paper, we consider two-sample non-parametric detection using dependent samples. By two-sample detectors we mean the number of sets of observations available at the detector input. The first set consists of M samples which may contain a signal, and the second set consists of MN reference noise samples. It is assumed that square-law envelope detection is used and that the signals fluctuate in accordance with a chi-square distribution.

Since the main purpose of this paper is to compare the performance of two-sample non-parametric detectors processing dependent observations with those processing independent observations, and not to compare non-parametric detectors with optimum parametric detectors, the background noise is assumed to be Gaussian. For the case in which the background noise is non-Gaussian and the observations are independent, the relative performances of the two-sample detectors considered here are compared in detail (**12**) with the corresponding performances of parametric detectors. By using these methods the results obtained previously (**12**) can be extended to the case in which the observations are statistically dependent. We shall consider the performance of the Generalized Sign (GS), Mann–Whitney (MW), Modified Savage (MS) and Modified Rank Squared (MRS) detectors expressed in terms of their asymptotic processing time relative to the asymptotic processing time of the same detectors when independent samples are used. In addition, we consider the asymptotic relative processing time (ARPT) performance of the two-sample GS and MW detectors when coherent rather than envelope detection is used and when the signal to be detected is constant and positive.

The results obtained indicate that the use of dependent samples, as compared with the use of independent samples, gives improved ARPT performance, and that the detector implementation becomes easier, particularly as the order of the ranking increases.

II. Detection Using Dependent Samples

Assume that at the output of an envelope detector samples are taken at T sec intervals and that there are $N+1$ such independent samples per pulse. Let it also be supposed that of the $N+1$ samples, N are noise samples, and that the remaining sample may contain signal. This situation for independent samples is illustrated in Fig. 1a. Further, let it be assumed that if samples are taken closer together than T sec then they become statistically dependent.

If the sampling rate is increased by a factor γ, that is, the sampling interval is decreased from T sec to T/γ sec then there are a total of $(N+1)\gamma$ dependent samples per sweep and if, in accordance with the method suggested by Kassam and Thomas (**7**) these dependent samples are divided into groups of n samples, with each group being separated by $\theta(\geqslant\gamma-1)$ samples that are discarded from the decision making process, then the variable formed by adding the samples within each group of n samples is statistically independent from group to group and can, therefore, be used in a non-parametric detector to obtain a CFAR probability. This situation for dependent samples is illustrated in Fig. 1b. It is important to note that in choosing n and γ, the value of n should be set equal to γ or $\gamma+1$ so as to ensure that, when a signal is actually present, all samples

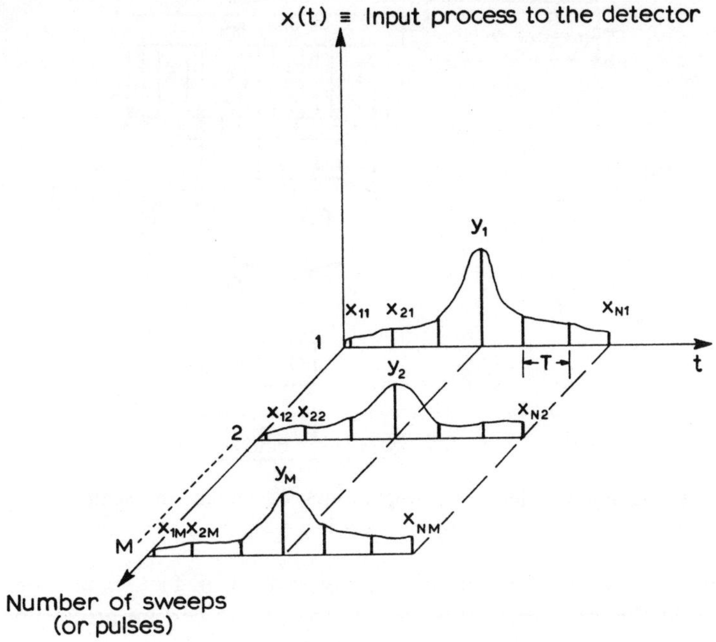

FIG. 1(a). Independent sampling case $x(t)$.

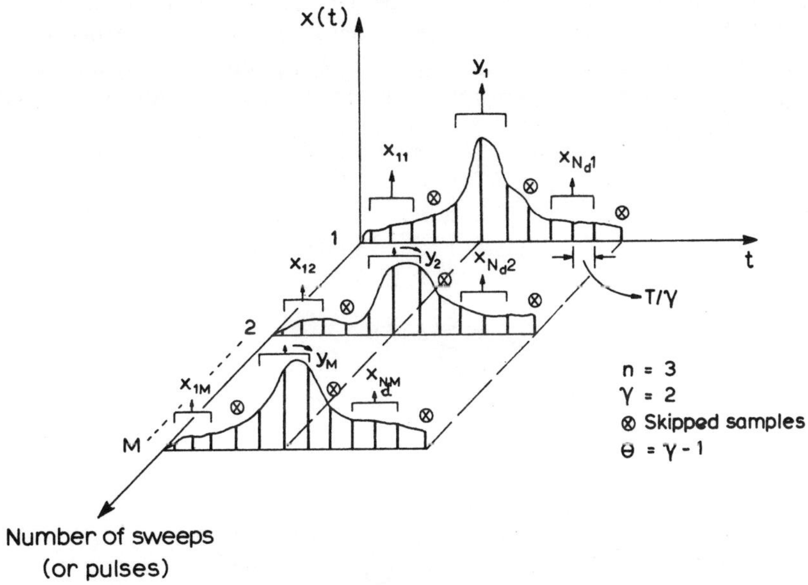

FIG. 1(b). Dependent sampling case.

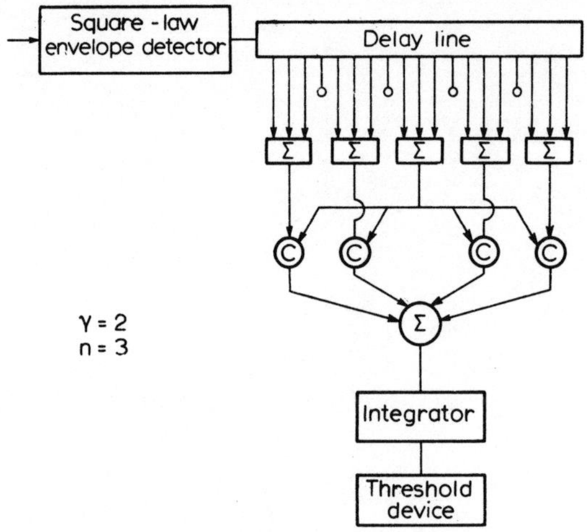

FIG. 2. A nonparametric detector using groups of dependent samples and sample skipping ($n = 3$, $\gamma = 2$).

associated with the group of samples under test are in fact signal samples. By choosing n in this manner it also ensures that, for a certain group under test, the situation in which all the samples in a group are signal samples occurs only once per sweep, and a unique maximum output is thereby achieved, (see Fig. 2 for an illustration of the scheme).

Let y_j denote the sum of the n samples associated with the group under test, and let x_{ij}; $i = 1, \ldots, N_d$; $j = 1, \ldots, M$ denote the sum of the n noise samples, associated with the (ij)th group of samples. Then, in terms of this notation, the GS, MW, MS and MRS non-parametric detectors* operate respectively by forming the statistics T_{GS}, T_{MW}, T_{MS} and T_{MRS}, and then comparing the appropriate statistic with an appropriate fixed threshold. The four statistics are:

$$T_{\mathrm{GS}} = \sum_{j=1}^{M} \sum_{i=1}^{N_d} U(y_j - x_{ij}) \tag{1}$$

$$T_{\mathrm{MW}} = \sum_{j=1}^{M} \sum_{k=1}^{M} \sum_{i=1}^{N_d} U(y_j - x_{ki}) \tag{2}$$

$$T_{\mathrm{MS}} = \sum_{j=1}^{M} a_j(r_j) \tag{3}$$

where

$$a_j(r_j) = \sum_{l_j = N_d + 1 - r_j}^{N_d + 1} 1/l_j$$

and

$$r_j = \sum_{i=1}^{N_d} U(y_j - x_{ij}); \quad 0 \leq r_j \leq N_d$$

* A detailed description of these detectors is to be found in Refs. (**8**) and (**9**).

and

$$T_{\text{MRS}} = \sum_{j=1}^{M} r_j^2 \tag{4}$$

where r_j is defined as for T_{MS}.

In these equations;

$$U(y_j - x_{ki}) = 1 \quad \text{if} \quad y_j \geqslant x_{ki}$$
$$= 0 \quad \text{otherwise.}$$

In this paper the number of discarded samples is kept to a minimum, i.e. θ is set equal to $\gamma - 1$: hence

$$N_d = \frac{(N+1)\gamma}{n+\gamma-1} - 1.$$

III. The ARPT of the Detectors

(a) *The ARPT when using envelope detection*

We now consider the performance of the above four detectors, when using dependent samples, by considering in each case the asymptotic processing time and comparing this time with the asymptotic processing time of the detector using independent samples. This is a comparison, of course, in terms of the familiar ARPT. In deriving the results we shall assume that the previously mentioned conditions of chi-square signal fluctuations, Gaussian background noise, and envelope detection apply. Also it is assumed that the signal and noise components are independent. In particular, we assume that the noise at the input to the envelope detector is a stationary T-dependent* Gaussian process, and that the target signal-to-noise power ratio at the detector input is a chi-square density function given (**10**) by

$$f_S(S) = \frac{k}{(k-1)!\,\bar{S}} \cdot \left(\frac{kS}{\bar{S}}\right)^{k-1} \cdot \exp\left(-\frac{kS}{\bar{S}}\right); \quad S > 0 \tag{5}$$

where S is the instantaneous single-pulse signal-to-noise ratio, $\bar{S}$ is the average single-pulse signal-to-noise ratio, and k is equal to half the number of degrees of freedom.

In Appendix 1, use is made of Kanter's method (**11**) to derive the density function of the statistic y_j formed from a group of n dependent samples. In the appendix it is shown that at the output of the envelope detector the pdf of any

* T-dependent means that samples taken T or more sec apart are statistically independent, and that if they are taken less than T sec apart then they are statistically dependent.

group sample, y is†

$$f_y(y)=\frac{1}{2\pi i}\int_{\alpha-i\infty}^{\alpha+i\infty}\frac{\frac{py}{e^{2\sigma^2}}\cdot[\prod_{j=1}^{n}(1+\lambda_j p)]^{k-1}}{\left[\prod_{j=1}^{n}(1+\lambda_j p)+\frac{p\bar{S}}{k}\sum_{l=1}^{n}G_l\prod_{j=1}^{\prime n}(1+\lambda_j p)\right]^k}\cdot dp/2\sigma^2 \quad (6)$$

where $i=\sqrt{-1}$,

the λ_j's are the eigenvalues of the normalized convariance matrix C,

$G_q=\sum_{l=1}^{n}\sum_{j=1}^{n}L_l^{(q)}L_j^{(q)}$ with the $L^{(q)}$'s being the corresponding orthogonal eigenvectors, and $L_l^{(q)}$ is the lth element of the vector $L^{(q)}$

$2\sigma^2$ denotes the average noise power in each of the input pulses,

and the prime denotes the deleted product.

The pdf at the output of the envelope detector, when noise alone is present at the detector input, can be obtained from Eq. (6) simply by putting $\bar{S}$ equal to zero.

In Appendix 1 expressions are derived for the ARPT of the GS and MW detectors, and two cases are considered. Initially, it is assumed that $n=2$, and it is found that in this case the ARPT's of the GS and MW detectors are independent of γ. In the second case, the ARPT's are derived when $n=3$ and $\gamma=2$. The results derived in Appendix 1 for the ARPT's are:

Case I: $n=2$, γ arbitrary

$$\text{ARPT}_{(GS)}=\frac{16(1+2/N)}{1+2/N_d}\cdot\left\{\frac{3-R(t)}{8}\right\}^2 \quad (7)$$

$$\text{ARTP}_{(MW)}=\frac{16(1+1/N)}{1+1/N_d}\cdot\left\{\frac{3-R(t)}{8}\right\}^2. \quad (8)$$

Case II: $n=3$, $\gamma=2$

$$\text{ARPT}_{(GS)}=\frac{16(1+2/N)}{1+2/N_d}\cdot\left[\frac{60-20R(T/2)-69R^2(T/2)+28R^3(T/2)+18R^4(T/2)-8R^5(T/2)}{32\{2-R^2(T/2)\}^2}\right]^2 \quad (9$$

$$\text{ARPT}_{(MW)}=\frac{16(1+1/N)}{(1+1/N_d)}\cdot\left[\frac{60-20R(T/2)-69R^2(T/2)+28R^3(T/2)+18R^4(T/2)-8R^5(T/2)}{32\{2-R^2(T/2)\}^2}\right]^2 \quad (10)$$

† It is important to note that in deriving Eq. (6) in Appendix 1 it was assumed that signal components in the same group were completely correlated. This is a situation which corresponds (**10**) to Swerling's models I and III.

where $R(\cdot)$ is the normalized autocorrelation function of the input process.

Although it is possible, as above, to derive explicit expressions for the ARPT's of the GS and MW detectors, it is not possible to derive similar expressions for the ARPT's of the MS and MRS detectors. With these detectors it is necessary to compute the ARPT's numerically. It is important to note that the Eqs. (7–10) and the numerically computed values of ARPT are independent of k (see Appendix 1, section B), that is to say are valid for all chi-square fluctuating signals, and thus for all cases of Marcum and Swerling (**10**).

(b) *The ARPT of two-sample detectors using coherent detection*

In addition to considering the ARPT's of the four two-sample detectors when envelope detection is employed, we have also considered the performance of the two-sample GS and MW detectors when coherent detection is used and the signals to be detected are positive and non-fluctuating. The derivations of the ARPT's is presented in Appendix 2, from which it can be seen that

$$\text{ARPT}_{(\text{GS})} = \frac{n^2}{n + 2\sum_{i=1}^{\gamma-1}(n-i)R(iT/\gamma)} \cdot \left(\frac{1+2/N}{1+2/N_d}\right) \tag{11}$$

and

$$\text{ARPT}_{(\text{MW})} = \frac{n^2}{n + 2\sum_{i=1}^{\gamma-1}(n-i)R(iT/\gamma)} \cdot \left(\frac{1+1/N}{1+1/N_d}\right). \tag{12}$$

It should be noted that Eqs. (11) and (12) are true for $n \geq \gamma$, but in our case, for the reasons mentioned earlier, n is restricted to have the value of γ or $\gamma+1$.

IV. Results

The ARPT's have been computed using Eqs. (7–12) and, where appropriate, numerical techniques, and the main results are shown in Figs. 3 and 4. Figure 3 shows the results obtained when envelope detection is employed and Fig. 4 the corresponding results for coherent detection. In Fig. 3, the ARPT's are given as a function of the autocorrelation function $R(T/2)$ and in Fig. 4, the ARPT's are given as a function of the group size n, for various autocorrelation functions.

In interpreting Figs. 3 and 4, it must be remembered that the ARPT's as given indicate the extent of the improvement in detector performance when using dependent as opposed to independent samples, and that they are not indicative of the absolute performance of a detector. Figure 3 shows that for $|R(T/2)|$ less than approximately 0.8 the ARPT's are greater than 1, which indicates that an actual improvement in detector performance takes place as a result of using dependent samples. The figure indicates that the improvement is greater for the GS and MW detectors than for the MS and MRS detectors, and furthermore, the improvement is most significant for negatively correlated samples. Figure 3 indicates that the greatest improvement in performance is obtained by using the GS and MW detectors with N tending to infinity. A rather special, but interesting and important result should be noted from Fig. 3 in connection with the performance of the GS and MW detectors. Figure 3 shows that, for the special case in which $n=2$, the GS and MW detectors operating without sample skipping and with $N=5$ have an identical ARPT to that obtained for the detectors when N is allowed to approach infinity. That this is a special case can be seen from Fig. 3b from which it is clear that, when $n=3$, the ARPT's obtained for the GS and MW detectors operating with $N=5$ and without the use of sample skipping are different from the ARPT obtained for the detectors when N tends to infinity. From Figs. 3, and 4, when comparing the ARPT performances for $N=5$ and $N=\infty$ respectively, it is clear that little gain is obtained in performance. This is based on the assumption that M is large. The number N of reference noise samples per pulse (N per sweep)

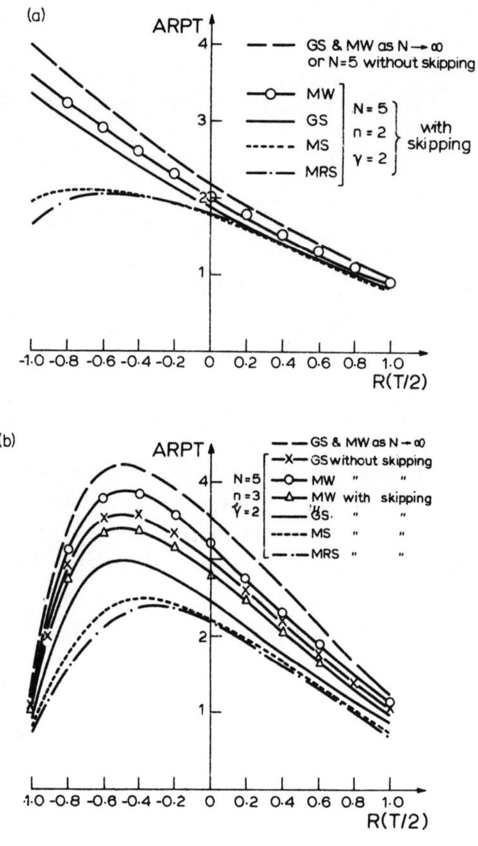

FIG. 3. Detector ARPT versus autocorrelation function $R(T/2)$.

(a) $N=5,\ n=2,\ \gamma=2$

(b) $N=5,\ n=3,\ \gamma=2$.

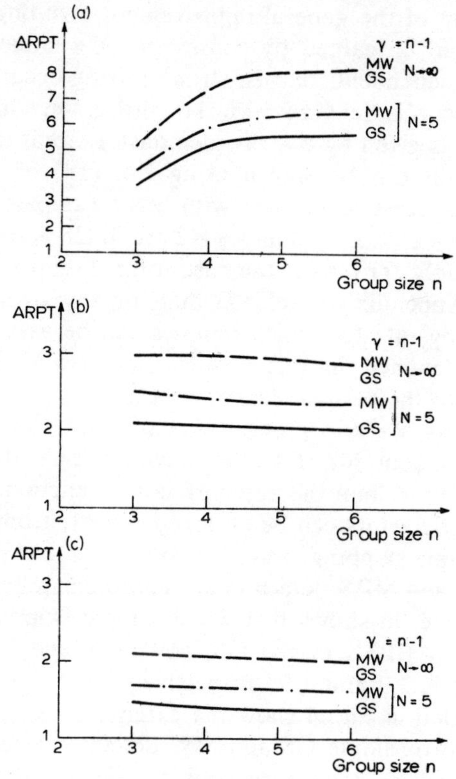

FIG. 4. Detector ARPT versus group size, n.

(a) $$R(t)=\left(1-\frac{|t|}{T}\right)^2\cdot\left\{1+\frac{|t|}{2T}\right\}\cos\left(\frac{2\pi t}{T}\right); \quad |t|\leq T$$
$$=0 \qquad ; \quad |t|>T$$

(b) $$R(t)=\left(1-\frac{|t|}{T}\right)\cos\left(\frac{\pi t}{T}\right) \qquad ; \quad |t|\leq T$$
$$=0 \qquad ; \quad |t|>T$$

(c) $$R(t)=\left(1-\frac{|t|}{T}\right)^2\left\{1+\frac{|t|}{2T}\right\} \qquad ; \quad |t|\leq T$$
$$=0 \qquad ; \quad |t|>T.$$

was chosen arbitrarily to be 5 for the following reasons:

(a) it is a representative example for small reference noise samples per sweep in comparison with the case of using an infinite number

(b) to give an integer number of group samples N_d associated with small values of γ and n which can be easily designed for practical systems.

A specific indication of the general improvement resulting from the use of dependent samples can be gained by considering the situation in which the samples are strongly dependent, that is, strong mixing occurs. An example of the strong mixing process is the Gauss–Markov process in which the autocorrelation function, $R(t)$, is given by $e^{-|t|}$. To demonstrate that the Gauss–Markov is not T dependent, it can be shown using Eq. (18) of Ref. (7) that the correlation between successive groups, with $\gamma = 1$ samples skipped between them, for $n = 2$ and $\gamma = 2$, and $n = 3$ and $\gamma = 2$ are 0.028 and 0.017 respectively which are approximately zero, as is the case for a T-dependent process. It is also proved in the Appendix of Ref. (7) that for any strong mixing process there exists a sampling rate for which samples can be assumed independent. Thus, the analysis obtained in this paper is also valid for strong mixing processes additional to the Gauss–Markov process.

Now, for the Gauss–Markov process, the samples can be considered as approximately independent for $T = 1/\pi$, in which case the autocorrelation function $R(T/2) = 0.208$. When the autocorrelation function has this value the ARPT's for the four detectors can be obtained directly from Fig. 3. Figure 3a shows that when sample skipping is used, $N = 5$, $n = 2$ and $\gamma = 2$; the ARPT's of the GS, MW, MS and MRS detectors are approximately 1.66, 1.8, 1.6 and 1.6 respectively. Figure 3b shows that when sample skipping is used, $N = 5$, $n = 3$ and $\gamma = 2$; the ARPT's of the GS, MW, MS and MRS detectors are approximately 2.2, 2.5, 1.9 and 1.95 respectively.

The results presented in Fig. 4 show the extent of the improvement in the performance of the two-sample GS and MW detectors when dependent samples are used in a system employing coherent detection, and in which the signals to be detected are positive and non-fluctuating. The main points that are clear from Fig. 4 are that the ARPT's for both detectors are greater than 1, which indicates that a definite improvement results from using dependent samples, and, secondly, a significant improvement results when using relatively small group sizes ($n = 4$), and that there is little or nothing to be gained from using larger group sizes.

V. Conclusions

The improvement in the performance of the two-sample non-parametric detectors as a result of using dependent rather than independent samples has been examined and it has been found that the performance of the GS, MW, MS and MRS detectors is improved, except when the samples are very highly correlated ($|R(T/2)| > 0.8$), by using dependent samples. It has also been found that significantly improved performance can be achieved when using relatively small ($N_d \doteq 2$) groups of samples.

In addition to leading to an improvement in detector performance, the use of groups of dependent samples can also result in a simplification of the implementation of the various detectors. For example, the ranking procedure is decreased for the GS and MS detectors from MN to MN_d and for the MW detector from M^2N to M^2N_d. If N, γ and n are 5, 2 and 3 respectively then the number of ranking operations needed is decreased to two-fifths of the number necessary when using independent samples.

References

(1) M. Kanefsky, and J. B. Thomas, "On polarity detection schemes with non-Gaussian inputs", *J. Franklin Inst.*, Vol. 280, pp. 120–138, Aug., 1965.

(2) L. D. Davisson, E. A. Feustel and J. W. Modestino, "The effects of dependence on nonparametric detection", *IEEE Trans. Info. Theory*, Vol. IT-16, pp. 32–41, Jan., 1970.

(3) E. P. Cunningham, "False alarm rate for rank quantizer when input range samples are not independent", *IEEE Trans. on Aerospace and Electronic Systems.* Vol. AES 12, Corresp., pp. 65–67, Jan., 1976.

(4) M. Kanefsky and J. B. Thomas, "On adaptive nonparametric detection systems using dependent samples", *IEEE Trans. Info. Theory*, Vol. IT-11, pp. 521–526, Oct., 1965.

(5) G. V. Trunk, B. H. Cantrell and F. D. Queen, "Modified Generalized sign test processor for 2-D radar", *IEEE Trans. Aerospace and Electronic Systems*", Vol. AES-10, pp. 574–582, Sept., 1974.

(6) M. N. Woinsky and L. Kurz, "Nonparametric detection using dependent samples", *IEEE Trans. Info. Theory*, Vol. IT-16, pp. 355–358, Corresp., May, 1970.

(7) S. A. Kassam and J. B. Thomas, "A class of nonparametric detectors for dependent input data", *IEEE Trans. Info. Theory*, vol. IT-21, pp. 431–437, July, 1975.

(8) V. G. Hansen and B. A. Olsen, "Nonparametric radar extraction using a generalized sign test", *IEEE Trans. on Aerospace and Electronic Systems*, Vol. AES-7, pp. 942–950, Sept., 1971.

(9) E. K. Al-Hussaini, F. M. Badran and L. F. Turner. "Modified Savage and Modified Rank Squared Non-Parametric Detectors", *IEEE Trans. on Aerospace and Electronic Systems*, Vol. AES-14, No. 1, Jan., 1978.

(10) P. Swerling, "Recent developments in target models for radar detection analysis", AGARD Avionics tech. Symp. proc., Istanbul, Turkey.

(11) I. Kanter, "A generalization of the detection theory of Swerling", EASCON-74, pp. 198–205, Oct., 1974, also in "Detection and Estimation applications to radar", S. S. Hykin, Ed. pp. 159–166, Dowden, Hutchinson and Ross, Inc., 1976.

(12) E. K. Al-Hussaini, F. M. Badran and L. F. Turner, "The Nonparametric Detection of Signals Embedded in Non-Gaussian Noise", *IEEE Trans. on Aerospace and Electronic Systems*", Vol. AES-14, No 2, March, 1978.

(13) G. E. Noether, "On a Theorem of Pitman", *Ann. Math. Stat.*, Vol. 26, pp. 64–68, 1955.

Appendix 1.

The probability density function of a univariate group at the output of a square-law envelope detector:

(A) *The Determination of the ARPT*

First, it is required to derive the pdf of y where:

$$y = \sum_{i=1}^{n} (y_{I_i}^2 + y_{Q_i}^2) \tag{A1-1}$$

and y_{I_i}, s and y_{Q_i}, s are the elements of the in-phase and quadrature phase vectors Y_I and Y_Q respectively; and each vector is composed of n dependent variables. The joint density function of Y_I and Y_Q is given by

$$f(Y_I, Y_Q)=\frac{\exp\left\{-\frac{1}{2\sigma^2}[(Y_I-\bar{Y}_I)^tC^{-1}(Y_I-\bar{Y}_I)+(Y_Q-\bar{Y}_Q)^tC^{-1}(Y_Q-\bar{Y}_Q)]\right\}}{(2\pi\sigma^2)^n\cdot|c|} \tag{A1-2}$$

where t denotes the transpose of the vector and

$$\bar{y}_{I_i}=\sqrt{\sigma_s}\cdot\cos\theta,\quad \bar{y}_{Q_i}=\sqrt{\sigma_s}\cdot\sin\theta$$

σ_s is proportional to the square of the signal amplitude, and θ is the phase of the signal.

Following Kanter's (**11**) method, the normalized covariance matrix C can be transformed to a diagonal matrix if we use the orthogonal transformation:

$$U=L^tY_I \quad \text{and} \quad V=L^tY_Q \tag{A1-3}$$

where

$$L^tCL=\begin{bmatrix}\lambda_1 & & & & 0\\ & \lambda_2 & & & \\ & & \cdot & & \\ & & & \cdot & \\ & & & & \cdot \\ 0 & & & & \lambda_n\end{bmatrix}$$

and $\lambda_1,\ldots,\lambda_n$ are the eigenvalues of the covariance matrix C and their n orthogonal eigenvectors $L^{(j)}$s, constituting the matrix L, are the solutions of:

$$CL^{(j)}=\lambda_jL^{(j)};\quad 1\leq j\leq n. \tag{A1-4}$$

The problem is then transformed to deriving the pdf of y where:

$$y=\sum_{i=1}^{n}(u_i^2+v_i^2) \tag{A1-5}$$

in which case the components of each vector (U and V) are independent. Equation (6) results from applying Kanter's (**11**) method.

If we consider the case when $k=1$, i.e. for Swerling case 1, then for the cases in which the group sample n equals 2 or 3, we obtain for the pdf of y:

(a) $k=1$, $n=2$ and γ is arbitrary

$$f(y)=\frac{1}{2\sigma^2}\left\{\begin{aligned}&\left[(1-R^2(t))\left(1+\frac{2\bar{S}}{1+R(t)}\right)\left(\frac{1}{1+R(t)+2\bar{S}}-\frac{1}{1-R(t)}\right)\right]^{-1}\\ &\times\exp\left[\frac{-y}{2\sigma^2(1-R(t))}\right]+\left[(1-R^2(t))\left(1+\frac{2\bar{S}}{1+R(t)}\right)\right.\\ &\left.\times\left(\frac{1}{1-R(t)}-\frac{1}{1+R(t)+2\bar{S}}\right)\right]^{-1}\exp\left[-\frac{y}{2\sigma^2(1+R(t)+2\bar{S})}\right]\end{aligned}\right\} \tag{A1-6}$$

If we put $\bar{S}$ equal to zero in the above equation we obtain, for noise alone, with dependent sampling

$$f(y)\Big|_{\bar{S}=0} = \frac{1}{4\sigma^2 R}\cdot\left[\exp\left(\frac{-y}{2\sigma^2(1+R)}\right) - \exp\left(-\frac{y}{2\sigma^2(1-R)}\right)\right]. \tag{A1-7}$$

If further, we put, R equal to zero in Eq. (A1-7), we obtain for noise alone with independent sampling

$$f(y)\Big|_{\substack{\bar{S}=0\\R=0}} = \frac{y}{(2\sigma^2)^2}\exp\left(\frac{-y}{2\sigma^2}\right) \tag{A1-8}$$

which is a chi-square distribution with four degrees of freedom resulting from summing two independent samples each with an exponential probability density function (i.e. chi-square with two degrees of freedom).

(b) $k=1$, $n=3$ and $\gamma=2$.

$$\begin{aligned} f(y) = \frac{1}{2\sigma^2}\Bigg\{&\left[\frac{d_1 d_3}{d_4}\left(1-\frac{d_2-d_3}{2d_4}\right)\right]^{-1}\exp\left[-\frac{y(d_2-d_3)}{4\sigma^2 d_4}\right] \\ &+\left[d_1\left(1-\frac{d_2-d_3}{2d_4}\right)\left(1-\frac{d_2+d_3}{2d_4}\right)\right]^{-1}\exp\left[-\frac{y}{2\sigma^2}\right] \\ &+\left[-\frac{d_1 d_3}{d_4}\left(1-\frac{d_2+d_3}{2d_4}\right)\right]^{-1}\exp\left[-\frac{-y(d_2+d_3)}{4d_4\sigma^2}\right]\Bigg\} \end{aligned} \tag{A1-9}$$

where

$$d_1 = [1-2R^2(T/2)]\left[1+\bar{S}\cdot\frac{(3-4R(T/2))}{1-2R^2(T/2)}\right]$$

$$d_2 = 3\bar{S}+2$$

$$d_3 = [9\bar{S}^2+16\bar{S}R(T/2)+8R^2(T/2)]^{1/2}$$

$$d_4 = [1-2R^2(T/2)+3\bar{S}-4\bar{S}R(T/2)]$$

The corresponding equations to [A1-7] and (A1-8) for noise alone conditions are given by:

$$\begin{aligned} f(y)|_{\bar{S}=0} = \frac{1}{2\sigma^2}\Bigg\{&\frac{1+\sqrt{(2)}R}{4R^2}\exp\left[-\frac{y}{2\sigma^2(1+\sqrt{(2)}R)}\right] - \frac{1}{2R^2}\exp\left[-\frac{y}{2\sigma^2}\right] \\ &+\frac{1-\sqrt{(2)}R}{4R^2}\exp\left[-\frac{y}{2\sigma^2(1-\sqrt{(2)}R)}\right]\Bigg\} \end{aligned} \tag{A1-10}$$

and

$$f(y)\Big|_{\substack{\bar{S}=0\\R=0}} = \frac{y^2\exp(-y/2\sigma^2)}{2!\,(2\sigma^2)^3}. \tag{A1-11}$$

As before, the last equation is a chi-square distribution with six degrees of freedom resulting from the summation of three independent samples each having a chi-square distribution with two degrees of freedom.

(B) *The ARPT can be derived as follows:*

The ARPT between two detectors D_1 and D_2 is defined (**7**) by:

$$\mathrm{ARPT}_{1,2} = \mathrm{ARE}_{1,2} \times \frac{t2}{t1}$$

where t_i ($i=1$ or 2) is the time between samples of detector D_i. So, the derivation of the ARPT (for the same detector) is equivalent to the derivation of the ARE between dependent and independent sampling cases when MN_d and MN reference noise samples are used respectively.

According to the Pitmen–Noether (**13**) theorem, the ARE between two detectors is defined as the ratio between their efficiencies. The efficacy of a two-sample test statistic T is defined by:

$$\varepsilon = \lim_{M\to\infty} \frac{1}{M\sigma_0^2(T)} \left[\frac{\partial E(T/\bar{S})}{\partial \bar{S}} \bigg|_{\bar{S}=0} \right]^2 \tag{A1-12}$$

where $\sigma_0^2(T)$ is the variance of T when noise alone is present, which is clear to be independent on k. Also $\partial E(T/\bar{S})/\partial \bar{S}|_{\bar{S}=0}$ is dependent mainly on $\partial f_y(y)/\partial \bar{S}|_{\bar{S}=0}$ of Eq. (6) which is again independent on k. Thus, the efficacies of the detectors under consideration and hence their ARE's are independent on k, i.e. the ARE results are valid for all signal fluctuations with a chi-squared distribution.

In section (A) of this Appendix, the pdf was derived for $k=1$, taken as an example, for simplicity to show how it can be derived for different γ and n. But, for the ARPT results there is no restriction on k.

Appendix 2

The ARPT of GS and MW Detectors when Coherent Detection is used and the signal is Positive and Non-fluctucting.

In this case the noise samples are assumed Gaussian with zero mean and unit variance. For a single group of size n when a constant positive signal with magnitude A is present we have, for a single group output:

$$Y = x_1 + x_2 + \cdots + x_n + nA. \tag{A2-1}$$

When noise alone is present, we have

$$f(y) = \frac{1}{\sqrt{(2\pi\sigma^2)}} \exp\left(-\frac{y}{2\sigma^2}\right) \tag{A2-2}$$

where

$$\sigma^2 = n + 2\sum_{i=1}^{\gamma-1}(n-i)R(i\,T/\gamma); \quad n \geq \gamma.$$

When a signal is present, we have

$$g(y) = \frac{1}{\sqrt{(2\pi\sigma^2)}} \exp\left[-\frac{(y-nA)^2}{2\sigma^2}\right] \tag{A2-3}$$

where σ^2 is as defined for Eq. (A2-2).

Following the same steps as in Ref. (**9**), the efficacies under dependent sampling can be obtained and the ARPT results are as given in Equations (11) and (12).

BIBLIOGRAPHY

List of sources and abbreviations used in the bibliography:

IRE/IEEE Transactions	Trans.
Automatic Control	Autom. Control
Aerospace and Electronic Systems	Aerosp. Electr. Syst.
Computers	Comput.
Communications	Commun.
Information Theory	Inf. Theory
Systems, Man and Cybernetics (formerly Systems Science and Cybernetics)	Syst., Man and Cybern. (Syst. Sci. and Cybern.)
Sonics and Ultrasonics	Sonics and Ultrason.
IRE/IEEE Proceedings	Proc.
IRE/IEEE Conferences, Conventions, and Symposia	
Canadian Communications and Power Conference	CCPC
Decision and Control Conference	DCC
Electronics and Aerospace Systems Convention	EASCON
International Conference on Acoustics, Speech and Signal Processing	ICASSP
International Conference on Communications	ICC
International Convention	ICONV
International Radar Conference	IRC
International Symposium on Information Theory	ISIT
National Aerospace Electronics Convention	NAECON
National Communications Symposium	NCS
National Electronics Conference	NEC
National Telecommunication Conference	NTC
Pattern Recognition and Image Processing Conference	PRIPC
Reliability and Maintainability Symposium	RMS
Symposium on Adaptive Processes	SAP
Southwestern IEEE Conference	SWIEEECO
Western Electronic Show and Convention	WESCON
Other Publications	
Allerton Conference on Communications, Control and Computing (formerly Allerton Conference on Circuit and System Theory)	Aller. Conf. Commun., Control and Comput. (Aller. Conf. Circuit Syst. Theory)

Asilomar Conference on Circuits and Systems	Asilomar Conf. Circuit Syst.
Automation and Remote Control; translation of Avtomatika i Telemekhanika (Consultants Bureau)	Autom. Remote Control
Computers and Electrical Engineering	Comput. Electr. Eng.
Conference on Information Sciences and Systems (Princeton or Johns Hopkins)	Conf. Inf. Sci. Syst.
Engineering Cybernetics; translation of Tekhnicheskaya Kibernetika (Scripta)	Eng. Cybern.
Hawaii International Conference on Systems Science	Hawaii Int. Conf. on Syst. Sci.
Institution of Electrical Engineers Conference Publication	IEE Conf. Publ.
Information and Control	Inf. and Control
Information Sciences	Inf. Sci.
International Telemetering Conference	ITC
Joint Conference on Mathematical and Computer Aids to Design (SIAM-ACM)	Joint Conf. Math. Comput. Aids to Design (SIAM-ACM)
Journal of the Acoustical Society of America	Acoust. Soc. Am. J.
Journal of the Franklin Institute	J. Franklin Inst.
M. J. Kelly Communications Conference (University of Missouri-Molla)	M. J. Kelly Commun. Conf. (Univ. of Missouri-Rolla)
Midwest Symposium on Circuits and Systems	Midwest Symp. Circuits and Syst.
Problems of Information Transmission; translation of Problemy Peredachi Informatsii (Consultants Bureau)	Probl. Inf. Transm.
Radio Engineering and Electronic Physics; translation of Radiotechnika i Electronika (Scripta)	Radio Eng. Electr. Phys.
Society for Industrial and Applied Mathematics: Journal on Applied Mathematics	SIAM J. Appl. Math.
Symposium on Computer Processing in Communications (Polytechnic Institute of Brooklyn)	Symp. Comput. Process. in Commun. (Polytech. Inst. Brook.)
Telecommunication and Radio Engineering; translation of Elektrosvyaz i Radiotechnika (Scripta)	Telecommun. Radio Eng.
Workshop on Communication Theory and Applications (Air Force Office of Scientific Research)	Workshop Commun. Theory and Appl. (AFOSR)

Bibliography

Abel, S. W., and J. R. Betten. 1969. Application of the Nonparametric Sign Test to Sequential Detection. *ISIT Abstr.*, pp. 41–42.

Abend, K. 1966. Compound Decision Procedures for Pattern Recognition. *NEC Proc.*, pp. 777–780.

Abuskhairoon, S. S., and M. Arozullah. 1972. Nonparametric Detection of Time-Varying Signals in the Presence of Intersymbol Interference. *ICC Proc.*, pp. 46.21–46.26.

Akimov, P. S. 1977. Nonparametric Detection of Signals. *Telecommun. Radio Eng.* **31/32**, pt. 1, 26–36.

Akimov, P. S., and V. S. Yefremov. 1975. Optimum Rank Sequential Detection of a Pulsed Signal. *Radio Eng. Electr. Phys.* **20**:44–53.

Akimov, P. S., and V. S. Yefremov. 1975. Ranked Sequential Procedure of Detection of Noncoherent Signals. *Radio Eng. Electr. Phys.* **20**:73–77.

Akimov, P. S., and V. S. Yefremov. 1977. An Adaptive Rank Sequential Detection Procedure. *Radio Eng. Electr. Phys.* **22**:56–61.

Akimov, P. S., V. S. Yefremov, and A. N. Kubasov. 1975. Sequential Detection of Signals Using Statistics Based on a Sum of Ranks. *Radio Eng. Electr. Phys.* **20**: 119–123.

Akimov, P. S., V. S. Yefremov, L. P. Zhurkov, and A. B. Kosolapov. 1976. Realization of a Parametric Rank Detector. *Telecommun. Radio Eng.* **30/31**, pt. 2, 127–128.

Al-Hussaini, E. K., and A. F. Hassan. 1974. Adaptive Nonparametric Detectors for Weak Signals in Noise. *ISIT Abstr.*, p. 34.

Al-Hussaini, E. K., F. M. Badran, and L. F. Turner. 1977. Two-Sample Non-parametric Detectors Using Dependent Samples. *J. Franklin Inst.* **304**:265–278.

Al-Hussaini, E. K., F. M. Badran, and L. F. Turner. 1978. Modified Savage and Modified Rank Squared Nonparametric Detectors. *IEEE Trans. Aerosp. Electr. Syst.* **AES-14**:242–251.

Al-Hussaini, E. K., F. M. Badran, and L. F. Turner. 1978. The Nonparametric Detection of Signals Embedded in Log-Normal Noise. *IEEE Trans. Aerosp. Electr. Syst.* **AES-14**:668–670.

Anderson, M. W., and R. D. Benning. 1970. A Distribution-Free Discrimination Procedure Based on Clustering. *IEEE Trans. Inf. Theory* **IT-16**:541–548.

Antoniak, C. E. 1974. The Use of Walsh Functions for Distribution-Free CFAR Signal Detection. *ISIT Abst.*, p. 33.

Antoniak, C. E., and G. M. Dillard. 1968. A Distribution-Free Sequential Probability-Ratio Test for Multiple-Resolution-Element Radars. *IEEE Trans. Inf. Theory* **IT-14**:822–825.

Antoniak, C. E., G. M. Dillard, and R. A. Shorack. 1967. Distribution-Free Detection in Radar with Multiple Resolution Elements. *SWIEEECO Rec.*, pp. 17.2.1–17.2.8.

Arkhipov, V. S., and B. R. Levin. 1975. Asymptotically Optimum Rank Algorithms of Detection of Signals in Correlated Noises. *Radio Eng. Electr. Phys.* **20**:133–135.

Armstrong, G. L. 1966. Performance of the Neyman-Pearson Detector for Correlated Input Samples. *IEEE Proc.* **54**:1190–1191.

Arozullah, M., and S. V. Whitten. 1976. Effect of Sample Dependence and Sample Size on Some Nonparametric Signal Detectors. *ICC Proc.*, pp. 43.12–43.19.

Baronkin, V. M., A. F. Kushnir, and A. I. Pinskii. 1974. Asymptotically Similar Tests of a Composite Hypothesis Versus a Nonstationary Alternative. *Probl. Inf. Transm.* **10**:239–339.

Beakley, G. W., and F. B. Tuteur. 1970. Comments on "A Nonparametric Partitioning Procedure for Pattern Classification." *IEEE Trans. Comput.* **C-19**:362–363.

Beakley, G. W., and F. B. Tuteur. 1972. Distribution-Free Pattern Verification Using Statistically Equivalent Blocks. *IEEE Trans. Comput.* **C-21**:1337–1347.

Bello, P., and W. Huggins. 1961. Effect of Hard Limiting on the Probabilities of Incorrect Dismissal and False Alarm at the Output of an Envelope Detector. *IRE Trans. Inf. Theory* **IT-7**:60–66.

Bershad, N. J., and P. L. Feintuch. 1974. Sonar Array Detection of Gaussian Signals in Gaussian Noise of Unknown Power. *IEEE Trans. Aerosp. Electr. Syst.* **AES-10**: 94–99.

Brownstein, H., and L. Kurz. 1969. An Adaptive Nonparametric Data Acquisition Method Based on a Modified Robbins-Monro Method. *Joint. Conf. Math. Comput. Aids to Design (SIAM-ACM) Digest*, p. 428.

Brownstein, H., and L. Kurz. 1969. A Sequential Detection Procedure Based on a Gambler's Ruin Problem. *Conf. Inf. Sci. Syst. (Princeton) Proc.*, pp. 188–192.

Brownstein, H., and L. Kurz. 1970. Parametric and Nonparametric Detectors Based on a Sequential Sample Median Test. *Inf. and Control* **17**:417–435.

Bushnell, W. J., and L. Kurz. 1974. The Optimization and Performance of Detectors Based on Partition Tests. *Aller. Conf. Circuit and Syst. Theory Proc.*, pp. 1016–1023.

Capon, J. 1959. *Nonparametric Methods for the Detection of Signals in Noise*. Tech. Rep. T-1/N, Columbia Univ., New York.

Capon, J. 1959. A Nonparametric Technique for the Detection of a Constant Signal in Additive Noise. *WESCON Rec.*, pt. 4, pp. 92–104.

Capon, J. 1960. Optimum Coincidence Procedures for Detecting Weak Signals in Noise. *ICONV Rec.*, pt. 4, pp. 154–166.

Capon, J. 1961. On the Asymptotic Efficiency of Locally Optimum Detectors. *IRE Trans. Inf. Theory* **IT-7**:67–71.

Capon, J. 1963. Locally Optimum Nonparametric Methods for the Detection of Signals in Noise. *SAP Proc.*, pp. 64–77.

Carlyle, J. W. 1967. A Nonparametric Coincidence-Array Detector. *Conf. Inf. Sci. Syst. (Princeton) Proc.*, p. 336.

Carlyle, J. W. 1968. Nonparametric Methods in Detection Theory. In *Communication Theory*, A. V. Balakrishnan, ed. New York: McGraw-Hill, pp. 293–319.

Chadwick, H. D., and L. Kurz. 1967. The Dual Test—A Sequential Nonparametric Detection Technique. *Conf. Inf. Sci. Syst. (Princeton) Proc.*, pp. 324–339.

Chadwick, H. D., and L. Kurz. 1968. Two Sequential Nonparametric Detection Procedures. *Inf. and Control* **13**:403–428.

Chen, Y. B., and A. Nadas. 1970. Nonparametric Bayes Approach to M-ary Detection. *Hawaii Int. Conf. on Syst. Sci. Proc.*, pp. 813–817.

Chien, Y. T., and K. S. Fu. 1967. A Nonparametric Sequential Ranking Procedure in Pattern Recognition. *Conf. Inf. Sci. Syst. (Princeton) Proc.*, pp. 98–102.

Ching, Y. C., and L. Kurz. 1969. Adaptive and Non-Adaptive Non-parametric Detectors Based on m-Interval Partitioning. *Joint Conf. Math. Comput. Aids to Design (SIAM-ACM) Digest*, p. 427.

Ching, Y. C., and L. Kurz. 1971. Lower Bounds on the Capacity of Permutation Channels. *IEEE Trans. Inf. Theory* **IT-17**:134–140.

Ching, Y. C., and L. Kurz. 1971. Two M-ary Nonparametric Detectors. *Conf. Inf. Sci. Syst. (Princeton) Proc.*, p. 337.

Ching, Y. C., and L. Kurz. 1972. Nonparametric Detectors Based on m-Interval Partitioning. *IEEE Trans. Inf. Theory* **IT-18**:251–257.

Cobben, W. 1976. A Nonparametric Detection Scheme for Regression Alternatives. *ISIT Abstr.*, p. 138.

Cochrane, J. I., and L. Kurz. 1972. Generalized Sign-Test Classifiers Within a Unified Framework of M-sample Nonparametric Tests. *Conf. Inf. Sci. Syst. (Princeton) Proc.*, pp. 479–483.

Cohen, S., and L. Kurz. 1969. An Efficient Nonparametric Detection Procedure Using a Computer Selection Technique. *Symp. Comput. Process. in Commun. (Polytech. Inst. Brook.) Proc.*, pp. 617–628.

Cohen, S., and L. Kurz. 1969. A Further Contribution to Nonparametric Detection Based on a Three-Sample Classification Model. *ISIT Abstr.*, p. 42.

Cohen, S., and L. Kurz. 1970. A Two-Sample Nonparametric Detection Procedure Based on a Split-Signal Transmission Technique. *Conf. Inf. Sci. Syst. (Princeton) Proc.*, p. 160.

Cole, L. G., and P.-W. Chen. 1978. Constant False Alarm Rate Detector for a Pulse Radar in a Maritime Environment. *NAECON Proc.*, pp. 1110–1113.

Cover, T. M., and T. J. Wagner. 1976. Topics in Statistical Pattern Recognition. In *Digital Pattern Recognition*, K. S. Fu, ed. Heidelberg: Springer-Verlag, pp. 15–46.

Cunningham, E. P. 1976. False Alarm Rate for Rank Quantizer When Input Range Samples Are Not Independent. *IEEE Trans. Aerosp. Electr. Syst.* **AES-12**:65–67.

Curry, T. J., and D. W. Tufts. 1976. Comparison of Transit Detectors in Ocean-like Environments, *ICASSP Proc.*, p. 379.

Dabkowski, J., and L. C. Peach. 1970. CFAR Operation in Non-Homogeneous and Point Clutter. *NEC Proc.*, pp. 611–615.

Daly, R. F. 1966. Nonparametric Detection Based on Rank Statistics. *IEEE Trans. Inf. Theory* **IT-12**:272 (ISIT Abstract).

Daly, R. F., and C. K. Rushforth. 1965. Nonparametric Detection of a Signal of Known Form in Additive Noise. *IEEE Trans. Inf. Theory* **IT-11**:70–76.

Davisson, L. D. 1967. Comments on "Performance of the Neyman Pearson Detector for Correlated Input Samples." *IEEE Proc.* **55**:239.

Davisson, L. D., and E. A. Feustel. 1968. Nonparametric Detection with Dependent Observations. *Hawaii Int. Conf. Syst. Sci. Proc.*, pp. 612–615.

Davisson, L. D., E. A. Feustel, and J. W. Modestino. 1970. The Effects of Dependence on Nonparametric Detection. *IEEE Trans. Inf. Theory* **IT-16**:32–41.

Davisson, L. D., and P. Papantoni-Kazados. 1972. Small Sample Efficiencies of Rank Tests. *ISIT Abstr.*, p. 57.

Devroye, L. P., and G. L. Wise. 1977. Nonparametric Detection of Changes in System Characteristics. *Midwest Symp. Circuits and Syst. Proc.*, pp. 735–739.

Dillard, G. M. 1972. Distribution-Free Sequential Signal Detection. *ISIT Abstr.*, p. 59.

Dillard, G. M. 1975. Application of Ranking Techniques to Data Compression for Image Transmission. *NTC Proc.*, pp. 22.18–22.22.

Dillard, G. M., and C. E. Antoniak. 1970. A Practical Distribution-Free Detection Procedure for Multiple-Range-Bin Radar. *IEEE Trans. Aerosp. Electr. Syst.* **AES-6**: 629–635.

Dillard, G. M., and J. T. Rickard. 1974. A Distribution-Free Doppler Processor. *IEEE Trans. Aerosp. Electr. Syst.* **AES-10**:579–586.

Dimitriadis, D., and T. Kotsidas. 1978. Sequential Detection of M-ary Data with Higher Than First-Order Markov Dependence. *ICC Rec.*, pp. 50.6.1–50.6.4.

Dimitrienko, A. N., and V. A. Korado. 1972. Nonparametric Properties of Detectors Optimized for Gaussian Interference of Unknown Intensity. *Radio Eng. Electr. Phys.* **17**:1649–1652.

Dorofeyuk, A. A. 1971. Automatic Classification Algorithms (Review). *Autom. Remote Control* **32**:1928–1958.

Dugundji, J., and E. Ackerland. 1957. Automatic Bias Control for a Threshold Detector. *IRE Trans. Inf. Theory* **IT-3**:65–70.

Dwyer, R. F. 1977. Robust Sequential Detection Using Acoustical Arrays in Undefined Noise. *Acoust. Soc. Am. J.* **62**(suppl. 1):550.

Dwyer, R. F., and L. Kurz. 1977. Sequential Partition Detectors with Dependent Sampling. *ISIT Abstr.*, p. 47.

Ekre, H. 1963. Polarity Coincidence Correlation Detection of a Weak Noise Source. *IRE Trans. Inf. Theory* **IT-9**:18–23.

Endresen, K., and R. Hedemar. 1961. Coincidence Techniques for Radar Receivers Employing a Double-Threshold Method of Detection. *IEEE Proc.* **49**:1561–1567.

Epstein, M. I., and L. Kurz. 1971. Nonparametric Estimators of Communication Channel Conditions. *Conf. Inf. Sci. Syst. (Princeton) Proc.*, pp. 340–343.

Feustel, E. A., and L. D. Davisson. 1966. A Nonparametric Detector More Efficient Than Polarity Coincidence Correlation. *IEEE Proc.* **54**:2019–2020.

Feustel, E. A., and L. D. Davisson. 1967. The Asymptotic Relative Efficiency of Mixed Statistical Tests. *IEEE Trans. Inf. Theory* **IT-13**:247–255.

Feustel, E. A., and L. D. Davisson. 1967. Nonparametric Detection by Rank Statistics. *ICONV Rec.*, pt. 11, pp. 34–39.

Feustel, E. A., and L. D. Davisson. 1968. On the Efficiency of Mixed Locally-Most-Powerful One- and Two-Sample Rank Tests. *IEEE Trans. Inf. Theory* **IT-14**:776–778.

Feustel, E. A., and L. D. Davisson. 1969. Nonparametric Sequential Testing Using Mixed Statistics. *ISIT Abstr.*, p. 42.

Figueiras-Vidal, A. R. 1978. Comments on "False Alarm Rate for Rank Quantizer When Input Range Samples Are Not Independent." *IEEE Trans. Aerosp. Electr. Syst.* **AES-14**:703–705.

Finn, H. M. 1966. Adaptive Detection in Clutter. *NEC Proc.*, pp. 562–567.

Finn, H. M. 1967. Adaptive Radar Detection with Regulated Error Probabilities in Extended Clutter Target Environments. *Aller. Conf. Circuit and Syst. Theory Proc.*, pp. 502–512.

Fowler, J. C., and J. R. Betten. 1968. Sequential Detection Using Nonparametric Statistics. *NEC Proc.*, pp. 405–509.

Friedman, J. H. 1977. A Recursive Partitioning Decision Rule for Nonparametric Classification. *IEEE Trans. Comput.* **C-26**:404–408.

Fu, K. S., and Y. T. Chien. 1967. Sequential Recognition Using a Nonparametric Ranking Procedure. *IEEE Trans. Inf. Theory.* **IT-13**:484–492.

Gibson, A. E. 1973. Adaptive Detection Probabilities for Fluctuating Target Models in Nonhomogeneous Gaussian Noise. *IEEE Trans. Aerosp. Electr. Syst.* **AES-9**: 113–115.

Gibson, J. D., and J. L. Melsa. 1975. *Introduction to Nonparametric Detection with Applications*. New York: Academic Press.

Goldstein, G. B. 1973. False-Alarm Regulation in Log-Normal and Weibull Clutter. *IEEE Trans. Aerosp. Electr. Syst.* **AES-9**:84–92.

Gonzales, R. C., and L. C. Howington. 1976. Machine Recognition of Abnormal Behavior in Nuclear Reactors. *PRIPC Proc.*, pp. 8–16.

Gotz, B., and L. Kurz. 1972. Non-Rank Distribution-Free Detectors Based on m-Interval Statistics. *ISIT Abstr.*, p. 58.

Grieve, P. G. 1977. The Optimum Constant False Alarm Probability Detector for Relatively Coherent Multichannel Signals in Gaussian Noise of Unknown Power. *IEEE Trans. Inf. Theory* **IT-23**:708–721.

Groeneveld, R. A. 1967. A Nonparametric Rank Correlation Method for Detecting Signals in Additive Noise. *IEEE Trans. Inf. Theory* **IT-13**:315–316.

Groeneveld, R. A., and M. J. Hinich. 1969. The Efficiencies of Level Tests in the Detection of Weak Stochastic Signals. *IEEE Trans. Inf. Theory* **IT-15**:502–504.

Groginsky, H. I., L. R. Wilson, and D. Middleton. 1966. Adaptive Detection of Statistical Signals in Noise. *IEEE Trans. Inf. Theory* **IT-12**:337–338.

Gupta, D. V., J. F. Vetelino, T. J. Curry, and J. T. Francis. 1977. An Adaptive Threshold System for Nonstationary Noise Backgrounds. *IEEE Trans. Aerosp. Electr. Syst.* **AES-13**:11–16.

Hancock, J. C., and D. G. Lainiotis. 1964. Adaptive and Distribution-Free Coincidence Detectors. *NCS Proc.*, pp. 433–439.

Hancock, J. C., and D. G. Lainiotis. 1965. On Learning and Distribution-Free Coincidence Detection Procedure. *IEEE Trans. Inf. Theory* **IT-11**:272–280.

Hansen, V. G. 1962. Coincidence Techniques for Radar Receivers. *IEEE Proc.* **50**:480.

Hansen, V. G. 1969. Detection Performance of Some One-Sample Nonparametric Tests and an Application to Radar. *ISIT Abstr.*, p. 49.

Hansen, V. G. 1970. Detection Performance of Some Nonparametric Rank Tests and an Application to Radar. *IEEE Trans. Inf. Theory* **IT-16**:309–318.

Hansen, V. G. 1970. Nonparametric Detection Procedures Applied to Radar. *M. J. Kelly Commun. Conf. (Univ. of Missouri-Rolla) Proc.*, pp. 24.3.1–24.3.6.

Hansen, V. G. 1972. Detection Performance of a Narrow-Band Wilcoxon Detector Against Gaussian Noise. *IEEE Trans. Inf. Theory* **IT-18**:664–667.

Hansen, V. G. 1972. Generalized Constant False Alarm Rate Processing and an Application to the Weibull Distribution. *ISIT Abstr.*, p. 68.

Hansen, V. G. 1973. Constant False Alarm Rate Processing in Search Radars. *Radar—Present and Future*, IEE Conf. Publ. 105, pp. 325–332.

Hansen, V. G. 1974. Importance Sampling in Computer Simulation of Signal Processors. *Comput. Electr. Eng.*, pp. 545–550.

Hansen, V. G., and B. A. Olsen. 1971. Nonparametric Radar Extraction Using a Generalized Sign Test. *IEEE Trans. Aerosp. Electr. Syst.* **AES-7**:942–950.

Hansen, V. G., and H. R. Ward. 1972. Detection Performance of the Cell Averaging LOG/CFAR Receiver. *IEEE Trans. Aerosp. Electr. Syst.* **AES-8**:648–652.

Hansen, V. G., and A. J. Zottl. 1971. The Detection Performance of the Siebert and Dicke-Fix CFAR Radar Detectors. *IEEE Trans. Aerosp. Electr. Syst.* **AES-7**: 706–709.

Henrichon, E. G., Jr., and K. S. Fu. 1969. On Calamity Detection Using Nonparametric Statistics. *IEEE Trans. Syst. Sci. and Cybern.* **SSC-5**:150–155.

Henrichon, E. G., Jr., and K. S. Fu. 1969. A Nonparametric Partitioning Procedure for Pattern Classification. *IEEE Trans. Comput.* **C-18**:614–624.

Ho, Y. C., and A. K. Agrawala. 1968. On Pattern Classification Algorithms: Introduction and Survey. *IEEE Proc.* **56**:2101–2114; *IEEE Trans. Autom. Control* **AC-13**: 676–690.

Kalyuzhnyy, A. Ya., and L. G. Krasnyy. 1977. Noise Immunity of Optimum and Quasioptimum Rank "Contrast" Detectors. *Radio Eng. Electr. Phys.* **22**:139–141.

Kanal, L. 1974. Patterns in Pattern Recognition:1968–1974. *IEEE Trans. Inf. Theory* **IT-20**:697–722.

Kanefsky, M. 1966. An Automatic Decision Threshold for Polarity Coincidence Arrays. *NEC Proc.*, pp. 2–9.

Kanefsky, M. 1966. Detection of Weak Signals with Polarity Coincidence Arrays. *IEEE Trans. Inf. Theory* **IT-12**:260–268.

Kanefsky, M. 1977. Book Review: *Introduction to Nonparametric Detection with Applications*, by J. D. Gibson and J. L. Melsa. *IEEE Trans. Inf. Theory* **IT-23**: 149–150.

Kanefsky, M., and J. B. Thomas. 1964. Polarity Coincidence Correlation Using Dependent Samples. *NEC Proc.*, pp. 714–719.

Kanefsky, M., and J. B. Thomas. 1965. On Adaptive Detectors for Two-Input Systems. *IEEE Trans. Autom. Control* **AC-10**:427–433.

Kanefsky, M., and J. B. Thomas. 1965. On Adaptive Nonparametric Detection Systems Using Dependent Samples. *IEEE Trans. Inf. Theory* **IT-11**:521–526.

Kanefsky, M., and J. B. Thomas. 1965. On Polarity Detection Schemes with Non-Gaussian Inputs. *J. Franklin Inst.* **280**:120–138.

Kassam, S. A. 1976. A Modified Nonparametric Wilcoxon Test for Signal Detection. *Conf. Inf. Sci. Syst. (Johns Hopkins) Proc.*, pp. 481–486.

Kassam, S. A. 1977. A Conditional Rank Test for Nonparametric Detection. *IEEE Trans. Inf. Theory* **IT-23**:315–320.

Kassam, S. A. 1977. Conditional Tests in Nonparametric Detection. In *Nonparametric Methods in Communications*, P. Papantoni-Kazakos and D. Kazakos, eds. New York: Marcel Dekker, pp. 145–201.

Kassam, S. A. 1978. Correlated-Input and Sequential Nonparametric Detection. *Workshop Commun. Theory and Appl. (AFOSR) Proc.*, pp. 58–60.

Kassam, S. A., and T. L. Lim. 1978. A Conditional-Test Sign Detector for Nonparametric Radar. *IEEE Trans. Aerosp. Electr. Syst.* **AES-14:668–692.**

Kassam, S. A., and J. B. Thomas. 1974. Dependent Sampling in Non-Adaptive Nonparametric Detectors. *Aller. Conf. Circuit Syst. Theory Proc.*, pp. 1007–1015.

Kassam, S. A., and J. B. Thomas. 1975. A Class of Nonparametric Detectors for Dependent Input Data. *IEEE Trans. Inf. Theory* **IT-21**:431–437.

Kassam, S. A., and J. B. Thomas. 1975. Detectors Based on Conditional Tests. *ITC Proc.*, pp. 26–33.

Kassam, S. A., and J. B. Thomas. 1975. A Modified Sign Detector Based on a Four-Level Input Quantization. *Aller. Conf. Circuits Syst. Theory Proc.*, pp. 204–210.

Kassam, S. A., and J. B. Thomas. 1976. Array Detectors for Random Signals in Noise. *IEEE Trans. Sonics and Ultrason.* **SU-23**:107–112.

Kassam, S. A., and J. B. Thomas. 1976. Dead-Zone Limiter: An Application of Conditional Tests in Nonparametric Detection. *Acoust. Soc. Am. J.* **60**:857–862.

Kassam, S. A., and J. B. Thomas. 1976. Generalizations of the Sign Detector Based on Conditional Tests. *IEEE Trans. Commun.* **COM-24**:481–487.

Kassam, S. A., and J. B. Thomas. 1977. Improved Nonparametric Coincidence Detectors. *J. Franklin Inst.* **303**:75–84.

Kassam, S. A., and J. B. Thomas. 1977. Nonparametric Weighted-Sign Tests for Location. *SIAM J. Appl. Math.* **32**:649–652.

Kazakos, D., and B. Dimitriadis. 1977. Sequential Nonparametric Detection with Dependent Data. *Conf. Inf. Sci. Syst. (Johns Hopkins) Proc.*, pp. 415–417.

Kersten, P. R., and L. Kurz. 1976. Robustized Vector Robbins-Monro Algorithm with Application to m-Interval Detection. *Inf. Sci.* **11**:121–140.

Kersten, P. R., and L. Kurz. 1977. Bivariate m-Interval Classifiers with Application to Edge Detection. *Inf. and Control* **34**:152–168.

Kersten, P. R., and L. Kurz. 1978. Improved Operation of m-Interval Detectors by Optimum Signal Selection. *IEEE Trans. Inf. Theory* **IT-24**:477–484.

Klamer, D. N. 1977. Nonparametric Maneuver Detection in Kalman Filtering. *Proc. DCC*, pp. 544–547.

Kobyashi, H. 1970. Distance Measures and Asymptotic Relative Efficiency. *IEEE Trans. Inf. Theory* **IT-16**:288–291.

Kobayshi, H., and J. B. Thomas. 1967. Distance Measures and Related Criteria. *Aller. Conf. Circuit Syst. Theory Proc.*, pp. 491–500.

Komo, J. J. 1969. Cherroff Bounds for the False Dismissal Probabilities of the Kolmogorov-Smirnov Detector. *IEEE Trans. Inf. Theory* **IT-15**:323–325.

Komo, J. J. 1970. An Application of Nonparametric Statistics of the Kolmogorov-Smirnov Type to the Measurement Selection Problem. *Hawaii Int. Conf. on Syst. Sci. Proc.*, pp. 735–738.

Komo, J. J., and G. W. Schroeder. 1971. Error Probabilities for Nonparametric Detectors of the Kolmogorov-Smirnov Type. *ICC Proc.*, pp. 13–19.

Kurz, L. 1970. Rank and Non-Rank Methods in Nonparametric Detection. *M. J. Kelly Commun. Conf. (Univ. of Missouri-Rolla) Proc.*, pp. 24.1.1–24.1.4.

Kurz, L. 1973. Rank and Nonrank Nonparametric Methodology in Classification Problems. *ISIT Abstr.*, p. E.1-3.

Kurz, L. 1977. Nonparametric Detectors Based on Partition Tests. In *Nonparametric Methods in Communications*, P. Papantoni-Kazakos and D. Kazakos, eds. New York: Marcel Dekker, pp. 73–143.

Kurz, L., and G. Pasternack. 1976. Nonparametric Classification Using Quadratic Quantile Statistics. *ISIT Abstr.*, p. 137.

Kushnir, A. F. 1971. Rank Algorithms for the Post-Detector Detection of Signals. *Telecommun. Radio Eng.* **25/26**, pt. 2, 80–83.

Kushnir, A. F., and B. R. Levin. 1968. Optimal Rank Algorithms for Detecting Signals in Noise. *Probl. Inf. Transm.* **4**:1–12.

Kuznetsov, V. P. 1971. Decision Invariance Under a Nuisance Parameter. *Probl. Inf. Transm.***7**:306–312.

Kuznetsov, V. P. 1972. Invariance of Maximum-Likelihood Decisions in the Presence of Nuisance Parameters. *Probl. Inf. Transm.* **8**:207–214.

Lainiotis, D. G. 1965. Generalized Distribution-Free Coincidence Detection Procedures. *ICONV Rec.*, pt. 7, pp. 214–219.

Lainiotis, D. G. 1967. On a Class of Generalized Polarity Coincidence Correlators. *Asilomar Conf. Circuit Syst. Proc.*, pp. 980–986.

Lainiotis, D. G. 1968. On a Class of Generalized Polarity Coincidence Correlators. *IEEE Trans. Inf. Theory* **IT-14**:603–604.

Lainiotis, D. G., and J. C. Hancock. 1964. Distribution-Free Coincidence Detection Procedures. *WESCON Rec.*, pt. 4, pp. 13.4.1–13.4.9.

Levin, B. R. 1971. Current Problems in the Statistical Synthesis of Information Systems. *Telecommun. Radio Eng.* **25/26**, pt. 2, 59–60.

Levin, B. R., and A. F. Kushnir. 1969. Asymptotically Optimum Rank Algorithms of Signal Detection in Noise. *Radio Eng. Electr. Phys.* **14**:221–228.

Levin, B. R., and A. F. Kushnir. 1971. Signal-Detection Rank Algorithms Using a Sequence of Correlated Observation Samples. *Telecommun. Radio Eng.* **25/26**, pt. 2, 83–86.

Levin, B. R., and A. K. Rybin. 1969. Nonparametric Amplitude and Phase Methods of Signal Detection. I. *Eng. Cybern. No. 5*, pp. 103–110.

Levin, B. R., and A. K. Rybin. 1970. Nonparametric Amplitude and Phase Methods for Signal Detection. II. *Eng. Cybern. No. 1*, pp. 140–147.

Lim, T. L., and S. A. Kassam. 1978. A Conditional-Test Sign Detector for Nonparametric Radar. *IEEE Trans. Aerosp. Electr. Syst.* **AES-14**:688–692.

Lomakin, D. V. 1972. Effectiveness of Certain Easily Realizable Rank Algorithms for Detections of Signals Against a Background of Noise. *Radio Eng. Electr. Phys.* **17**:1564–1566.

Lytle, D. W., P. H. Moose, and L. L. Scharf. 1971. Comments on "Nonparametric Detection." *IEEE Proc.* **59**:631–632.

Marks, R. J., II, G. L. Wise, and D. G. Haldeman. 1977. Further Results on Detection in Laplace Noise. *Midwest Symp. Circuits and Syst. Proc.*, pp. 735–739.

Marks, R. J., II, G. L. Wise, D. G. Haldeman, and J. L. Whited. 1977. Some Preliminary Results on Detection in Laplace Noise. *Conf. Inf. Sci. Syst. (Johns Hopkins) Proc.*, pp. 541–546.

McCormick, D. J., and S. S. Wolff. 1975. Beamforming with a Linear Array. *Conf. Inf. Sci. Syst. (Johns Hopkins) Proc.*, pp. 45–48.

Millard, J. B., and L. Kurz. 1966. Adaptive Threshold Detection of M-ary Signals in Statistically Undefined Noise. *ICONV Rec.*, pt. 7, pp. 131–142; *IEEE Trans. Commun.* **COM-14**:601–610.

Millard, J. B., and L. Kurz. 1966. Nonparametric Signal Detection—An Application of the Kolmogorov-Smirnov, Cramer-von Mises Tests. *IEEE Trans. Inf. Theory* **IT-12**:275 (ISIT Abstract).

Millard, J. B., and L. Kurz. 1967. The Kolmogorov-Smirnov Tests in Signal Detection. *IEEE Trans. Inf. Theory* **IT-13**:341–342.

Miller, J. H., and J. B. Thomas. 1972. Detectors for Discrete-Time Signals in Non-Gaussian Noise. *IEEE Trans. Inf. Theory* **IT-18**:241–250.

Miller, J. H., and J. B. Thomas. 1972. Small Sample Performance of Detectors in Non-Gaussian Noise. *Asilomar Conf. Circuits and Syst. Proc.*, pp. 438–443.

Miller, J. H., and J. B. Thomas. 1975. Numerical Results on the Convergence of Relative Efficiencies. *IEEE Trans. Aerosp. Electr. Syst.* **AES-11**:204–209.

Miller, R. S., and M. M. Rochwarger. 1978. Applications of Invariant Hypothesis Testing Techniques to Radar Signal Classification. *PRIPC Proc.*, pp. 78–84.

Mizogenchi, R., and M. Shimara. 1976. Nonparametric Learning Without a Teacher Based on Mode Estimation. *IEEE Trans. Comput.* **C-25**:1109–1117.

Modestino, J. W. 1977. Adaptive Nonparametric Detection Techniques. In *Nonparametric Methods in Communications*, P. Papantoni-Kazakos and D. Kazakos, eds. New York: Marcel Dekker, pp. 29–71.

Modestino, J. W., and L. D. Davisson. 1969. Adaptive Detection of Weak Signals in Statistically Undefined Noise Backgrounds. *Symp. Comput. Process. in Commun. (Polytech. Inst. Brook.) Proc.*, pp. 629–651.

Moser, J. N. 1971. Detection Probabilities for a Rank Statistic. *IEEE Trans. Inf. Theory* **IT-17**:346–349.

Mucci, R. A., and D. H. Reynhout. 1976. Limiting Performance of Generalized Sign Test Detector. *IEEE Trans. Aerosp. Electr. Syst.* **AES-12**:505–509.

Nagy, G. 1968. State of the Art in Pattern Recognition. *IEEE Proc.* **56**:836–862.

Nitzberg, R. 1972. Constant-False-Alarm-Rate Signal Processors for Several Types of Interference. *IEEE Trans. Aerosp. Electr. Syst.* **AES-8**:27–34.

Nitzberg, R. 1973. Constant-False-Alarm-Rate Processors for Locally Non-Stationary Clutter. *IEEE Trans. Aerosp. Electr. Syst.* **AES-9**:399–405.

Nitzberg, R. 1978. Analysis of the Arithmetic Mean CFAR Normalizer for Fluctuating Targets. *IEEE Trans. Aerosp. Electr. Syst.* **AES-14**:44–48.

Novak, L. 1974. On the Detection Performance of a Cell-Averaging CFAR in Non-stationary Weibull Clutter. *ISIT Abstr.*, p. 32.

Odell, P. L., and B. S. Duran. 1974. Comparison of Some Classification Techniques. *IEEE Trans. Comput.* **C-23**:591–596.

Papantoni-Kazakos, P. 1975. Small-Sample Efficiencies of Rank Tests. *IEEE Trans. Inf. Theory* **IT-21**:150–157.

Papantoni-Kazakos, P. 1977. Robustness in Parameter Estimation. *IEEE Trans. Inf. Theory* **IT-23**:223–231.

Papantoni-Kazakos, P. 1977. Small Sample Results. In *Nonparametric Methods in Communications*, P. Papantoni-Kazakos and D. Kazakos, eds. New York: Marcel Dekker, pp. 203–263.

Papantoni-Kazakos, P., and D. Kazakos, eds. 1977. *Nonparametric Methods in Communications.* New York: Marcel Dekker.

Picinbono, B., and G. Vezzosi. 1972. New Results Concerning the Signal Detection in a Noise of Unknown Power. *ISIT Abstr.*, p. 105.

Poage, J. L. 1975. A Sequential Nonparametric Pattern Classification Algorithm Based on the Wald SPRT. *IEEE Trans. Syst., Man and Cybern.* **SMC-5**:128–134.

Prewitt, J. 1972. Parametric and Nonparametric Recognition by Computer: An Application to Leukocyte Image Processing. In *Advances in Computers*, vol. 2. New York: Academic Press, pp. 285–414.

Prokof'yev, V. N. 1970. Signal Detection in Unknown Noise Using a Set of Non-Equivalent Solutions. *Radio Eng. Electr. Phys.* **15**:705–707.

Prokof'yev, V. N. 1976. Double-Sample Rank Criteria for Detection of a Signal on the Basis of Extremal Observations. *Radio Eng. Electr. Phys.* **21**:144–148.

Pyle, H. W., and L. E. Vogel. 1975. The Utilization of Functionally Dedicated Micro Processors in Radar Video Processing. *IRC Proc.*, pp. 268–273.

Reid, W. S., M. R. Saltsman, and L. E. Vogel. 1975. Technique for Clutter False Alarm Control in Automated Radar. *IEEE Trans. Aerosp. Electr. Syst.* **AES-11**:424 (IRC Abstract).

Ribin, A. K. 1975. Classification of Weak Signals in Non-Gaussian Noise. *Eng. Cybern. No. 5*, pp. 901–909.

Rickard, J. T., and G. M. Dillard. 1976. Adaptive Detection Algorithms for Multiple-Target Situations. *Asilomar Conf. Circuits and Syst. Proc.*, pp. 303–305.

Rickard, J. T., and G. M. Dillard. 1977. Adaptive Detection Algorithms for Multiple-Target Spectral Data. *DCC Proc.*, pp. 515–518.

Robertson, G. H. 1976. Computation of the Noncentral F Distribution (CFAR Distribution). *IEEE Trans. Aerosp. Electr. Syst.* **AES-12**:568–571.

Rossi, A., and M. Arozullah. 1975. A Ranking Procedure for Nonparametric Detection of M-ary Signals. *NTC Proc.*, pp. 30.23–30.27.

Rowlands, R. O. 1962. The Relative Efficiencies of Various Binary Detection Systems. *ICONV Rec.*, pt. 4, pp. 185–189.

Sasser, G. E., and D. O. Richards. 1973. The Kolmogorov-Smirnov Test Modified for Censored Data. *RMS Proc.*, pp. 177–180.

Scharf, L. L., and D. W. Lytle. 1971. Signal Detection in Gaussian Noise of Unknown Level: An Invariance Application. *IEEE Trans. Inf. Theory* **IT-17**:404–411.

Schroeder, G. W., and J. J. Komo. 1972. A Nonparametric Detector with Applications to Noncoherent Receivers. *ISIT Abstr.*, p. 58.

Schultze, H. M., and M. Arozullah. 1975. Nonparametric Detectors Using Intersymbol Interference for Signal Detection. *ICC Proc.*, pp. 15.10–15.14.

Schwartz, L. S. 1965. Proposed Signal Detection Techniques for Ionospheric Sounding. *ICC Proc.*, pp. 61–63.

Schwartz, M. I. 1970. Detection Performance Measures and Approximate Evaluation of Detection Probabilities. *ISIT Abstr.*, p. 68.

Schwartz, R. E. 1969. Minimax and CFAR Detection in Additive Gaussian Noise of Unknown Covariance. *IEEE Trans. Inf. Theory* **IT-15**:722–725.

Shapiro, A. 1970. A Nonparametric Detector for a Narrow-Band Signal. *Acoust. Soc. Am. J.* **47**:694–700.

Shin, J. G., and S. A. Kassam. 1977. Two-Input Correlator Detectors for Random Signals. *Aller. Conf. Commun., Control and Comput. Proc.*, pp. 517–526.

Shin, J. G., and S. A. Kassam. 1978. Three-Level Coincidence-Correlator Detectors. *Acoust. Soc. Am. J.* **63**:1389–1395.

Shin, J. G., and S. A. Kassam. 1979. Multilevel Coincidence Correlators for Random Signal Detection. *IEEE Trans. Inf. Theory* **IT-25**:47–53.

Shloma, A. M. 1974. Nonparametric Multichannel Sequential Analysis on Extremal Statistics. *Radio Eng. Electr. Phys.* **19**:39–45.

Sidorov, E. Yu. 1972. Nonbiased Rule for Detection of Radar Signals in Noise of Unknown Power. *Radio Eng. Electr. Phys.* **17**:1747–1750.

Silber, D. 1966. Probabilities of Detection and False Alarm for a Coherent Detector with Amplitude Limiting of Arbitrary Hardness. *ICONV Rec.*, pt. 7, pp. 107–123.

Simakhin, V. A. 1973. Nonparametric Signal Detectors When the Signal Samples Are Dependent. *Telecommun. Radio Eng.* **27/28**, pt. 2, 66–72.

Sirvanci, M. B. 1971. Asymptotic Normality of Certain Two-Sample Nonparametric Test Statistics. *Conf. Inf. Sci. Syst. (Princeton) Proc.*, p. 430.

Sirvanci, M. B., and S. S. Wolff. 1976. Nonparametric Detection with Autoregressive Data. *IEEE Trans. Inf. Theory* **IT-22**:725–731.

Skomal, E. N. 1967. Application of the Nonparametric c_1 Statistical Test to Radar Cross Section Signature Comparisons. *IEEE Proc.* **55**:1628–1629.

Smith, J. M. 1977. Problems Solved and Unsolved in Radar Detection Theory. *Radar-77*, IEE Conf. Publ. 155, pp. 263–265.

Spooner, R. L. 1968. On the Detection of a Known Signal in a Non-Gaussian Noise Process. *Acoust. Soc. Am. J.* **44**:141–147.

Srinivasan, V. S. 1976. Distribution-Free Processors Based on the Theory of Runs. *IEEE Trans. Aerosp. Electr. Syst.* **AES-12**:77–80.

Stoll, E. D., and L. Kurz. 1967. Most Powerful Rank Detection—An Extension and Application of Hoeffding's Theorem. *Conf. Inf. Sci. Syst. (Princeton) Proc.*, pp. 330–334.

Stoll, E. D., and L. Kurz. 1967. Suboptimum Rank-Detection Procedures Using Rank-Vector Codes. *ICONV Rec.*, pt. 11, pp. 22–33.

Stoll, E. D., and L. Kurz. 1968. A Contribution to the Multiple-Polarity Coincidence Detection Schemes. *IEEE Trans. Commun.* **COM-16**:676–682.

Stoll, E. D., and L. Kurz. 1968. Suboptimum Rank Detection Procedure Using Rank Vector Codes. *IEEE Trans. Commun.* **COM-16**:402–410.

Tamm, Yu. A. 1967. Noise Immunity of a Binary Signal Receiver Using Optimal Weighted Integration of the Signal "Sign." *Telecommun. Radio Eng.* **21/22**, pt. 1, 26–31.

Tantarantana, S., and J. B. Thomas. 1976. Performance of a Sequential Sign Detector in Antipodal Signal Detection. *Conf. Inf. Sci. Syst. (Johns Hopkins) Proc.*, pp. 475–480.

Tantarantana, S., and J. B. Thomas. 1977. A Modified Sequential Rank Test for Real Time Implementation. *Conf. Inf. Sci. Syst. (Johns Hopkins) Proc.*, pp. 395–400.

Tantaratana, S., and J. B. Thomas. 1977. On Sequential Sign Detection of a Constant Signal. *IEEE Trans. Inf. Theory* **IT-23**:304–315.

Tarasenko, F. P. 1971. Properties of Sampling Blocks as the Basis for Nonparametric Procedures. *Telecommun. Radio Eng.* **25/26**, pt. 2, 89–92.

Tarasenko, F. P., and V. P. Shalenin. 1971. Some Properties of a Nonparametric Procedure for Detecting Signals in Noise. *Telecommun. Radio Eng.* **25/26**, pt. 2, 87–88.

Thomas, J. B. 1970. Nonparametric Detection. *IEEE Proc.* **58**:623–631.

Thomas, J. B., and T. R. Williams. 1959. On the Detection of Signals in Nonstationary Noise by Product Arrays. *Acoust. Soc. Am. J.* **31**:453–462.

Tong, P. S., and B. Liu. 1966. Application of Stochastic Approximation Techniques to Two Signal Detection Problems. *NEC Proc.*, pp. 804–809.

Trunk, G. V. 1970. Small and Large Sample Behavior of Two Detectors Against Envelope-Detected Sea Clutter. *IEEE Trans. Inf. Theory* **IT-16**:95–98.

Trunk, G. V. 1971. Further Results on Detection of Targets in Non-Gaussian Sea Clutter. *IEEE Trans. Aerosp. Electr. Syst.* **AES-7**:553–556.

Trunk, G. V., and S. F. George. 1970. Detection of Targets in Non-Gaussian Sea Clutter. *IEEE Trans. Aerosp. Electr. Syst.* **AES-6**:620–628.

Trunk, G. V., B. H. Cantrell, and M. E. Owens. 1974. Modified Generalized Sign Test Processor. *ISIT Abstr.*, p. 33.

Trunk, G. V., B. H. Cantrell, and F. D. Queen. 1974. Modified Generalized Sign Test Processor for 2-D Radar. *IEEE Trans. Aerosp. Electr. Syst.* **AES-10**: 574–582.

Tugnait, J. K., and S. Prasad. 1977. Adaptive Radar Detection with Asymptotically Regulated False-Alarm Rate. *IEEE Trans. Aerosp. Electr. Syst.* **AES-13**:390–394.

Tuteur, S. B. 1965. Optimum Detection of a Stochastic Signal in a Gaussian Noise of Unknown Power Level. *ICC Proc.*, pp. 63–66.

Tuteur, F. B. 1965. Optimum Detection of a Stochastic Signal in a Gaussian Noise of Unknown Strength. *IEEE Proc.* **53**:487–488.

Tuteur, F. B. 1967. On the Detection of Transiting Broadband Targets in Noise of Uncertain Level. *IEEE Trans. Commun.* **COM-15**:61–69.

Viterbi, A. J. 1975. Asymptotic Performance of Three Hard-Limiting Non-Coherent Correlator-Envelope Detectors. *ITC Proc.*, pp. 196–205.

Vogel, L. E., W. S. Reid, and P. E. Steichen. 1975. An Examination of Radar Processing Via Nonparametric Techniques. *IRC Proc.*, pp. 533–537.

Ward, H. R. 1972. Dispersive Constant False Alarm Rate Receiver. *IEEE Proc.* **60**: 735–736.

Woerrlein, H. H. 1977. CFAR-Detection Derived from Correlation Processing in Adjacent Range Bins. *Radar-77,* IEE Conf. Publ. 155, pp. 266–269.

Woinsky, M. N. 1970. Nonparametric Target Detection Using Spectral Data. *CCPC Dig.*, pp. 21–22.

Woinsky, M. N. 1972. Nonparametric Detection Using Spectral Data. *IEEE Trans. Inf. Theory* **IT-18**:110–118.

Woinsky, M. N. 1972. Nonpararmetric Versus Limit-Sum Detection. *ISIT Abstr.*, p. 60.

Woinsky, M. N. 1972. Nonparametric Detection Performance in Radar, Sonar Applications. *ISIT Abstr.*, p. F2-6.

Woinsky, N. M., and L. Kurz. An Efficient Nonparametric Detector Based on a Three-Sample Classification Model. *IEEE Trans. Inf. Theory* **IT-14**:744–752.

Woinsky, M. N., and L. Kurz. 1968. A New Approach to Nonparametric Detection Using Dependent Samples and Time Continuous Processing. *Conf. Inf. Sci. Syst. (Princeton) Proc.*, pp. 178–183.

Woinsky, M. N., and L. Kurz. 1968. A Noncoherent Detector with Improved Immunity to Fading. *EASCON Rec.*, pp. 290–295.

Woinsky, M. N., and L. Kurz. 1968. Sequentially Adaptive Nonparametric Detection with Prescribed Total Error Probability. *Hawaii Int. Conf. Syst. Sci. Proc.*, pp. 605–608.

Woinsky, M. N., and L. Kurz. 1969. Sufficient Sample Sizes for a Three-Sample Nonparametric Detector. *ISIT Abstr.*, p. 41.

Woinsky, M. N., and L. Kurz. 1970. Nonparametric Detection Using Dependent Samples. *IEEE Trans. Inf. Theory* **IT-16**:355–358.

Wolff, S. S., and R. E. Keelen. 1970. Rank Tests with Shared Samples. *M. J. Kelly Commun. Conf. (Univ. of Missouri-Rolla) Proc.*, pp. 24.2.1–24.2.4.

Wolff, S. S., and M. B. Sirvanci. 1973. "Studentizing" Rank Tests for Dependent Data. *Conf. Inf. Sci. Syst. (Princeton) Proc.*, pp. 63–66.

Wolff, S. S., J. L. Gastwirth, and H. Rubin. 1966. The Sign Test on Autoregressive Data. *IEEE Trans. Inf. Theory* **IT-12**:279 (ISIT Abstract).

Wolff, S. S., J. L. Gastwirth, and H. Rubin. 1967. The Effect of Autoregressive Dependence on a Nonparametric Test. *IEEE Trans. Inf. Theory* **IT-13**:311–313.

Wolff, S. S., J. B. Thomas, and T. R. Williams. 1962. The Polarity-Coincidence Correlator: A Nonparametric Detection Device. *IRE Trans. Inf. Theory* **IT-8**:5–9.

Zeoli, G. W. IF Versus Video Limiting for Two-Channel Coherent Signal Processors. *IEEE Trans. Inf. Theory* **IT-17**:579–586.

Zeoli, G. W., and T. S. Fong. 1971. Performance of a Two Sample Mann-Whitney Nonparametric Detector in a Radar Application. *IEEE Trans. Aerosp. Electr. Syst.* **AES-7**:951–959.

Zigangirov, K. Sh. 1965. A Nonparametric Criterion for Comparison of Samples. *Probl. Inf. Transm.* **1**:94–96.

Zoldi, J. M., and M. Arozullah. 1971. A Class of M-ary Distribution-Free Detectors for Channels with Unknown Additive and Multiplicative Noises. *ICC Proc.*, pp. 13.13–13.18.

AUTHOR CITATION INDEX

SUBJECT INDEX

About the Editors

SALEEM A. KASSAM received the B.S. degree in engineering from Swarthmore College, Swarthmore, Pennsylvania, in 1972, the M.S.E. and M.A. degrees in electrical engineering from Princeton University, Princeton, New Jersey, in 1974, and the Ph.D. degree in elecrical engineering from Princeton in 1975. Since September 1975 he has been an assistant professor in the Department of Systems Engineering of the Moore School of Electrical Engineering, University of Pennsylvania, Philadelphia. His major areas of research are signal processing and statistical communication theory.

Dr. Kassam is a member of Phi Beta Kappa and Sigma Xi.

JOHN B. THOMAS received the A. B. degree from Gettysburg College, Gettysburg, Pennsylvania, in 1944, the B.S. degree from the Johns Hopkins University, Baltimore, Maryland, in 1952, and the M.S. and Ph.D. degrees from Stanford University, Stanford, California, in 1953 and 1955, respectively.

During World War II he served in the U. S. Army. From 1946 to 1952 he was employed by Koppers Co., Baltimore, Maryland. Since 1955 he has been a member of the Department of Electrical Engineering of Princeton University, Princeton, New Jersey. In 1967–1968 he was a senior postdoctoral National Science Foundation fellow and visiting professor at Stanford University and the University of California at Berkeley.

Dr. Thomas is a fellow of the Institute of Electrical and Electronics Engineers and a member of Tau Beta Pi, Sigma Xi, and Eta Kappa Nu.